Fundamentals of Internet of Things

IEEE Press
445 Hoes Lane
Piscataway, NJ 08854

Fundamentals of Internet of Things

For Students and Professionals

Farzin John Dian
British Columbia Institute of Technology
Canada

Published by John Wiley & Sons, Inc., Hoboken, New Jersey.
Published simultaneously in Canada.

Library of Congress Cataloging-in-Publication Data Applied for:

Hardback ISBN: 9781119847298

Cover Design: Wiley
Cover Image: © BAIVECTOR/Shutterstock

Set in 9.5/12.5pt STIXTwoText by Straive, Chennai, India

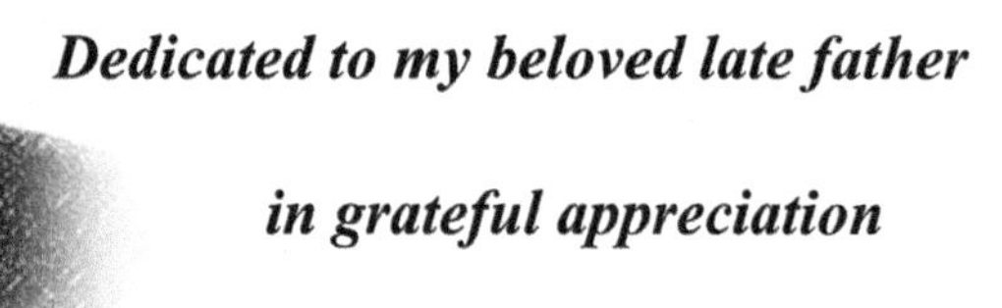

Dedicated to my beloved late father

in grateful appreciation

Contents

About the Author

Dr. F. John Dian is a faculty in the Department of Electrical and Computer Engineering at the British Columbia Institute of Technology in Vancouver, Canada. He received his Ph.D. degree from Concordia University, Canada, in Electrical and Computer Engineering. Dr. Dian has extensive experience in designing and implementing telecommunication systems and IoT networks. He holds a certificate in business analytics from Harvard Business School, USA, and co-chairs the center of excellence in analytics at BCIT. He has received numerous awards for his outstanding teaching and research, and has been an invited speaker at many forums and conferences. Dr. Dian is a senior member of the Institute of Electrical and Electronics Engineers (IEEE) and an active member of the Association of Professional Engineers and Geoscientists of British Columbia (APEGBC). He is the author of several books such as *IoT Use Cases and Technologies*, *Cellular IoT for Practitioners*, and *Physical System Modelling Using MATLAB®*.

Preface

The Internet of Things (IoT) is a fascinating and new technology. IoT enables the power of the Internet and analytics to be provided to physical objects in order to make them smart. Smart city, smart home, smart factory, and smart meter are all examples of objects that have become smart by implementing IoT. IoT is becoming a mature technology and is growing at a fast pace. It brings endless opportunities for various industries such as healthcare, energy, agriculture, and transportation. As a result, more and more IoT applications are getting implemented every day.

In the era of any new technology, training becomes one of the crucial tasks that prepares the workforce with the knowledge and skill sets that they need to design, implement, deploy, and test the new technology. Leaders and decision makers must be trained to be able to make appropriate decisions on how and when to implement a new technology in order to take advantage of the opportunities that it can provide to their organizations. Customers of a new technology also need to be trained to know what to expect from that technology and how to use it. IoT, as a new technology, is not an exception.

This textbook has been prepared with instructors and students in mind. It presents a clear, comprehensive introduction to IoT. The textbook emphasizes the fundamentals of IoT including applications, use cases, existing connectivity schemes, protocols, analytics, security, solution development, and cloud connection. It is suitable for undergraduate engineering students studying in universities or colleges, as well as students studying in various diploma programs in engineering disciplines. To accommodate students with limited knowledge in the area of data communications and networks, the first chapter of this book is dedicated to the fundamentals of data communications and networks. The book also discusses some more advanced topics in Chapter 8, which is suited to engineering students who have a deeper knowledge of data communications and networks. The textbook has been designed to be used during a term of 15 weeks. During a term, students who have taken courses related to data communications and networks in the past, may skip the first chapter and study Chapter 8, while other engineering students can study the first chapter thoroughly and skip Chapter 8.

My approach in writing this textbook is straightforward. It should convey the information through explanation of the concepts and discussion of practical applications. The complex concepts should be explained in simple language, and the content should flow smoothly from one chapter to another. The primary audience for this textbook is undergraduate engineering students, with the secondary audience consisting of IoT practitioners, network architects, software and hardware engineers, consultants, managers, and entrepreneurs.

This textbook focuses on the fundamentals of IoT and intends to give undergraduate students the basic knowledge they need about IoT. It does not discuss any specific platform, service provider, or embedded environment. To prepare their students for performing hands-on experiments with IoT technology, instructors may choose to introduce them to certain platforms, embedded systems or simulation tools. This can help students to gain practical knowledge and have a better understanding of the theoretical concepts discussed in this book. For this purpose, several practical assignments have been defined at the end of this book.

After reading this book, readers will be able to:

- Recognize the principles of smart objects and the potential for IoT to revolutionize people's lives, as well as understanding the impact it has on businesses
- Comprehend IoT architecture from sensors to the cloud, including edge gateways
- Choose among cloud computing, fog computing, and edge computing for an IoT system
- Determine the static and dynamic performance metrics for core sensors
- Explain the architecture of smart sensors and the concept of sensor fusion in IoT system design
- Learn how to use various wired and wireless technologies to provide connectivity for IoT applications
- Select a suitable application-layer protocol among the ones used for IoT applications
- Acquire deep understanding of cellular IoT in 4G and 5G eras
- Evaluate various cellular IoT modules
- Recognize how analytics and artificial intelligence algorithms are used in the IoT ecosystem
- Address IoT security and privacy concerns, as well as appropriate security and privacy protection schemes
- Become familiar with IoT solution development methodologies
- Conduct hands-on experiments to design IoT system
- Become inspired by the endless opportunities that IoT can provide

I would like to establish a line of communication with the readers of this book. I encourage all readers to send me their comments and suggestions. I will use the readers' feedback and suggestions for future editions of this book.

F. John Dian

John_dian @bcit.ca
https://www.linkedin.com/in/johndian/

Chapter 1 Data Communications and Networks

1.1 Introduction

The Internet of Things (IoT) is a network of physical objects that are embedded with sensors and electronics and are connected to each other over the Internet. To begin reading this book, you must have a basic grasp of networks, particularly Internet-based networks. To accommodate readers who have limited knowledge in the area of data communications, this chapter is dedicated to the fundamentals of data communications and networks. There are a vast number of topics, technologies, systems, and standards that are fundamental to this field of study. This chapter intends to provide materials that prepare readers with the required information to study the next chapters of this book. Readers who already have a good grasp of these materials may skip this chapter and start the book from Chapter 2. We will begin by defining what a network is, and then work our way toward other required topics related to data communications and Internet-based networks in this chapter.

A network is a collection of interconnected nodes that can communicate with one another. A computer network is defined as a network in which all the nodes are computers. A sensor network is a network in which each of its nodes is a sensor. Your network is a collection of your friends, and therefore, each node in your network is one of your friends. The most important characteristic of a network is the capability of its nodes to somehow communicate with each other. In other words, a node must be able to send data and exchange information with other nodes of the network. Data communication among the nodes of a network is possible only and only if all the nodes can talk in the same language and can communicate according to the same protocol. A protocol defines the required rules that determine the syntax, timing, and the method of communication, which all the nodes understand. In addition, each network node should have its own address. A source node cannot transfer data to a destination node if it does not know its address.

When designing a network, we should consider many factors such as the number of nodes in a network, the distance among the nodes, the environment in which the nodes are located, the type of data that they exchange, and the required rate of data transmission.

A network may consist of a large number of nodes or only a small set of nodes. However, a network should have at least two nodes. The nodes can communicate and send their data at a fast speed or have a slow rate of data transmission. In other words, a node may send gigabits of data per second or only transmit several bits in one second. The data rate of a network is determined by the network application. A car engine, for example, is a network of many nodes (various parts of the engine) that can interact with one another or with a diagnostic device connected to the engine. This network may not require a high data rate, whereas a network transmitting high-definition video needs to support a high data rate.

Fundamentals of Internet of Things: For Students and Professionals, First Edition. F. John Dian.

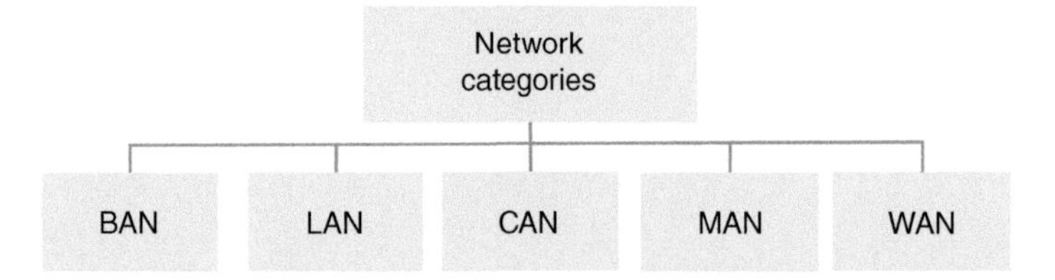

Figure 1.1 Network classification based on the maximum distance among the nodes.

The nodes of a network might be close to or far from each other. The Body Area Network (BAN) is a network with a limited number of nodes that are connected to the human body. Therefore, there is a very short distance among the nodes in BAN. Local Area Network (LAN) is a network with local nodes. A network of computers, printers, and other electronic equipment that can be found in a home or an office is a clear example of a LAN. The same goes with the network of many nodes inside a building. However, local area networks today can cover larger geographical areas, making the concept of local nodes to some extent fuzzy. Campus Area Networks (CAN) is a network where its nodes are located in a campus area. Metropolitan Area Network (MAN) covers a neighborhood, while a Wide Area Network (WAN) is a network with nodes that are spread across a large geographical area such as a city, country, or continent. Figure 1.1 shows different types of networks based on the maximum distance between the nodes of the network.

The environment in which a network will be deployed is also an important design consideration. A network may be deployed in a noisy environment or in an area in which the communication does not experience a huge amount of interference. For example, the design requirements for a network where its nodes are located in an urban area might be different than the requirements for a network in rural areas. Similarly, the design of an underwater network is different from a network on the ground, and the design of a network installed in an industrial facility is different from the one installed in a home or an office.

Generally speaking, a network can be classified based on many factors. For example, networks can be divided into wired or wireless networks. In a wired network, the nodes are connected to each other using a wire such as twisted pairs or fiber optic cables. In a wireless network, the nodes should have an antenna for data transmission. Similarly, a network can be divided based on the mobility of its nodes. In a non-mobile network, the nodes are stationary, while in a mobile network, the nodes can exchange data while they are moving. A network can also be categorized based on the amount of energy that it consumes. For example, in a low-power network, a great deal of effort has been spent on power saving strategies in order to reduce power consumption. Networks can also be classified based on their topology. Topology is the arrangement of the nodes and links in a network that shows how the nodes in a network are connected to each other. Figure 1.2 shows some of the most popular network topologies that are mesh, star, bus, ring, and tree topologies. Figure 1.2a shows a mesh topology. In a mesh topology, every node has a point-to-point connection to every other node in the network. This guarantees that each node has a dedicated link to carry its traffic to another node. Since nodes are not sharing any link for data transmission, it eliminates the possibility of unpredicted network traffic that causes packet loss and delay in a network. The data transmission also is very secure, since a dedicated link is used for the transmission of data between two nodes. The mesh topology is reliable and robust; if a link becomes faulty, it does not have any effect on the rest of the network. However, the amount of cabling used in a wired mesh network, or the wireless resources required for a wireless mesh network, is excessive. Figure 1.2b shows a star topology. In a star topology, each node is directly connected to a central node in a network, and therefore, each node requires only one link to connect to any other node in the network. Installation

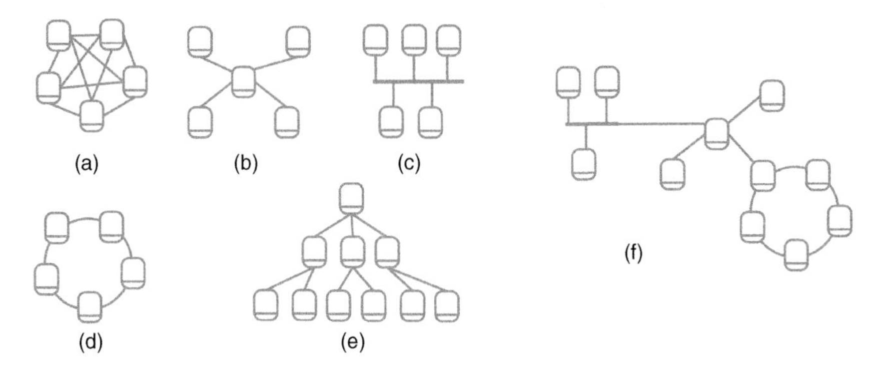

Figure 1.2 Different network topologies: (a) mesh; (b) star; (c) bus; (d) ring; (e) tree; (f) hybrid.

of a network based on the star topology is easy. The network is reliable and robust in the event that any node or any link becomes faulty, except for the central node. The entire star-based network is dependent on the central node, and if this node fails to operate, the entire network goes down. Figure 1.2c shows a bus topology. In this topology, all the nodes are connected to a shared link. Therefore, the deployment of the networks with bus topology is easy and the network uses less number of links as compared to other topologies. However, if the shared link becomes faulty, the entire network goes down. Figure 1.2d shows a ring topology. In this topology, each node has a point-to-point connection with the nodes on either side of it. For two nodes that are far from each other, the data travels along the ring in one direction passing through many nodes until it gets to the destination node. Ease of installation is one of the advantages of this topology. However, a break in a ring can bring down part of the network. Figure 1.2e shows a tree topology. In this topology, there is a parent-child hierarchy of a smaller star topology connection. Several child nodes are connected to a central parent node in a star topology, and the central node is a child node that is connected to its parent central node. This makes the arrangement of the nodes and links to be similar to a tree structure. The tree topology has a flexible and scalable structure. Figure 1.2f shows a hybrid topology. Hybrid topologies combine two or more different topologies. It should be noted that a network topology should be chosen based on the design requirements. There is no perfect, ideal topology that can satisfy the requirements of all applications.

A node must convert its data into an equivalent signal that can pass through a cable or be sent to an antenna. Assume a situation where a node intends to send a file to another node. The file consists of many bits of "0" and "1," and to transmit these bits to another node, the bits need to be changed to an electrical or optical signal. There are many methods to convert data bits to signals. The amount of data bits sent in one second is known as data rate, whereas the number of signal elements sent in one second is known as signal rate.

A network can be interconnected to other networks in order to create a larger network. The devices that are used for this purpose are called internetworking devices. In this chapter, we explain various types of internetworking devices.

1.2 OSI Model

When designing a network, we need to design the nodes of the network or the internetworking devices that connect networks together. The International Standardization Organization (ISO) published the Open Systems Interconnection (OSI) networking model in 1984, which discusses at a high level how a network node or an internetworking device should be designed in terms of its functionalities. In order to separate various functionalities that must be performed by a node or an internetworking device, the OSI model uses a layered approach. Let us start with some clarifications before we get into the layered approach that the OSI model introduces:

- One of the models that has been used for the design and understanding of a network is the OSI model. There are different networking models that can be used for this purpose. For example, the Transmission Control Protocol (TCP)/Internet Protocol (IP) networking model, which we will explain later in this chapter, is a model that has been used widely in the design of many networks. Also, the OSI model was published in 1984, and therefore, all the networks designed and deployed before this year are not based on the OSI model.
- The OSI model is used as a reference for describing and understanding a network. Due to the popularity and robustness of the TCP/IP networking model, many networks are designed based on this model. However, when there is a need to explain network functionalities, the OSI model is usually used. Most educational and research communities use the OSI model in order to explain various functionalities of a network, even if the network is not designed based on the OSI model. You might even see the TCP/IP networking model explained using the OSI model in some publications. Therefore, readers should understand that the OSI model is a reference model, which can be used for describing the functionalities of the nodes of a network, even if the design of that network is not based on this model.
- The OSI model does not discuss a specific protocol, communication standard, an electronic circuit, or a particular hardware or software. It only discusses how all the required functionalities of a node can be separated in a logical manner.
- The OSI model can be used in the design and explanation of different types of networks such as BAN, LAN, CAN, MAN, and WAN. In other words, wired or wireless networks, stationary or mobile networks, all can be modeled based on the OSI model.

The OSI model introduced a seven-layer approach for the design of each node in a network as shown in Figure 1.3. In this section, we explain each of these seven layers in simple language and without going into details.

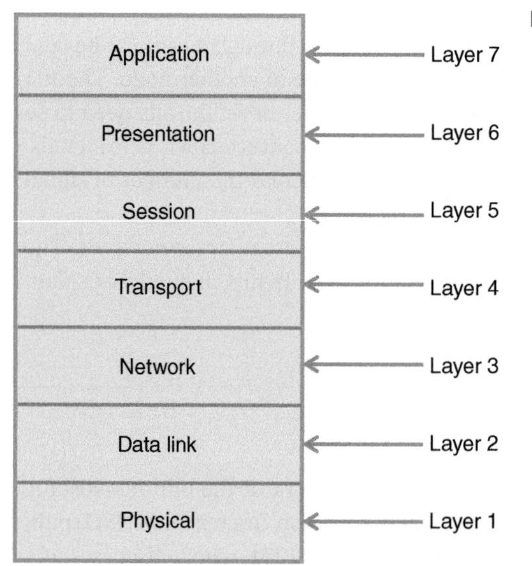

Figure 1.3 Seven layers of the OSI model.

1.2.1 Layer 1 – Physical Layer

The physical layer is the first layer of the OSI model, and it is responsible for transmitting data bits from one node to another. In other words, the physical layer discusses how a "0" or "1" is converted into a signal that is sent to a cable for wired communication or to an antenna for wireless transmission. The physical layer of the transmitter and receiver must be synchronized in order to send data from one node to another. Therefore, synchronization becomes a topic that needs to be addressed at the physical layer. As previously stated, the OSI model is a high-level model, and therefore, it does not discuss the types of synchronization schemes that should be employed for a given network. The physical layer defines how data needs to be transformed into signal based on the maximum distance between nodes in a network. As a result, we may conclude that the physical layer of the OSI model is in charge of data to signal conversion. Working on the physical layer design requires a deep understanding of telecommunication circuits and systems, synchronization, modulation, duplexing, and the design of wire and wireless interfaces. In a very robust physical layer, each "0" gets to the destination node as "0" and each "1" arrives at the destination as "1," even if the environment between the two nodes is highly noisy and there are interferences from numerous sources. It is important for readers to understand that designing a perfectly reliable physical layer is impractical. Because the design of a highly reliable physical layer can be complex and expensive, the physical layer should be designed depending on the requirements of an application. In other words, a node's physical layer should ensure that a "0" or "1" arrives at its destination safely in accordance with the design criteria and by considering the specifications of the communication channel.

1.2.2 Layer 2 – Data Link Layer

The second layer of the OSI model is called the data link layer. This layer is responsible for five important tasks as shown in Figure 1.4. These tasks are addressing, framing, error control, flow control, and access control. We explain each of these tasks in this section.

1.2.2.1 Addressing

Assume a node in a network wants to send data to another node. As previously stated, each node in a network must have a unique address in order to be identified among all other nodes in the network. Actually, most networks use two addresses for each node: a local address and a global one. Local addresses are used in a local setting, while a global address is used to send data through wide area networks. Global addresses must be unique globally, while local addresses need to be unique only locally. Even though it is not required, many networks employ local addresses that are also globally unique. Defining and using a local address for a node is a responsibility of the data link layer.

1.2.2.2 Framing

When a node intends to send data to another node, it makes a packet that has several data fields. Assume a network node with address 5 wants to send a large file to a node with a local address

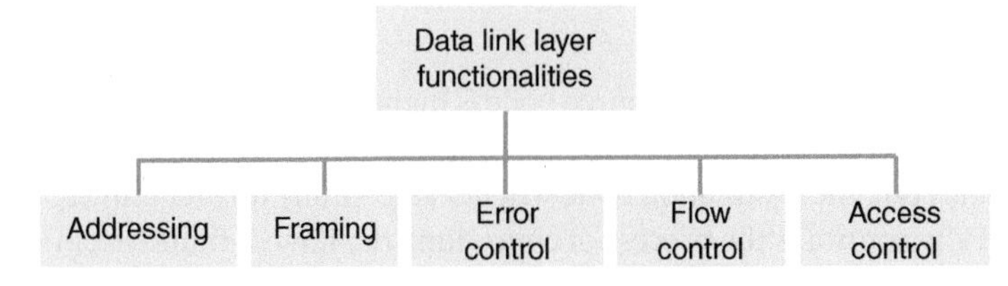

Figure 1.4 The functionalities of Layer 2 in the OSI model.

of 3. Layer 2 of Node #5 would get some portions of this file as its data field, and it adds other fields to it in order to construct a packet. A packet consists of several fields such as the address of a source node, the address of a destination node, and possibly other fields, in addition to its data field. A packet in Layer 2 is called a frame. In our example, Layer 2 of Node #5 gives the frame to its physical layer, bit by bit. Keep in mind that the physical layer is unaware of the concept of a frame. It only converts the frame's bits into signals and sends them to the physical layer of Node #3 via the communication link between the two nodes.

1.2.2.3 Error Control

Due to noise and interference, the transmitted "0" or "1" by a physical layer of a source node might not be received and detected by the physical layer of the destination node as a "0" and "1," respectively. Detecting a "1" as "0" or a "0" as "1" is an error. Detecting several consecutive corrupted bits is called a burst error, while a single error happens when only one bit among many bits becomes corrupted. The possibility of a burst error is usually higher as compared to a single bit error in most networks.

To enable the destination node to detect a corrupted frame, we can add some extra bits to a Layer 2 frame. For example, we can mathematically add all the bytes of a frame together and use the least significant 16 bits of the addition as a checksum value. When the source node intends to create a frame in this case, it adds a checksum field to its frame. When the data is received by Layer 2 of the destination node, this layer calculates the checksum of the frame. If the value of the checksum is not the same as the value in the checksum field, then this is an indication that an error has occurred. Clearly, the destination node does not know which bits are corrupted, or how many bits in a frame are wrong. It does not understand if the error is a single error or a burst error. However, it understands that the frame is not exactly the same as the transmitted frame.

Error detection is the process of detecting errors in a frame. Besides checksum, there are other methods to detect errors in a frame. One of the methods that has a strong property in detecting both single bit and burst errors is called Cyclic Redundancy Check (CRC). CRC has been used widely in various networking applications.

Error correction is the process of correcting errors in a frame. We can add a field to the frame in such a way that can help us to find out which bits are corrupted in a frame. Even though error correction has been utilized in some networks to fix a limited number of errors in a small frame, it is not possible to correct numerous corrupted bits inside a long frame.

We need a way to tackle the problem when an error is detected in which no error correction is used or can correct the frame. The solution to this problem is an error control mechanism. For example, when the destination node receives a corrupted frame, it can ask for the retransmission of the corrupted frame as an error control mechanism. Dropping the corrupted frames without taking any further action is a simple error control approach. In this case, it is obvious that the destination node does not have the corrupted frames. Layer 4 can remedy this problem by asking the source node to resend the erroneous frames, as we will see later in this chapter. In other words, an error control mechanism at Layer 2 may be as simple as detecting and dropping corrupted frames, but this places the burden of having a strong error control strategy on Layer 4.

1.2.2.4 Flow Control

It would be problematic if the speed of transmission of a source node is higher than the speed of the reception or processing of the destination node. In this situation, the destination node may need to buffer the data. If the buffer becomes full, the destination node will not receive any further frames, requiring frame retransmission. Flow control is the process of controlling the flow of frames from a source node to a destination node. For example, a source node can transmit more frames to a destination node only when the destination node allows the source node to do so. Taking no action would be a basic flow control mechanism. It is possible that you lose frames as a result of this. If this occurs, a higher layer must address the issue of lost or dropped frames.

1.2.2.5 Access Control

The method to determine which node in a network is allowed to access the network at a specific time is called access control. Many access control methods have been designed and used in different communication systems. Each of these methods uses a different strategy to determine which node has the right to use network resources and transmit data. We explain three strategies that have been used for access control: random access, master/slave strategy, and channelization.

Multiple Access (MA) is the simplest method based on the random access strategy, in which each node can access the network at any time. This may cause collisions and can result in corrupted frames. Another method based on the random access strategy is Carrier Sense Multiple Access (CSMA), in which each node checks the communication link before transmission. A node can access the network, if and only if the communication link is not busy. This method reduces the possibility of collisions, but it does not eliminate it completely. Consider a scenario in which multiple nodes begin transmitting at the same time. Carrier Sense Multiple Access/Collision Detection (CSMA/CD), which is similar to CSMA but checks for collisions during data transmission and stops the transmission of the rest of the frame if a collision is detected, is another way of access control which utilizes the random access strategy.

Another strategy for access control is based on a master and slave approach. In this method, a node in a network plays the role of a central unit (master node) and other nodes are the peripheral units (slave nodes). In this access control strategy, access is granted to a peripheral node by a central node.

Channelization is another strategy that has been used as an access control method in many networks. Time Division Multiple Access (TDMA) is an example of an access control method in this class. In TDMA, each node of a network is granted a specific time to access the network. Frequency Division Multiple Access (FDMA) is another method in which each node is given a part of the spectrum to access the communication channel at all times.

1.2.3 Layer 3 – Network Layer

Layer 3 is called the network layer. It is responsible for the delivery of packets outside a local setting and across possibly multiple networks. Layer 3 ensures that a packet originating from a node in a local network reaches a destination node that resides in another network. As mentioned earlier, we need a global address for this purpose. Layer 3 provides the addressing and handles the routing of packets in a network. Remember, a Layer 2 address is only locally usable. A Layer 3 address, also called a logical address, is a unique global address. Generally speaking, local networks can be connected to each other to make a larger network. This can be done by using internetworking devices such as routers that route the packets toward a destination node. Therefore, the network layer is responsible for the source-to-destination transmission of data. One of the most popular Layer 3 addresses is an IP address. We discuss IP addresses and IP routers in more detail in this chapter.

1.2.4 Layer 4 – Transport Layer

The transport layer is responsible for five important tasks: port addressing, end-to-end error control, end-to-end flow control, connection control, and congestion control. In this section, we look over these tasks.

1.2.4.1 Port Addressing

A node may run many processes simultaneously. Simply put, a process is an application program running on a node. The network layer ensures that a packet reaches its destination node; however, if a node runs several processes, the network layer does not specify which process in the node should receive the packet. For this reason, we need another address, called a port address or Layer 4 address. This address guarantees that a packet sent to a specified process on a specific destination address reaches the proper process on the correct node.

1.2.4.2 End-to-end Error Control

Similar to Layer 2, Layer 4 is responsible for error control. However, Layer 4 provides an end-to-end error control mechanism. The error control in Layer 2 is for a single frame across a single link, while the error control in Layer 4 is for the packets across the entire source-to-destination path. A packet in Layer 4 usually contains several Layer 2 frames. The end-to-end error control handles any corrupted, lost, or duplicated packet.

1.2.4.3 End-to-end Flow Control

Similar to Layer 2, Layer 4 is responsible for flow control. However, Layer 4 provides an end-to-end flow control mechanism, rather than flow control across a single link.

1.2.4.4 Connection Control

Generally speaking, communication between two nodes can be either connection oriented or connectionless oriented. In a connection-oriented transmission, a source node establishes a connection with a destination node before starting data transmission. During connection establishment, the network reserves the resources for the communication, verifies that the connection with the destination can be established, and ensures that the destination is ready to receive data. After a connection is established, then the nodes can exchange data. In the end, the connection is required to be released. In other words, a connection-oriented communication has three phases: connection establishment, data transfer, and connection tear down. In connectionless-oriented communication, the source node sends its data toward the destination node without any connection establishment. For example, making a phone call is an example of a connection-oriented communication, while sending a letter through the mail is considered as a connectionless-oriented communication. A transport layer can be designed to be connectionless or have a connection-oriented architecture.

1.2.4.5 Congestion Control

The traffic between two nodes in a wide area network is unpredictable. Due to the high traffic in the network, the links between any two nodes of the network can become congested at times. Network congestion results in packet loss or delay in arrival of packets at the destination. Congestion control is the process of controlling the flow of traffic in order to avoid or reduce the possibility of congestion, as well as taking proper actions when the network is congested. For example, if the source node knows that the network is congested, it may postpone the transmission of its data to another time in order to reduce network congestion. Overall, there is no benefit in transmitting a packet that most likely will be lost. In other words, when the network is congested, a source node can assist the network in better managing the situation by eliminating the need to initiate any additional transmissions.

1.2.5 Layer 5 – Session Layer

The session layer is required when there is a need for a dialog control during data communication between two nodes. In many networking applications, there is no need to have a dialog control during communication, and therefore, the session layer is not needed. But it is possible for a node to have a dialog with another node during data transmission. For example, a node that is sending a file containing a main document and some amendments may want to maintain a session with the other node to make sure that the main document has been delivered before sending the amendments.

1.2.6 Layer 6 – Presentation Layer

The presentation layer is concerned with data formatting. A node that wants to deliver a file to a destination node, for example, might change the file's format, compress it, or encrypt the data inside the file. This can be done for several different reasons. For instance, compressing a file reduces the amount of traffic associated with the transmission of the file. When compressed data reaches the destination node's presentation layer, it should be decompressed. Data encryption can be performed to increase the security of data transmission. Clearly, when encrypted data arrives at the presentation layer of the destination node, it should be decrypted. The format of data may also need to be changed from a sender-dependent format to a common format that is acceptable by the destination node. In this case, the presentation layer performs translation from one format to another so that the destination node can understand it.

1.2.7 Layer 7 – Application Layer

Application layer enables users or programs to access the network and interface with it in order to get access to services such as file transfer, web access, or electronic email. There are many popular application-layer protocols that have been used by users and programs in order to provide them access to various network services.

1.3 Header Encapsulation

Let us see how data is exchanged between two nodes on the OSI network. Assume a node intends to send a large file to another node on the network. The source node breaks down the large file into multiple pieces. Each piece becomes a Layer 7 data unit, represented by D_7, as shown in Figure 1.5. In this section, we discuss how D_7 goes through different layers of the OSI model until it is transmitted through a communication link and goes toward the destination node.

The application layer of the source node adds a header to its data unit (D_7). Let us represent the header of the application layer by H_7. The combination of $D_7 + H_7$ makes D_6, which is a data unit for Layer 6. The presentation layer also adds a header, H_6, to the presentation layer's data unit and sends it to Layer 5. For example, the header, H_6, has information about the format of data, whether the data is encrypted or compressed, or the type of encryption or compression that has been used. The presentation layer at the destination side must use this information to take proper actions in order to decrypt or decompress the data. As can be seen from Figure 1.5, at each layer, a header is added to the data unit. At Layer 2, a trailer is also added to the data. Then the data at the physical layer changes to an electromagnetic or optical signal and goes through the communication link.

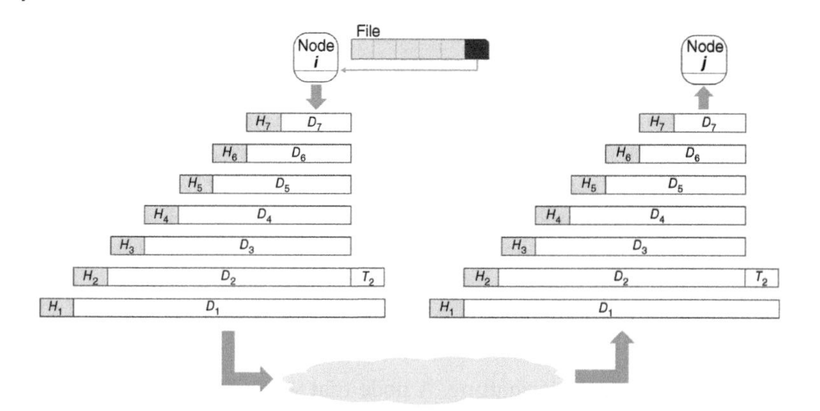

Figure 1.5 Header encapsulation.

When the transmitted signal arrives at the destination, it passes through the physical layer of the destination node, and it is transformed back into "0" or "1." The data then goes upwards through OSI layers. At each layer, the associated header or possibly the trailer is removed from the data, and based on the information inside the header, appropriate actions take place.

A packet in Layer 2 is called a frame. A frame consists of Layer 2 header, Layer 2 data unit, and a trailer. A packet in Layer 3 and Layer 4 may be called a datagram or a segment, respectively. Therefore, whenever we talk about a frame, we usually mean a packet in Layer 2.

1.4 Layer 2 – Ethernet

In the 1980s, IEEE initiated a project to build standards for communication among equipment from different manufacturers of local area networks. This project was called Project 802. The IEEE has defined several standards for local area networks. Among them Ethernet was defined as the IEEE 802.3 standard. Ethernet evolved during several decades to provide a communication standard that could be used in faster and more efficient local settings and became the most popular standard for local area networks.

Since its creation, Ethernet has gone through many generations: Legacy Ethernet (10 Mbps), Fast Ethernet (100 Mbps), Gigabit Ethernet (1 Gbps), Ten-Gigabit Ethernet (10 Gbps), Forty-Gigabit Ethernet (40 Gbps), Hundred-Gigabit Ethernet (100 Gbps), and beyond 100 Gbps Ethernet. At the beginning, Ethernet intended to make its new generations always 10 times faster than the previous ones. For example, 1 Gbps Ethernet was 10 times faster than Fast Ethernet, and Fast Ethernet was 10 times faster than legacy Ethernet. However, Ethernet agreed to make some changes from its 10 Gbps standard, which actually had two modes of operation at 10 Gbps and 9.5 Gbps. Also, 40 Gbps was just four times faster than 10 Gbps, not 10 times faster. The reason for this change was that Ethernet intends to become a standard not only for local area networks but also for metropolitan area networks and wide area networks. For example, 9.5 Gbps and 40 Gbps are the data rates that are used by wide area network technologies.

Ethernet is one of the most important Layer 2 standards and has been widely used to establish local area networks. It performs all the functionalities of the second layer of the OSI model. As discussed earlier, Layer 2 is responsible for framing, addressing, error control, flow control, and access control. We discuss how Ethernet performs each of these tasks in this section.

1.4.1 Framing

An Ethernet frame has a very simple frame format that consists of the following fields: destination address, source address, length of data, user data, and CRC. The length of an Ethernet frame is variable. However, the maximum length of an Ethernet frame is 1518 bytes. The length of the destination and source addresses are six bytes. The third field of each Ethernet frame indicates the length of user data in the frame. This field has 2 bytes. The CRC field has 4 bytes, and user data field can be up to 1500 bytes. Ethernet tried to maintain the structure of its frame as it evolved from legacy Ethernet to the newer versions of Ethernet. It should be noted that there are also jumbo Ethernet frames that can carry up to 9000 bytes, which is more than the 1500 bytes originally set by the Ethernet standard. Many Gigabit Ethernet switches support jumbo frames as well as the standard frame size.

1.4.2 Addressing

In order for two Ethernet-enabled nodes to communicate, each one should have an Ethernet interface, with an address assigned to it. Each Ethernet address is 6-bytes long, and it is also known as a physical address, Layer 2 address, or Media Access Control (MAC) address. The Ethernet address is represented in hexadecimal notation in which each byte is separated by a colon. For example, an Ethernet address can be expressed as 05:03:13:18:2B:1F.

In a local area network, we may need to send unicast, multicast, or broadcast messages. Therefore, it is necessary to define Ethernet multicast and broadcast addresses in addition to the unicast addresses. An Ethernet device is always assigned to a unicast address. However, if an Ethernet device wants to send a frame to a group of Ethernet devices in a network, the multicast address assigned to that group should be used as the destination address. In this case, the sender should use a unicast Ethernet address as its source address and a multicast Ethernet address as its destination address. If the least significant bit of the first byte in an Ethernet address is 1, the address is a multicast address; otherwise, the address is a unicast Ethernet address. A broadcast address is a multicast address that includes all the nodes of the network. An Ethernet broadcast address consists of 48 bits of "1" and can be expressed as FF:FF:FF:FF:FF:FF.

Since an Ethernet address has 6 bytes of hexadecimal numbers (48 bits), it can represent $(2)^{48}$ unique addresses. This is a huge address space, and therefore, Ethernet did not need to increase its addressing space as it evolved during the years. One of the interesting aspects of Ethernet is its addressing scheme. Even though, originally Ethernet was used in local area networks with a limited number of nodes, its address space was large. It has been shown that a small address space can create huge problems. For example, due to a limitation in their phone address space, many telephone companies were forced to change the phone numbers of their customers and use phone numbers with additional digits, as the number of their subscribers in a region increased. The Y2K problem also happened as a result of limitations in recording date information, where a four-digit number representing a year had been contracted to a two-digit number. It is more convenient to use a wider address space, especially if the number of nodes in a network is likely to grow significantly. Using a large address space, however, increases the overhead, since each frame must carry the extra bits of a larger address space.

1.4.3 Error Control

Ethernet uses a very simple error control mechanism. It only performs error detection on each frame. To do that, each frame has a CRC field. By checking the CRC field, it can detect whether an error has occurred or not. In the case of an error, the frame will be dropped and no further action will be performed. In this case, it is the duty of Layer 4 to have an error control mechanism for the retransmission of the dropped frame. Due to robust communication schemes used in the physical layer, and good quality of cables used in the industry today, there is no need for Ethernet to have an extensive error control mechanism. If a frame becomes corrupted, Ethernet only drops the frame, and Layer 4 is responsible for taking care of the missing frame.

1.4.4 Flow Control

Similar to error control, the flow control used in Ethernet is simple, and a source node does not need to wait for any acknowledgment from a destination node before sending the next frame. It gives the responsibility to Layer 4 to perform end-to-end flow control.

1.4.5 Access Control

As discussed earlier, access control defines how different nodes of a network access the shared medium. Ethernet, originally was a CSMA/CD technology. In this technology, a node needs to listen to the shared medium channel to make sure no other node is transmitting data and using the channel. If the channel is free, then the node can transmit its data, and at the same time, it should listen to the channel to detect any possible collision. In case of a collision, the node should stop its transmission and try transmission of its frame at a later time. After each consecutive collision, the node should usually wait for a longer period of time before it is allowed to transmit again. The waiting period is randomly selected from a range, which will be increased after each consecutive collision.

The legacy Ethernet became famous for its CSMA/CD access control scheme and maintained this method when it published Fast Ethernet and Gigabit Ethernet standards. Actually, CSMA/CD was not often used with the faster networks. The Fast Ethernet standard and faster Ethernet standards use full duplex switched Ethernet technology, a peer-to-peer connection between an Ethernet device, and an Ethernet switch with no possibility of collision. However, Ethernet marketed itself as a CSMA/CD technology and supported this technology as part of its newer standards until finally dropped it in 10 Gbps standard. Therefore, 10 Gbps Ethernet and faster ones do not support CSMA/CD anymore.

1.5 Layer 3 – IP

Many Layer 3 protocols have been designed and used as part of various communication systems. However, the Internet Protocol (IP) is by far the most popular one, which has been used widely as a network-layer protocol in order to provide the addressing and routing requirements of various communication systems. We can divide the networks into two types based on their network layers: IP-based and non-IP-based networks. As the name suggests, IP-based networks are the ones whose network layer uses the IP protocol. In other words, IP-based networks use IP addresses and each of their packets has an IP header. In this section, we discuss IP addresses, the assignment of IP addresses in a network, and the IP header structure.

An IP address is a unique global number that is used to identify a node or a network device. There are two types of IP address formats available today: IP Version 4 (IPV4) and IP Version 6 (IPV6). There is an organization called Internet Assigned Numbers Authority (IANA), which is responsible for tracking IP addresses, domain names, and protocol parameters that are used by the Internet standards.

When an organization needs IP addresses, it can ask IANA to provide the organization with the required IP addresses. Actually, organizations do not go directly to IANA to get their IP addresses and instead obtain them from their service providers, which provides the organization with a block of IP addresses. This block consists of several consecutive IP addresses. The length of the block is based on the needs of the organization. For example, consider an organization that has 1600 computers, printers, and other IP-based electronic devices. Assume that each device requires only one IP address for its operation. The organization also has many internetworking devices (switches and routers) that require 200 IP addresses. In this case, the organization needs a total of 1800 IP addresses for its operation. It is a good practice to obtain some extra IP addresses considering the possibility of a network expansion. Therefore, the organization may decide to ask for $2^{12} = 2048$ IP addresses. In this scenario, the organization receives a block of 2048 IP addresses, and assigns the IP address among the network nodes and internetworking devices. Generally speaking, a computer may need one or several IP addresses. You can install several Network Interface Cards (NIC) on a computer, and give each interface card an IP address. In this case, the computer plays the role of several nodes of the network where all these nodes are located in a physical device (computer). Also, some internetworking devices such as routers usually need several IP addresses for their operations.

Initially, IANA assigned IP addresses as blocks of 256, 65 536, and 16 777 216. This method was called classful addressing, in which an organization could ask only for a block of 256 IP addresses (also called a class C address), 65 536 IP addresses (also class a class B address), or 16 777 216 IP addresses (also called a class A address). For example, a small company that needed 100 IP addresses would receive 256 addresses. Similarly, if a service provider would ask for one million IP addresses, they would be given a block of class A addresses that had a lot more than what was requested. The assignment of classful addressing was easy and allowed the organizations to have enough IP addresses in case of their network expansion. As the Internet grew and there was a need for more and more IP addresses, it became clear that classful addressing was not efficient. In classless addressing, the length of a block of IP addresses can be expressed as 2^x ($x = 1, 2, \ldots$). In this case, a home that needs a small number of IP addresses may get four IP addresses(2^2), a small organization may be provided with 128 IP addresses (2^7), a small local service provider may ask for 65 536 IP addresses (2^{16}), and a larger service provider can request for 4 million IP addresses (2^{22}).

IPV4 identifies almost 4 billion numbers (2^{32}). Therefore, each IP address is a number from zero to four billion. IPV6 has a larger address space and can identify (2^{128}) IP addresses. With the exponential growth of the Internet, there were some concerns that the number of IPV4 addresses is not enough to satisfy the needs of the increasing number of nodes on the Internet. These concerns started in the early 1990s, but IPV4 addresses are still in use, since there are a number of techniques that are used in order to assign IPV4 addresses more efficiently. We will discuss some of these techniques later in this section. IPV6 addressing is also in use today and all IPV4 addresses will be transitioned to IPV6 eventually. Due to its large address space, IPV6 can guarantee that there is no need for extending the IP address space for some time.

The notation for representing an IPV4 address is a.b.c.d, where a, b, c, and d are decimal numbers between 0 and 255. As mentioned earlier, the IPV4 address space contains almost 4 billion addresses. Here, 0.0.0.0 represents the first number that is 0, 0.0.0.1 represents 1, 0.0.0.255 represents 255, and 0.0.1.0 represents 256. The number associated with a.b.c.d can be expressed as (1.1):

$$a \times 256^3 + b \times 256^2 + c \times 256^1 + d \tag{1.1}$$

Since a, b, c, and d are numbers from 0 to 255, each can be represented by one byte, and therefore an IPV4 address is a 4-byte address. The notation representing the IPV6 address is a:b:c:d:e:f::g:h, where a, b, c, d, e, f, g, and h are 16-bit hexadecimal numbers. As can be seen, these hexadecimal numbers are separated by colons. For example, an IPV6 address can be expressed as 002F:0000:0001:0000:0000:00FF:1000:01CD. This representation is long, and there is a method to abbreviate an IPV6 address. The rules for IPV6 abbreviation are as follows:

- The leading zeros of a hexadecimal number can be omitted. It should be noted that trailing zeros cannot be omitted. Therefore, the IPV6 address discussed earlier can be expressed as 2F:0:1:0:0:FF:1000:1CD.
- Consecutive hexadecimal numbers consisting of zeros can be replaced by :: symbol (this can be done just once). Therefore, the IPV6 address discussed earlier can be expressed as 2F:0:1::FF:1000:1CD.

An IPV4 or IPV6 address can be represented in Class Inter Domain Routing (CIDR)) format. In CIDR notation, there is a prefix that is an IP address and a suffix that indicates how many bits of the address are not part of the entire address block. The prefix and the suffix are separated by a slash (/) mark. For example, 192.20.10.0/31 is an IPV4 address in CIDR notation, whose block has just 1 bit (32-31) or 2 addresses. In other words, 31 bits out of 32 possible bits of an IPV4 address are not part of this block of addresses. Therefore, there are only two addresses(2^1) in this block. Even though the CIDR notation represents an IP address, we can look at it as a representation of a block of addresses as well. We can say that a block of IPV4 addresses can be expressed as a.b.c.d/n, where the length of the block is 2^{32-n}. For example, 32.100.21.0/30 represents a block of four IP addresses 32.100.21.0, 32.100.21.1, 32.100.21.2, and 32.100.21.3. Similarly, the same can be said about an IPV6 block of addresses. For example, 2001::DB00/120 is a block of 256 IPV6 addresses ($2^{128-120}$), starting from 2001::DB00 and ending with 2001::DBFF.

When an organization receives a block of IP addresses, it may want to distribute these addresses among its different divisions. This process is called subnetting. Similarly, a large Internet Service Provider (ISP) may have a huge number of IP addresses and need to give its addresses to different homes, offices, companies, and even smaller ISPs. Let us demonstrate this with simple examples.

Example 1.1 A Simple Subnetting Exercise

A company receives a block of 256 IPV6 addresses F0::25:0/120 and needs to make three subnets. The first subnet has 128 addresses, while each of the other two subnets needs 64 IP addresses. We can make these three subnets as:

- Subnet 1: F0::25:0 - F0::25:7F expressed as F0::25:0/121
- Subnet 2: F0::25:80 - F0::25:BF expressed as F0::25:80/122
- Subnet 3: F0::25:C0 - F0::25:FF expressed as F0::25:C0/122

Example 1.2 Another Subnetting Exercise

An ISP has a block of addresses starting with 110.10.0.0/16. These addresses must be given to two groups of customers. The first group contains 64 customers and each customer needs 256 IPV4 addresses. The second group contains 128 customers and each customer needs 128 IPV4 addresses. Let us design these subnets.

The first group requires 16,384 (64 × 256) IP addresses. We can design 64 blocks for this group as:

- Customer 1: 110.10.0.0 --- 110.10.0.255 (110.10.0.0/24)
- Customer 2: 110.10.1.0 --- 110.10.1.255 (110.10.1.0/24)

⋮

- Customer 64:110.10.63.0 ---110.10.63.255 (110.10.63.0/24)

The second group also requires 16,384 (128 × 128) IP addresses. We can design 128 blocks for this group as:

- Customer 1: 110.10.64.0 --- 110.10.64.127 (110.10.64.0/25)
- Customer 2: 110.10.64.128 --- 110.10.64.255 (110.10.64.128/25)

⋮

- Customer 128: 110.10.127.128 ---110.10.127.255 (110.10.127.128/25)

In the next section, we discuss IPV4 and IPV6 headers and the techniques that made the assignment of IPV4 addresses more efficient.

1.5.1 IPV4 and IPV6 headers

As mentioned in Section 1.3, each packet in Layer 3 contains two parts: a header and a data section. The IP header is an important part of a packet and contains information for routing and delivery of a packet across a network. Routers read the IP header of a packet and decide how to route the packet. The length of an IPV4 or IPV6 packet can be up to 65 536 bytes. Header has a variable size and contains a fixed length section, called base header, and a variable section. The length of a header in IPV4 is from 20 to 60 bytes. The base header is 20 bytes, and the variable section can be up to 40 bytes.

A header consists of several fields. Here, we explain briefly the information in the base header for both IPV4 and IPV6. Let us start with the IPV4 header as shown in Figure 1.6. As can be seen from this figure, the base header for IPV4 has 12 fields. We now provide a short description of each of these fields:

- Version: This field defines the version of the IP protocol. For IPV4, this field is 4.
- Header length: This field shows the length of the header in 4-byte words. For example, a value of 5 in this field represents that the header has a length of 20 bytes (5 × 4). This means that the header does not have any variable section. A value of 6 in this field shows that the header consists of a 20-byte base header and a variable section that is 4 bytes. The variable section is optional, and its length can be up to 40 bytes.
- Type of Service (ToS): This field can be used to differentiate among different types of data that is carried by an IP packet. By giving different service types, different priorities can be defined. A router reads the service field of an IP header and provides better service to packets with higher priorities.

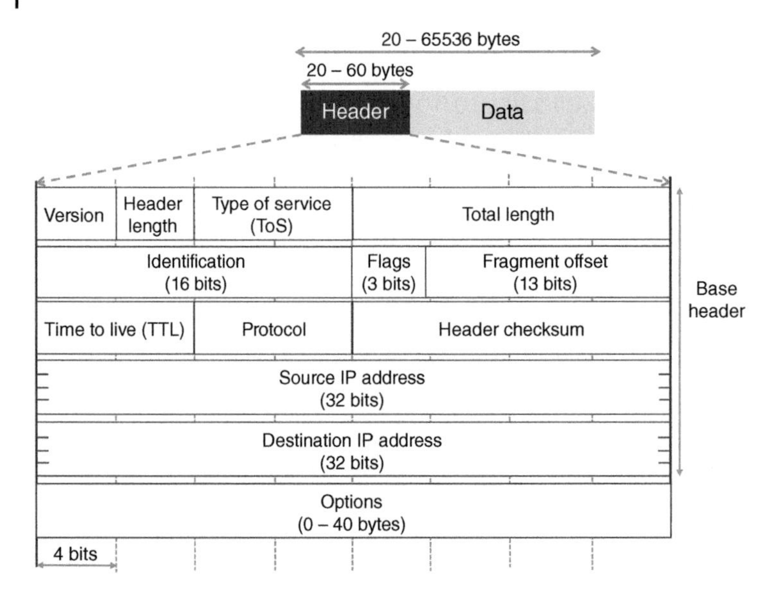

Figure 1.6 IPV4 header format.

- Total length: This field shows the total length of the header and data.
- Identification, flags, and fragment offset: These three fields are mostly related to the fragmentation of an IP packet. An IP packet might need to be fragmented in order to pass through a Layer 2 frame. For example, an IP packet can be up to 65,536 bytes, while an Ethernet frame is only 1500 bytes. To be able to send an IP datagram that is larger than 1500 bytes on an Ethernet layer, the IP datagram needs to be fragmented. Layer 3 on the destination node should be able to assemble all the fragments together and make the original IP datagram.
- Time to Live (TTL): As an IP packet travels from a source to a destination node, each router along the way reads this field and reduces its value by one. If the value of this field becomes zero, the router drops the packet. In other words, this field limits the number of routers that a packet can go through. This ensures that a packet does not loop on the Internet without reaching its destination. Without this field, there is a possibility that a packet becomes corrupted and stays on the network, going from one router to another without getting to its final destination. The value inside this field limits the lifetime of a packet and the number of routers that it can go through.
- Protocol: This field defines the type of protocol used in Layer 4. There exists many Layer 4 protocols. One of the most important Layer 4 protocols is TCP, which we will discuss in the next section.
- Header checksum: This field is used to ensure that the header information is not corrupted. The source calculates the sum of all the bytes in a header and stores the least significant 4 bytes in this field. The destination node as well as all the routers along the way calculate the sum of all the bytes in the header and compare it to the value of the checksum field. When a router gets a packet and notices that the checksum value does not match the least significant 4 bytes of the sum of all header bytes, the router drops the packet.
- Source IP address: This is a 4-byte field that contains the IP address of the source node.
- Destination IP address: This is a 4-byte field that contains the IP address of the destination node.

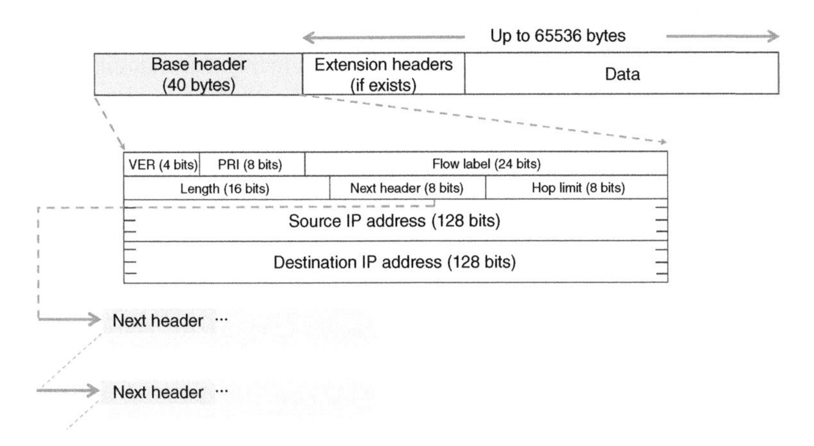

Figure 1.7 IPV6 header format.

We now explain the header section of an IPV6 packet, as shown in Figure 1.7. The length of the base section in IPV6 is 40 bytes. The IPV6 base header contains several fields. Let us give a short description of each of these fields.

VER: This field shows the version of the IP, and therefore, the value of this field for IPV6 is 6.

PRI (Priority): This field shows the priority of a packet with respect to other packets. If the network is congested, the higher priority packets have a better chance of going through the network without being dropped as compared to the lower priority packets.

Flow label: This field has 20 bits, and it indicates how a particular flow of traffic needs to be handled. A flow of traffic is a collection of packets that share certain characteristics. For example, they may require the same security specifications; they may go to the same destination address or at least part of their traveling path are the same, or they may have the same Quality of Service (QoS) requirements. To support different flows of traffic, routers should have a flow table in addition to their routing table. Routers can look at their flow table to decide how to route a packet that belongs to a specific flow of traffic.

Payload length: This field shows the length of a packet excluding the length of the base header.

Next header: This is used to point to another item of the header that defines an optional functionality. Each optional item can point to another one. In IPV6, we can define many optional functionalities. Depending on the application, the required functions can be added by a sequence of pointers. If there is no need for any extra option, this field is similar to IPV4's protocol field.

Hop limit: This is similar to the TTL field that we explained for the IPV4 header.

Source IP address: This is a 16-byte field that contains the IP address of a source node.

Destination IP address: This is a 16-byte field that contains the IP address of a destination node.

1.5.2 Improving IPV4 Address Assignments

As mentioned earlier, IPV4 has an address space of around 4 billion addresses. With the growth of the Internet, IPV4 seemed to be short in providing enough IP addresses for an increasing number of IP nodes. However, there were a few techniques that made IPV4 address assignment more efficient. Because classful addressing was not an efficient approach for allocating IP addresses, one of these techniques was to adopt classless addressing instead of classful addressing, as we discussed before. We now discuss two more techniques that have paved the way for more efficient assignment of IP addresses.

So far, we have discussed how IP addresses are assigned in a static manner. When a node is assigned a static IP address, it is given an IP address that it can keep permanently. For example, if a device is turned off for a long time and no longer requires an IP address, the IP address is retained. It is obvious that keeping an IP address, while it is not in use, is inefficient. Dynamic Host Configuration Protocol (DHCP) is a protocol that brings the concept of assigning IP addresses dynamically. In other words, an IP address is assigned to a node, when the node requests to receive an IP address. A DHCP server provides the node with an IP address for a limited amount of time. If the node needs to use the IP address for a longer period of time, it can renew its lease with the DHCP server. Otherwise, the server can assign the same IP address to another node. A company may prefer to have a smaller set of IP addresses and assign them dynamically than to have a larger set of addresses and assign them statically. This certainly allows for more efficient use of IP addresses.

The second technique is Network Address Translation (NAT), which can substantially save IP addresses. The idea behind NAT is that most organizations usually need a large set of addresses internally and can work with a small set of IP addresses externally. There are many messages exchanged inside an organization and those messages can use an internal IP address, while when the nodes need to communicate with the outside world, they need to use external IP addresses. As a result, the Internet authorities have set aside three sets of IP addresses for internal use. These addresses are called internal or private addresses. Private IP addresses are intended for internal use and cannot be used on the Internet. The private addresses are free to use locally and can be used by anyone. IANA assigns the following three sets of addresses as private IP addresses:

- 10.0.0.0 ---- 10.255.255.255 (2^{24} addresses)
- 172.16.0.0 --- 172.31.255.255 (2^{20} addresses)
- 192.168.0.0 ---- 192.168.255.255 (2^{16} addresses)

These addresses can be used internally by any organization. When an organization needs to send a packet through the Internet, they need non-private IP addresses. To accomplish this task, organizations need to set up an NAT router to do the translation between the internal and external addresses. This is shown in Figure 1.8, in which three different companies all use the same block of private IP addresses (192.168.1.0/24) internally. However, each company has acquired some IP addresses for its external use. The external addresses are assigned to the NAT router. When an NAT router receives a packet whose source address is a private address, and its destination address is an address on the Internet, the NAT router translates the source address to one of its external

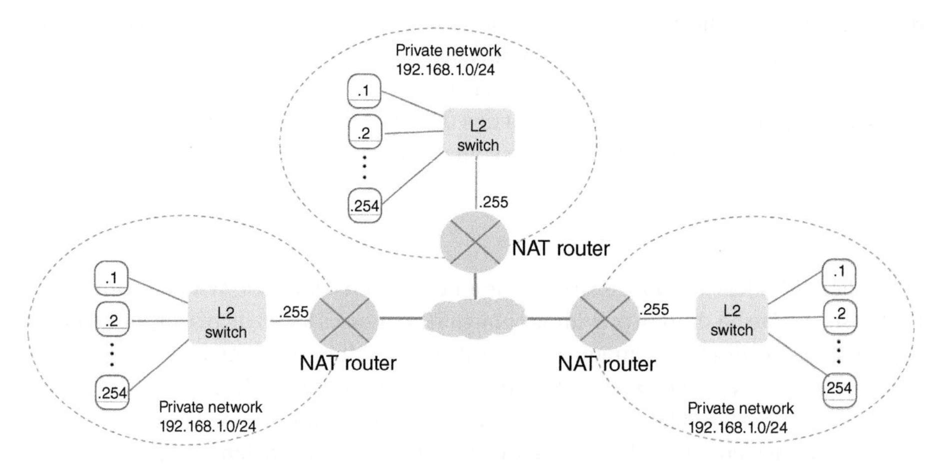

Figure 1.8 An example of using NAT in a network.

addresses and writes this information inside an NAT translation table. This table shows how a private IP address has been changed to an external IP address. When the NAT router receives a packet from the Internet, it can determine the internal IP address of the node by using its translation table.

1.6 Layer 4 – TCP and UDP

In this section, we explain TCP and User Datagram Protocol (UDP). These two important and popular Layer 4 protocols are very robust and have been widely used during the years. TCP is used mainly for non-real-time applications, while UDP is simpler and mostly used in real-time applications where the retransmission of packets is not needed. Both TCP and UDP use a client-server architecture, in which a node of the network that plays the role of a client communicates with another node that has the role of a server.

A node of a network can run several processes at the same time, and therefore, the network should be able to send the packets not only to a destination node but also to a specific process within the node. The network layer can route a packet based on its destination IP address to a destination node, but it does not deliver the packet to a specific process inside the node. To identify a specific process within a node, there is a need for another address. TCP calls this address a port number. A TCP port number is a 2-byte integer number, and therefore, a TCP port number can be a number between 0 and 65 535. TCP port numbers are divided into three groups by the Internet Corporation for Assigned Names and Numbers (ICANN):

- Well-known port numbers: These port numbers are between 0 and 1023. They are assigned to specific servers by Internet addressing authorities. For example, web servers usually use port number 80 or FTP servers use port number 20.
- Registered port numbers: A port number in the range of 1024 and 49 151 is a registered port number. They are not specifically assigned by Internet addressing authorities, but they have been registered by vendors for their own server applications.
- Dynamic port numbers: These port numbers are between 49 152 and 65 535. Each dynamic port number can be used by a TCP client at any time.

When a node uses TCP as its Layer 4 protocol to communicate with another node, it selects a port number at random and uses this number as its source port number. A destination port number is also required by the node to indicate the server side. Consider the following scenario: a node tries to communicate with two separate web servers on the Internet. The node needs to choose a source port number at random and use 80 (because the server side is a web server) as the destination port number when communicating with the first web server. Keep in mind that the TCP header includes the source and destination port numbers. Later in this section, we will go over the TCP header and its fields. In this scenario, let us say the node has chosen 50 000 as its source port number. In addition, the node must connect with the second web server. The node also uses another dynamic port number (let us say 51 000) as its source port number for this connection and the port number 80 as its destination port number. The first web server uses port 80 as its source port and port 50 000 as its destination port when sending a packet back to the node. When the second web server sends a packet to the node, it uses port number 51 000 as the destination port number.

The combination of an IP address and a port number is called a socket. A socket represents a certain process inside a specific node. The concept of port number is the same for both TCP and UDP protocols. We start with the TCP and move on to UDP at the end of this section.

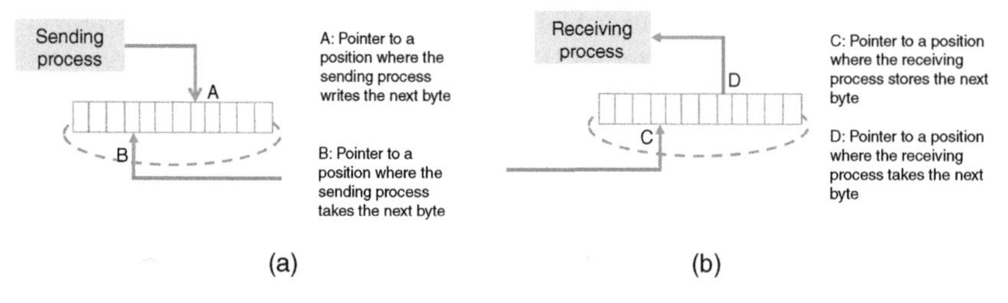

Figure 1.9 TCP buffers: (a) sending process and (b) receiving process.

To understand TCP, we should understand how TCP manages the sending and receiving processes. Since the rate of sending packets by a sending process may not be exactly the same as the rate of receiving and processing of a packet by a receiving process, TCP uses buffers on both sides to store the packets. Therefore, there is a buffer for a sending process and a buffer for a receiving process.

TCP is called a stream-oriented protocol that sends and receives data as a stream of bytes. The data is sent to the sending process buffer byte by byte. Then, in order to create a TCP packet, TCP takes several bytes from this buffer and adds a header to them. A TCP packet is also called a segment. The length of a segment is not fixed. We discuss several reasons that affect the length of a TCP segment. In other words, the number of bytes TCP takes from a sending process buffer to make a segment is variable, depending on many factors. The segment is sent to Layer 3 after the sending process adds the TCP header to the data bytes taken from the sending process buffer. The sending and receiving buffers are shown in Figure 1.9a, b, respectively.

1.6.1 TCP Header

TCP headers can range from 20 to 60 bytes in length. The main part of the header is the first 20 bytes, which are not optional. If optional fields are used, TCP headers can be up to 60 bytes long. The explanation of the optional part of the header is outside the scope of this book. The TCP header is shown in Figure 1.10. The fields associated with the first 20 bytes are discussed here.

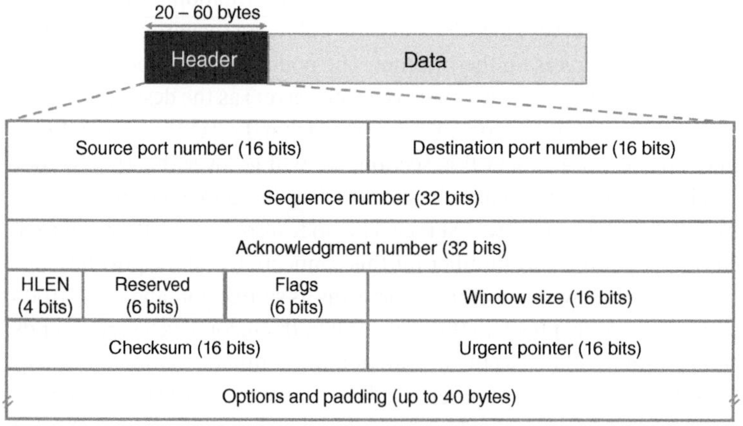

Figure 1.10 TCP header.

- Source port number: This 2-byte field shows the source port number. As discussed, a node chooses the source port numbers randomly.
- Destination port number: This 2-byte field shows the port number of the process at the destination node.

Sequence number: TCP counts the number of the transmitted bytes. The sequence number informs the destination about the number of bytes that have been transmitted so far and the order of segments. TCP chooses a random value as the initial value for the sequence number and adds the number of data bytes that are transmitted in each segment to the initial value. For example, let us consider a situation where a sending process randomly chooses the value of 100, as its initial value, and sends three segments of 100, 250, and 300 bytes. In this case, the sequence number of these three segments is 201(101 + 100), 451 (201 + 250), and 751 (451 + 300), respectively. It is clear that the sequence number identifies the number of bytes sent by the sending process so far. The initial value of the sequence number is also called Initial Sequence Number (ISN), and it is created using a random generator during connection establishment. A sequence number has 32 bits. The sequence numbers are wrapped around when they reach to $2^{32} - 1$.

- Acknowledgment number: This field defines the byte number that the receiving process is expecting to receive. Let us consider a situation that we explained earlier, in which a sending process with an ISN of 100 sends three packets of 100, 250, and 300 bytes. If the receiving process receives the first packet and wants to send an acknowledgment packet, it puts the value of 201 in its acknowledgment field. This means that the receiving process has received the first 100 bytes and expects to receive a packet with the sequence number 201. Now assume that the receiving process receives two more segments and wants to acknowledge these two segments. In this case, the acknowledgment has the value of 751. It should be mentioned that there is no need for a receiving process to send an acknowledgment for every segment that it receives.
- Header LENgth (HLEN): This field shows the length of a TCP header in a 4-byte format. This is a 4-bit field, where a value of 5 indicates that the header has 20 bytes, and a value of 15 shows that the length of the header is 60 bytes. A TCP header is between 20 and 60 bytes.
- Reserved: This is a 6-bit field and it is reserved for future use.
- Control: This field defines six different control bits or flags. These six flags are urgent pointer (URG), acknowledgment (ACK), request to push (PSH), reset (RST), synchronize sequence number (SYN), and terminate the connection (FIN). The detailed descriptions of these flags are outside the scope of this book. However, readers should know that these flags are used to indicate and control some of TCP operations such as connection establishment (SYN+ACK), connection termination (FIN+ACK), or connection reset (RST).
- Window size: This field is used by the receiving process to advertise the number of bytes that it can accept in its buffer before the buffer becomes completely full. This is a 2-byte field, and therefore, the maximum value for this field is 65 536 bytes. This value is also called the receiving window (rwnd).
- Checksum: This 16-bit field contains the checksum of the header and is used to detect the corrupted headers.
- Urgent pointer: There is a possibility that a node wants to send an urgent or critical data ahead of the ones that are in the sending process buffer. In this situation, TCP sets the urgent flag and inserts the urgent data in front of the buffer. Urgent pointer points to the last urgent byte in the data section of a segment. It should be noted that the value of this field is meaningful, only when the urgent flag is set.

1.6.2 TCP Functionalities

TCP has five main functionalities: process-to-process communication, connection control, end-to-end flow control, end-to-end error control, and congestion control. In this section, we discuss these functionalities.

1.6.2.1 Process-to-process Communication

TCP performs process-to-process communication using port numbers. As we explained earlier, TCP needs a source and a destination port number in order to establish communication between the two processes.

1.6.2.2 Connection Control

TCP is a connection-oriented protocol. This means that when two nodes intend to communicate, a virtual connection between the transport layers of these two nodes needs to be created. Generally speaking, a connection-oriented transmission consists of three phases: connection establishment, data transfer, and connection termination. Therefore, TCP needs to establish a connection with the destination node, and after the connection is established, bidirectional data transfer can take place. TCP uses a client-server model in which both the client and server can send data and acknowledgments. The acknowledgments can be sent inside the data packets. Figure 1.11a shows the connection establishment in TCP and the TCP flags that are involved in this process. Figure 1.11b shows the data transfer phase, in which the TCP client sends 500 bytes of data with the sequence number of 3501, and expects to receive packets with the sequence number of 2001. The two numbers of 3500 and 2000 are ISN values for the client and servers. These values have been chosen at the time of connection establishment by the sending and receiving processes, respectively. The client then sends another 500 bytes of data. The server acknowledges receipt of these two segments (Ack:4501) and sends 1000 bytes of data to the client node. Figure 1.11c shows the connection disconnection and the flags that are involved in this process.

1.6.2.3 Flow Control

To control the flow of packets from a source node to a destination node, and in order to ensure that the buffer associated with the receiving process does not become full to disrupt the flow of traffic, TCP should not allow the sending process to send out more data bytes than what the receiver

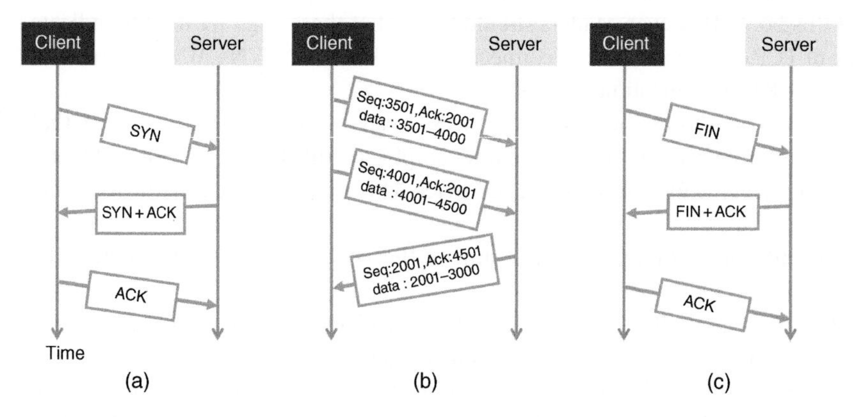

Figure 1.11 TCP connection-oriented phases: (a) connection establishment; (b) data transfer; (c) connection tear down.

buffer can handle. To do this, the sending process uses a sliding window that covers part of its buffer. The sending process is only allowed to send out the bytes that are inside this window. The receiving process advertises the size of the available section of its buffer inside the window size field in the TCP header, and the sending process uses this information to adjust its sliding window accordingly.

1.6.2.4 Error Control

To ensure that a segment has arrived at the receiving process, TCP uses acknowledgments. If a segment is not acknowledged during a defined period, TCP assumes that the segment is lost, or corrupted, and therefore, it needs to be retransmitted. TCP retransmits a segment in two cases: When the sending process sends a segment, it starts a timer at the same time. If the timer expires while the receiving process has not acknowledged the receipt of the segment, TCP transmits the unacknowledged segment again. This is the first case for retransmission of a segment by TCP and is shown in Figure 1.12a. This figure shows a situation where the TCP client transmits segments of 1000 bytes. The client sends the first segment (numbered as 401–1400) and the second segment (numbered as 1401–2400). These two segments go through the network and become acknowledged by the server (Ack:2401). Then the client sends a segment (numbered as 2401–3400) with the sequence number of 2401, which is not received by the server. When the timer associated with this segment times out, the client sends the segment again.

The second case, called fast retransmission, happens when the sending process sends a segment and receives three packets from the receiving process, in which none of them acknowledges the sent segment. An example of the second case is shown in Figure 1.12b. This figure shows a situation where the client transmits segments of 1000 bytes. The client sends a segment with the sequence number of 2401 that becomes corrupted. The client continues sending more segments and receives three acknowledgments from the server during this period. However, the server does not acknowledge receipt of the segment with the sequence number of 2401. Therefore, the client

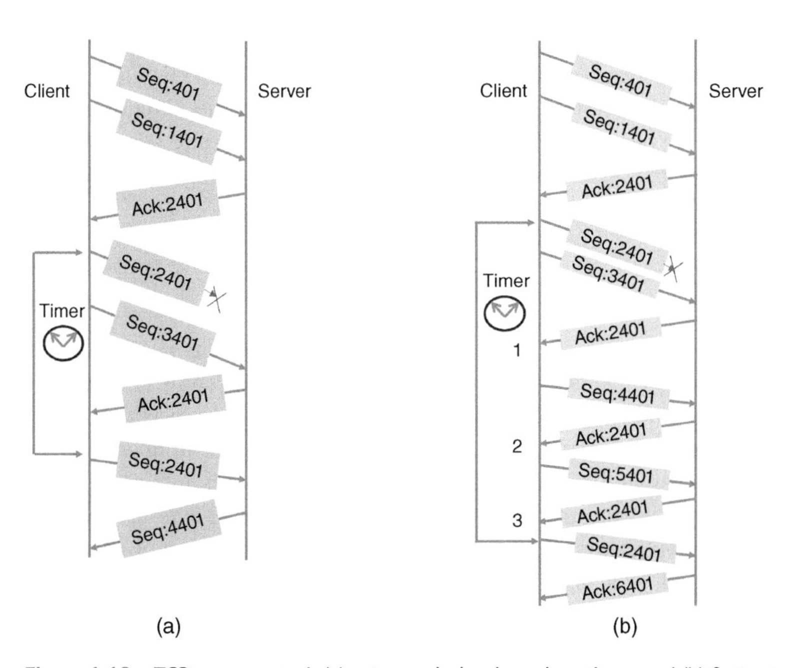

Figure 1.12 TCP error control. (a) retransmission based on timer and (b) fast retransmission.

resends this segment again. Since all other transmitted segments have been received by the server without any issues, the server acknowledges the reception of all other segments (by sending a packet with ACK:6401) as shown in Figure 1.12b.

1.6.2.5 Congestion Control

A receiving process determines the available space of its buffer, also called the receiver window or rwnd, writes this information inside the window size field in the TCP header, and sends the segment to the sending process. The value of rwnd can help the sending process to accordingly adjust the length of its segments. This certainly helps to control the flow of traffic in situations where the speed of processing of the receiving process is slower than the rate of transmission of the sending process. However, this does not solve the problem when the network is congested. Network congestion slows down traffic and results in packet loss. If similar to the receiving process, the network was also able to send the number of bytes that could handle, referred to as congestion window or cwnd, then the sending process could make a better decision on how many bytes to transmit in a segment to satisfy both the network and the receiving process. If that was possible, the sending process could adjust the size of its segment to be less than the minimum of rwnd and cwnd values. Unfortunately, the network is not able to give this information to the sending process, and therefore, TCP has three policies to determine the network conditions and estimate a value as cwnd. The three TCP policies used for congestion control are called slow start, congestion avoidance, and congestion detection. We briefly explain these three policies.

Slow start: When TCP wants to start its transmission, it does not have any idea about the network situation, and therefore, assumes that the network might be congested. TCP chooses a very small value as cwnd at this point, which is called Minimum Segment Size (MSS), and if no congestion is observed, TCP exponentially increases the value of cwnd.

Congestion avoidance: To slow down the exponential growth of cwnd value, TCP starts an additive increase of the cwnd value after cwnd reaches a threshold.

Congestion detection: When TCP begins to detect signs of congestion (unacknowledged segments), it starts to react to the situation. TCP reacts strongly when it does not receive an acknowledgment for a segment within the duration of a timer that is set for that segment. In this situation, TCP goes back to slow start and sets the cwnd value to MSS. In case of not receiving any acknowledgment, when it has received three packets from the receiving process, TCP changes the value of cwnd to half of its value.

1.6.3 UDP

The User Datagram Protocol (UDP) is a simple Layer 4 protocol. As we mentioned earlier, each Layer 4 (L4) protocol is responsible for five tasks: process-to-process communication, connection control, error control, flow control, and congestion control. To handle process-to-process communication, UDP performs exactly similar to TCP (by defining and handling port numbers). UDP is a connectionless-oriented protocol, and therefore, does not need to establish any connection before data transmission. UDP is used mostly for real-time applications in which applications are time sensitive. The real-time applications do not tolerate a large delay and do not need the retransmission of the lost, corrupted, or delayed packets. For this reason, UDP does not perform any error control, flow control, or congestion control. This makes UDP faster and simpler. The UDP header is shown in Figure 1.13, which has only four fields: source port number, destination port number, total header length (header + data), and checksum.

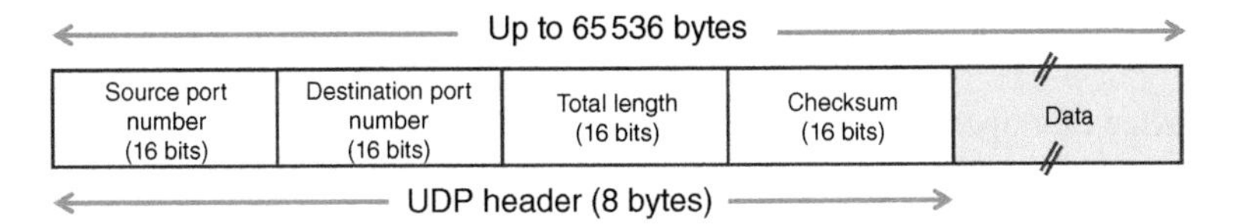

Figure 1.13 UDP header and packet.

1.7 TCP/IP Networking Model

Similar to the OSI model, TCP/IP is a networking model that consists of several layers. This networking model does not consist of only two protocols (IP and TCP), as its name suggests. TCP/IP model has a layered approach. It originally introduced four layers: the network interface layer, Internet layer, the transport layer, and the application layer. It changed to a five-layer model later on. These five layers were the physical layer, data link layer, Internet layer, TCP layer, and application layer. The TCP/IP model can be compared to the OSI model as shown in Figure 1.14. Practically, TCP/IP has been an extremely successful networking model and the five-layer version aligns well with the OSI model.

A protocol suite is a set of protocols that have been designed to work together. The TCP/IP protocol suite consists of many protocols that can operate at one of these five layers of the TCP/IP model. Some of the most important protocols in the TCP/IP protocol suite are shown in Figure 1.14. Some of the application-layer protocols in the TCP/IP protocol suite are Teletype network (Telnet) protocol, File Transfer Protocol (FTP), Simple Mail Transfer Protocol (SMTP), Domain Name System (DNS) protocol, and Simple Network Management Protocol (SNMP). The explanation of these protocols is outside the scope of this book.

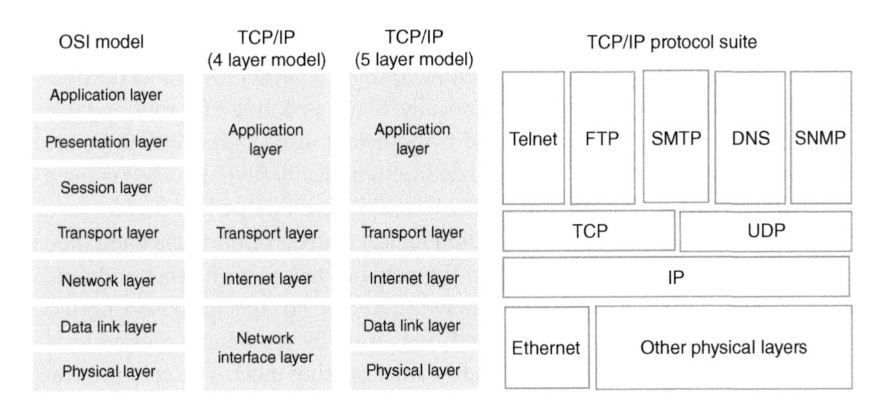

Figure 1.14 TCP/IP networking model and its simplified protocol suite.

1.8 Internetworking Devices

We can define four internetworking devices based on their functionalities as it relates to the OSI model. These four devices are hubs or repeaters that operate in Layer 1 (L1); bridges or switches that operate in L1 and Layer 2 (L2); routers or Layer 3 (L3) switches that operate in L1, L2, and L3; and finally gateways that operate in all layers.

A hub operates only on the physical layer. When this device receives a signal at its input, it regenerates the original signal bit pattern. It does not amplify the signal, rather, it refreshes the signal. A hub forwards every bit of a packet that it receives, and it does not have any filtering capability.

A bridge or switch is a device that operates in both the physical and the data link layer. Similar to a hub, it regenerates the signal that it receives. But it can also check the physical addresses (L2 addresses) contained in a packet. A bridge has filtering capability and can check the destination physical address of a packet and decide whether the frame should be forwarded to a destination physical port of the bridge or dropped. A bridge has a table that maps addresses to the interface ports. To clarify, we consider a switch as a device that works in two layers: the physical layer and data link layer. The term "bridge" has been used in IEEE documentations, but it is not a common name used by people in the industry. People usually use the term "switch" to identify a bridge. However, the term "switch" has been used in different contexts, and therefore, the term "Layer 2" switch, or L2 switch, is a better representation of a bridge. An L2 switch has several physical ports. Each port might be connected to a node of the network or to a port of another L2 switch. When a switch port receives a packet, the switch stores the source physical address of the packet in a table, which is called a MAC table. This process is called MAC learning and during this process, an L2 switch dynamically learns about the network nodes. The MAC table associates one or several physical addresses to a port. In other words, a MAC table associates a port with the nodes connected to it. An L2 switch uses the MAC table in order to forward a packet to a destination node through the physical port of the switch that is associated with the destination node. Therefore, an L2 switch has filtering capability. When it receives a packet on one physical port, it does not send the packet out of all other physical ports. Instead, it looks at the destination L2 address, and only sends the packet out of the port that is associated with that L2 address.

A router is a three-layer device that routes packets based on their Layer 3 addresses. That is the main difference between a router and an L2 switch. Router uses the L3 destination address to route a packet, while an L2 switch uses the L2 destination address for forwarding a packet. A router is the main building block of WANs and is also used to connect LANs to the Internet. Each router has a routing table that is used for making decisions on how to route a packet toward its destination. A routing table can be updated statically or dynamically. Static updating of a routing table means updating a routing table manually. This method is inefficient, especially when there are many routers in a network. A routing table can be updated automatically (dynamically) using a routing protocol. Generally speaking, a router has two responsibilities. First, when it receives a packet, it searches its routing table based on the destination logical address of the packet and then accordingly forwards the packets. Second, the routers in the network talk to each other and send signaling packets to each other in order to find out the network routes. For this purpose, they use a routing protocol. A routing protocol is a set of defined rules used by routers to communicate with each other in such a way that enables them to update their routing table. The communication among routers for updating the routing tables is considered as the signaling information, since no user data is transferred during this communication. The term "switch" has also been used to identify devices that forward the packets based on the logical addresses as well as those that transmit the packets based on the physical addresses. Therefore, the term "three-layer switch," or L3 switch, better represents a switch that works at the network layer and performs the functionalities of a router.

A gateway is normally an internetworking device that operates on all seven layers of the OSI model. It is usually a software program installed within a router. A Layer 2 device can read the L2 header, but not the higher layer header. Similarly, a router can read the L2 and L3 headers, but it does not read Layer 4 and higher headers. A gateway can read all the headers. For example, a mail gateway can be used to connect two systems that are using two different mail protocols.

It should be noted that the best way to name an internetworking device is based on its functionality or in terms of the OSI layers that it supports. Many devices on the market today are programmable, and they can be configured to play different roles. For instance, they can be configured as an L2 switch, or a router, or even a gateway. Also, a device may support several functionalities at the same time. For example, a device may have an L2 switch module and several interfaces that route the packets based on the L3 address.

In Software-Defined Networking (SDN), we try to separate the signaling and the data transfer of an internetworking device. The idea is to use hardware devices that can forward packers at a fast rate. In this situation, a device only forwards the packets, and it is not involved in signaling operations such as MAC learning or updating the routing table. This way, the device can play the role of any type of internetworking device. In other words, the internetworking device is only responsible for packet forwarding, and it receives its table from a controller on the network.

1.8.1 VLAN

Virtual networks play an important role in networking. When something is virtual, it means that it is not physical, but plays the role of the physical one. In other words, a virtual object can perform the same tasks as a physical object. Think about virtual memory. You are given 10 G bytes of memory that you can use, but you do not know where physically the memory is located. You cannot touch it. Actually, it might not be a single 10 G byte storage element. You might have been given two 5 G bytes of a physical memory, each located in different locations. It is also possible that there is a 20 G byte physical device that a section of it is partitioned for you to store your data.

Similarly, a virtual LAN is not a physical local area network, but it performs the same functions as a LAN. To make a LAN, we need to connect all nodes of a network to an L2 switch. For example, for making a LAN with 24 nodes, you need an L2 switch that has 24 ports. In other words, a physical switch can represent only one physical LAN. If you have five nodes in a network, you need a 5-port switch. If you use a 24-port switch for this network, certainly, many ports of the switch are not being used.

A Virtual Local Area Network (VLAN) is a local area network that is configured by software. There is no need to have a physical switch per a network, and a physical switch can be divided into several virtual switches. In addition, a virtual switch with more ports can be created by combining multiple physical switches. To be able to do this, an L2 switch should have a VLAN software. Using VLAN software, any port of a physical switch can be configured to be part of a specific network. VLAN allows the local area network to be reconfigured to meet the demands of an enterprise at any time. The network can be modified without affecting the physical wiring by changing the switch configurations. Simply put, the basic purpose of a VLAN is to group nodes of a network logically.

Let us use a simple example to demonstrate the concept of VLAN. Figure 1.15 shows a network with three L2 switches labeled SW1, SW2, and SW3. This network is utilized by two divisions inside an organization: engineering and manufacturing. You can assign the ports of a physical switch that are connected to the staff in the engineering division to VLAN 1 and the ports that are connected to the staff in the manufacturing division to VLAN 2 in order to isolate these two divisions virtually. If someone from the engineering division (VLAN 1) sends a broadcast message, the message will only come out of the VLAN 1-assigned ports. Some ports can be assigned to both VLANs. These ports are called trunks or tag ports. When a node connected to SW1 sends a broadcast packet to VLAN 1 ports, the switch sends the packet out of all VLAN 1 ports on SW1, as well as the trunk ports. The same happens when a node that is connected to SW1 sends a packet to ports assigned

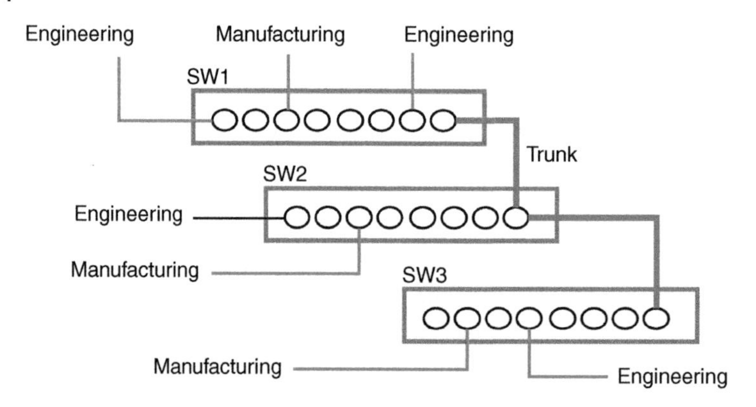

Figure 1.15 A multi-switch VLAN example.

Destination address	Source address	4 Byte Tag	Data	Frame check sequence

Figure 1.16 802.1Q tagging protocol.

to VLAN 2. When the VLAN 1 traffic and VLAN 2 traffic arrive at SW2, there must be a way for SW2 to understand to which VLAN these packets belong to. To solve this problem, SW1 adds a tag to the packet that indicates which VLAN it belongs to and sends it to SW2, which removes the tag and distributes the packet to all of the ports belonging to that VLAN.

Tagging is done just between switches. The trunking protocol, IEEE 802.1Q, is an industry standard for this purpose, which adds a 4-byte tag field within an Ethernet frame as shown in Figure 1.16.

1.8.2 Quality of Service (QoS)

As mentioned earlier, when a router receives a packet, it searches its routing table and accordingly forwards the packet out of one of its interfaces. This interface is usually connected to another router that performs a similar task of forwarding the packet toward its final destination. This method of delivery of packets is called best-effort services. Historically, the Internet relied on best-effort service delivery. Despite the fact that the IPV4 header included a field for type of services, the traditional Internet only supported a single class of traffic, and all communications among the nodes in the network was handled in the same way. The best-effort service delivery does not ensure that packets will be delivered to the destination, and so merely states that it will route the packets to the best of its ability. For many applications, this may not be a realistic option.

To provide better service to customers and applications, the network should ensure the quality of its delivery services or at the very least, differentiate between different network traffics. Besides the best-effort service delivery, differentiated services and integrated services are other types of service delivery that can be provided in a network. The routers in the differentiated services architecture, as the name implies, differentiate between different types of traffic. For example, if a router requires to drop a packet due to network congestion, it drops the low-quality packets, while forwarding the higher priority ones. To be able to implement differentiated services in a network, routers should support differentiated services. It is clear that best-effort routers cannot provide differentiated services. An integrated services router, on the other hand, guarantees that the network supports specific QoS metrics. In the integrated services model, applications ask the network for an explicit reservation of resources. Routers keep track of all traffic and reservations, ensuring that there are sufficient network resources available before accepting new packets.

QoS is typically measured in terms of parameters such as bandwidth, delay, jitter, and packet loss. Different applications may need different QoS requirements. If a service provider can guarantee the requested QoS, the node or application that requested that specific QoS will be able to exchange data in a manner that meets its requirements. For example, an application may require constant bandwidth for its operation. If internetworking devices can reserve the required bandwidth for this application, then the network guarantees that it can provide the quality requested by the application. It is clear that the required bandwidth differs from one application to another. In a videoconferencing application, we may need several Mbps to stream a video signal, while another application may require low bandwidth for its operation. To provide QoS, a service provider may also need to guarantee the amount of packet loss to be lower than a threshold amount. Delay is another essential QoS metric. Some applications are sensitive to delays, while others can tolerate large delays. Videoconferencing or emergency services require very low delay, while performing a file transfer or sending an email might not be sensitive to delay. A voice over IP application may require strict delay values defined by standards. Another important QoS metric is the maximum jitter value that an application can tolerate. The variation in delay for packets belonging to the same flow is called jitter. Let us explain jitter with a simple example. If three packets leave a node at times 0, 1, and 2, and arrive at a destination node at times 10, 11, and 12, then the delays are always 10 time units, and therefore, there is no jitter. However, if these packets arrive at the destination at times 10, 12, and 15, then each of these packets have experienced 10, 11, and 13 time units of delay, respectively.

1.9 Summary

This chapter provides readers with basic knowledge about data communications and networks. There are a vast number of topics, technologies, systems, and standards that are fundamental to the field of data communications and networks. This chapter intends to provide the material that prepares readers with the required information to study the next chapters of this book.

Networks can be classified in terms of many factors. For example, they can be classified based on the maximum distance among the nodes into BAN, LAN, CAN, MAN, or WAN networks. They also can be classified based on their topologies, into mesh, ring, star, bus, or tree networks. Networks can be divided into wired or wireless networks. In a wired network, the nodes are connected to other nodes using wires such as twisted pairs or fiber optic cables. In a wireless network, each node should have an antenna for data transmission. A network can be divided based on the mobility of its nodes. In a non-mobile network, the nodes are stationary, while in a mobile network, the nodes can exchange data while they are moving. The network can be categorized based on the amount of energy that it consumes. For example, in a low-power network, power saving strategies are utilized in order to reduce power consumption.

OSI and TCP/IP are two of the most widespread networking models. The OSI model has become a theoretical model that is mostly used for understanding and the study of networks, while TCP/IP is a practical model and most data communication networks are designed based on this model.

The OSI model introduced a seven-layer approach for the design of each node in a network. These seven layers are the physical, data link, network, transport, session, presentation, and application layers. Each of these layers has its own responsibilities in the network. In this chapter, we briefly explained the most important Layer 2, Layer 3, and Layer 4 protocols that are Ethernet, IP, and TCP, respectively.

TCP/IP provides a five-layer approach for the design of each node in the network. These five layers are physical, data link, Internet, transport, and application layers. TCP/IP suite introduced a collection of popular protocols that are designed to work together under the umbrella of this networking model.

The internetworking devices play an important role in a network. We can define four internetworking devices based on their functionalities as they relate to the OSI model. These four devices are hubs or repeaters that operate in L1; bridges or L2 switches that operate in L1 and L2; routers or L3 switches that operate in L1, L2, and L3; and finally gateways that operate in all layers. We discussed how a packet forwarding action on a network is performed by an L2 switch or a router. In this chapter, the concepts of virtual LAN and QoS-enabled routers were also briefly explained.

As previously stated, the purpose of this chapter is to provide a fundamental understanding of data communications and networks to those readers who are unfamiliar with the subject. This understanding is essential for studying the remaining chapters of this book. This chapter can be used as a fast introduction of the main concepts in this discipline for students who are familiar with data communication systems and technologies.

Readers who are interested are encouraged to read [1–3] for a more in-depth examination of the subjects covered in this chapter or for more advanced topics pertaining to data communications and networks.

References

1 Forouzan, B.A. (2021). *Data communications and networking with TCP/IP Protocol suite.* New York, NY: McGraw Hill.

2 Stallings, W. (2014). *Data and Computer Communications*, 10e. Upper Saddle River, NJ: Prentice Hall.

3 Comer, D. (2004). *Computer Networks*. Upper Saddle River NJ: Prentice Hall.

Exercises

1.1 A network, which is designed based on a mesh topology, has eight nodes. How many links are needed to establish a network with this topology? What if the number of nodes becomes 10?

1.2 There are six nodes in a network. Assume one link of this network fails to operate. Discuss different scenarios that can happen if the topology of the network is (a) mesh, (b) star, (c) unidirectional ring, (d) tree, and (e) bus.

1.3 Draw a network with a hybrid topology that has a star backbone, with two ring networks and one tree network.

1.4 Name the OSI layer responsible for the following functions: duplexing, routing, access control, end-to-end error control, and compression.

1.5 Name the functionalities of the data link layer.

1.6 What is access control?

1.7 How does a burst error differ from a single-bit error?

1.8 Why in the CSMA control access method, the possibility of a collision still exist?

1.9 An IPV4 router receives two packets that their IP headers start with the following byte (in binary format). Explain whether the router discards these packets or not.

a. 0100 0011
b. 0110 1000

1.10 The total length of an IPV4 packet is 1100 bytes, of which 1072 bytes is data from the higher layers. What is the value of HLEN?

1.11 Find the number of addresses, the first address, and the last address in a block of IP addresses given to an organization, if one of the IP addresses of this block is 140.120.84.24/20.

1.12 An organization uses private addresses in the range of 192.168.1.0–192.168.1.255 for its internal addresses and has the block of 176.10.10.0–176.10.10.7 as its non-private IP addresses. Can the organization allocate one of the private IP addresses as its website IP address? What about non-private ones? Explain why.

1.13 Explain the functionalities of Layer 4 of the OSI model.

1.14 Explain the difference between physical addressing, logical addressing, and port addressing.

1.15 Explain how TCP assigns the destination and source port addresses.

1.16 What is a socket number?

1.17 Explain how error control is managed in TCP?

1.18 Two nodes in a network use TCP for their Layer 4 communication. If the receiving buffer has the capacity for 7000 bytes, and 1000 bytes of this buffer is not processed yet, what is the value that the receiving process advertises in the window size field of the next TCP segment?

1.19 A company has two separate office buildings. In some projects, some people from the first office building and some people from the second office building need to be in the same work group. We are not interested in changing the physical configurations of the network that is installed in these buildings. Provide a solution to this situation.

1.20 Explain the advantages of using VLAN.

1.21 How can VLAN traffic be sent from one L2 switch to another L2 switch? Does the frame change when it goes from one switch to another?

1.22 Router A receives an IP packet. The source and destination IP addresses in the IP header of this packet are S-IP and D-IP, respectively. The source and destination physical addresses in the MAC header of this packet are S-MAC and D-MAC, respectively. Which of these addresses will be changed (if any) when the router sends the IP packet to the next router? Explain how they will be changed, or why they will not get changed.

1.23 Can a hub read or change the Ethernet address of a packet? How about a router?

1.24 What do you call
- **a.** an internetworking device that receives a packet and can read its transport layer (L4) header?
- **b.** a packet in Layer 2, Layer 3, and Layer 4?
- **c.** a router that does not support any QoS and does not differentiate between traffics?
- **d.** an internetworking device that you can set up a VLAN on?

1.25 The following blocks of IP addresses 10.0.0.0/8, 172.16.0.0/12, and 192.168.0.0/16 are private addresses, and they cannot be used on the Internet. How many private addresses exist in total?

1.26 Why is UDP better for real-time applications than TCP?

Advanced Exercises

1.27 Do most networking systems today are designed based on the OSI networking model? If not, why is it important to learn about the OSI model?

1.28 Ethernet addresses are 6 bytes long. How many nodes can use this address space in a local area environment? Is it efficient to use this address space in a local setting?

1.29 Assume a network does not perform error control in its Layer 2.
- **a.** What happens if a frame becomes corrupted?
- **b.** Why does a network decide to avoid supporting error control in its Layer 2?

1.30 If the communication links of a network are error-free, is there a need for the data link layer? If yes, what is it needed for?

1.31 Why is there a maximum length for an Ethernet frame?

1.32 Calculate the total time required to transmit a 10 Mbps IEEE 802.3 MAC frame whose length field is 180 in hexadecimal.

1.33 An organization is given a block of addresses as 132.19.219.64/24. The organization has three offices, each one needs 32, 16, and 16 addresses, respectively. We need to make five subnetworks. Three subnetworks are needed for the offices and two subnetworks each having four addresses for the interconnection between routers. Design the block of addresses for each subnetwork from the provided block of addresses.

1.34 In Figure 1.E34, the client sends 425 bytes of data in its last packet. Find the values of x, y, z, and t of the last packet. ($z - t$ shows the range of data from z to t).

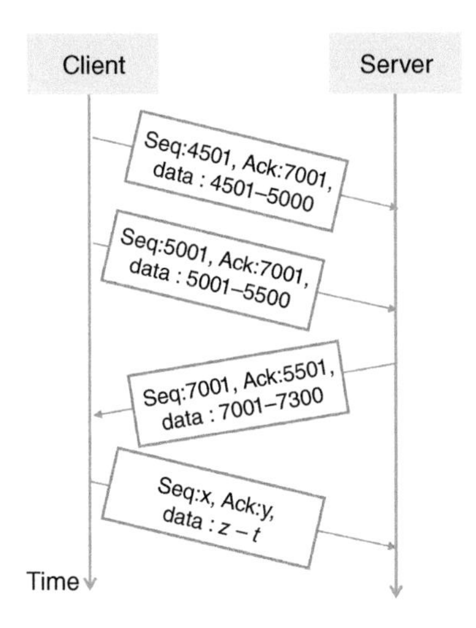

Figure 1.E34 TCP data transfer.

1.35 If UDP does not support any error control, flow control, or congestion control mechanism, what happens to corrupted, delayed, or lost packets?

1.36 Which of these communications are connection oriented, and which ones are connectionless oriented?

a. Communication between Layer 2 of two nodes using Ethernet.
b. Communication between Layer 3 of two nodes using IP.
c. Communication between Layer 4 of two nodes using TCP.
d. Communication between Layer 4 of two nodes using UDP.

1.37 Explain the difference between the delivery of a packet in a local area network and delivery of a packet through a wide area network.

1.38 There are many organizations that publish standards in the area of telecommunications and networks. The most important ones are ISO, ITU-T, ANSI, IEEE, and IETF.

a. Find out what does the name of each of these organizations stand for?
b. Which organization published the OSI model? What about Ethernet?

Chapter 2 Introduction to IoT

2.1 Introduction

The Internet of Things (IoT) is a network of physical objects that are embedded with sensors and electronics, and are connected to each other over the Internet. This architecture enables physical objects to exchange data through the Internet in order to increase their performance or make smart decisions. Physical objects range from simple household devices to sophisticated industrial ones. Many breakthroughs in numerous technologies, including telecommunications, low-power sensor design, cloud computing, and analytics, have made IoT technology a reality. In an IoT ecosystem, after an IoT-based object sends its data to the cloud or a database system, the data is analyzed and usually it is used in one of the following three possible scenarios, as shown in Figure 2.1.

1. The analyzed data is used to enhance that object's performance or the performance of other objects in the system in a non-real-time manner (Figure 2.1a). Consider a scenario in which all visitors to a painting art gallery are needed to wear an IoT-based wristband that tracks their movements inside the gallery. The wristband tracks the movement of each visitor and sends the location data to the Internet. After collecting data for a period of time, the data can be utilized to determine which paintings are more popular and which ones draw less attention. The gallery can then decide to remove unpopular paintings from the gallery.
2. The collected data from a physical object is analyzed in real time, and the results are sent to the object in order to increase its performance or help the object to make a better decision on its operation (Figure 2.1b). Consider the case of a patient who is wearing a medical device that transmits his blood glucose level to the Internet. The data can be used to track the patient's blood concentration in relation to his activity or food consumption. Assume the IoT-connected medical device includes an insulin infusion pump. After real-time data processing, a command to regulate insulin dosage can be sent to this pump.
3. The sensor data is analyzed in real time, and the results are sent to one or more objects in the network to give them appropriate commands regarding their operations or help those objects to make better decisions in order to increase their performance (Figure 2.1c). Consider the case of a person who is wearing a wearable device. The device transmits information about skeletal muscle movements to the Internet. This data can be used to determine the sleep pattern of the individual who is wearing the wearable device. Assume this individual wakes up every morning and turns on the light in his room and the coffee machine in the kitchen. In this case, the wearable device sends data to the Internet, which is evaluated to determine when the individual wakes up in the morning. Following the analysis of the data, the necessary commands are delivered to that object or other objects. In this situation, for example, a command is sent to turn on the light and a second command is sent to turn on the coffee machine.

Fundamentals of Internet of Things: For Students and Professionals, First Edition. F. John Dian.

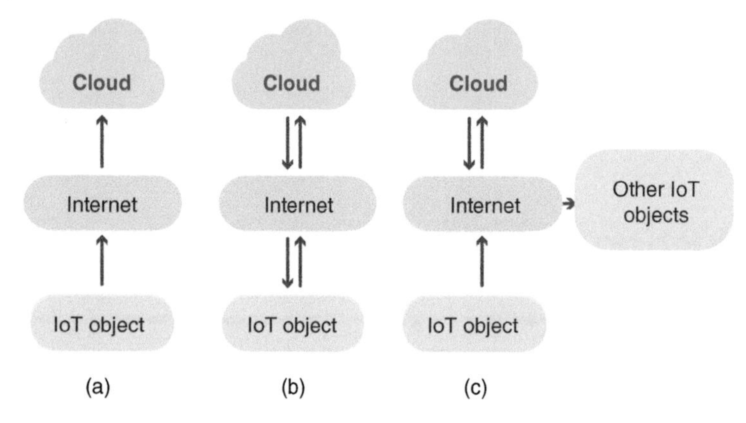

Figure 2.1 IoT device and cloud connection scenarios. (a) Scenario 1, (b) Scenario 2, (c) Scenario 3.

IoT has come to impact every aspect of our lives. There are so many organizations that can benefit from IoT technology. Manufacturing, healthcare, automotive, retail, transportation and logistics, and agriculture are just a few examples of these sectors. The use cases for IoT are endless. Smart homes, smart utility meters, smart cars, and smart wearables are some of these use cases. Some use cases may need specific requirements in terms of connectivity, the amount of power consumption, or the physical size of the IoT system. In the future, every home becomes smarter and will be equipped with multiple sensors. This requires IoT service providers to have enough capacity to provide connectivity to a large number of smart homes. Smart utility meters such as gas, electricity, or water are other use cases of IoT. Smart meters may be installed in environments such as basements with poor wireless coverage. This requires IoT service providers to provide good coverage for IoT devices that are located in challenging areas. Autonomous vehicles and self-driving cars, which use hundreds of sensors, are becoming a reality. These sensors generate very large amounts of data. Processing, analyzing, and transmission of huge amounts of data require a very fast network with high processing capacity. Also, IoT system developers should consider processing time and communication delay as part of their design specifications. Smart wearables have many applications in healthcare, sports, safety, and location tracking. These IoT-based wearables need to be small in size to be integrated into clothing. In addition, many smart systems are battery powered, and therefore, the IoT technology needs to support low power consumption.

2.2 IoT Traffic Model

The traffic and data flow of most IoT applications are somehow different from the traffic generated by humans when they access the Internet. For example, in many IoT applications, the IoT device transmits its data after the detection of an event, while human-generated traffic is not usually event based. There are also many IoT applications that send their collected data through the day periodically, while the human-generated traffic during the day is substantially higher as compared to the traffic during the night. Another important difference is the amount of traffic generated by IoT devices as compared to the data generated by humans. Overall, the amount of generated traffic by an IoT device might be significantly lower or higher as compared to the generated traffic by humans. There exist IoT-based systems that generate vast amounts of data, far more than the data that a person can generate. An example of these systems is the machineries used in smart factories.

Traffic generated by humans	Traffic generated by IoT devices
Higher traffic during the day compared to the night	Almost uniform traffic all the time
Higher downlink traffic compared to the uplink	Substantially higher or lower traffic as compared to the traffic generated by humans
High data transmission rate	Low or high data transmission rate

Figure 2.2 Comparison between the traffic generated by humans and generated by IoT devices.

But in many IoT applications, an IoT device generates a very small amount of data. For example, consider a water meter that sends its data every week to a server. A user may receive a large amount of data every day, while it might be enough for a water meter to access the Internet daily or even weekly in order to send the value of its counter, which is just a few bytes. Also, IoT applications have higher uplink data to transmit as compared to downlink, while human's traffic is often in the downlink direction. In terms of mobility, many IoT devices are stationary and have mobility that is lower than humans. However, there are applications where the IoT device experiences the same mobility as humans. For example, the mobility of an IoT system, such as a wearable device, is the same as the person who wears it. You should consider that humans send and receive data while they are traveling with cars and high-speed trains, and therefore, the mobility of the smart car and the passengers of this smart car are the same. Figure 2.2 shows a comparison between traffic generated by humans and traffic generated by IoT devices.

Many IoT devices monitor a physical condition and send some information when an event is detected. The probability of the occurrence of these events is usually rare. An alarm from a smoke detector or a power failure notification from a smart meter are examples of these events. In these examples, an IoT device reports the event and expects the report to be received with a low latency. Due to the importance of these events, the IoT device may also expect to receive an acknowledgment from the network. It should be highlighted that the traffic generated by an IoT device is application oriented. While a power outage notice might generate a relatively small amount of data, an IoT-based camera surveillance system may transmit a large amount of data upon the detection of an incident.

There are IoT devices that monitor a system constantly and need to send their sensor data periodically. Utility meters or environmental data collection systems are examples of IoT systems with periodic data transmissions. Some IoT devices send data neither periodically nor upon the detection of an event. These IoT devices perform a task or send data after receiving a command. For instance, a command can be sent from a server to turn on an IoT-based street lighting system, or to ask a reading from a smart utility meter.

Another type of IoT-related traffic is software update traffic, which is sent to an IoT device in order to add new features or improve the existing ones, enhance its performance, or fix existing security bugs. IoT devices are expected to go through software updates occasionally. During a software update process, large amounts of data might be sent to an IoT device.

2.3 IoT Connectivity

There are many IoT connectivity methods that are used to connect an IoT device to the Internet. Wired connections are fast, secure, and extremely reliable. However, using wired connections is only practical if IoT devices are located close to a wired Internet access point and also close to each

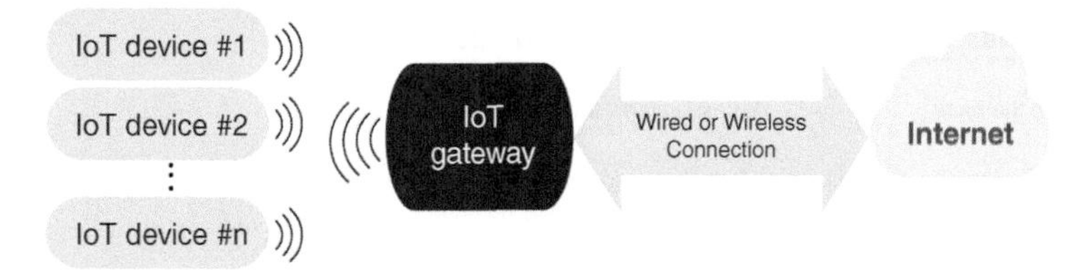

Figure 2.3 IoT gateway connection.

other in order to reduce cabling costs. For most IoT applications, the wired connectivity method is not very practical. Short-range wireless technologies such as Bluetooth and Zigbee are attractive connectivity schemes, particularly for low-power IoT devices. But these technologies are not designed for providing long-range communications to the Internet. The same goes for the WiFi technology that enables connectivity by adding wireless access points, which consequently connect IoT devices to a wired or cellular connection. To connect IoT devices to the Internet using short-range technologies such as Bluetooth, ZigBee, or WiFi, there is a need for an IoT gateway as shown in Figure 2.3. In general, a gateway provides a bridge between different types of communication technologies. An IoT gateway can act as a bridge and receive data through technologies such as Bluetooth, ZigBee, or WiFi from IoT devices. It can also send the data to the Internet using a wired or wireless broadband technology. Since an IoT gateway is located at the edge of the network and may have limited processing power, it might be able to perform some processing, if needed. Figure 2.3 shows a situation where several IoT devices are connected to an IoT gateway.

It is also feasible for one or more of the network's IoT devices to act as an IoT gateway. Figure 2.4 depicts a mesh local area network with two IoT devices serving as gateways. A mesh network, as described in Chapter 1, is a network in which all the network's nodes can communicate with one another.

Cellular IoT (CIoT) networks are highly reliable and available with almost ubiquitous coverage. A CIoT technology may operate in an unlicensed band where there is no need to acquire a license for transmission, or in a licensed band where it requires to have a license for its data transmission. While there exist some IoT-enabling technologies in unlicensed bands today that may be able to address the wide area coverage requirements of IoT devices, they fall short in terms of coverage, scalability, interoperability, Quality of Service (QoS), and security as compared to licensed-band CIoT technologies. A Low Power Wide Area Network (LPWAN) is defined as a network that can

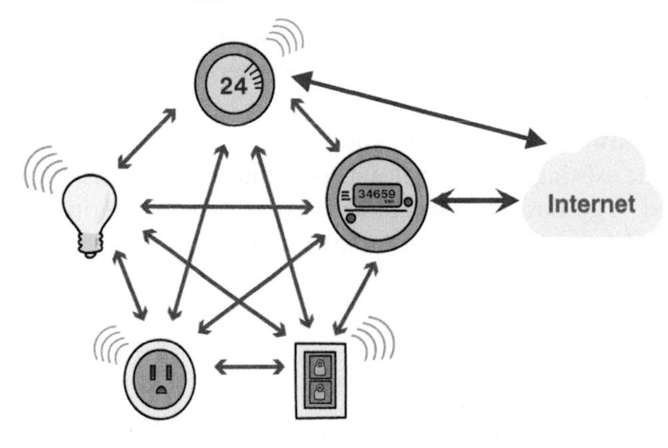

Figure 2.4 An IoT mesh network with two IoT devices serving as gateways.

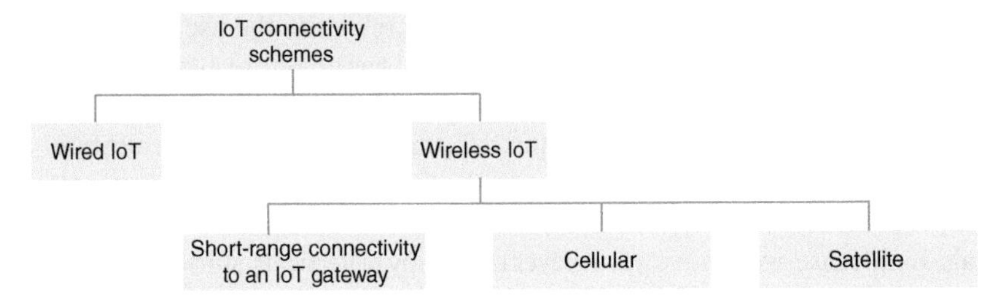

Figure 2.5 IoT connectivity schemes.

provide high coverage while having low power consumption. The CIoT technologies usually belong to the category of LPWAN systems.

Another technology that can be utilized to enable connectivity for IoT devices is satellite technology. This is especially important when IoT systems are intended to operate in places where terrestrial network connections are not available. For example, IoT devices might be distributed in remote areas such as forests, deserts, or oceans. In such cases, using satellites to provide wide area coverage could be a realistic option. Consider cargo containers being transported across an ocean, which are equipped with tracking devices. There is no cellular coverage in the region; therefore, satellites are the only means to track these containers. In other words, asset management applications that need to maintain connectivity to their assets, no matter where they are, can benefit from combining satellite technology with terrestrial IoT. While satellite technology is an excellent complement to terrestrial IoT networks, previous adaptations of satellite technology were not built for IoT and were thus very expensive and complex. Traditional satellite systems could not provide connectivity for IoT applications that required low latency, low power consumption, and reliable availability. They were also prohibitively expensive, making satellite technology unsuitable for most IoT applications. However, with recent advancements in satellite technology, satellites are now able or will soon be able to provide affordable, available, and accessible services to IoT devices. The use of satellite technology for IoT applications is not covered in this book. However, a brief discussion on satellite IoT is provided in Appendix C.

Figure 2.5 shows different classifications of IoT connectivity schemes. We will discuss the wired connectivity schemes in Chapter 5, short-range wireless technology in Chapter 6, and LPWAN technologies in Chapters 7 and 10. In Appendix C, there is a brief description of satellite IoT.

2.4 IoT Verticals, Use Cases, and Applications

As illustrated in Figure 2.6, the utilization of IoT may be classified into three different levels of verticals, use cases, and applications. A vertical can have multiple use cases, each of which can have multiple applications.

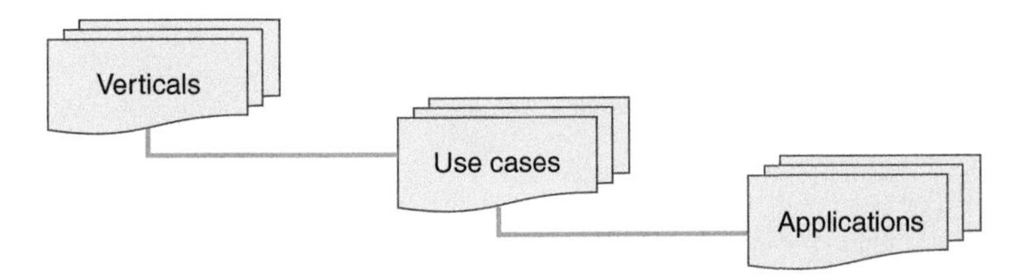

Figure 2.6 Categorization of IoT usage into three segments of verticals, use cases, and applications.

The first layer in our taxonomy is the vertical layer, which often relates to the use of IoT in a specific industry segment or a certain class of users. A specific vertical has unique regulatory bodies and supports a certain set of standards, specialized policies, procedures, and protocols. Examples of these IoT verticals include transportation, energy, industrial, smart cities, retail, or healthcare. As IoT grows and is implemented in more industry segments, the list of IoT verticals grows as well. There is no common agreement on the list of IoT verticals. A large vertical can be further divided into several subverticals. In this case, we can use a four-layer taxonomy of verticals, subverticals, use cases, and applications. For example, smart homes can be considered as an individual IoT vertical or as a subvertical under smart cities.

In our taxonomy, use cases are the second layer. A use case subdivides a vertical into sections that can often be served by the same platform and usually require similar types of processing. IoT use cases of a vertical usually have similar specifications in terms of factors such as throughput and latency. For example, in the transportation vertical, there may be use cases related to autonomous vehicles or smart highways. It is clear that there would be more similarities between the use cases within a single vertical and the ones that belong to different verticals.

The third layer in our taxonomy is the application layer. Applications further subdivide a use case into sections that can often be served by similar solutions and software programs. Developers with knowledge in a specific application of a particular use case belonging to a certain vertical are expected to be able to apply that expertise to effectively develop other applications in that specific use case and vertical.

Hereunder, we discuss some of the possible use cases and their corresponding verticals. Some of the existing IoT verticals and use cases are also shown in Figure 2.7.

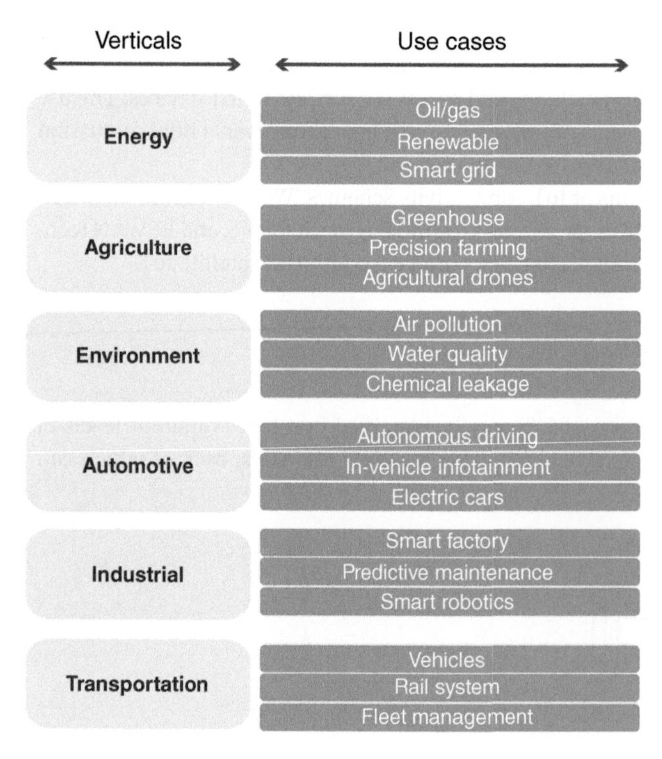

Figure 2.7 Some of the existing IoT verticals and examples of possible use cases.

- Smart city vertical: Examples of use cases in this vertical would be smart parking, public transport, smart street lighting, smart waste management, route optimization for garbage pickup trucks, public safety surveillance, and structural health monitoring systems for roads and bridges.
- Healthcare vertical: IoT can revolutionize patient monitoring as well as access to treatment and medical services. Examples of use cases in this vertical are remote patient monitoring, fall detection, patient surveillance, sleep control, and robotic surgery.
- Retail vertical: Examples of use cases of this vertical are supply chain monitoring, inventory management, predictive equipment maintenance, demand-aware warehouse, connected consumer, smart store, smart payment, smart marketing, smart display, and fraud detection.
- Transportation vertical: Examples of use cases in this vertical are fuel consumption optimization, fleet maintenance, inventory tracking and warehousing, predictive analytics, supply chain management, location management systems, quality control for transportation goods, fleet tracking, vehicle auto diagnostic, and road pricing.
- Energy vertical: Examples of use cases in this vertical are oil and gas, renewables, and smart grid. Utilities are building smart grids that can help detect and isolate outages and integrate power generated by customers into the grid. Energy monitoring in different renewable formats such as wind, solar, or wave is an important use case of this vertical.
- Environment vertical: Fire detection, air pollution monitoring, landslide and avalanche detection, meteorological station, marine and coastal surveillance, and chemical leakage detection can be considered as important use cases of this vertical.

2.5 IoT Value Chain

The IoT value chain explains how IoT creates value, identifies the players involved, and shows how they interact with each other. It also explains the way that these different players deliver value. The IoT value chain consists of the following sections as shown in Figure 2.8.

- IoT chip/module: This section of the value chain includes sensors, actuators, and chipsets.
- IoT device host: This is an application-specific environment that hosts the sensors and IoT communication module.
- Connectivity: This section provides end-to-end connectivity of IoT devices installed in the network. It includes connection to the Internet through a gateway by utilizing technologies such as Bluetooth, Zigbee, or WiFi, as well as those wired or wireless technologies that provide a direct connection to the Internet.

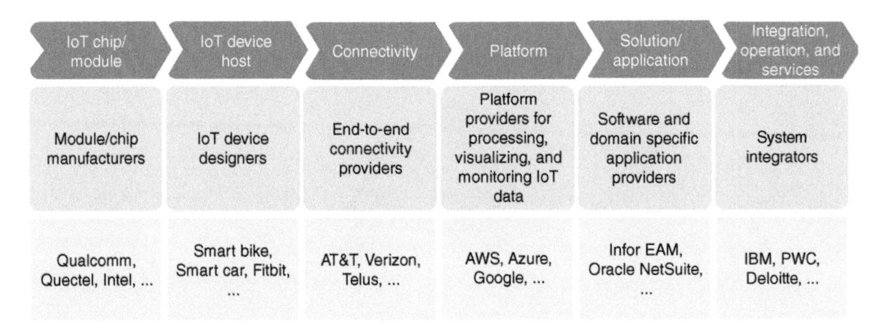

Figure 2.8 IoT value chain.

- Platform: This section of the value chain belongs to software platforms that are used for aggregating, processing, securing, storing, analyzing, controlling, and monitoring of IoT-based data.
- Solution and applications: This section covers the software programs and services that make use of IoT data.
- Integration, operations, and services: This section provides system integration in order to provide seamless connection between different systems, integrate various components of the IoT ecosystem, and provide managed services to the clients.

We defined the IoT value chain based on their functionalities. Traditionally, a specific player was in charge of a specific function. However, most players like to move beyond their main functions and offer other functions across the value chain. For example, telecommunication operators try to move along the value chain beyond connectivity. Clearly, operators have a huge presence in the connectivity sections of the IoT value chain. However, connectivity only generates a small portion of the revenue that can be gained through the entire value chain. The traditional telecommunication operators have competition from new LPWAN operators that only provide connectivity for IoT devices. Therefore, traditional operators must provide efficient connectivity schemes in order to ensure that they have a strong presence in the connectivity market. In addition, these operators must understand to what extent they can move along the value chain, which verticals are better to go in, and what new roles they can play. One may think that an operator may be able to provide end-to-end solutions for verticals such as agriculture, healthcare, or automotive. However, becoming an expert in a specific vertical that has various markets and use cases requires a huge amount of expertise, skill sets, and investment that makes entering that specific vertical not profitable for an operator. For this reason, operators often like to build platform services or application services that can be used for many verticals. Today, many telecommunication operators offer services in various areas such as fall detection, smart lock, or security services.

2.6 Examples of IoT Use Cases and Applications

2.6.1 IoT-based Structural Health Monitoring System

You may have heard in the news about a bridge that has collapsed, a wind turbine with broken blades, or a crane that has toppled on top of a building as shown in Figure 2.9. All these structures are usually designed and built based on standards. So, one may wonder why these structures collapse, get broken, or are fallen apart. We can blame a heavy load, a strong wind, or large vibrations

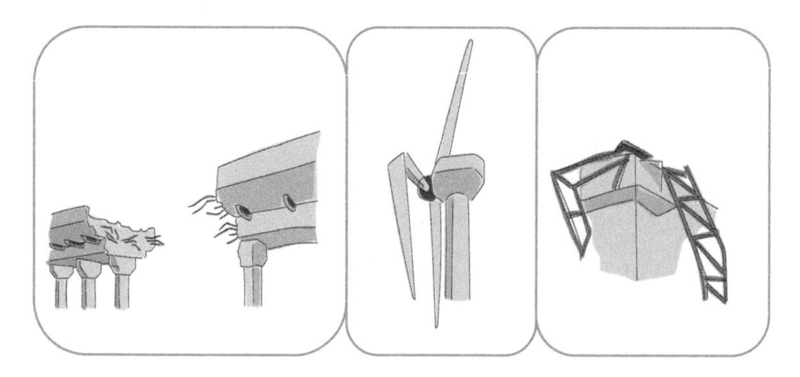

Figure 2.9 Collapsed and broken structures.

as a reason for these problems. It is possible that a heavy load, heavier as set by the standard, or a strong wind, stronger than the maximum allowable wind, will be the main cause. In most cases, however, these are not the root cause of the problem. It should be noted that structures such as a building, a wind turbine, or a crane, are vulnerable as they are exposed to harsh environments. They experience constant varying loads by wind, operate in extreme temperature and humidity variations, and suffer from erosion and corrosion. Therefore, to find the existing conditions of a structure and prevent accidents, the structure needs to be monitored at all times.

An IoT-based Structural Health Monitoring System (SHMS) can be an excellent solution for understanding the current condition of a structure. Knowledge about the status of a structure enables us to prevent many problems that could occur in the future. Figure 2.10 shows how different structures can be monitored using an IoT-based SHMS system. As shown in this figure, the collected data from these structures needs to be sent to a cloud computing platform through the Internet. The data is analyzed in the cloud and the results might be sent to the structure or to an IoT management system in order to determine a suitable course of action upon the detection of a problem.

SHMS can be considered for new or existing structures. The reason behind monitoring a new structure is to collect real-time data related to structural components used in the manufacturing processes, in order to examine safety risks involved in the fabrication of new structures. The objective of using structural health monitoring for existing structures is to find the condition of the structure, calculate its remaining lifetime, or estimate the maximum load the structure can tolerate with its current condition. The monitoring system follows any changes in the structural components, as well as the environment, in order to detect possible events or the deterioration rate of the structure. This information can be used for maintenance, repair, and safety purposes. In general, an SHMS consists of three subsystems. These three subsystems are data acquisition, data transmission, and data analysis. The first subsystem relates to the sensors that need to be installed on the structure in order to collect the necessary data for processing. The second subsystem provides connectivity between the sensors and the cloud. The third subsystem relates to the computing and calculating of the condition of a structure by analyzing the collected data.

In some situations, time synchronization between the sensors installed on the structure would be crucial. This means that the application needs accurate time stamping of the events for

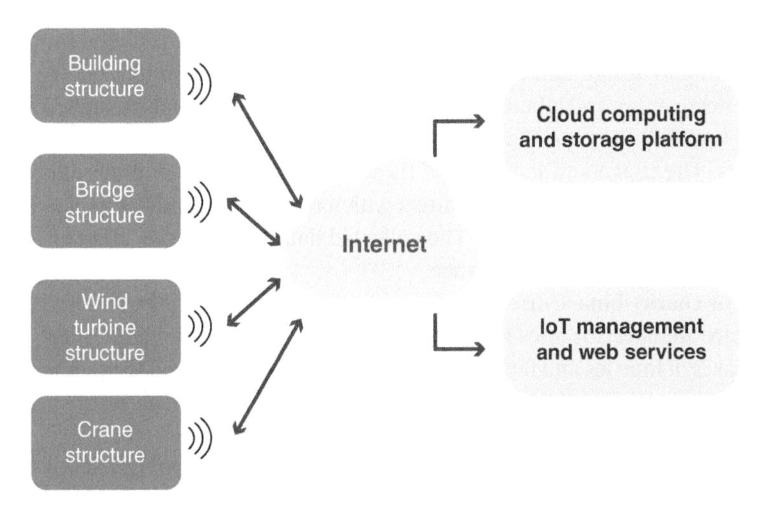

Figure 2.10 IoT-based SHMS.

further analysis. In other words, the collected data from various sensors, which are installed on a structure, is meaningful if and only if we have an accurate understanding of the time among all the sensors. For example, if there is a need to determine the structural integrity of a building after an earthquake, the sensors installed on different parts of the building need to be synchronized in order to enable us to study the effect of forces applied to the structure accurately at a given point of time.

2.6.2 IoT-based Electric Meter

To reduce energy consumption and carbon footprints, governments around the world as well as utility companies would like to take advantage of IoT technologies to update existing energy infrastructure into smart grids. This is beneficial for customers, since customers can have more visibility over their energy usage, manage their energy usage patterns more efficiently, reduce their power consumption, and benefit from various pricing models. It is also beneficial for utility companies to reduce the cost of their operations and maintenance by managing operations remotely, improving power consumption forecasts, reducing energy theft, improving customer satisfaction, shifting power consumption peak hours, and tracking renewable power.

To achieve this objective, the first step is to transform the metering system in such a way that an electric meter can send its data to the cloud for processing and storage. CIoT might be considered as a good connectivity solution for smart metering. While most homes or buildings have WiFi connections, utility companies usually do not have access to this type of connection. Also, utility companies cannot force all homeowners to have a WiFi connection. Cellular connectivity schemes are suitable options if they can provide good coverage for all locations where electric meters are installed. The smart meters might be installed in challenging locations inside a house, such as basements. Therefore, cellular connectivity methods need to provide the required coverage for communication with these meters in order to be considered as a suitable connectivity solution for this use case.

2.6.3 IoT-based Waste Management System

To keep our cities and neighborhoods free from unwanted waste materials, waste management services exist everywhere from populated cities to rural areas. Many of the current waste management operating standards are not only resource intensive but also extremely inefficient. The inefficiency is caused by outdated collection methods as well as logistical processes. The waste management industry is beginning to design and deploy IoT-based solutions to solve its existing problems and to enhance the effectiveness and efficiency of its operations. Sensors can be installed inside the waste bins to track how full they are. The IoT system can notify the crew when they need to empty these waste bins. The usage and locations of these bins can also be monitored in order to determine if there are enough bins in a particular area, which can consequently help plan to increase or decrease the number of bins accordingly. The collected data can help city planners to optimize their operations and improve their performance.

The same idea can be used for charity bins. Currently, the methods used for the collection and management of donation bins are archaic. Charities rely upon a scheduled pickup, much like a city garbage pickup. Due to many cases of injuries and fatalities, media coverage has alarmed the public about donation bins, where a person can get into a donation bin, with no possibility of getting out, which can cause permanent damage, and even death.

Some cities are expected to save money by using a more efficient waste collection system, as shown in Figure 2.11. For example, a city may decide to adopt wireless autonomous sensor technology in order to determine the status of a trash waste bin. The information is transmitted to a

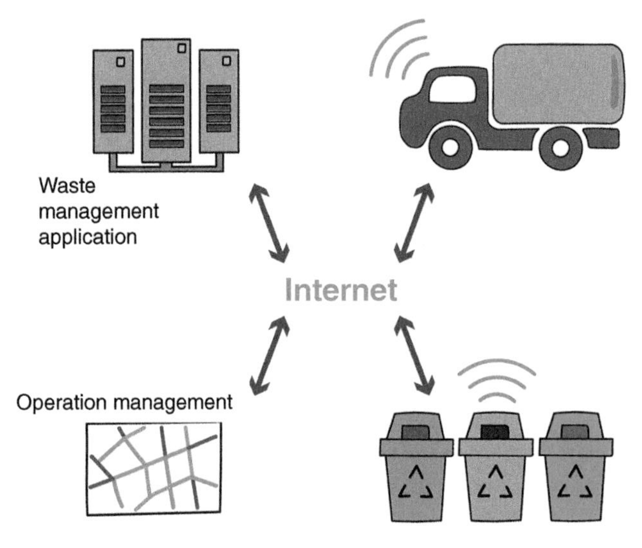

Figure 2.11 IoT-based waste management system.

software platform that plans the scheduling process for emptying the bins and optimizes the trash collection routes. Optimizing routes also results in less traffic and fewer truck emissions.

2.6.4 IoT-based Earthquake Detection

Due to devastating effects of a large earthquake, there is a need for predicting earthquakes in advance. Despite considerable efforts by scientists, it is still impossible to predict the location or the time of an earthquake yet.

Some people believe that certain animals can predict earthquakes ahead of time by exhibiting unique behavior. If their theory is correct, then tracking the behavior of these animals could lead to the prediction of an earthquake's location and time. For example, it has been observed that lizards crawl out of their cracks, wolves leave the forests, or foxes yell some time before an earthquake. Despite the fact that these unique behaviors have been observed, animals may express these behaviors for other reasons as well.

Using the concept of IoT, the design of a smart earthquake prediction system might be possible. Monitoring a large number of animals simultaneously and analyzing the data that represents the behavior of animals might be a good start for finding a solution for earthquake detection. Let us assume that all animals are equipped with IoT-based devices that connect them to the Internet. If a large number of animals exhibit odd behaviors in a given geographic location, it could be a sign of an earthquake, and it could help us predict the event. Clearly, a false alarm in predicting an earthquake may have many damaging consequences. Therefore, data should be analyzed carefully in order to correctly classify an earthquake from other possible reasons for unusual behaviors. The distribution of the animal's unusual behavior across a geographical area could also assist in locating the earthquake's epicenter.

2.6.5 IoT-based Car Software Update

Modern cars have many components that can be controlled through software, and therefore, new software programs are designed regularly to enhance the functionalities, add new features,

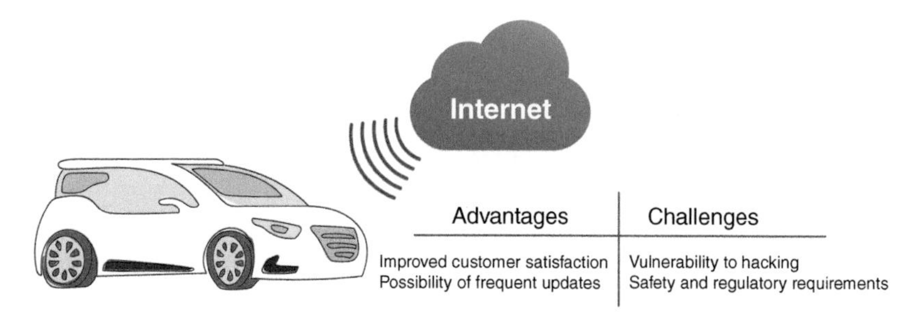

Internet
Advantages
Challenges
Improved customer satisfaction
Possibility of frequent updates
Vulnerability to hacking
Safety and regulatory requirements

Figure 2.12 Advantages and challenges of OTA software updates for a smart car.

or fix the existing bugs. Therefore, the software update is regularly required to be performed on these components. It would be very efficient to update these software programs remotely and Over The Air (OTA). For instance, the updates can be sent to improve cruise control accuracy, enhance battery performance, improve the brake system, make transmission adjustments, tweak fuel efficiency, or be used to better recognize stop signs and traffic lights for autonomous cars. Today, many vehicles perform OTA updates on a regular basis. However, these updates are only done for non-essential components such as a sound system or a navigation map. In general, most automakers do not perform software updates on components that have a direct relation to the driving of a car or the ones that may impose any risk. For any critical software update, owners need to take their cars to the dealership in order to perform the software update. The main reason for not updating any critical software updates over the air is security issues. Even though it is efficient and certainly convenient to update the latest software over the air, OTA updates might be vulnerable to cyberattacks in which a hacker can send malicious software updates to critical, driving-related components of a car or a large group of cars. Figure 2.12 shows the advantages and challenges of OTA software updates for a smart car.

2.6.6 IoT-based Mountain Climbing Information System

An IoT system can be installed along the tracks in specific locations of a mountain to collect data such as temperature and humidity, as well as the number of people who are passing through those locations. The IoT system makes the collected data available to anyone who visits a dedicated website designed for this purpose. The climbers can use the information on the website to plan their activities and bring appropriate clothing and equipment. The information can be used to make further improvements to the trails and the other nearby facilities, or be utilized to plan appropriate actions that need to be done during peak times. The climbers can use the information to understand the status of the tracks and find out the periods of time that the tracks become crowded.

2.6.7 IoT-based Agriculture – Pest Management

Pest management is a crucial part of the agricultural industry. Farmers are usually seeking efficient ways to apply precise pest management schemes in order to ensure their crops stay healthy. Since pesticides are harmful to humans, animals, and the environment, they should be used carefully. Traditional pest management schemes use calendar-based or prescription-based pest management methods that are not efficient. Due to the lack of visibility on data such as exact location of pests at a specific time, or type and number of pests in the farm, a lot of pesticides need to be sprayed on the whole farm at specific times during the year. IoT-based intelligent devices allow farmers to reduce pesticide usage substantially by precisely spotting the pest. In other words, pest management in non-IoT-based systems is a guessing game, while IoT-based pest management systems are precise. Crop health monitoring and pest management depend on finding the type, amount, and locations of the disease at a specific point in time, and treating the disease by applying a precise amount of pesticides on exact spots at appropriate times throughout the crop cycle.

Pest management is only a small portion of smart agriculture. Smart agriculture also includes use cases such as smart irrigation, smart fertilization, and yield monitoring, to name a few. In smart irrigation, instead of schedule-based watering, soil conditions are monitored using soil moisture sensors. The data related to the soil condition along with the weather information and other factors are used in order to determine the time and location that watering is needed. Smart fertilization precisely estimates the required dose of nutrients by site-specific soil nutrient level measurements according to various conditions such as the type of the crop, type and absorption capability of the soil, and weather conditions. It is clear that applying extra amounts of fertilization has negative effects on the environment. An IoT-based system can estimate crop yield in real time based on various factors such as crop mass, moisture content, and harvested crop quality.

2.6.8 IoT-based Wearable in Sports

IoT has the potential to open up a world of possibilities for sports in general. This includes improving services for athletes, coaches, fans, and sports organizations. IoT-based systems can increase the efficiency and offerings in sports by creating an environment in which athletes can receive better training or have access to the data that can help to keep them healthier. Coaches are able to analyze injuries or find metrics to evaluate the performance of players. The organizations can offer fan engagement strategies or allow fans to receive personalized offerings from their favorite team. Smart wearable technology is a game changer in sport activities. However, the use of wearable IoT is currently limited due to some regulations and some challenges that need to be resolved. Yet, as more and more organizations recognize the advantages that smart wearables can provide to players, coaches, and fans, wearable IoT devices will eventually become more and more prevalent in sports [1].

2.6.9 IoT-based Healthcare System

Today, the healthcare system is a clinic-oriented or hospital-oriented process in which a patient needs to go to a clinic or a hospital for diagnosis. IoT technology can change this orientation toward home-oriented diagnosis and treatment as shown in Figure 2.13. Using IoT technology, remote real-time monitoring data of a patient can be stored in a health cloud data center or sent to a local healthcare service provider. The health data can be collected from sensors connected

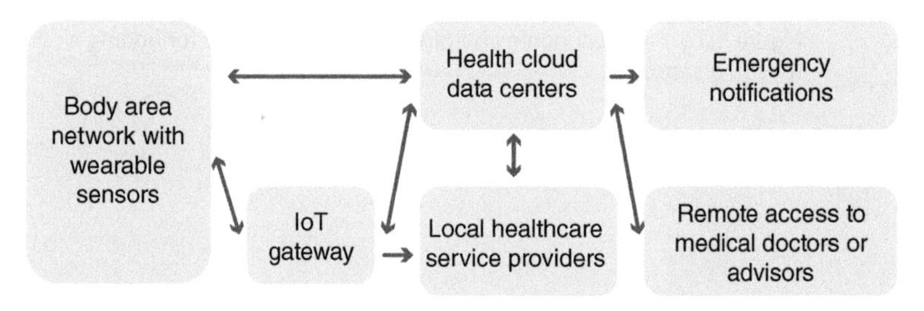

Figure 2.13 The architecture of a home-based healthcare system.

to the body, installed in smart clothing, or obtained from ingestible pills. This data can be shared with all authorized parties including physicians, insurance companies, and medical research centers. The data can be analyzed, and in critical and life-threatening circumstances, an alert can be sent to the appropriate medical personnel. In an emergency situation, patients can contact doctors remotely, an ambulance might be sent to their home, or medicine might be ordered for them through a pharmacy.

IoT technology can also be used to improve medical knowledge through performing analytics on huge amounts of collected data. It is clear that collecting data manually or conducting randomized controlled trials would be very time consuming, expensive, and sometimes not feasible. However, applying analytics and machine learning algorithms to massively collected data would be a good solution in finding the factors that cause a specific disease. This knowledge can also be used to develop new medications or provide knowledge about the types of patients who are eligible to receive specific medications.

2.6.10 IoT-based Augmented Reality (AR) System

The real-world environment can be complemented by using AR technology, which uses computers to generate imaginary virtual objects and animations, and augments them to the real-world environment. Although AR applications are very interesting, it would be more realistic if it would be possible to physically interact with the virtual objects. You can add a hologram of a dog that moves like a real dog to your real-world environment, but it would be more interesting if you could pet this virtual dog. This challenge can be solved by using wearable IoT haptic devices that are comfortable to wear and are capable of providing sensations. A haptic device can provide stimulations such as vibrations or the sensation of touching a material, as well as a sense of object temperature. Therefore, IoT technology can help us to have a better AR environment. An example is shown in Figure 2.14 in which upon touching a virtual dog, real-time data is sent to the cloud and you receive some information on your haptic interface that provides the stimulation of touching a real dog.

Generally speaking, haptic devices can create a real sense of touch. Assume you are wearing haptic gloves and you are interacting with a virtual object. The gloves pull your fingers back as you contact the virtual object, allowing you to feel the shape of the object. You can feel not only the object's presence but also its texture. This allows you to interact with the virtual object more effectively and gives you the impression that you have touched the object. This is possible if real-time data related to the location of the virtual object and your hand is sent to a processing unit. When your hand approaches the virtual object, the haptic gloves receive the analyzed data, which specifies the feedback force that should be applied to your hand and each of your fingers.

Figure 2.14 IoT technology helping augmented reality for adding a touching sense.

2.6.11 IoT-based Food Supply Chain

IoT can increase the efficiency of the supply chain substantially. For instance, transporting food from a farm to warehouses, processing plants, and finally to stores can be completely monitored using IoT technology. The collected data through the supply chain segments can increase the transparency and visibility of information in such a way that food manufacturers, transportation companies, retail stores, restaurants, regulatory agencies, and consumers can better collaborate in the food supply chain.

Using IoT, tracking shipment locations correlated with factors such as temperature or moisture that can have an impact on food quality, becomes feasible throughout the supply chain. Therefore, in transporting temperature-sensitive food products, IoT technology can better manage the product movement in real time, discover issues in the supply chain faster, and even better respond to possible issues before they result in product contamination or recall.

It is clear that food recall is a very important public safety issue, which has a significant economic impact and causes customer dissatisfaction. The collected data from all segments of the supply chain can provide more visibility and consequently enable us to narrow the scope of a recall. Due to the lack of visibility, in the traditional food supply chain, there is no way to understand exactly which final products are made from the original food products of a specific farm. Also, it would be very difficult to identify the root of a food contamination, due to the absence of real-time data from consistent tracking and tracing of the entire supply chain.

There are standards and regulations in the food industry that set the required temperature for different products such as refrigerated foods at stores and during transportation. Traditionally, this was a labor-intensive task, since it was done manually. The correlation between the real-time temperature and location data of a product can give the confidence that foods are continuously kept at their required temperatures. The products at each stage of the supply chain can also be tagged using barcodes. The tag can show the type of the food product and any other data required by regulatory bodies such as the farm of origin, the transportation method, or time of transportation. The tagged information can be read by various people involved in the supply chain in different formats and levels of detail. A consumer of a product might not be interested or does not need to know about the length of time that it took for the product to move from a farm to a logistic center, or if the truck carrying the product had an accident, while this information might be crucial for another person with more authority in the supply chain.

2.6.12 Smart Grid System

Smart grid provides many advantages to utility companies as well as their customers. To demonstrate this, let us compare how an outage caused by a short circuit is handled by the old grid and the smart grid systems. This simple example shows how utility companies can take advantage of IoT to perform their operations more efficiently. Figure 2.15a shows a simplified version of what is called one line diagram of a utility company, which is essentially the blueprint for the analysis of power systems. The lines in this figure are called feeders, which are transmission lines through which electricity is transmitted in power grids. Typically, feeders have three cables and there are switches in series with these cables. In traditional grid systems, these switches do not measure voltage or current, do not perform any processing, and do not have the capability to send data or receive information. If the utility company needs to open or close them, it must send its technicians to change them manually. These switches are expressed as $SW_i(i = 1, \ldots, 15)$ in Figure 2.15a.

The feeders in Figure 2.15a are connected to four substations. Each substation is usually represented by a specific color in the diagram. Also, each feeder is represented by the color that matches

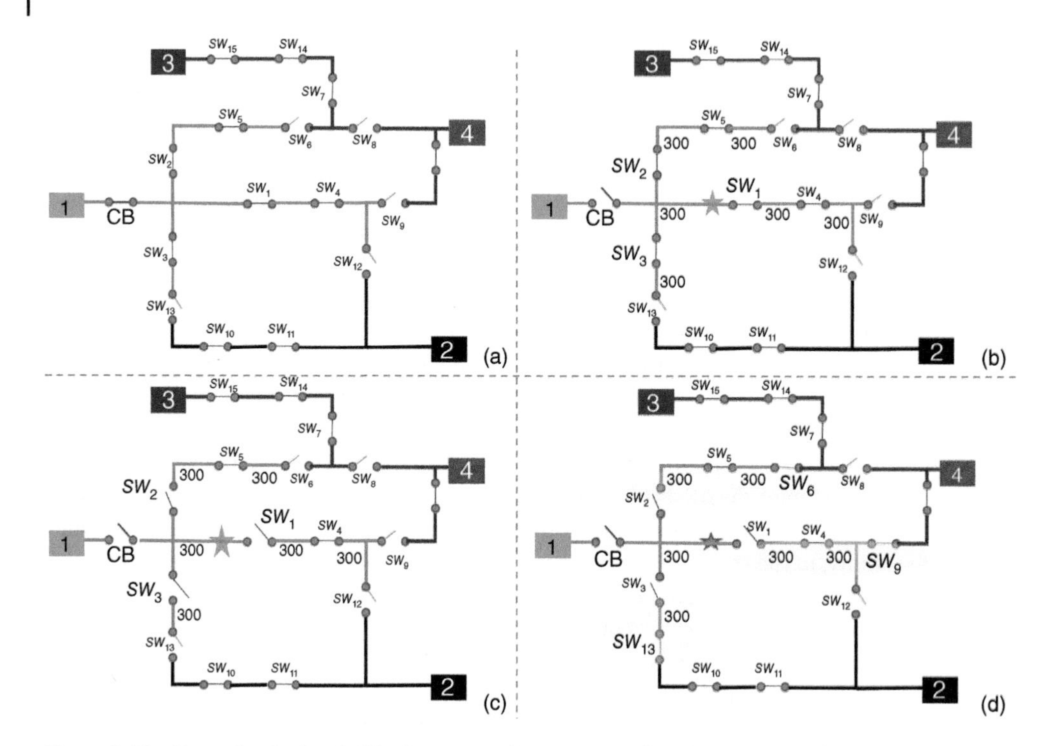

Figure 2.15 Example of using IoT in the smart grid system. (a) One line diagram of a utility company; (b) one line diagram showing the effect of a short circuit; (c) one line diagram showing the isolation of the problem caused by the short circuit; (d) one line diagram showing how the other substations' extra capacity is used to get the power back to at least some of the affected customers

the color of the substation that it is drawing power from. In this example, the only feeder that draws power is the one that it is connected to substation 1. This feeder provides power to 1800 customers that can be residential or commercial customers. Let us assume that the customers are spread out evenly. Therefore, in this example, each link has 300 customers as shown in Figure 2.15b. When a tree falls on a feeder, it can cause a short circuit. Let us consider a situation where a tree falls on a feeder close to SW_1. The location of the short circuit is shown with a star sign in Figure 2.15b. We would like to understand how many customers, and for how long, have been affected in the old grid system as well as the smart grid system.

As a result of the short circuit, the upstream circuit breaker opens to interrupt the flow of the current as shown in Figure 2.15b. The circuit breaker closes after a short time to check whether it was only a transient issue or not. In our case, let us assume the short circuit does not go away by itself, and therefore the circuit breaker stays open. As a result, many customers lose electricity and start calling the utility company about the problem. In the old grid system, the utility companies would learn about the outage from their customers. After the utility company finds out about the outage, it dispatches a crew and a truck toward the location of the problem. It usually takes longer than usual for the truck to reach the area, since the traffic lights of the affected areas do not work either. The truck then needs to patrol the area around the feeders to find the location where the problem has occurred. Remember, the feeder lines are usually around 20–50 miles, and therefore, it takes time to find the location of the problem. When the crew finds the location of the problem, they need to call their control center to discuss the nature and the location of the short circuit. The control center tells the crew which switches need to open in order to isolate the problem. In this example, switches SW_1, SW_2, and SW_3 must be opened to isolate the problem caused by the short circuit as shown in Figure 2.15c. Therefore, the crew needs to travel to each of these switches to manually open them. Then, the control center needs to determine whether the other substations in the area have extra capacity and whether it is possible to get the power back to at least some of the affected customers. Let us assume that:

1. Station 4 has some extra capacity, and therefore, SW_9 can be closed in order to restore power for 600 customers. It is clear that the crew should drive to this switch to close the switch manually.
2. Substation 3 has some extra capacity, and therefore, SW_6 can be closed in order to restore power for another 600 customers.
3. Substation 2 has some extra capacity, and therefore, SW_{13} can be closed in order to restore power for another 300 customers.

After all is done, 1500 customers out of 1800 affected customers have their power back as shown in Figure 2.15d. At this point, the crew can travel to the location of the problem and start their repair work. As you can see, it takes a long time to travel to the affected area and find the location of the problem. It also takes a long time to isolate the switches around the problem area and connect the area to other substations that have extra capacity. This simple example shows an isolated problem; you can imagine how long it takes to restore power when a storm has caused many problems in a network.

In a smart grid system, we use smart switches that can measure voltage and current. Smart switches have computing power to process the data, and transceivers which enable them to communicate not only with each other but also with circuit breakers and control systems. This means that the upstream circuit breaker can listen to communication between switches. Therefore, the voltage at each switch, the current of each link, the customer's load, and the capacity of the other substations are known to the network. If a short circuit happens in a smart grid system, the switches around the affected area talk to each other and can isolate the network in a matter of seconds. In other words, while it took a long time for SW_1, SW_2, and SW_3 to become open manually by the crew, it happens within seconds in a smart grid system. Since there is a good understanding of the substations that have extra capacity, the power of 1500 out of 1800 affected customers can also be restored within seconds. The location of the problem is somewhere close to the SW_1, SW_2, and SW_3 switches, and therefore, this information can help the crew to find the location of the problem faster. Also, since the power is restored for many areas after several seconds, the traffic lights in most of the affected areas are working, and therefore, it takes less time for the crew to travel to the affected area. It is clear that the amount of time that is required to repair the real problem is the same in the old grid and smart grid systems.

This example shows the business impact of a smart grid as compared to the old grid system. When it is possible for many customers of an old grid system to be affected by a simple short circuit for hours, it takes only several seconds for these customers to have their power back in a smart grid system.

2.7 IoT Project Implementation

When an organization decides to start an IoT project, the organization should include an assessment on how to plan on proceeding with an IoT initiative. If there is already a solution available in the market for the project, the organization can implement the existing solution. However, if there is no such solution, the organization needs to build a solution in-house or needs to work with a partner to develop the solution. Generally speaking, most IoT initiatives are difficult to implement without the help of partners, since IoT solutions require various expertise. This means that organizations require many skilled IoT designers to develop a solution. For example, an IoT solution may include an electronic design solution from one company, a connectivity solution for data transmission and reception from a telecommunication provider, a cloud software that performs sophisticated analytics from a cloud provider, and a solid customer-based portal that is designed by another company. An organization can hire IoT system integrators to perform these complex tasks. IoT system integrators work with the organization and all other necessary companies and technology players on behalf of the organization. We will discuss IoT solution development methodologies in Chapter 13.

2.8 IoT Standards

One company alone cannot cover all of the IoT market and its value chain. As more and more IoT devices become available, interoperability becomes an important issue. Therefore, there is a need for various embedded devices and sensors from a variety of manufactures to interact with each other seamlessly and securely. Also, there must be an organization that unifies IoT standards and makes specifications for protocols and open source projects. The Open Connectivity Foundation (OCF) is an industry organization established to develop specification standards and promote a set of interoperability guidelines. OCF also provides a certification program for IoT devices as well as the certified labs to do the IoT testing. There are many companies that have joined OCF as a member, such as Intel, Cisco, Qualcomm, Samsung Electronics, Microsoft, and Electrolux. OCF has become one of the most important connectivity standards organizations for IoT and seeks to include a broader range of companies in its membership. IoT requires easy discovery as well as trusted and reliable connectivity among IoT devices. OCF has developed IoTivity, an open source framework that allows seamless connectivity among IoT devices. OCF has also established Universal Plug and Play (UPnP), which is a collection of networking protocols that provides seamless device discovery and connectivity establishment.

2.9 Summary

The material in this chapter is an introduction to IoT technology, which provides background for our subsequent discussions. Explanation of the IoT traffic model, although brief, provides a basic idea that the requirements needed for IoT applications might be different from other types of networking applications. The brief discussion on various connectivity schemes available for IoT in Section 2.3 also provides readers with some basic understanding of connectivity schemes used for IoT. We will discuss these technologies in more detail in the next chapters.

The IoT value chain explained in Section 2.5 shows how IoT creates value, identifies the players involved, and their interaction with each other. As IoT becomes more mature, more companies and businesses will be entering the IoT value chain.

We have introduced a taxonomy that categorizes the use of IoT into a hierarchy of three different layers of verticals, use cases, and applications. A vertical has several use cases and each use case can have several applications. There is no common understanding of how the phrases vertical, use case, and application are used, causing misunderstanding.

Section 2.6 is dedicated to examples of IoT use cases and applications in various verticals. These examples provide a big picture of the applicability of IoT in various industries and serve the readers well in their journey through this book. The examples, even though explained at a high level and are strictly introductory, give a good understanding of how different industries can benefit from IoT implementations. For further examples of IoT use cases and applications, interested readers are encouraged to look at [2] and [3].

The implementation of IoT solutions requires various skill sets that organizations are usually lacking. It is important to understand that one company alone cannot cover the entire IoT market. As more and more IoT devices become available, interoperability becomes an important issue. This has been briefly addressed in Section 2.8.

References

1 Dian, F.J., Vahidnia, R., and Rahmati, A. (2020). Wearables and the internet of things (IoT), applications, opportunities, and challenges: a survey. *IEEE Access*. 8: 69200–69211. https://doi.org/10.1109/ACCESS.2020.2986329.

2 Dian, F.J. and Vahidnia, R. (2020). *IoT Uses Cases and Technologies*. Vancouver: BCcampus https://pressbooks.bccampus.ca/iotbook/.

3 Behmann, F. and Kwok, W. (2015). *Internet of Things (IoT): For Future Smart Connected Life and Business*. Hoboken, NJ: Wiley.

Exercises

2.1 For each of the following IoT verticals, find a list of some use cases and applications:
- a. Energy
- b. Agriculture

2.2 Discuss which IoT vertical and use case each of the examples of Section 2.6 belongs to.

2.3 For an urban area, the household density per square km is 1517. Assuming that there are 40 devices per household and the area of a cell is 0.866 km^2. What is the CIoT device density per cell for this urban area?

2.4 A welding machine is used as part of the manufacturing of a product. This machine welds 50 spots in order to complete this product. During the operation in each welding spot, the welding machine collects the voltage, current, and temperature values. The welding process for each spot takes 10 seconds, and a welding machine transmits data at a data rate of 2 Kbps during this period. The welding machines make 800 products per day. How many mega bytes of data is generated by this welding machine daily?

2.5 Bombardier has used 5000 sensors in its C-Series jetliner. These sensors produce as much as 10 GB of data each second. Why does a jet engine require so many sensors? How many pitta bytes of data gets generated after one day? How about after one month?

2.6 Give three use case examples for the industrial vertical.

2.7 How can fleet management use IoT to improve efficiency, reduce costs, and provide compliance with government regulations?

2.8 Low cost sensors can be used for detecting forest fires using IoT technology. Sensors can be placed in the forest for measurement of some parameters such as temperature, relative humidity, atmospheric pressure, carbon monoxide, and carbon dioxide. If each of these measured parameters goes above a configured threshold, the system sends an alarm. For firefighters to be able to go and extinguish the fire, what other parameters are needed? Do you think the IoT system designed for this application needs to be battery powered? Is the power consumption of each IoT system important in this application? Why?

2.9 In 2015, Bosch launched the Track and Trace program. The idea behind this program was that workers would spend a lot of their time trying to find their tools. So, the company added sensors and IoT technology to its tools to track them. The first tool was a cordless nutrunner. What effects do you think the use of IoT had on the performance of workers? Do you think this method can be used for assembly operations?

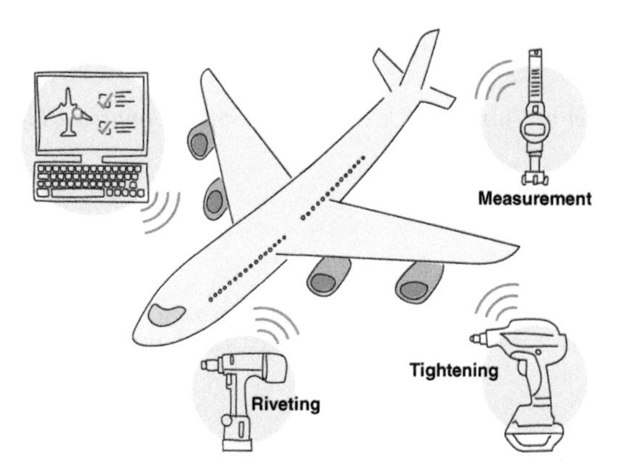

Figure 2.E9 Bosch track and trace program. https://blog.bosch-si.com/industry40/airbus-factory-future/track-and-trace-connected-tools-in-manufacturing-bosch/.

Advanced Exercises

2.10 Early or late frost can damage the year-end harvest or plants that are ready to bloom. Explain how IoT technology can be useful in this application.

2.11 IoT technology could help beekeepers in situations where honeybee colonies have collapsed and honeybees have died off at a fast rate. Examples of these situations are when a hive has tipped over or when the bees are having trouble gathering pollen from the surrounding areas. Why do researchers like to know everything about the hive including hive temperature and humidity, noise, or even light levels?

2.12 For its printers, HP has transformed part of its business model as well as its supply chain into a subscription-based revenue model. HP smart printers monitor the amount of ink of the printer's cartridge and send their data to the cloud. Using this information, the company can ship new ink cartridges before the printer runs out of ink. This makes HP to have a more accurate forecast, but now HP needs to modify its distribution strategy for ink shipment. Discuss the effect of this IoT-based model for HP and its customers. Does this IoT application require low power consumption?

2.13 Due to the fact that food sources are becoming more available in urban areas, the rodent population is increasing. Rodents transmit many types of diseases, and they are one of the main causes of electrical fires in buildings. Are traditional methods (using traps) for getting rid of rodents still effective? Explain how an IoT technology can help.

2.14 Amazon Key is a new smart access technology developed by Amazon to allow customers to give access to delivery personnel who need to put delivered goods inside a customer's house or garage. The operations performed by Amazon Key are shown in Figure 2.E14. Explain how Amazon key uses IoT technology and how this application is secure.

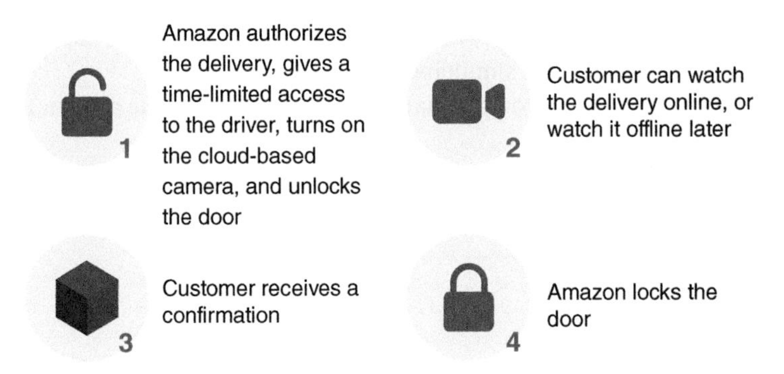

Figure 2.E14 Amazon key smart access technology.

2.15 IoT can enable the existence of digital twins. A digital twin is a virtual replica of a physical object. The object uses different sensors related to vital areas of its functionalities. These sensors generate data about various aspects of the physical objects' performance. The data is then sent to a processing system and applied to the digital copy.

a. What is the difference between digital twins and simulation?
b. Why are digital twins popular within the industrial world?
c. How can building a digital twin for a car help in car diagnostics?
d. Some say that Tesla is making a digital twin of every sold car. Do you think this could be true?

2.16 An urban area in Vancouver, Canada, is connected to transformers from four substations that provide hundreds of megawatts to power the area as shown in Figure 2.E16. Also, there is a fifth transformer that is in standby mode. This transformer will be used when one of the main transformers fails to operate. In other words, when a transformer fails, the power

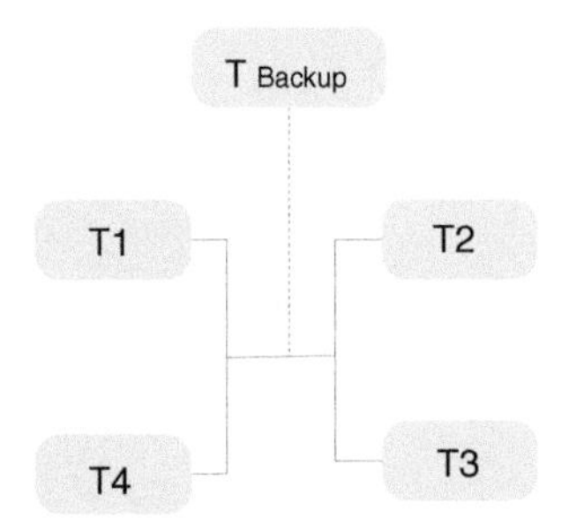

Figure 2.E16 Utility grid connection for the area in Exercise 2.16.

capacity of the system does not decrease, since the backup transformer will be switched on. However, if two transformers fail, the power capacity of the system decreases, and it causes grid equipment to operate above their maximum power capacity. Therefore, in this situation, the operator should disconnect some of the customers to avoid equipment failure.

a. In the traditional grid, how does the utility company avoid disconnecting customers in the event that two transformers fail?
b. Knowing that the price of real estate is high in Vancouver, what is the problem with the method in (a)?
c. In a smart grid system, how can a utility company avoid disconnecting the customers in the event that two transformers fail simultaneously?
d. Based on your answer in (c), can you say that the smart grid can create a virtual transformer for this area?

Chapter 3 IoT Architecture

3.1 Introduction

To be able to design and develop robust and efficient Internet of Things (IoT) systems, there must be a well-thought architecture that perfectly connects an IoT solution to the requirements of an IoT application. This architecture can be used to describe IoT systems in a systematic way, compare IoT protocols based on their functionalities, and most importantly be a reference model for not only the design and development but also understanding of IoT systems. This is extremely important in the IoT ecosystem where a design may involve the integration of many different kinds of physical objects, devices, technologies, and services. Despite the need for a well-accepted architecture for IoT systems, one of the challenges in both the design of an IoT system and describing the functionalities of various protocols and services used in the IoT domain is the absence of a general architecture that can simplify the high-level design. Various architecture models have been published by different companies, organizations, and research communities. But there is no specific model that is agreed by everyone or can handle the requirements of all types of IoT applications.

The terminology and classification of functionalities of IoT architecture models published by different players in the IoT ecosystem is quite different. For example, Microsoft uses a three-layer model of things, insight, and action; Intel has published a three-layer model of things, network, and cloud; Google introduced a three-layer model of device, gateway, and cloud; and Ericson has a five-layer model of device, gateway, connectivity, cloud, and users. Also, many architecture models have been published to be suitable for specific industries. For example, the Industrial Internet Consortium (IIC) has published an open architecture model for Industrial Internet Reference Architecture (IIRA), which is specific to industrial IoT systems.

Even though different organizations may use different terminologies, or use architectures that group the IoT functionalities in more or less layers, they usually all have the same objective. They try to classify the functionalities and operations involved in the collection, transmission, storage, and processing of the collected data from various IoT devices in a modular manner and clear format, and integrate all these functionalities in such a way that an organization can benefit from the IoT implementation.

There are a number of important factors to consider when designing an IoT architecture model. These factors are discussed at the beginning of this chapter, followed by an explanation of a seven-layer IoT architecture model published by the Internet of Things World Forum (IoTWF), and finally a comparison of the IoTWF model with other IoT architectural models published by other players in the IoT domain.

Fundamentals of Internet of Things: For Students and Professionals, First Edition. F. John Dian.

The most important role in an IoT architecture is carried out by data processing and analysis, which can be performed in various locations across an IoT system. In this chapter, we become familiar with three different types of processing in IoT systems, which are cloud computing, fog computing, and edge computing.

3.2 Factors Affecting an IoT Architectural Model

There are several factors that are specific to IoT systems, and therefore, should be considered as the basis of any IoT architectural design. The most important factors that should be considered in the design of any architecture model for IoT systems are:

Sensors and devices: An IoT system requires one or multiple sensors to collect data. These sensors are different in terms of functionalities and specifications, and some of them might have limited processing capabilities. Clearly, sensors should be part of any architectural model that is designed for IoT systems.

Scalability: It is clear that the design of a small network is not necessarily the same as the design of a large network. However, the architectural model for IoT should be scalable and consider small networks with a limited number of nodes as well as large networks with large numbers of IoT devices. Since an IoT network can support a massive number of devices, the IoT architecture must be designed to be able to handle many IP-based endpoint devices. For example, the addressing scheme used for IoT endpoints should be large enough to be used by a massive number of IoT devices in the network, especially as this number grows. Device management also becomes important when there are a huge number of devices in the network, and therefore, there must be efficient methods to connect to each node or update any device in the network, when needed.

Handling devices with different requirements: Many IoT applications have constraints in terms of their cost, size, and power consumption. This means that IoT devices in these applications should be low cost, small in size, and able to work with limited processing power, storage capacity, and transmission power. For example, some IoT devices might be battery powered, transmit small amounts of data at very low bit rates, or transmit infrequently. Therefore, the network should be able to support a large number of devices that access the network infrequently. On the other hand, we have IoT devices that are more expensive and more complex. These IoT devices usually have large processing power and storage. The IoT architecture should be designed in such a way that it works with the devices that have lots of constraints in terms of their processing power, cost, and size, as well as those devices that do not have any limitations.

Data size: IoT devices generate huge amounts of data. The data can be structured or unstructured. Structured data has a specific format, and therefore, it can easily be stored in relational databases. Unstructured data is often qualitative information that is usually stored in its native format. Audio and video data are examples of unstructured data. In many applications, some parts of the collected data might not be needed to be stored and can be filtered out in order to reduce the size of data. The IoT architecture model should consider that the entire collected data might not need to be stored unnecessarily.

Latency: IoT applications may have various latency requirements. We can define latency as the time needed for an IoT device to gain access to the network and send its data, until the time that the data is transferred to the cloud, or even until the time that an action is taken after analyzing the data. Understanding that there are IoT applications that are sensitive to large delays may affect the architectural design of an IoT system.

Connectivity and availability: To connect IoT devices to a processing center, such as the cloud, they require an IP connection and a connectivity strategy. Many IoT applications need the Internet connection as well as the cloud to be always available. For time-critical IoT applications, the availability of the Internet and the cloud is a necessity.

Computation and analytics: Various IoT applications require different levels of computational complexity and processing sophistication. The IoT architecture model should consider the appropriate resources that are needed to be allocated to various types of IoT applications in order to enable them to perform the required computation.

Security: To have a secure architecture, security must be embedded in each of the functionalities defined in an IoT architecture model. This includes the device security, connection security, and cloud security. As data needs to be secured during transmission, it should also be secured when stored in databases, and sensitive information needs to be secured and kept private.

Business integration: The relationship between a business goal and the IoT implementation needs to be integrated as part of an IoT architecture model. For an organization, the objective of an IoT implementation is to achieve its goals in terms of business efficiency, process optimization, and economic growth. Business integration highlights the management requirements that need to be coordinated by an IoT architecture in order to add value to the business.

3.3 IoT Architectural Model

Generally speaking, an architecture model permits visualization of high-level information regarding a system, and can be used by system designers who intend to develop the detailed design of a system based on a specific architecture model. An IoT architectural model involves how the data is collected, transported, analyzed, and finally acted upon.

Even though every IoT system is unique, the principle of the IoT architecture for these systems is almost the same. The same goes with the flow of data in IoT ecosystem. An IoT system consists of a physical object connected to the Internet directly or via an IoT gateway. The transmitted data by IoT devices need to be preprocessed to become ready to be analyzed. After the data is transferred to storage devices, the data needs to be processed in order to be transformed into actionable information.

3.4 IoTWF Architectural Model

The IoTWF architectural model consists of seven layers as shown in Figure 3.1 [1]. From bottom to top, these seven layers are: physical devices and controllers (Layer 1), connectivity and communications (Layer 2), edge or fog computing (Layer 3), data accumulation (Layer 4), data abstraction (Layer 5), application (Layer 6), and collaboration and processes (Layer 7).

Layer 1 is related to IoT physical devices that generate data or endpoint devices that can be controlled by the IoT network. A Layer 1 device can be a simple sensor used in an IoT system or a very complex machinery installed in a smart factory. Each IoT physical device generates real-time data at a specific rate based on application requirements and design specifications and transfers the data to higher layers. This layer represents "Things" and can include not only sensors and controllers but also smart sensors that have processing capabilities. We will discuss in detail about sensors, smart sensors, and their applications in the IoT system in Chapter 4.

Layer 2 is related to connectivity schemes that are used among Layer 1 devices as well as the communication of IoT devices with Layer 3. This layer encompasses all connectivity elements

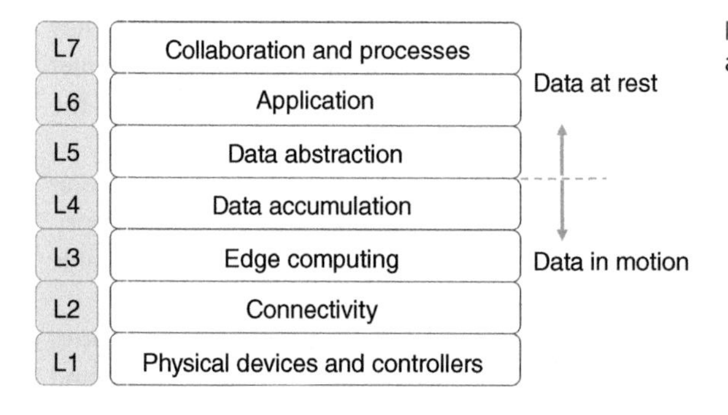

Figure 3.1 IoTWF seven-layer IoT architecture.

and spans from IoT devices up to the cloud. Layer 2 provides a transport layer using a single telecommunication technology or a combination of technologies to establish required connectivity. It includes the connection of IoT physical devices to each other, or directly to the cloud, as well as the connectivity between IoT devices and IoT gateways, and IoT gateways and the cloud. It also includes the processing that can be performed using the devices in this layer.

Layer 3 is related to edge computing or, more precisely, cloud edge computing. This layer interfaces the data to higher layers in the cloud and can perform limited computations at the cloud edge gateway. Examples of the computations that can be performed at this layer are reformatting, protocol conversion, and fast decision-making analysis that can be beneficial for delay-sensitive applications. One of the main objectives of this layer is to reduce the size of data. This layer prepares the data in such a way that it is more efficient for storage or more suitable for processing at higher layers. This includes filtering and reformatting of data. It is also possible to perform real-time computations at the cloud edge in order to analyze the data, take appropriate actions, or send the required notifications.

Layer 4 is data accumulation that is related to the storage and preparation of data for processing in higher layers. The data before Layer 4 is called data in motion. Since the data at Layer 4 is stored in databases, the data after Layer 4 is called data at rest. At this layer, data is stored in databases for further processing. In other words, the data can be queried. As mentioned earlier, physical IoT devices may perform their measurements at different frequencies at Layer 1 based on the requirements of their applications. When data arrives at Layer 4, it needs to be prepared for processing in higher layers, and therefore, data needs to be buffered in such a way that higher layers can handle the data at their own desired speed.

Layer 5 is related to data abstraction. This layer is responsible for accommodating multiple data formats coming from various sources in order to ensure consistent interpretation of data. This layer ensures that the data is complete and stores the data in one single storage location or multiple locations. Using storage virtualization, the data can be pooled from several storage devices while it appears to be from a single storage device as one logical device. This layer is responsible for organizing upstream data (from IoT devices toward the cloud) into appropriate schemas as well as sending the data for further processing. In a similar manner, downstream data (from higher layers toward IoT devices) can be reformatted for interaction with IoT devices or stored for more processing.

Layer 6 is the application layer. In this layer, the data is analyzed and interpreted. After analysis of data, this layer can represent the data according to an IoT applications objective. Examples of IoT applications include visualization of statistical analysis, process optimization, and system monitoring.

Layer 7 is the collaboration and processes. This layer focuses on sharing the analyzed information. Collaboration and communication of the IoT-analyzed information among multiple stakeholders is what makes IoT useful. This layer can be beneficial to the business community in order to improve the efficiency of their processes. It is about leveraging the value of IoT to provide business efficiency, process optimization, enhanced productivity, social benefit, and economic growth. It involves human interaction with other layers of IoT at a high level to initiate collaboration among stakeholders and to achieve business goals. In other words, this layer provides humans with the business intelligence and insights to make decisions based on facts that are obtained from data and presented by visualization tools. It should be noted that the application layer cannot provide decision-making capabilities that are needed, and that is why Layer 7 is defined as a separate level on top of Layer 6.

Example 3.1 IoT-based Donation Bin System

Let us discuss the seven-layer IoTWF architecture model using a practical example. Several incidences of injuries and fatalities caused by people who climbed inside donation bins have been reported in previous years, where people became trapped and could not get out of these bins. A researcher designs an IoT system to solve the safety problems associated with these donation bins. In his design, the donation bin is equipped with sensors to detect the presence of a human or any living being inside the bin. There are also several sensors that are installed inside the donation bin in order to determine how full the bin is. The data from these sensors is sent to an IoT gateway, which subsequently sends the data to the cloud for processing. We now explain this system based on the seven-layer of the IoTWF architecture model.

Layer 1: Layer 1 is related to IoT devices that are installed inside each donation bin. These IoT devices should have sensing devices to accurately detect the motion inside the donation bins as well as other sensors in order to detect the level at which a donation bin is filled up.

Layer 2: The IoT system is connected to an IoT gateway. An IoT gateway is usually installed somewhere close to the donation bin, where there is a connection to the Internet. The data from the donation bin can be sent using short range wireless technology to the IoT gateway. The IoT gateway should have a wired or wireless connection to the cloud.

Layer 3: At the cloud edge, the data that represents the presence of humans inside the bin might be processed in real time. This information is delay sensitive. If data indicates that there is no movement inside the bin, the data might not be necessary to be stored, since it does not contain any valuable information. Otherwise, appropriate action needs to be taken in real time. The other information must be sent to the higher layer.

Layer 4: This layer prepares the data that needs to be stored in a database. The data might be aggregated with the received data from other donation bins in the neighborhood, or received data from other IoT-based systems, if needed.

Layer 5: This layer ensures that the data is complete and has the correct format. This is especially important when data comes from various sources.

Layer 6: The bin management application analyzes the data. This application can improve the efficiency of the processes. The non-IoT-based method of collection of items from the donation bins is time based in which the donation bins are emptied at certain intervals such as daily or weekly. Valuable time and money are wasted using this method. The

(Continued)

Example 3.1 (Continued)

current method can be costly if some or most of the bins are not full, since the charity pays for the fuel and transportation costs. On the other hand, there is a possibility that a donation bin is full and stays full for several days. Since major stakeholders in this case are charitable organizations, the money that is saved from improving the current management method could potentially be used for charitable causes.

Layer 7: The processed data from these donation bins is shared with those who need to make decisions about possible changes to the management of this system. This layer provides decision-makers with the business intelligence and insights they need to make informed decisions.

Example 3.2 Comparing Microsoft IoT Architecture Model with IoTWF Architecture Model

The three-layer IoT architectural model published by Microsoft is shown in Figure 3.2. The model consists of three layers of things, insights, and actions. The "things" layer consists of IoT devices as well as IoT gateways that are able to connect to the cloud in order to send or receive data. This layer is connected to the "insights" layer using a cloud gateway service (IoT hub). There are several subsystems inside the "insights" layer including stream processing, reporting tools, and storage. Stream processors analyze data, integrate it with business processes, and store the data on a storage device. A User Interface (UI) and reporting tool is used to visualize the analyzed data. The last layer is the "actions" layer that includes a business integration unit that creates the appropriate actions that need to be taken based on the analyzed data.

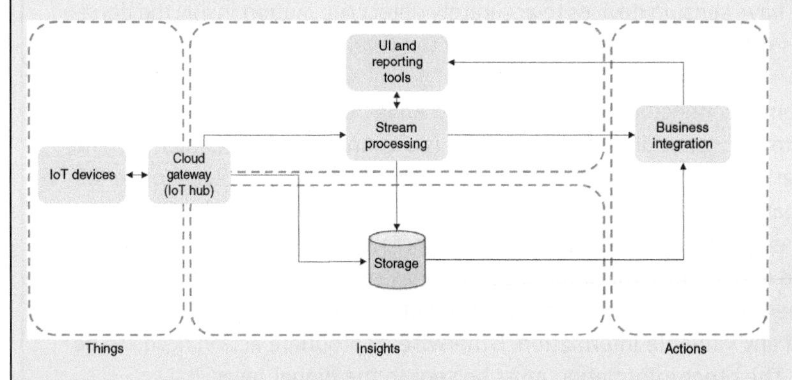

Figure 3.2 Microsoft IoT architecture model. Source: Adapted from Microsoft Azure [2].

Now let us compare this architecture with the IoTWF architecture model. The layer that is called "things" in the Microsoft architecture model covers the first and second layers of the IoTWF model. The insights layer covers Layers 3–6 of the IoTWF model, and the actions layer is the same as Layer 7 of the IoTWF architecture model. Microsoft also has another architectural model that further subdivides each sublayer of its original architecture model. This architecture model is shown in Figure 3.3. This architecture is based on Lambda architecture, which is a data-processing architecture for handling large amounts of data. Lambda architecture takes advantage of both online and off-line processing of data and tries to optimize latency and throughput by using non-real-time and real-time processing at the same time. In this model, the storage element is divided into two sections. The cold path section is related to a part of data that is stored for a

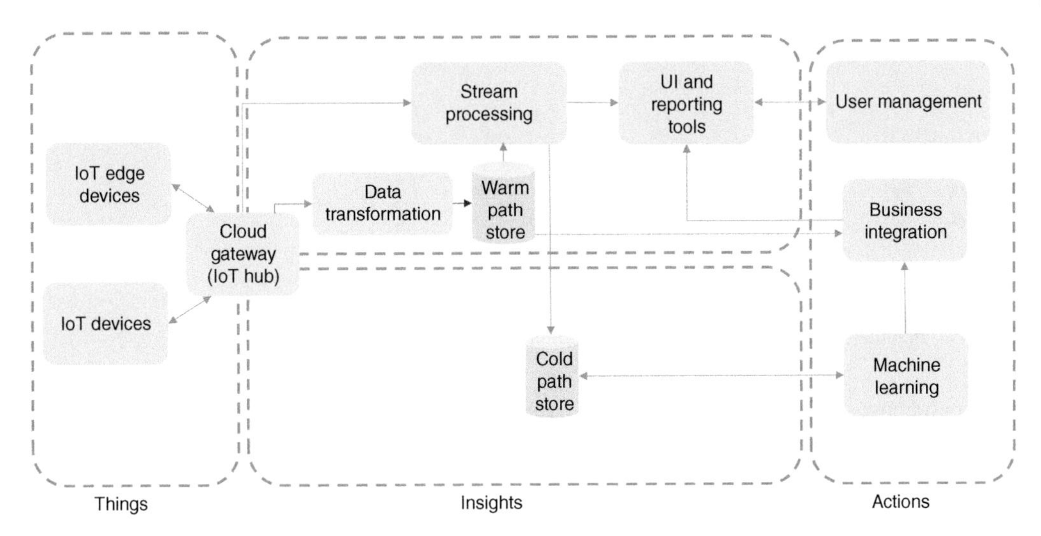

Figure 3.3 Microsoft architectural model with more detailed sublayers. Source: Adapted from Microsoft Azure [2].

longer period and is used for off-line processing. The warm path section is the storage element that is used for real-time processing. The user management subsystem permits the specification of different capabilities for various users and groups. Data transformation subsystem performs data aggregation and data manipulation.

3.5 Data Center and Cloud

A data center is a facility with several servers, storage devices, network infrastructure (switches and routers), and a cabling system for establishing connectivity among servers and storage elements. The facility must be equipped with redundant and backup power supplies as well as redundant data communication connections to improve its availability. A very important part of a data center is the environmental control equipment, which is responsible for performing operations such as air conditioning and fire suppression in order to establish a suitable condition for the data center to operate. A data center should also be equipped with security devices such as firewalls, Virtual Private Network (VPN) gateways, and intrusion detection systems.

You may virtually visit a data center to get an idea of how it looks on the inside. Many companies provide virtual tours of their data centers. For example, Google has published pictures of some of its data centers and allows you to virtually move inside different parts of these data centers.

Traditionally, a large company would build a data center for its own use. Today, data centers are managed by Internet Service Providers (ISPs) for hosting their own servers as well as third-party servers. There are modular data centers that can be used by service providers to make a functional data center. For this purpose, the service provider must purchase or acquire land to install the modular data centers, ask a local utility company to provide the power that is needed to run the data center, and ask a telecommunication service provider to establish the required connectivity to the data center. The modular data centers have everything including the racks, servers, computers, cooling systems, and the cabling system. Examples of companies that build modular data centers are Cisco, IBM, and SGI. These modular data centers are usually designed in three different categories as shown in Figure 3.4. Small modular data centers need less than 1 MW of power and have up to 4 racks per unit, where each rack can have 20–40 servers. Medium data centers need 1–4 MW

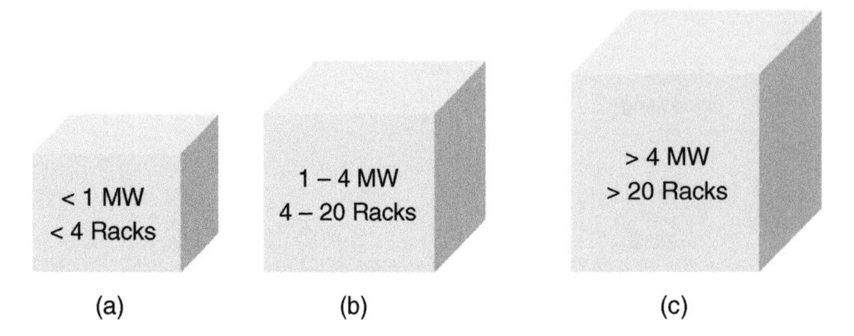

Figure 3.4 Modular data centers: (a) small; (b) medium; (c) large.

of power and have from 4 to 20 racks per unit. Large modular data centers are the ones that require more than 4 MW of power and have more than 20 racks per unit. By connecting several modular data centers to each other, an organization or a service provider can build a larger data center or a data park. In other words, this makes the design of a data center scalable. The power efficiency of a data center is expressed as Power Usage Effectiveness (PUE). It is defined as the power entered into a data center divided by the power used for servers and storage devices. For example, if PUE is 1.05, it means that there is a need for 1.05 MW to run a 1 MW data center. Unfortunately, PUE of data centers is usually not close to one. This means that data centers need to utilize a large amount of electricity for other operations.

As previously mentioned, the cabling system is one of the most important parts of a data center. To have an efficient cabling system, structured cabling is used in data centers, and all the cables are numbered. Different colors are assigned to various types of cables, which are neatly organized within the data center. Most data centers have false floors that have been used for heating and cooling as well as cabling. Within a data center, different cables may be separated from one another and organized in different locations. For example, fiber optic cables can be arranged on top of racks, while power cables can be placed in the middle, and signaling cables can pass through the bottom of racks.

Generally speaking, data centers can be classified into various levels based on their performance. For example, Uptime institute classifies data centers into four tiers. Simply put, Tier I data centers have the lowest performance. The annual downtime for Tier I data centers is around 30 hours. Tier II data centers have better performance in terms of availability. This is achieved by using redundant components and raised floors. Tier III data centers reduce downtime to around two hours per year. This is achieved by using multiple paths for the cabling distribution as well as the power distribution. Also, Tier III data centers are connected to more than two telecommunication service providers. Tier IV data centers are the best in terms of performance and can achieve less than 30 minutes of downtime each year. Table 3.1 outlines the four tiers of data centers, as well as the specifications that each tier is defined by.

Table 3.1 Classification of data centers into four tiers.

	Downtime per year (hours)	Redundant components	Dual power for servers	Number of Telecomm. providers	Redundant cabling	Power cable distribution	Raised floor
Tier I	~30	No	No	≥1	No	Single path	No
Tier II	~22	Yes	No	≥1	No	Single path	Yes
Tier III	~2	Yes	No	≥2	No	Multiple paths	Yes
Tier IV	~0.4	Yes, multiple	Yes	≥2	Yes	Multiple paths	Yes

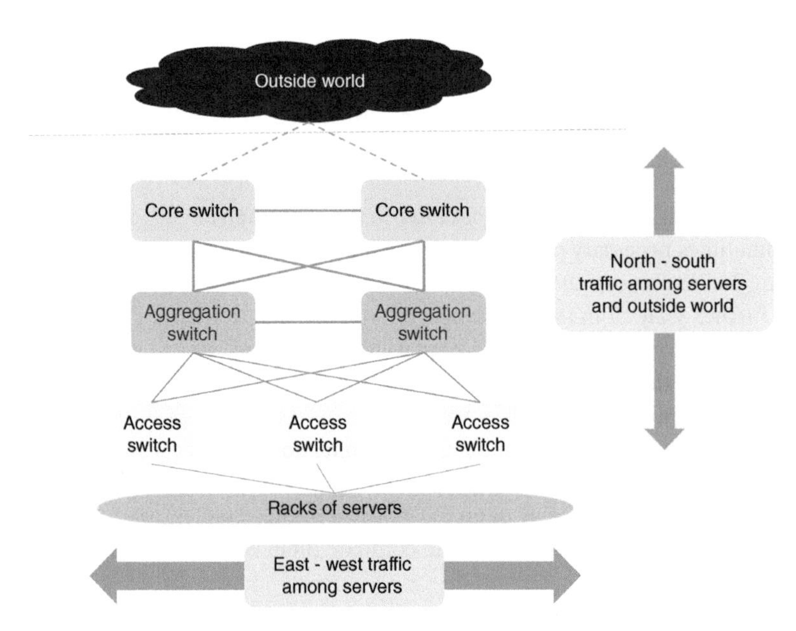

Figure 3.5 East-west traffic versus north-south traffic on a data center.

One important factor in the design of a data center pertains to the amount of traffic that comes in or goes out of the data center as compared to the amount of traffic among the servers and storage devices inside a data center, as shown in Figure 3.5. Generally speaking, servers that often communicate with one another to deliver a specific service are typically placed close to one another inside the data center. For many IoT applications, the east-west traffic (server-to-server) is significantly higher as compared to the north-south traffic (from IoT devices toward servers or vice versa). If there are x servers underneath of a core router at the edge of a data center, where each server has a y Gbps connection, and assuming that the core router has z Gbps connection to the outside world, then we can say that the core router is: $z:x \times y$ over subscripted. The value of z is usually substantially smaller than the value of $x \times y$ in most data centers. In other words, to provide a service to the outside world, there is a large amount of traffic among servers in the east-west direction. Therefore, most data centers are designed based on the fact that north-south traffic is lower than east-west traffic.

Simply put, the cloud is basically composed of a group of multi-tenant data centers. In the data communication world, servers and networking infrastructure are usually represented by clouds in technical diagrams. That is why the term *cloud* or *cloud computing* indicates the location where the networking infrastructure and computing power resides. It is possible to move the applications or services provided by a data center to another data center. For example, if latency is not a big issue, computing can be performed in any data center that has available resources, or the energy cost is lower. Since the energy consumption and cost of electricity is usually lower at night, cloud computing can follow the moon. For example, during the day, part of North American cloud computing can be moved to be performed in data centers in Japan where it would be nighttime. In a similar way, computing from the east coast can be moved to the west coast and vice versa. Even though this idea sounds interesting, security and privacy factors must also be taken into account. The ability to transport data or applications to be processed in other data centers resembles the movement of clouds and relates to the definition of the term “cloud.”

3.6 Computing (Cloud, Fog, and Edge)

In regard to the required processing of IoT applications, there exists three terms called the *cloud computing*, *fog computing*, and *edge* (mist) *computing* as shown in Figure 3.6. In this section, we explain these terms and their applications in IoT.

In nature, mist is closer to the earth than fog, and fog is closer to the earth than clouds. Similarly, for IoT applications, it is sometimes necessary or perhaps more beneficial to process data closer to IoT devices. Edge computing means processing the IoT data close to the IoT device. Fog processing means processing the data further away from the IoT device, but not in a data center or cloud. For example, processing data inside an IoT gateway is considered as fog computing.

Depending on the amount of data, the complexity of processing, the storage and security requirements, and the sensitivity of an IoT application to delays, we can determine the appropriate processing location for performing the required computation for an IoT application. In other words, these factors can help us to see which processing scheme can be used to fulfill the design requirements of an IoT application or can improve the efficiency of an IoT system. In this chapter, we evaluate the cloud, fog, and edge (mist) processing in more detail and compare them with each other.

3.6.1 Cloud Computing

For the development of an IoT application, cloud computing can offer many advantages. It allows for fast implementation of IoT applications, with no upfront implementation cost in an environment that is scalable and accessible from anywhere that an Internet connection is available. Cloud platform is elastic, and therefore, IoT system designers can easily scale up or down the required resources on the cloud, especially when the number of IoT nodes changes. IoT developers also do not have to worry about performing system backups and maintenance.

Cloud computing technology provides different types of services that can be categorized into three groups as shown in Figure 3.7:

- Infrastructure as a Service (IaaS) is an environment that provides resources such as processing power (servers), data storage capacity, and networking devices. IaaS is also called Hardware as a Service (HaaS), since the service provider only provides the hardware resources. Therefore, the service provider manages the hardware resources, and the users are responsible for operating systems, databases, application programs, and virtualization programs.
- Platform as a Service (PaaS) is an environment that provides the necessary tools for designing, testing, and launching applications. The service provider manages hardware resources as well as the operating system, virtualization programs, and software resources for creating applications.
- Software as a Service (SaaS) is an environment that provides software programs tailored to various business needs. Therefore, the service provider manages the entire application.

In terms of deployment, cloud can be divided into three categories: public, private, and hybrid clouds. In a public cloud, resources are shared among many customers, while a private cloud is

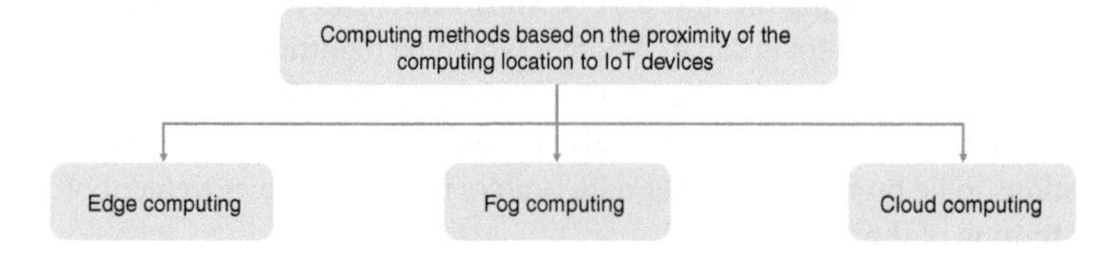

Figure 3.6 Computing methods based on the proximity of the computing location to IoT devices.

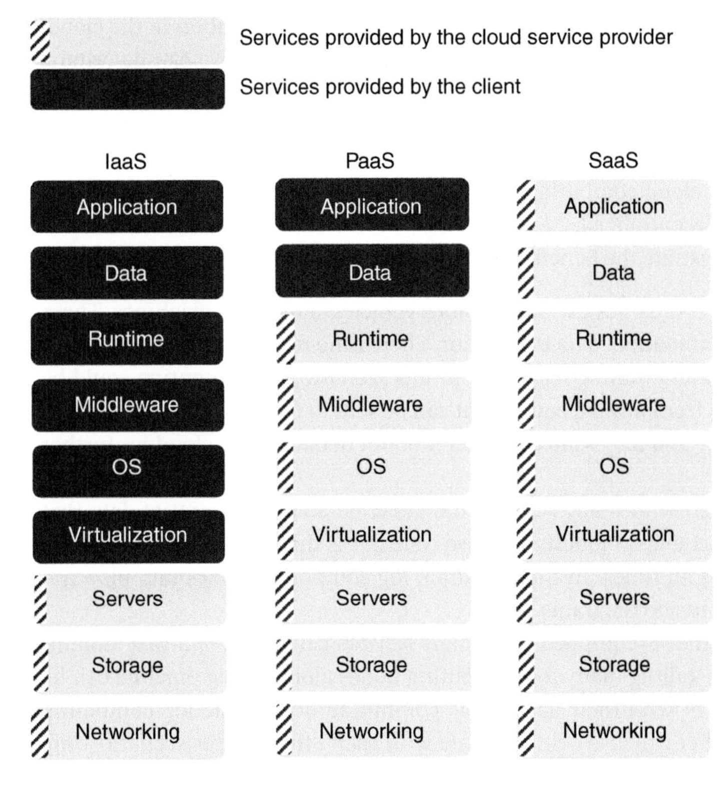

Figure 3.7 Various types of services in cloud computing technology: IaaS, PaaS, and SaaS. Source: Adapted and modified from [3].

dedicated solely to one customer. Therefore, a private cloud can provide a higher level of data security. Despite the fact that a public cloud shares its resources with a large number of clients, the data and applications that are hosted in the public cloud are still very secure. Hybrid clouds can use both public and private clouds.

Clouds, in general, provide an environment that is suitable for a wide range of non-delay-sensitive IoT applications, particularly those that demand complex processing and a considerable amount of storage capacity. Cloud computing cannot ensure that it will be a good option for applications needing low latency due to the considerable distance between an IoT device and the cloud, as well as network traffic unpredictability. Cloud availability is another issue that can be considered a drawback of cloud computing. Even the data centers with the highest performance ratings (data center tier IV), as we previously stated, experience brief periods of outage. This problem could be handled by utilizing two or more cloud service providers at the same time. However, this technique increases implementation costs. Security and privacy are other problems with public cloud computing. Even though the cloud environment is very secure, you should still consider that the cloud uses a centralized model and data passes through wide area networks alongside the data from other sources.

3.6.2 Fog Computing

Fog computing nodes are geographically closer to IoT devices, and therefore, they can provide faster responses as compared to cloud computing servers. For this reason, fog computing can be very beneficial for some IoT applications.

For many IoT applications, fog computing is not a solution replacing computation in the cloud, rather, it is a complement to cloud computing. For a specific IoT application, we can determine which part of data should be transferred to the cloud for processing and which part should be processed at fog computing nodes. In other words, a fog computing node is an intelligent device that offloads a portion of traffic and processing from the cloud. One main difference between cloud computing and fog computing is that cloud computing uses a centralized model, while fog computing uses a distributed model that is decentralized.

The following examples demonstrate the benefits of fog computing for IoT applications:

- To ensure safety, many mining companies use autonomous vehicles and boring machines. These machines can generate a large amount of data every hour. There is no reliable cloud connectivity available in deep underground and mining locations. In this scenario, fog computing could be an excellent solution. The data from mining equipment can be sent to fog computing nodes that perform their processing locally and may send a smaller amount of data to the cloud for further processing.
- Fast moving trains are equipped with many sensors and generate a large amount of data that needs to be processed. Due to the fast mobility of these trains, it is difficult to maintain a reliable connection to the cloud at all times. In this situation, fog computing comes into play and processes the data somewhere inside the trains.
- A long smart pipeline system that is equipped with many sensors can generate a vast amount of data during its operation. Installing many fog computing nodes along a long pipeline can be a viable solution, where sensors send their data to fog computing nodes. The fog computing nodes that are close to each other can also communicate with each other. In this scenario, only high-level information will be sent to the cloud for storage and processing. This also reduces the cost associated with cloud computing.
- A drone takes videos of a surveillance area to track various targets. Sending live raw video signals taken by the drone to the cloud for processing would demand a lot of bandwidth and would be very expensive. It may be a good solution to send video signals to a fog computing node for processing and then deliver the analyzed data to the cloud. This technique not only minimizes the quantity of data that needs to be transferred to the cloud but it also reduces the latency as compared to when video signals are processed in the cloud.

3.6.3 Edge Computing

Edge (mist) processing places computing power as close as possible to IoT devices. The idea is to process directly on the device without the need to transmit the entire data to a fog computing node or to the cloud. Processing at the device improves data security, and reduces the amount of data that is sent to a fog computing node or to the cloud, if needed.

Edge computing means processing data at the source instead of sending the data to the cloud and waiting for the response from the cloud. It is a perfect solution for very delay-sensitive IoT applications. Edge computing is useful when processing is simple, large data storage is not needed, or where providing connectivity is difficult. For example, the power output of a wind turbine that is installed on a remote wind farm can be optimized based on real-time values of wind speed and wind direction. Edge computing can be considered as a potential solution in this scenario, where real-time wind data can be used by each wind turbine to compute the best settings of the wind turbine in order to maximize the output power. Edge computing allows the data to be processed locally without relying on a fog computing node or a cloud connection.

3.7 Summary

An architecture model permits visualization of high-level information about a system. It can be used by designers to develop the detailed design of a system based on a specific architecture model.

There are several IoT architecture models published by various IoT players. In this chapter, we discussed the IoTWF architecture model and compared this model with the model published by Microsoft.

One important part of any IoT architecture model is related to physical objects and sensors. We will discuss sensors in Chapter 4. Connectivity is another important part of any IoT architecture. We will discuss various technologies used for providing connectivity in Chapters 5–8, as well as the important IoT protocols in Chapter 9. Another part of any IoT architecture model is related to computation of data. In this chapter, we reviewed several computing techniques depending on the proximity of computing nodes to IoT devices. We defined cloud computing, fog computing, and edge computing as three computing strategies. We also addressed the benefits and drawbacks of each of these computing methods. Cloud computing, as its name suggests, means the processing of data in the cloud. Fog computing nodes are geographically closer to IoT devices, and therefore, they can provide faster responses as compared to cloud computing servers. Edge computing allows the data to be processed locally without relying on a fog computing node or a cloud connection. Various methods of analyzing and processing data will be discussed in Chapter 11.

Architectural models can incorporate vertical layers as well as horizontal layers. For example, security can be considered as a vertical layer. This means that security should be considered for all the horizontal layers of an architectural model. We will discuss IoT security in Chapter 12.

An important layer in most IoT architecture models is related to business integration and collaboration among business stakeholders. This layer can help decision-makers to make better decisions on how the IoT system can add value to their organization. IoT solution development is discussed in Chapter 13.

References

1 Hanes, D., Salgueiro, G., Grossetete, P. et al. (2017). *IoT Fundamentals: Networking Technologies, Protocols, and use Cases for the Internet of Things*. Cisco Press.

2 Microsoft Azure (2020). Microsoft Azure IoT Reference Architecture, version 2.1.

3 Fitzgerald, J., Dennis, A., and Durcikova, A. (2014). *Business Data Communications and Networking*. NJ: Wiley.

Exercises

3.1 A simple IoT architecture uses three layers of perception, network, and application. The perception layer presents things, the network layer is responsible for connectivity, and the application layer represents the processing of data. What is the most important layer that is not considered in this IoT architecture?

3.2 Is it correct to say an architectural model that is used to make the architecture of an IoT system is the same as the blueprint of a building?

3.3 What advantages edge computing can provide for an IoT system?

3.4 A smart city uses smart hydrant systems. Each hydrant measures the water pressure and is able to transmit the pressure value, if needed. Explain how edge computing and fog computing can be useful for this application.

3.5 Among edge, fog, and cloud computing, compare each computing scheme based on the application sensitivity to the delay.

3.6 Give an example of a fog computing node.

3.7 What is the main reason for data abstraction in the cloud?

3.8 What happens if a massive number of IoT devices send simultaneously their data to the cloud?

3.9 What is the reason behind data accumulation in the cloud?

3.10 IBM has a five-layer architectural model of user, proximity network, public network, provider cloud, and enterprise network. The user layer is related to IoT devices. The proximity network layer defines the IoT gateway. The public network layer defines the connection to the cloud. The provider cloud layer is related to cloud processing and application. The enterprise network layer defines the network that receives the analyzed data. This information can be used for decision-making. Compare the IBM model to the IoTWF architecture model.

3.11 The IoT data can be structured or unstructured. Give an example of structured and unstructured data.

3.12 Describe an IoT application that contains unnecessary collected data that must be filtered out in the cloud.

3.13 There are four different tiers of data centers. Describe how these tiers differ.

Advanced Exercises

3.14 What do you think PUE of the existing data centers is?

3.15 Do you believe that fog is scalable in the same way that cloud is?

3.16 Smart electric pumps used in the oil and gas industry are usually installed at the center of an oil well, which is located in deep underground locations in order to pump oil to the surface. These smart pumps have several sensors and generate a large amount of data. The process for monitoring the condition of a pump and performing predictive maintenance is not extremely complex but is not simple either. Explain what computing scheme you suggest for this application.

3.17 Smart meters generate large amounts of data. Meter Data Management Systems (MDMS) use the data in order to forecast future energy demand. For this purpose, data from smart meters needs to be aggregated. The data aggregation process is very slow. There is also a large amount of data that is not necessary to be sent to the cloud. Explain a hierarchy of two sets of fog computing nodes: one for data aggregation, and one for data reduction that can be used in this smart grid application.

3.18 In a cloud environment, the core router is 1 : 24 over subscribed. There are 240 servers underneath the core router, and each server has a 1 Gbps connection. What is the core router's link speed?

3.19 Can we say that a fog is a small cloud (mini cloud)?

3.20 Do we need fog computing, if the cloud is able to provide extremely low latency, unlimited processing power, unlimited storage capacity, and outstanding security and privacy protection?

Chapter 4 IoT Sensors

4.1 Introduction

Internet of Things (IoT) and sensors help each other to exist and grow. Simply put, IoT does not exist without sensors. In Chapter 3, we learned that one of the most important parts of any IoT architecture model is the "things" layer, which consists of sensors and physical objects. An IoT system needs its sensors to satisfy design requirements in terms of accuracy, power consumption, size, weight, cost, and other design parameters. In order for IoT to grow and increase the number of its verticals, use cases, and applications, there must be suitable sensors to fulfill the requirements of various applications.

On the other hand, the sensor industry has found a golden opportunity to grow with the rise of IoT. The forecast of massive numbers of IoT devices in the near future can be translated into the need for a great range and a large number of sensors to be designed, developed, and manufactured.

The performance of an IoT system is greatly dependent on the accuracy and performance of its sensors. If the quality of the sensed data is poor, IoT systems usually do not perform well, and therefore, cannot provide smart decisions. This does not mean that every IoT application requires a sensor with the highest resolution or best accuracy. It means that each IoT application requires sensors with specific performance specifications for its various attributes such as size, power, weight, cost, accuracy, frequency response, or sensitivity. Therefore, the sensor industry has been given the opportunity to design and develop various types of sensors to fulfill the requirements of numerous types of IoT applications. For example, the size of a sensor is an important factor in the design of wearable IoT devices. However, the size attribute might not be a very important factor in sensors that are used in the machineries of a smart factory. In a similar context, the amount of power that a sensor consumes might be a very important factor in battery-powered IoT devices, while this attribute might not be an essential factor in systems that are directly connected to electricity.

There are many external factors such as noise or temperature that may affect the performance of a sensor and the quality of sensed data. A sensor may have excellent performance under normal conditions, but experiences performance degradation in the presence of such external factors. Some IoT systems may need to operate in harsh environments, and therefore, their sensors have to be resilient to tolerate such harsh conditions. With advancements in sensing materials, especially nanomaterials, it is possible to design sensors that are resilient and can perform well under extremely harsh conditions.

To ensure that a sensor can provide its required performance, it should be calibrated. In-factory calibration, even though costly, only ensures that the sensor meets its specified performance for a specific period of time. Therefore, sensors need to be recalibrated at later times as suggested by the manufacturers. Performing calibration on massive numbers of IoT devices would be a tedious

Fundamentals of Internet of Things: For Students and Professionals, First Edition. F. John Dian.

task and almost impossible. Quick and inexpensive self-calibration techniques might be needed to avoid producing sensor data with large uncertainties and unknown accuracy.

In this chapter, we briefly explain the sensors' performance metrics and the application of sensors in IoT systems. We then define the concept of smart sensors by a brief discussion on the general architecture of these sensors and their important components. It should be noted that detailed discussions on sensors and their performance are outside the scope of this book. The objective of this chapter is to give readers a basic understanding about sensors as well as smart sensors. By bringing some practical examples, we highlight the important factors on utilizing sensors and smart sensors in IoT systems.

4.2 Sensor and Its Performance Metrics

Sensors and actuators are important parts of IoT systems. A sensor is a device that measures a physical quantity and provides an output that is usually represented by an electrical signal. An actuator changes an electrical signal to a physical quantity. For example, a stepper motor is an actuator that changes an electrical pulse signal to a mechanical force that causes the motor to move. In IoT systems, sensors and actuators are both important and may work together. For example, an IoT wearable medical device may have a sensor to measure the glucose level in the blood and an insulin pump (actuator) in order to adjust the insulin dosage in patients with diabetes. In this chapter, our focus is on sensors.

There are many metrics that show the performance of a sensor. The performance metrics can be classified into two main groups. The first group demonstrates the steady-state response of a sensor to a stimulus input. The second group represents the dynamic characteristics of a sensor that are based on the sensor's transient response to an input.

4.2.1 Static Performance Metrics

Following are some of the most essential metrics related to the static characteristics of a sensor:

Range: A sensor can measure limited variations of a physical quantity. The limited variations indicate the range of a sensor. For example, a temperature sensor may be able to measure temperatures between −10 and 70 °C. In this case, the range of this sensor is [−10, 70]. A temperature sensor is one of the sensors that has been widely used in IoT applications.

Span: The span of a sensor is the difference between the maximum and minimum values of the range of a sensor. For example, a pressure sensor may have a range from 0 to 25 bar, whereas a bar is the metric unit of pressure equivalent to 100 kiloPascal (kPa). In this case, we can say that the pressure sensor has a span of 25 bars. A pressure sensor is used to measure the magnitude of the physical pressure that is being exerted on a sensor. It can also indirectly measure other variables such as the flow, speed, or level of liquids and gases. Pressure sensors are used widely in medical, automotive, and industrial IoT applications. Generally speaking, there are two types of pressure sensors. The first type is absolute pressure sensors, which are used to measure pressure against a pressure of 0 bar (vacuum). The second type is called relative pressure sensors, which are used to compare a pressure with the ambient pressure.

Accuracy: The accuracy of a sensor is the maximum difference between the actual true value of a physical quantity and the one measured by the sensor. Accuracy can be expressed as an absolute value or as a percentage of the full scale. For example, if a tilt sensor measures the slopes from 0 to 90° and if the accuracy of this sensor is ±1%, then the accuracy in absolute term is ±0.9°.

A tilt sensor (slope sensor) measures the angle of the slope of an object with respect to gravity's direction. Tilt sensors are used in various industrial applications [1]. An application of using a tilt sensor is in IoT-based structural health monitoring systems in order to determine whether the slope of a structure has been changed, or not.

Sensitivity: The sensitivity of a sensor indicates the ratio of the change of the sensor's output to the per unit change of the input. For example, the sensitivity of a sound sensor (microphone) is expressed as the ratio of the change of audio voltage (mV) to the change of sound pressure (Pa). Therefore, the sensitivity of a sound sensor is expressed as mV/Pa. Sound sensors are very popular in a variety of IoT applications to sense audio signals from people or surrounding environments.

Resolution: The resolution of a sensor is the smallest change of the input signal that is detectable in the output. For example, if a position sensor measures a displacement up to 50 mm, and its output has 10 distinct levels, then its resolution is 5 mm. Position sensors can detect the movement of an object as compared to a reference point and provide an output that is proportional to the displacement. They can also be used as occupancy sensors to detect the presence or absence of an object.

Nonlinearity: The nonlinearity of a sensor determines the maximum deviation of the output of a sensor from an ideal line that shows the linear relationship of a sensor's input and output. It can be expressed as the ratio of the maximum deviation of the sensor's output from an ideal line to the maximum full scale of its input expressed in percentage. The nonlinearity error is shown in Figure 4.1a. For example, for an imaging sensor such as Charged-Coupled Devices (CCDs), the input–output relationship over a wide range is almost linear. However, under high illumination intensity, the input–output relationship of CCD sensors becomes nonlinear. Imaging sensors such as CCDs have been used in many IoT applications to capture visual content.

Hysteresis: The hysteresis of a sensor is defined as the maximum difference between the output values of the sensor at any constant input within the sensor's input range, when the input signal is approached from the smaller values of the input toward the larger values of the input and vice versa. The hysteresis error is shown in Figure 4.1b.

Repeatability: The repeatability of a sensor is the ability of a sensor to generate the same output under the same input and the same conditions. Repeatability is defined as the variance of the data values that are collected by repeating the same measurements multiple times. In other words, repeatability represents the range of values the sensor can generate, relevant to previous measurements for the same input and under exactly the same conditions. Sometimes, the standard deviation of the mean is used to express repeatability. The standard deviation of the mean can be calculated by dividing the standard deviation by the square root of the number of measurements.

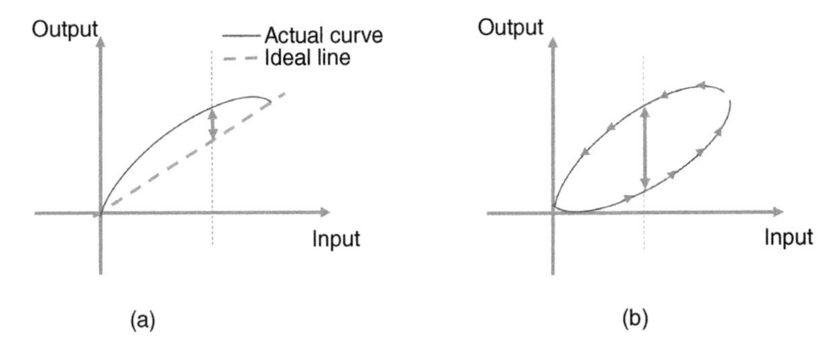

Figure 4.1 Examples of uncertainty in sensor measurement: (a) nonlinearity error and (b) hysteresis error.

Drift: The drift of a sensor is the degree to which the sensor output remains constant when applied to the same input over time. This metric is also called stability. For example, if a temperature sensor is drifting at the rate of 0.002 °C/year, the device is slowly changing its performance over several years. If we have good knowledge regarding the drift's rate, then the output can be corrected using processing techniques.

4.2.2 Dynamic Performance Metrics

Everything in engineering is viewed as a system, and sensors are no exception. Engineers would like to model everything as linear systems because of their simplicity. Many systems, in general, are either inherently linear or may be approximated as linear systems. Considering a sensor as a linear system, we can model it mathematically as an nth-order linear differential equation expressed as Eq. (4.1)

$$a_0 \frac{d^N y}{dt^N} + a_1 \frac{d^{N-1} y}{dt^{N-1}} + \ldots + a_{N-1} \frac{dy}{dt} + a_N y(t) = b_{N-M} \frac{d^M x}{dt^M} + b_{N-M+1} \frac{d^{M-1} x}{dt^{M-1}} + \ldots + b_{N-1} \frac{dx}{dt} + b_N x(t) \tag{4.1}$$

where x and y are the input and output of a sensor, respectively. M and N are the highest order of derivatives for input and output. For practical systems, the value of M is usually less than or equal to N. The values of a_i ($i = 0, N$) and b_j ($j = 0, M$) are the coefficients of the linear differential equation. If a sensor is modeled as a zero-order system, then there is no delay in the output response. For example, a potentiometer that is used as a position sensor, to indicate the linear or rotary displacement, can be represented by a zero-order system. This system can be expressed as $a_0 y(t) = b_0 x(t)$. The output of a sensor, modeled as a first-order system, changes with a specific time constant to reach its steady state. The first-order system can be represented by:

$$a_0 \frac{dy}{dt} + a_1 y(t) = b_0 \frac{dx}{dt} + b_1 x(t) \tag{4.2}$$

If the value of $b_0 = 0$, then the first-order system can be expressed as:

$$a_0 \frac{dy}{dt} + a_1 y(t) = b_1 x(t) \tag{4.3}$$

Solving the differential equation of a system such as the one expressed in Eq. (4.3) provides us with the time-domain output response of the system to any input. Without going through the mathematical calculation on how to solve Eq. (4.3), we provide the final solution to this equation in Eq. (4.4):

$$y(t) = y(0)\, e^{-\frac{a_1}{a_0} t} + \frac{b_1}{a_0} e^{-\frac{a_1}{a_0} t} * x(t) \tag{4.4}$$

where $y(0)$ represents the initial value of the system at $t = 0$, and the operator * represents the convolution operation. For example, the output of this system to a step function ($x(t) = Au(t)$) can be calculated using Eq. (4.4). The final result can be simplified as:

$$y(t) = \frac{b_1}{a_1} A + \left(y(0) - \frac{b_1}{a_1} A \right) e^{-\frac{a_1}{a_0} t} \tag{4.5}$$

The mathematical modeling of a sensor as a linear system is outside the scope of this book. However, readers should understand that when the input to a sensor changes, the output of the sensor may vary in time as it reaches its steady state. For example, it is possible for second-order systems or higher that the output exceeds its final steady-state values (overshoot) or oscillates above and below its final value during its transient time. For the first-order systems as modeled using Eq. (4.3), the output of the system does not have any overshoot or oscillations in reaching its steady state. Some of the performance metrics based on dynamic characteristics of a sensor are:

Dead time: The dead time of a sensor is the duration of time from when an input is applied until the time that the output begins to respond to the applied input.

Response time: The response time of a sensor defines the rate of change of the output on a step change of the input. This can be measured in terms of rise time, settling time, and overshoot.

Frequency response: The frequency response of a sensor shows the amplitude and phase of a sensor in the steady-state condition, when the frequency of the input signal changes within a specified frequency range, which is defined for the sensor.

Bandwidth: The bandwidth of a sensor is the frequency at which the magnitude of the output response falls 3 dB below the Direct Current (DC) gain value.

4.2.3 Sensor Selection

To select an appropriate sensor for an IoT application, we should look at the static and dynamic characteristics of possible sensor choices as well as the design requirements. To explain this further, we discuss some of the characteristics of the two sensors. The first one is related to different types of temperature sensors and their different characteristics. The second one discusses piezoelectric pressure sensors and their suitability for certain applications.

The most common sensors for measuring temperature are thermocouples, Resistance Temperature Detectors (RTDs), Thermistors, and Integrated Circuit (IC) temperature sensors. Thermocouples are made by connecting two dissimilar metals. The point of contact between these metals generates a voltage that is approximately proportional to temperature. Thermocouples support a wide temperature range of up to approximately 1200 °C, and generate a low output voltage for a unit change of the input temperature. RTDs are basically resistors whose resistance varies with temperature. They support temperatures up to 750 °C. They have excellent accuracy, repeatability, and an approximately linear input–output relationship. Thermistors are temperature-dependent resistors. The most common thermistors are called Negative Temperature Coefficients (NTC). The span of NTCs is 150 °C, and they demonstrate poor linearity and very high sensitivity. IC temperature sensors are silicon-based sensors. These low-cost temperature sensors have excellent linearity and support temperature values of up to 150 °C. We briefly explained some of the static characteristics of different temperature sensors. These characteristics may help us to choose a suitable type of temperature sensor for an IoT application. However, when there is a need to choose a specific sensor for an application, we should look at the complete static and dynamic characteristics to choose a suitable sensor for an application.

To illustrate with another example, let us consider the use of piezoelectric pressure sensors that have been widely used in many industrial applications and work extremely well in harsh environments. When a force is applied to a piezoelectric material, an electric charge is generated across it that is proportional to the pressure. In many piezoelectric sensors, a static force results in a charge across the sensor that leaks away over time for many reasons. Therefore, piezoelectric pressure sensors do not usually perform accurately when constant pressure is applied. However, these sensors are robust and very sensitive to changes in pressure across a wide range of frequencies. Therefore, many piezoelectric pressure sensors can detect small changes in pressure, even when the pressure is high. Even though piezoelectric pressure sensors have many advantages, the electric charge is lost over time, and long-term stable measurement is not possible. On the other hand, using a strain gauge pressure sensor, a very long-term stable measurement can be achieved. The strain gauge is basically a spring. When force is applied to the strain gauge, it deforms proportionally to the applied force.

Example 4.1 Accelerometer

An accelerometer is an electromechanical sensor that can sense motion, vibrations, and shocks and convert these quantities to an electrical signal. The basic principle of an accelerometer is based on a mass-spring system as shown in Figure 4.2. As seen from this figure, the mass, M, is attached to a wall by a spring that has a constant coefficient of k. When a force, F, is applied to the mass, it causes the mass to move x units in the direction of the applied force. Using Newton's law, we can find that acceleration on the mass is:

$$F = Ma = kx \rightarrow a = \frac{kx}{M} \tag{4.6}$$

In other words, if we can measure x, we can find acceleration on the mass. There are different ways to calculate the displacement x. Capacitive accelerometers, for example, use a capacitor to determine displacement. The capacitance of a capacitor is a function of the distance between the two plates of the capacitor. Therefore, a change in the distance between the two plates of a capacitor can change its capacitance, which can be measured electronically using a simple circuit. Assume that one of the capacitor plates is connected to the mass M, while the other remains stationary. When the mass moves as a result of the applied force, the distance between the two plates changes. By measuring the capacitance, an accelerometer can determine the amount of displacement.

Assume the range of a 16-bit digital accelerometer is $\pm 100g$, where g is the acceleration due to gravity that is equal to 9.81 m/s^2. The frequency response of this sensor is shown in Figure 4.3.

a. Can this sensor be used to measure acceleration of 50 m/s^2?
b. Can this sensor distinguish between 10 and 10.001g?
c. Can this sensor measure slow vibrations less than 2 kHz?

Solution

The range of this accelerometer is $\pm 100g = \pm 981$ m/s^2. Therefore, it should be able to measure acceleration of 50 m/s^2.

a. The resolution of this 16-bit digital accelerometer is $\frac{200g}{2^{16}} = 0.03g$. Therefore, it cannot distinguish between 10 and 10.01g.
b. The frequency response of this sensor shows that the sensor cannot detect very slow vibrations.

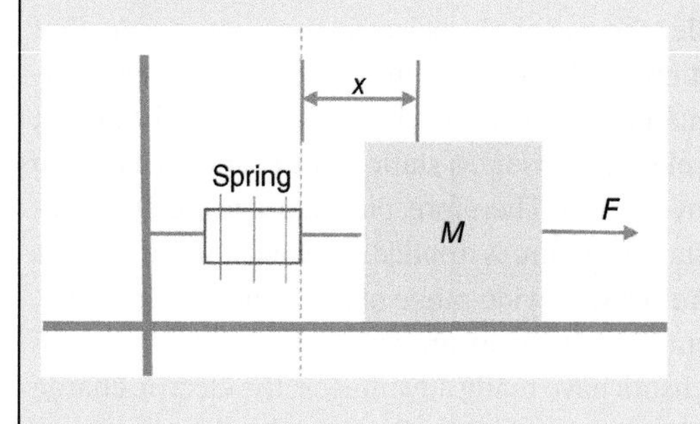

Figure 4.2 A basic model for an accelerometer.

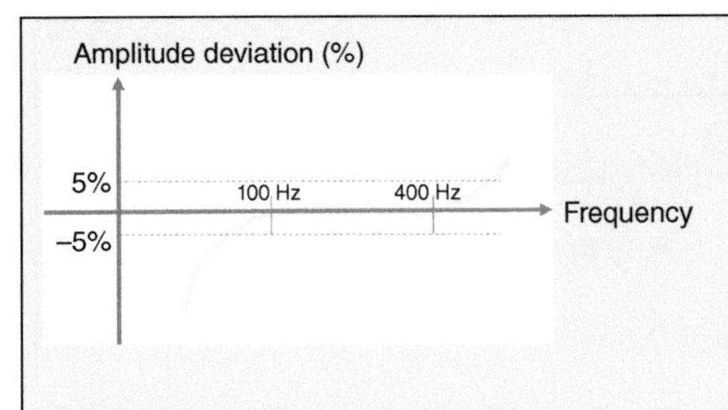

Figure 4.3 Frequency response of the accelerometer in Example 4.1.

Example 4.2 Imaging Sensor

An IoT-based device uses an imaging sensor with the size of 1.5 × 1.5 mm, containing a two-dimensional array of about 360 000 single sensors. The space between these sensors is equal to the width of these sensors. A lens is also used to focus the image onto the sensor array.

a. If the lens has a focal length of 15 mm, what is the field of view (in degrees) of this imaging sensor?
b. Can this IoT device detect an object that is approximately one mile away and has a size of 80 cm? For the sake of simplicity, assume that the image's object must cover at least four CCD sensors in order to be able to detect the image.

Solution

a. If the sensor array covers an area of 1.5 × 1.5 mm, and contains 360 000 CCDs, then the sensor array has 600 × 600 CCD sensors. Assuming equal spacing between the CCD sensors, this results in 600 CCD sensors and 600 spaces on a line of 1.5 mm long. Therefore, each CCD sensor is 1.5 mm/1200 = 0.00125 mm in diameter. Since the focal length of the lens is 15 mm, then the field of view can be calculated based on Figure 4.4.

$$\text{Field of view} = 2\,tan^{-1}\left(\frac{\frac{1.5}{2}}{15}\right) = 5.8^{\circ} \tag{4.7}$$

b. The object is at a distance of approximately 1609 m (1 mile). If we assume that the device has a focal length of 15 mm, then we can say:

$$\frac{f}{d} = \frac{s}{h} \tag{4.8}$$

where d is the distance between the lens and the object, f is the focal length, h is the length of the object, and s is the height of the object's image on the imaging sensor. Considering that the object has a height of 80 cm or 0.8 m:

$$s = \frac{h \times f}{d} = \frac{(0.8\text{ m}) \times 15 \times 10^{-3}}{1609\text{ m}} = 0.0074\text{ mm} \tag{4.9}$$

(Continued)

Example 4.2 (Continued)

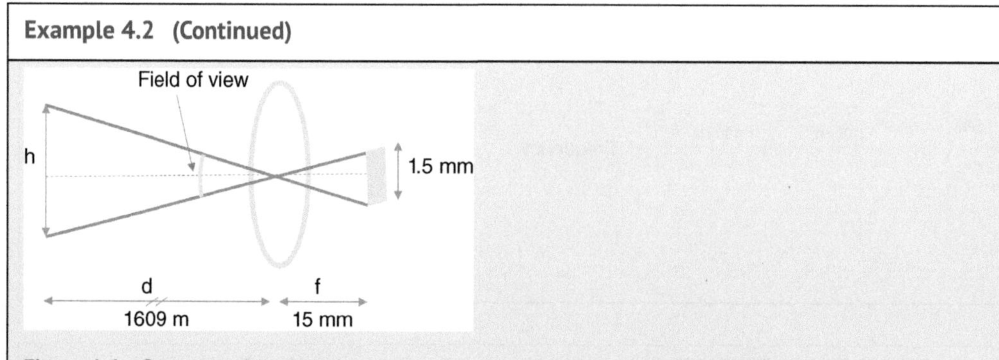

Figure 4.4 Structure of the imaging sensor, lens, and the object in Example 4.2.

Earlier we found that the length of each CCD sensor is about 0.00125 mm. Therefore, the image of this object covers about five CCD sensors, and therefore, the sensor array can detect this object at the specified distance.

4.3 Smart Sensors

The sensors that we explained in the previous section are basically sensing devices that only measure a physical quantity and provide outputs proportional to their inputs. We can define these sensors as core sensors or base sensors. A base sensor does not have any intelligence and only performs sensing measurements.

A smart sensor, also called an intelligent sensor, is a sensing device that performs many other operations besides sensing [2]. The architecture of a smart sensor is shown in Figure 4.5. As can be seen from this figure, a smart sensor consists of a base sensor and several other functional blocks to perform filtering, signal conditioning and Analog/Digital (A/D) conversion, calibration, data processing, power management, diagnostic, and communication. A smart sensor has the computing power, storage element, and connectivity capability in addition to one or more base sensing units.

A base sensor, as previously said, normally produces an electrical output after sensing a physical quantity. The filter block reduces noise from the range of frequencies that the base sensor supports. For example, a low pass filter can be applied to the output of a tilt sensor to reduce high frequency noise. A band pass filter can be applied to the output of a vibration sensor to separate random vibrations from the main vibrations that may indicate a problem with a machine in a smart factory. For applications that need to sense fast input changes, noise filtering should be applied in such a way that it does not affect the response of the system to the fast transitions. To make the output signal of a base sensor band limited, low pass filtering can be applied. This can be beneficial if we want to sample the sensor's output signal and encode it to a digital format whereby making the signal band limited, we can reduce the possibility of aliasing. Simply put, aliasing happens as a result of low sampling rates. It is the distortion that occurs when a continuous signal is under-sampled. In other words, aliasing causes a reconstructed signal from its samples to be different from the

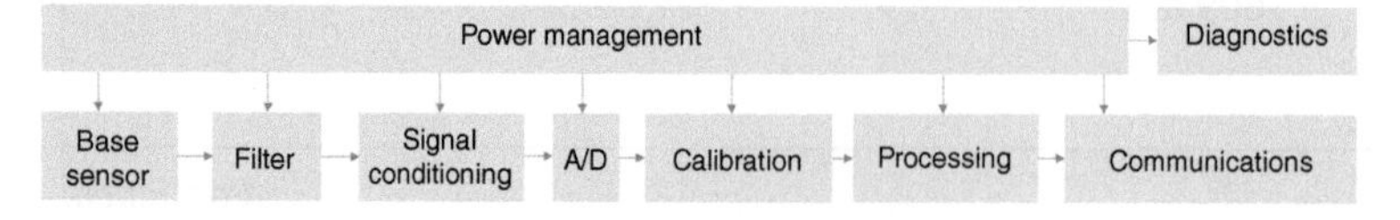

Figure 4.5 A general architecture for smart sensors.

original continuous signal. Generally speaking, aliasing causes the high frequency components of a signal to become part of the lower frequency components of the sampled signal. Since the proper rate of sampling has a direct relationship with the highest frequency of a signal, by filtering the high frequency components of a signal, the lower sampling rate can be applied.

Signal conditioning is the process of preparing a sensor's output signal so that it can be processed by a microprocessor or a data acquisition system in the next steps. A signal conditioner is a circuit that performs signal conditioning. For example, a sensor's output might be too low or high to be used by the next system. Signal conditioning ensures that the sensor's output is suitable for a system that is connected to the sensor. The A/D block converts the analog signal to a digital signal. The calibration block enhances the accuracy of measurements by applying a correction formula. Calibration is an important task that can increase the accuracy of a sensor. The data processing block prepares the data to support IoT applications. Edge computing can be performed in this block. The communication block provides connectivity to the Internet or to an IoT gateway. We discuss several important wired and wireless technologies used in the communication block in the next chapters. The power management block is responsible for optimizing the power consumption of the smart sensor by providing power to different blocks of the sensor in an efficient manner. For example, the power management block may introduce a low power state to be utilized between measurement periods to conserve energy. The diagnostic block is responsible for self-assessment. For example, it can be used to detect sensor contamination or sensor failure.

4.4 MEMS

By using similar technology as the one used in Very Large-Scale Integration (VLSI) and semiconductor fabrication, Micro-Electro-Mechanical Systems (MEMS) technology can enable the production of very small mechanical and electro-mechanical sensors inside a chip. MEMS sensors are very accurate and highly sensitive. The size of MEMS sensors is extremely small, their production costs on large scale as well as their power consumption are very low, and they can demonstrate high repeatability. For these reasons, MEMS sensors are ideal for IoT applications. An MEMS sensor may have a simple or complex architecture with no or many moving parts and can vary in size from less than a micrometer to a few millimeters.

Adding an MEMS sensor inside an IC with other elements for processing, storage, A/D conversion, signal conditioning, and communication in the same package enables us to have a smart sensor.

Most IoT applications would not benefit from building an accelerometer using a mass and spring combination, as shown in Figure 4.2. MEMS sensors contain both electrical and mechanical components, but they are fabricated on a micromillimeter scale. Using MEMS inside an IC makes the size of the sensor become very small. For example, the structure of an MEMS accelerometer is shown in Figure 4.6, in which a series of fixed plates are assembled in the outer frame. There are also a series of movable plates. These movable plates are connected to a mass and four springs. When the system moves due to acceleration, it makes the movable plates move which consequently changes the capacitance. By measuring the capacitance, we can determine the acceleration to the mass. Applying this sensor in different directions of x, y, and z can enable us to measure acceleration in directions of x, y, and z, respectively.

The frequency response of the accelerometer in Figure 4.3 indicated that the sensor does not perform well at 0 Hz (DC response) as well as at low frequencies. Figure 4.7 shows a typical frequency response for MEMS accelerometers. As can be observed, the MEMS accelerometer supports static acceleration such as gravity and slow vibrations, as well as higher frequency vibrations.

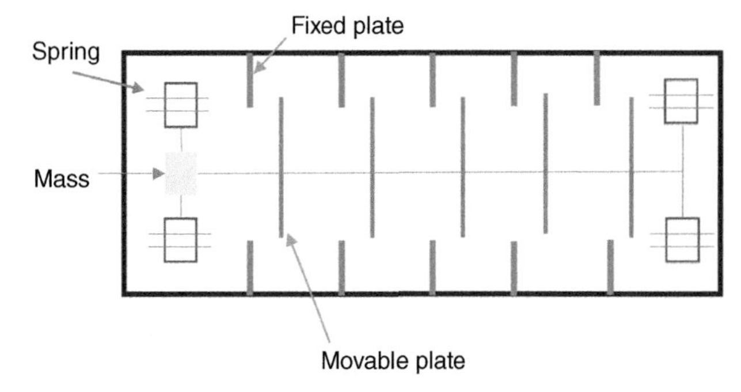

Figure 4.6 MEMS accelerometer.

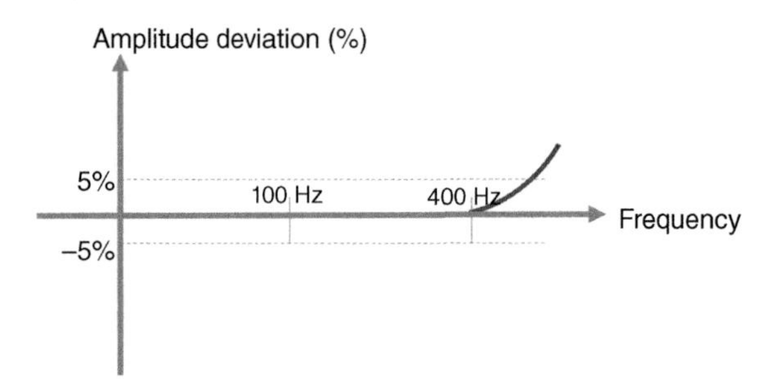

Figure 4.7 A typical frequency response of an MEMS accelerometer.

Gyroscopes are sensors that measure angular rotation. The principal operation of a single-axis MEMS gyroscope is shown in Figure 4.8a. Depending on the angular rotation, the movable mass inside the gyroscope moves away or toward the center of rotation, which changes the location of the movable plates of the capacitors and consequently the value of the capacitance. This movement is the result of Coriolis acceleration, which applies a force on the MEMS frame. Figure 4.8b shows the direction of motion on mass M, when the direction of the force applied on the inner frame is in $-x$ direction. Using a three-axis gyroscope, we can measure changes in angular rotations in three dimensions.

An Inertial Measurement Unit (IMU) is composed of a three-axis accelerometer, a three-axis gyroscope, and a magnetometer. Therefore, IMUs use a combination of sensors and can measure motion. For example, a robot might use an IMU to measure changes in angular rate and direction as well as vibrations in its joints. The average output signal from an accelerometer or a gyroscope has a small offset even when there are no movements or vibrations. This is called sensor bias.

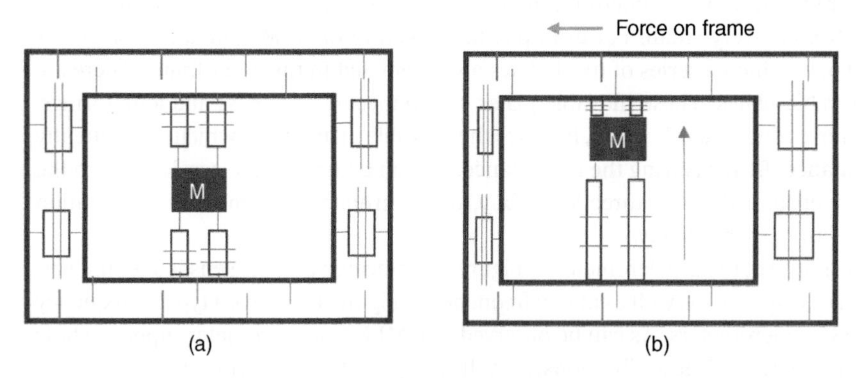

Figure 4.8 MEMS gyroscope. (a) Without applying any force (b) when force is applied to the frame.

MEMS is an enabling technology for IoT that makes the existence of small, low-cost, and high-quality sensors and actuators a reality.

4.5 Sensor Fusion

Sensor fusion plays an important role in IoT systems. Generally speaking, for a system, sensor fusion is the process of combining two or more sensors' output to get a better understanding of the system or make a better model of the system. We can obtain data from several similar or different sensors and apply them to a fusion algorithm to find a result that is more accurate, more reliable, and more functional as compared to the output of a single sensor. We can also look at it from another angle and say that through sensor fusion, and by fusing several sensors together, we can build better sensors. Sensor fusion can be used to achieve the following four objectives. We use practical examples to explain these objectives.

4.5.1 Improving the Quality and Accuracy of a Sensor

By fusing outputs from several sensors, we can improve the quality of data and the accuracy of a sensor. For example, the output of an accelerometer is usually noisy and the amount of noise depends on the quality of the accelerometer. We can make a better sensor if we use another accelerometer and average the data from two accelerometers. If the noise has a mean of zero, adding the output signals of several sensors can substantially reduce the noise. In this example, the fusion algorithm is a simple averaging of the two outputs. Therefore, by simple fusion algorithm and the use of two sensors, we have made a sensor that has improved performance. If two similar sensor measurements x_1 and x_2 have different noise variances of ${\delta_1}^2$ and ${\delta_2}^2$, respectively, then the sensor fusion output x can be calculated as:

$$x = \frac{{\delta_2}^2 x_1 + {\delta_1}^2 x_2}{{\delta_1}^2 + {\delta_2}^2} \tag{4.10}$$

If in Eq. (4.10) both sensors have the same noise variance (${\delta_1}^2 = {\delta_2}^2$), then this equation can be written as:

$$x = \frac{x_1 + x_2}{2} \tag{4.11}$$

We can also fuse two or more dissimilar sensors to reduce noise and enhance the quality as compared to the quality of a single sensor. For example, an IMU uses a magnetometer in addition to an accelerometer and a gyroscope. Generally speaking, the use of a combination of two dissimilar sensors (a magnetometer and a gyroscope) is better than using two similar magnetometers or two similar gyroscopes, since the noise of similar sensors might be correlated. Therefore, fusing a magnetometer and a gyroscope can produce better results if the noise signals of these two sensors are not correlated. Actually, that is the reason behind using three types of sensors in an IMU, which is used for movement measurements.

4.5.2 Improving the Reliability of a Sensor

By fusing the data from several similar sensors, a more reliable sensor can be created. Consider a scenario in which several similar sensors are utilized instead of a single sensor, and one of the sensors malfunctions either temporarily or permanently. The presence of numerous sensors ensures that the sensor fusion method can produce reliable sensor output in this case. It should be noted that failing a sensor that is used as part of the sensor fusion process may reduce accuracy. In addition, if one of the sensors malfunctions and produces data that is significantly different from the other sensors, the fusion algorithm may discard the data generated by that sensor. For example, if three wind sensors are providing data on wind speed and direction, and one of them is providing data that differs from the other two, the fusion algorithm can use the data from the two sensors that produced comparable outputs and ignore the data from the other sensor.

4.5.3 Improving the Capability of a Sensor

By fusing the outputs from several sensors, we can make a sensor with improved capabilities as compared to the capabilities of each of the single sensors. For example, an ultrasonic sensor has a narrow field of view. With fusing several ultrasonic sensors, positioned at different angles, we can build a sensor that has a wider field of view.

4.5.4 Measuring a Different Physical Quantity

By fusing the data of several sensors, we may be able to measure a physical quantity that none of the sensors involved in the fusing process is able to measure. For example, a we are unable to discern the size of an object inside an image which is taken by a camera, but combining data from a camera with an ultrasonic sensor that provides a sense of distance between the sensor and the object allows us to determine the size of the object within the image.

4.6 Self-calibration

To ensure that a sensor is providing its best possible performance, it needs to be calibrated. Calibration can minimize uncertainty in measurements and provide data that is more accurate. When sensors are out of calibration, they can produce data that might not have the required accuracy, and therefore, the IoT system might not be able to make smart decisions. A sensor might go out of calibration as a result of aging, drift, or excessive noise and vibrations, which are applied to the sensor. For example, a sensor installed on a smart car may go out of calibration after an accident.

Sensor calibration for IoT systems presents significant challenges. Consider a scenario in which a company attempts to measure the wind data of a potential wind park in an offshore location. The company builds a met mast structure and installs a first-class anemometer sensor to accurately measure wind speed and direction. The wind data will be used to determine the amount of energy that the wind park can generate. In an offshore wind park, the connection between the anemometer and the cloud is often established using satellite communications. Assume that the anemometer has come with a first-class certification verifying that the anemometer has been calibrated. But the sensor needs to be recalibrated every 18 months. Traveling to the met mast location is weather dependent and the cost and labor entailed in the calibration process is high. In this simple example, you can see the importance of self-calibration that can reduce the cost and ensure that the wind data provides the required accuracy. This issue becomes more problematic when there are massive numbers of IoT sensors installed in different locations, and there is a need for periodic calibration of these sensors.

Self-calibration means that the sensor can automatically calibrate itself. Generally speaking, we would like to determine the best possible parameters and configurations of a sensor or those parameters related to the synchronization of multiple sensors. Some of the techniques for self-calibration include utilizing additional sensors to compensate for cross-sensitivity and using an actuator to generate a local calibration signal. Detailed discussion on self-calibration techniques is outside the scope of this book.

4.7 Sensors of the Future

The sensor industry is growing rapidly and sensor manufacturers are producing a variety of sensors that can be used in various IoT applications.

To demonstrate some of the new developments in sensor research, we discuss three interesting ideas among many new research activities in designing smart sensors.

1. The use of ingestible sensors is becoming an ever-growing reality. These small, pill-sized sensors can be swallowed to enter the patient's stomach. They can then collect and send health data outside the body. For example, these ingestible pills can monitor the pH level or temperature inside the body. Recently, MIT researchers designed an expandable ingestible pill that can stay in the stomach for up to a month and can track many health conditions [3]. By drinking a specific solution, the pill can shrink in size and pass out of the body. These ingestible sensors can find environmental data in the stomach, measure the pH level, look for signs of bacteria, and monitor different gas levels in the gut. Additionally, in future, a camera might be added to observe the progress of tumors inside the body. The ingestible sensors get their power from the environment. For example, zinc and copper electrodes might be attached to an ingestible sensor. The zinc electrode emits ions into the acid inside the stomach to power the electronic circuit inside the ingestible sensor.
2. Smart fabrics and textiles are garments with embedded smart sensors. Adding sensors into the fabric can be done using several methods, such as multilayer 3D printing or conductive fibers. Fabric sensors can measure many physical quantities such as force, pressure, chemicals, and temperature variations, and therefore, they play an integral role in wearable IoT.
3. Optical chemical sensing attempts to monitor many chemical characteristics by transmitting the optical signal and finding specific molecular events. This area of research has many applications for determining water or air quality as well as monitoring quality of food. Generally speaking, a chemical sensor measures the chemical information such as absorption, concentration, pressure, or activities of particles of a chemical substance. Optical chemical sensing is performed based on the interaction of light with the chemical substance, and subsequently the conversion of the optical signal to an electrical signal that represents a chemical property of the substance. The wavelength of the optical light is an important factor in optical chemical sensing. It has been shown that many molecules have their maximum vibrational absorption in infrared wavelengths. Building an optical chemical sensor on a silicon chip at infrared wavelength introduces both many challenges and opportunities that can improve the quality of chemical sensors.

4.8 Summary

A sensor is a device that measures a physical quantity and provides an output that is usually represented by an electrical signal. Sensors are the building block of IoT systems, and without suitable sensors IoT systems cannot provide smart decisions. In this chapter, we briefly discussed the static and dynamic characteristics of a sensor. To select an appropriate sensor for an IoT application, we should look at these characteristics in order to find a good option that satisfies the design requirements.

Explanation of smart sensors, although brief, provides a basic idea about the functionalities of various components of these sensors. As explained in this chapter, a smart sensor has the computing power, storage capacity, and communication capability in addition to one or more base sensing units.

MEMS technology has enabled the production of very small mechanical and electro-mechanical sensors inside a chip. MEMS sensors are very accurate and highly sensitive. The size of MEMS sensors is extremely small, their production cost on large scale as well as their power consumption are very low, and they can demonstrate high repeatability. For these reasons, MEMS sensors are ideal for many IoT applications.

Sensor fusion is the process of combining the data obtained from several sensors of a system in order to get a better understanding of the system, make a better model of the system, or make a sensor that is more accurate, more reliable, and more functional as compared to a single sensor. Sensor fusion can be used to improve the accuracy, reliability, or capability of a sensor. Also, sensor fusion can enable us to measure a physical quantity that none of the single sensors used in the sensor fusing process can measure individually.

We emphasized that calibration is an important process that can ensure the accuracy of a sensor. Due to the challenges associated with calibration of a massive number of sensors in the IoT ecosystem, self-calibration of IoT sensors is an important task.

References

1 STMicroelectronics (2020). Precise and accurate tilt sensing in industrial applications, Application note, AN555- Rev 1.

2 Device, Analog (2016). The power of low noise in IoT smart sensors. https://www.analog.com/media/en/technical-documentation/tech-articles/The-Power-of-Low-Noise-in-IoT-Smart-Sensors.pdf (accessed 30 June 2022).

3 Trafton, A. (2018). Ingestible capsule can be controlled wirelessly. MIT news. https://news.mit.edu/2018/ingestible-pill-controlled-wirelessly-bluetooth-1213. (accessed 30 June, 2022).

Exercises

4.1 Which of the following devices are sensors, and which ones are actuators? a speaker, a DC motor, a microphone, a Light-Emitting Diode (LED), a gyroscope, and a camera.

4.2 What is signal conditioning?

4.3 A position sensor that is used in an IoT application has an input range of 0–2 inches, provides an output of 0–10 V DC, and its linearity error is specified as ±0.25% of full scale. The IoT application needs to find the position with the accuracy of 0.003. Is this sensor suitable for this application?

4.4 Explain a disadvantage of MEMS sensors.

4.5 Explain the accuracy and repeatability of each of the measurement results in Figure 4.E5. Use poor accuracy, good accuracy, poor repeatability, and good repeatability to describe each situation.

4.6 What is the difference between a base sensor and a smart sensor?

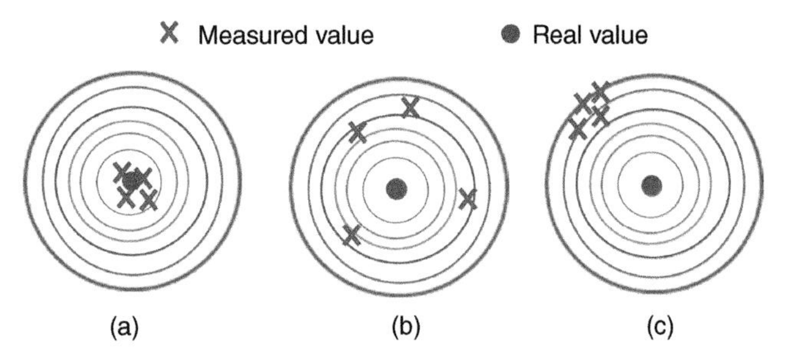

Figure 4.E5 Three different measurement results shown as a, b, and c.

4.7 A manufacturer of MEMS accelerometers has three different accelerometers listed in Table 4.E7. Another manufacturer has a new design with a noise density of 20 $\mu g/\sqrt{Hz}$ and power consumption of 350 μW. Compare the new design with devices A, B, and C.

Table 4.E7 Specifications of different accelerometers.

Device	Noise density ($\mu g/\sqrt{Hz}$)	Power (μW)
A	180	30
B	400	6
C	560	3

4.8 In an IoT application, there is a need for a temperature sensor.
a. Even though most NTC thermistors do not have excellent accuracy, what do you think might be the reasons that a designer uses an NTC thermistor in designing an IoT system?
b. A temperature sensor needs to measure the temperature of an object that can vary between 200 and 500 °C. What sensor do you suggest for this application?

4.9 What is the advantage of having a low hysteresis pressure sensor?

4.10 In an IoT application, we want to find the medicine taken by a patient during the day. Suggest a sensor for this application.

4.11 What is a self-powered sensor? How can a sensor operate without any energy source?

4.12 What is the difference between an absolute pressure sensor and a relative pressure sensor?

4.13 What is sensor fusion and why does it play an important role in IoT systems?

Advanced Exercises

4.14 A temperature sensor is modeled as a first-order system expressed as:

$$a_0 \frac{dy}{dt} + a_1 y(t) = b_1 x(t)$$

Find out after how many milliseconds the output of the system reaches 25 °C, if the input suddenly changes from 20 °C ($y(0) = 20$ °C) to 30 °C. Assume that $a_0 = 1$, $a_1 = 2$, and $b_1 = 2$.

4.15 A camera cannot measure distance from an object.
  a. Why is not a camera capable of distance measurement?
  b. Explain how fusing the data from this camera with the data from another camera, which is positioned at a different angle, can enable us to measure the distance.

4.16 What is a magnetometer and why is it used as part of IMU?

4.17 In the architecture of a smart sensor, why low pass filter before the conversion of the signal to digital format has been used?

4.18 What is chemical sensing? Briefly explain how optical chemical sensors work.

4.19 We can classify sensors as active and passive sensors. An active sensor requires an external source of power, while passive sensors do not need any external source of energy for their operation.
  a. Do you think a piezoelectric pressure sensor is an active sensor or a passive sensor?
  b. A Light Detection and Ranging (LIDAR) is a sensing device that transmits light waves into the surrounding areas. The transmitted waves bounce off surrounding objects and come to the sensing device. The device uses the reflected light to measure different physical quantities. Do you think LIDAR is an active or a passive sensor?

4.20 What is the gyroscope's bias?

4.21 Figure 4.8b shows the direction of the motion on mass M and the state of different movable springs when the direction of the applied force on the inner frame of an MEMS sensor is horizontal and in $-x$ direction. When a horizontal force in x direction is applied to the gyroscope, draw a picture similar to Figure 4.8b to demonstrate this situation.

4.22 After repeating an experiment for several times, the measurement values, which were obtained from a sensor (under exactly the same conditions), were 2, 2.05, 2.3, 2.2, and 2.1 mv. What is the repeatability of this sensor?

4.23 Explain how a capacitor can play a role in chemical sensing.

4.24 What are some of the reasons that a sensor can go out of calibration?

4.25 Two temperature sensors have different noise variances. The sensor that shows the temperature as 21° has a noise variance of 4 and the one showing the temperature as 24° has a noise variance of 2.
  a. If you fuse these two sensors, what would the output of the sensor fusion algorithm be?
  b. If these two sensors have the same noise valence, what would output of the sensor fusion algorithm be?

Chapter 5 IoT Wired Connectivity

5.1 Introduction

Connectivity for a Internet of Things (IoT) application can be designed in many ways, and there is not a single solution that is suitable for all IoT applications. Connectivity to the Internet can be wired or wireless. We use the term wired IoT, when IoT devices are connected to the Internet using cables. Similarly, we use the term wireless IoT when IoT devices are connected to the Internet using wireless technologies. An IoT device can connect to the Internet using both wired and wireless technology. For example, an IoT device might be connected to an IoT gateway, using a wire, while the gateway is connected to the Internet using a wireless technology. Similarly, IoT devices might be connected to the IoT gateway using a wireless technology, and the gateway might be connected to the Internet using a wired wide area network technology such as Asymmetric Digital Subscriber Line (ADSL) or cable TV.

Wired connections can offer many advantages for IoT applications. They can provide high data rates, low packet loss, reliable communication, and high security. However, they usually have higher costs of installation as compared to their wireless counterparts. They have challenges with scalability, and they cannot be used for mobile IoT devices. Besides wired wide area network technologies, the two main technologies that can be used for wired IoT are Ethernet and Power Line Communications (PLCs).

Standard Ethernet is a very popular method for connecting computers in a local area network and certainly can be used for some IoT applications. However, standard Ethernet does not have the capabilities required for most Industrial IoT (IIoT) applications, which need to handle time-critical and/or real-time applications. For example, in manufacturing operations of a smart factory in which tight coordination of sensors and actuators for creating an efficient closed-loop control system is essential, standard Ethernet might not be a suitable candidate. Since standard Ethernet cannot satisfy the requirements of automation and manufacturing applications in terms of factors such as guaranteed latency, packet loss, or throughput, several Ethernet-based protocols were designed that are called industrial Ethernet. Examples of these protocols are Process FIeld NETwork (PROFINET) and Ethernet for Control Automation Technology (EtherCAT) [1]. These protocols have modified standard Ethernet by adding some functionalities to standard Ethernet in order to become suitable for certain tasks in specific areas, and therefore, their applications are limited to those tasks or those specific areas. To address the needs for a suitable wired connection that can be used for all applications including IIoT applications, IEEE extended standard Ethernet to Ethernet Time-sensitive Networking (TSN). We use the term TSN or Ethernet TSN interchangeably to represent this new Ethernet standard.

Fundamentals of Internet of Things: For Students and Professionals, First Edition. F. John Dian.

PLC is another type of wired connectivity scheme for IoT devices. PLC technologies use the power lines and power distribution cables, as well as wiring inside houses and offices to provide a wired connectivity scheme. PLC can also provide broadband connectivity to rural and remote locations where other types of wired technologies are not available. In other words, power lines are infrastructures that can be used to transport data in addition to transporting energy. This makes them an attractive choice for providing connectivity in some IoT use cases.

In this chapter, we discuss Ethernet, Ethernet TSN, and PLC technologies. These technologies can provide wired connectivity for IoT applications.

5.2 Ethernet

As we discussed in Chapter 1, Ethernet is the most popular standard for local area networks. It has evolved from legacy Ethernet (10M bps) to above 100 Gbps Ethernet during four decades. We discuss some of the practical issues regarding the operation of IoT devices in an Ethernet-based local area network in this section.

Figure 5.1 shows a local area network in which several IoT devices are connected to an Ethernet switch using wired cables. The IoT devices can communicate with each other through the Ethernet switch in a star topology. These IoT devices usually need to be connected to the Internet. For this purpose, the switch should be connected to a router that is also connected to the Internet. Since this router sits at the edge of a local area network, it should also have an Ethernet port that is connected to the switch. Each IoT device should have an IP address. Since each IoT device has an Ethernet interface that is represented by a Media Access Control (MAC) address, an IoT device also has an MAC address.

As explained in Chapter 1, packets within a local area network are transmitted based on MAC addresses, while packets within a wide area network are routed using their IP addresses. Assume one of the IoT devices in the network shown in Figure 5.1, let us say Device #*i*, intends to send data to the cloud through the Internet. The packet generated by this IoT device should be sent to the router, which is a network entity capable of routing packets to the cloud through an Internet connection. In this scenario, the router is the network gateway to the Internet, and therefore, we should configure these IoT devices to have their default gateway as the IP address of the router (IP #*p*). If the IoT device #*i* intends to send its data to an IP address that is not part of the IP address of the local area network, the packet is automatically sent to the IP address defined as its default gateway, which in this scenario should be the IP address assigned to the Ethernet port of the router (IP #*p*).

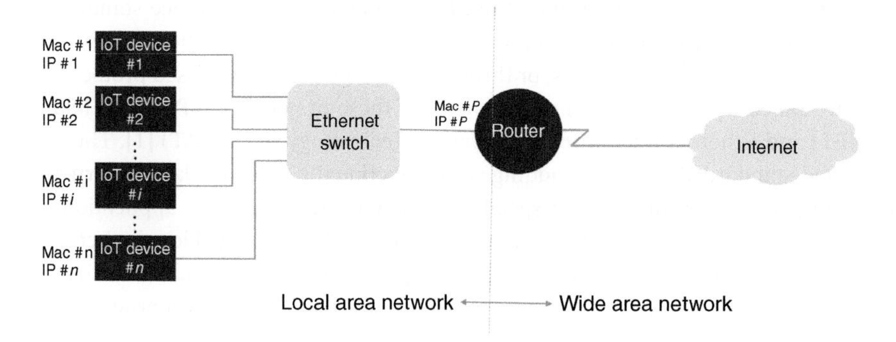

Figure 5.1 A local area network of IoT devices.

There is another issue that needs to be solved before the IoT device in our example can send its data to the router. To send data frames within a local area network, we need to know the MAC address of the IoT device as well as the MAC address of the Ethernet port of the router. Each Ethernet frame consists of a source Ethernet address and a destination Ethernet address. There is a protocol called Address Resolution Protocol (ARP), which is one of the supporting protocols for IP. This protocol maps an IP address to an MAC address. In other words, a device can send an ARP request in order to request for the MAC address associated with the IP address of a specific device. The router and each of the IoT devices know their own IP address as well as their own MAC address. If a device intends to find out about the MAC address of another device, knowing its IP address, the device can broadcast an ARP packet to all other devices in the network. When the broadcasted ARP packet reaches all devices in the network, then the device that has the same IP address as the IP address that is part of the ARP packet, can send back its MAC address to the device requesting it. That is exactly what happens in the network in Figure 5.1. The IoT device #*i* needs to broadcast an ARP request packet with the IP address of the default gateway (IP #*p*) and waits to receive an ARP response that contains the MAC address of the router (MAC #*p*). The ARP packet that is sent by the IoT device #*i* should have a source Ethernet address as MAC #*i* and a destination Ethernet address that is a broadcast MAC address expressed as FF:FF:FF:FF:FF:FF. When the IoT device #*i* finds out the MAC address of the Ethernet port of the router, it sends its data frame to the router. The router then sends the data to the cloud through the Internet.

The mapping ARP data (mapping the MAC and IP addresses used in the network) can also be kept somewhere in the network. For example, data related to the mapping of IP and MAC addresses of all nodes in the network can be kept on a proxy server. This proxy server can be a server connected to the network or even a router in the network. In this situation, IoT devices can send their ARP request packets to the ARP proxy server directly instead of sending a broadcast request. It is clear that the MAC address of the ARP proxy server should be given to all IoT devices.

5.2.1 Power over Ethernet (PoE)

PoE is a technology that allows network cables, which are used for data communication, to carry electrical power. Generally speaking, an IoT device requires a data connection to the network and a power connection to electricity in order to power the device. If an IoT device uses PoE technology, the network cables can carry both the electrical power and data, and therefore, the IoT device only needs one connection.

There are two ways to add PoE to an Ethernet network. The most important one is to use a PoE switch. Each port of a PoE switch has interfaces for both data and power. By connecting an IoT device to a PoE switch, the switch detects whether the IoT device is PoE enabled or not. If it is a PoE-enabled device, then it connects to the power automatically. Another method to add PoE capability to an Ethernet network is to use a device called a midspan or PoE injector. Midspans can be used to upgrade existing LAN installations to PoE. These two methods are shown in Figure 5.2.

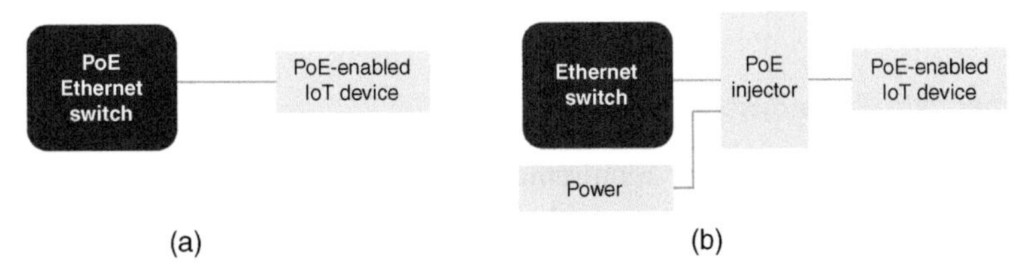

Figure 5.2 (a) A PoE-enabled network using PoE switch. (b) A PoE-enabled network using PoE injector.

Table 5.1 Different PoE standards.

		PoE	PoE+	PoE++	
Standards	IEEE standard	IEEE 802.3af	IEEE 802.3at	IEEE 802.3bt	
	PoE type	Type 1	Type 2	Type 3	Type 4
Switch port	Max. power per port	15.4W	30W	60W	100W
	Port voltage range	44–57V	50–57V	50–57V	52–57V
PoE-enabled device	Max. power to device	12.95W	25.5W	51W	71W
	Voltage range at the device	37–57V	42.5–57V	42.5–57V	41.1–57V
Cables and wirings	Number of used twisted pairs	2 pairs	2 pairs	2 or 4 pairs	4 pairs
	Supported cables	At least Cat3	At least Cat5	At least Cat5	At least Cat5

A PoE switch supplies power over two twisted pairs of an Ethernet wire and can deliver power up to 12.95W to a PoE-enabled device.

IEEE 802.3at standard, published in 2009, introduced PoE+ technology that increases the capabilities of PoE. A PoE+ switch supplies power over two twisted pairs of an Ethernet wire and is able to deliver power up to 25.5W to a PoE+-enabled device with a voltage range from 42.5 to 57V.

The IEEE 802.3bt standard, published in 2018, introduced PoE++, which has been classified into two types. The first one is called Type 3 and the second one is Type 4. PoE++ Type 3 enables two or all four twisted pairs in an Ethernet cable to deliver power up to 51W to a PoE++-enabled device, while Type 4 can send up to 71W over four twisted pairs of an Ethernet cable. The comparison of different PoE standards in terms of their power and voltages at the PoE switch and PoE-enabled device as well as the required class of the Ethernet cable for each standard is listed in Table 5.1.

5.3 Ethernet TSN

Ethernet TSN provides deterministic communication capability to standard Ethernet [2]. In deterministic communication, the duration of time that it takes for data to move between two nodes of a network is predictable and can be determined. To be able to have deterministic communications, all nodes of the network including IoT devices, non-IoT devices, and networking elements should have a common sense of time.

TSN is an abbreviation for time-sensitive networking. As the name indicates, there is an emphasis on time. Therefore, TSN uses time synchronization and time scheduling to guarantee on-time delivery of data and to minimize latency and jitter. In this section, we discuss TSN technology in more detail.

5.3.1 Challenges of Connectivity for Industrial IoT

Since standard Ethernet did not provide the requirements of IIoT applications, many Ethernet-based solutions are designed by different vendors to satisfy the various essential requirements of industrial applications. Industrial Ethernet is the common name used for these solutions. The solutions are provided by different vendors in order to control or monitor applications in many industries such as automation, automotive, and manufacturing [3]. These solutions extended standard Ethernet to provide deterministic latency, reliable communication, and real-time operations. The equipment and devices that are built based on these solutions should work in harsh environments such as smart factories and industrial manufacturing plants. One of the most important challenges in using industrial Ethernet protocols and equipment is that they are vendor dependent. This means that there is no compatibility among the equipment from different vendors. This forces the customers either to buy all the equipment from one vendor or to solve compatibility issues by designing and implementing protocol conversions or hardware adaptors. Solving compatibility issues is time consuming and tedious. Another challenging issue is that many industrial Ethernet solutions are designed for specific tasks and built for a specific purpose, and therefore, they might not be used for general IIoT applications.

One of the most popular Industrial Ethernet standards is PROFINET. It is used by many manufacturers such as General Electric (GE) and Siemens. It is actually an application-layer protocol over Ethernet. PROFINET runs over TCP, IP, and Ethernet for diagnostics and configuration. However, it skips TCP and IP and runs directly on Ethernet to provide fast and reliable real-time process data operations in a local industrial environment.

Ethernet TSN intends to provide a flexible standard in order to solve interoperability issues. Therefore, it can be considered as a generalized solution for various types of IIoT applications, where various industrial equipment from different vendors can be used in a network.

5.3.2 Ethernet TSN Features and Key Technologies

Before data is exchanged between two nodes in a TSN network, TSN switches should be configured. During the configuration, the required bandwidth needs to be reserved, and they should agree on the time synchronization strategy, the time scheduling method, and required metrics for the Quality of Service (QoS). The key technologies and important features of a TSN network are time synchronization, bandwidth reservation, redundant transmission, traffic shaping and scheduling, and latency minimization. In this section, we briefly explain each of these features and technologies.

5.3.2.1 Time Synchronization

To synchronize time across a network, each network element should have a clock that must be synchronized to a common clock. Time synchronization in a TSN network means that all devices and switches have the same time. This can be done by synchronizing the clock of all nodes to the clock of a specific node in the network. In other words, time needs to be distributed from a central source through the TSN network. This central node is called a grandmaster. It is the responsibility of the grandmaster to periodically send its clock to other nodes and also measures the roundtrip delay on each network node. To perform these tasks, TSN needs to use a time synchronization protocol. Precision Time Protocol (PTP) is a popular protocol for synchronization of clocks in a network and has been widely used to perform time synchronization on different types of telecommunication networks, such as cellular networks. Three versions of PTP protocols have been published as IEEE 1588-2002, IEEE 1588–2008, and IEEE 1588–2019. TSN uses an adaptation of PTP for providing time synchronization in a TSN network.

Generally speaking, a node that has a more accurate clock system must be selected as a grandmaster. In practice, this selection might be performed using an algorithm that is called Best Master Clock Algorithm (BMCA). In this algorithm, each node that intends to become a grandmaster must be configured to have two priority numbers: priority 1 and priority 2. These priority numbers accept any number from 0 to 255. There is also a clock accuracy parameter that can be configured to indicate the level of accuracy of a clock system. The BMCA algorithm selects the grandmaster based on the following steps:

1. The node that its clock has the lowest priority 1 value is chosen as a grandmaster.
2. If two or more nodes have the same lowest priority 1 number, the one with the lowest clock accuracy number among them is chosen as the grandmaster.
3. If two or more nodes satisfy the conditions in steps 1 and 2, then among these nodes, the algorithm chooses the one with the lowest priority 2 value as the grandmaster.
4. If there are two or more nodes that satisfy the conditions in steps 1, 2, and 3, then the one with the lowest MAC address among them is chosen as the grandmaster. MAC addresses are device dependent, and they are unique numbers. Therefore, it is not possible that two nodes share the same MAC address.

5.3.2.2 Bandwidth and QoS Reservation

To be able to reserve the required bandwidth for the communication between a source and a destination node, the source node should advertise a message with its required bandwidth and other QoS requirements to a connected switch. The switch then considers the reservation of the bandwidth and QoS parameters and sends the message to the next switch until it reaches the destination node. The destination node sends back a message toward the source node, and as the message comes back toward the source node, the bandwidth will be reserved on switches along the way.

5.3.2.3 Redundant Transmission

In a TSN network, it is important to prevent any loss of data. To be able to provide reliable communications, TSN allows a frame to be transmitted over several links at the same time. For applications with low latency requirements, this ensures reliable transmission and guarantees timely delivery of data. It should be noted that in case of a packet loss, the retransmission process can increase latency substantially. TSN tries to avoid retransmission by establishing a redundant communication scheme that increases the reliability for applications with tight latency requirements.

5.3.2.4 Traffic Shaping and Scheduling

Time scheduling is the process of forwarding the queued packets at specific points of time in order to ensure they arrive at their destination on time and without unexpected delays. It can be achieved by considering latency, jitter, bandwidth, and the size of the application's buffer at the destination node, in order to make delivering deterministic communication possible.

TSN allows different classes of traffic to coexist in its network. It should be noted that each class of traffic can have a different set of requirements in terms of bandwidth or desired latency. To allow Ethernet TSN to be backward compatible with standard Ethernet, TSN considers eight different priorities for Virtual Local Area Networks (VLANs). Therefore, each TSN switch maintains eight queues on each outgoing port. The data goes to an appropriate queue according to its traffic class and leaves the switch port eventually. This mechanism is called strict-priority traffic shaping. Even though this mechanism is simple, it has several problems. First, if there are many high priority packets, then a TSN switch might not be able to serve low priority frames at all. Second, if the switch receives many high priority frames belonging to the same class, these frames may still need to wait for a long time before leaving the switch, and therefore, the meaning of priority is lost.

It should be noted that when configuring a VLAN, we can assign it to a priority class value from 0 to 7. The lower the traffic class value, the lower is the priority. For example, a VLAN assigned to level 0 is sent as best-effort traffic. A VLAN that is assigned to level 7 indicates that its packets have high priority and they belong to time-sensitive applications.

5.3.2.5 Latency Minimization

Besides the simple VLAN priority scheme, TSN also introduces more complex traffic shaping mechanisms in order to provide low latency transmission. Examples of these mechanisms are Time-Aware Shaper (TAS) or Credit-Based Shaper (CBS) schemes.

TAS divides communication time into fixed length and repetitive cycles. In each cycle, different time slices are defined, where each time slice is assigned to one or more possible priorities. Figure 5.3 shows a situation where each cycle is divided into two time slices: slice 1 is assigned to the highest priority frames, and slice 2 is assigned to any other frames. A problem can arise when a low priority frame starts at the end of time slice 2, which causes the frame transmission to continue to the next cycle. This situation is shown in Figure 5.3 where a low priority frame has occupied a time slice that was assigned to high priority and critical traffic.

To address this issue, TAS implements a guard band that prevents the problem from occurring. During the guard band period, no node is allowed to send any Ethernet frames. The aforementioned problem is overcome if the guard band's duration is greater than an Ethernet frame. Figure 5.4 shows the introduction of the guard band for the TAS mechanism. Since the maximum size of an Ethernet frame is approximately 1500 bytes, then having a guard band equivalent to the duration of a 1500-byte Ethernet frame should be used.

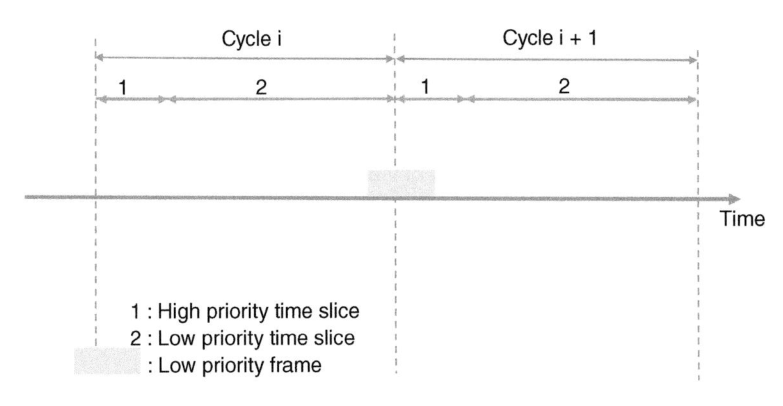

Figure 5.3 TAS cycles and a possible issue of TAS.

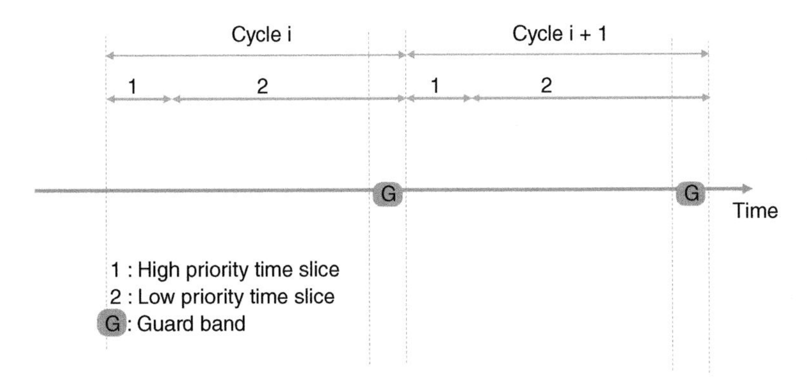

Figure 5.4 Use of a guard band to solve the problem with TAS.

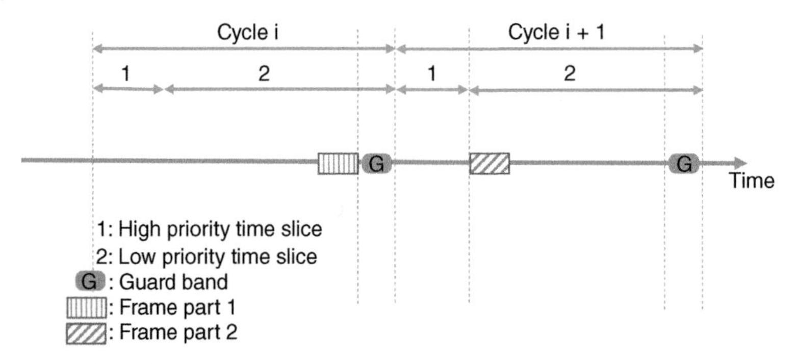

Figure 5.5 Pre-emption.

This solution has a major drawback. The introduction of the guard band can result in a reduction in the available time period for data transmission. For example, in Figure 5.4, the guard band has affected the available bandwidth for transmission of lower priority traffic. In other words, since during the guard band, no transmission can be started, the guard band reduces the available bandwidth. To be able to solve this problem, a new concept called pre-emption is introduced by TSN, which reduces the size of the guard band substantially. The pre-emption interrupts the transmission of low priority Ethernet frames before the guard band and continues the rest of the transmission in the next low priority cycle as shown in Figure 5.5. The concept of guard band in pre-emption is not exactly the same as the one used in TAS. In pre-emption, the guard band can be smaller as compared to the guard band defined in TAS, and it is limited to the maximum size of the Ethernet fragments, with a minimum size of 64 bytes. As mentioned earlier, the size of the guard band in TAS is the maximum size of an Ethernet frame.

In order to minimize jitter, CBS can be used in addition to TAS and pre-emption. In some applications the quality is degraded in case of jitter. For example, in video transmission, jitter may cause video frame skipping. In CBS, data transmission is controlled by a credit value, and an Ethernet node can only start transmitting data if the credit is positive or zero. The credit decreases during data transmission, and it increases when frames are waiting in the queue. This credit-based technique removes bursts from the traffic and generates an almost constant bitrate traffic that results in a reduction of frame loss and provides a lower jitter value.

5.3.3 A Simple Example

Let us explain time synchronization and scheduling with a simple example as shown in Figure 5.6a. In this example, there are three IoT-based cameras and a computer that are connected to a four-port switch. The clocks of these cameras, the switch, and the computer are all synchronized. To schedule the transmission time for each node in this network, we ask each camera to send its bursts of data at a specific point in time in such a way that there is no overlap in the transmissions of these cameras. For example, camera 1 may begin transmission at T and end at $T + 20$, camera 2 may begin transmission at $T + 20$ and end at $T + 40$, and the third camera may begin transmission at $T + 40$. If there is no traffic from the computer within these time periods, then the provided solution requires no queuing, there would be no latency inside the switch, and the jitter would be zero. The packets from each camera come to the switch and leave the switch without any interaction with the traffic generated by other devices, as shown in Figure 5.6b.

It is clear that the traffic pattern of each node of the network is not as simple as the traffic pattern that we discussed in this example. Now assume that there is some traffic from the computer that is low priority traffic as shown in Figure 5.6c. It would be interesting to queue the traffic and shape

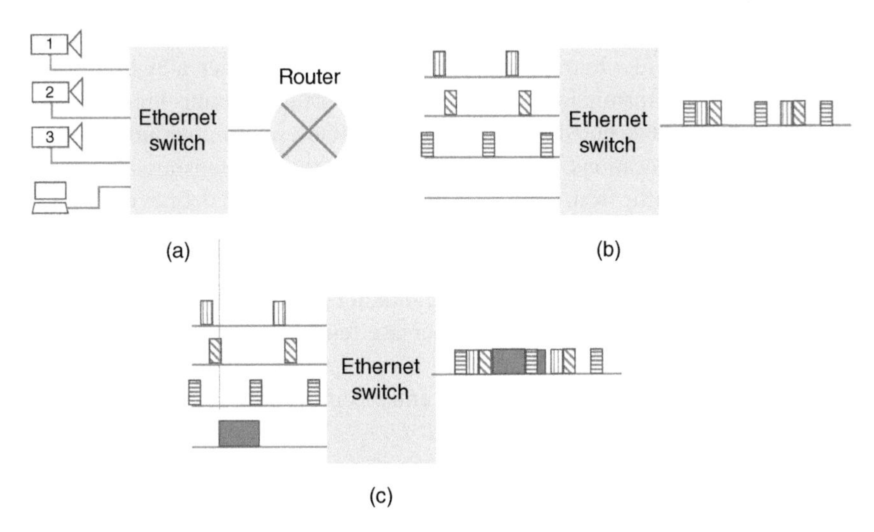

Figure 5.6 (a) Example of a local area network that contains three cameras and a computer. (b) Inputs and output of the switch when the computer is not sending any packet. (c) Inputs and output of the switch when the computer is sending low priority packets.

it accordingly in order to be able to stream the packets as smoothly as possible and with minimal delay. The TSN network uses simple traffic queuing and shaping techniques such as VLAN priority or complex techniques such as TAS, pre-emption, or CBS, as well as bandwidth reservation and redundant transmission to be able to smoothly forward the frames through the switches in its network.

5.3.4 Ethernet TSN Substandards

TSN features and key technologies that were explained earlier have been added to the standard Ethernet and have been published in a number of IEEE standard extensions [4]. These extensions discuss features such as time synchronization, redundancy, bandwidth reservation, queuing, and time scheduling. A list of some of these IEEE extensions that built Ethernet TSN is shown in Table 5.2.

Hereunder, we briefly explain these substandards. The main TSN features for providing deterministic communication are addressed by 802.1AS and 802.1Qbv standards. 802.1AS discusses timing and time synchronization, which is a profile of the IEEE 1588 PTP synchronization protocol. 802.1Qbv discusses traffic scheduling by defining how the flow of a queued traffic should be forwarded through gates at each output port of a TSN switch [5]. In other words, 802.1Qbv determines which queued traffic needs to be forwarded during a specific time window and which queued traffic should be blocked at that time. Therefore, 802.1Qbv has a vital role in controlling message latency.

Table 5.2 Some of TSN substandards.

TSN Sub-Standard	Description
IEEE 802.1AS	Enhanced timing protocol and time synchronization
IEEE 802.1Qbv	Time aware shaping
IEEE 802.1Qbu	Pre-emption
IEEE 802.1Qch	Cyclic queuing and forwarding
IEEE 802.1Qca	Path set up and control
IEEE 802.1CB	Frame replication
IEEE 802.1Qcc	Bandwidth reservation

As mentioned before, pre-emption interrupts the transmission of a standard Ethernet frame as well as an Ethernet jumbo frame in order to transmit a high priority frame. When a high priority frame is transmitted, it resumes the transmission of the previously interrupted frame. Pre-emption is defined in IEEE 802.1Qbu. It should be noted that with using 802.1Qbv, we cannot minimize the latency or optimize the bandwidth allocations. 802.1Qbv only ensures that the transmission of the critical messages is not affected by traffic from other network devices. 802.1Qch defines cycles for forwarding queued traffic and traffic shaping.

802.1Qca defines how disjoint paths in a network can be identified and how such paths can be set up. 802.1CB defines the mechanism to transmit the redundant copy of a message in parallel through some specific paths in the network in order to increase reliability. It uses 802.1Qca for path control.

802.1Qcc defines bandwidth reservation, and it is an enhancement to Stream Reservation Protocol (SRP), which has been published as IEEE 802.1Qat.

5.4 Power Line Communications (PLCs)

PLC brings a cost-effective approach to data transmission by utilizing the existing power line wiring. In other words, data is carried on cables that are used simultaneously for the distribution of electric power to customers. To be able to use power lines for data transmission, a broad range of PLC technologies have been developed to satisfy different requirements of various applications. Many PLC technologies use the wirings inside a single home or building, while there are other technologies that use both the distribution lines and building wiring.

PLC technologies can be divided into three main categories: Narrowband PLC (NB-PLC), Midband PLC (MB-PLC), and Broadband PLC (BB-PLC). Generally speaking, NB-PLC technologies operate at lower frequencies as compared to other PLC technologies. The operational frequency of NB-PLC is between 3 and 500 KHz, while BB-PLC operates at a higher frequency range between 18 and 250 MHZ. MB-PLC operates at frequencies below 12 MHz. Even though we used the term midband PLC technology, we can also consider it as a medium frequency BB-PLC technology. NB-PLC can be used for applications that require lower data rates of up to hundreds of kbps, while BB-PLC can be used for applications that need higher data rates of up to several Mbps. Due to the fact that lower frequencies can pass through transformers more efficiently, BB-PLC can operate at shorter distances as compared to NB-PLC. Figure 5.7 shows different types of PLC and their characteristics. The NB-PLC is also called the Distribution Line Carrier (DLC) technology due to its application on the distribution side of the smart grid.

Due to the high attenuation of wireless signal in metropolitan and dense areas, PLC technologies can be used as suitable wired connectivity schemes. For example, in a smart city use case, real-time readings of various smart meters, which are usually located in basements, can be provided using NB-PLC technologies.

Standardization organizations and their allocated frequency bands for the NB-PLC in different regions of the world are listed in Table 5.3. As it can be seen, the most important standardization bodies for NB-PLC are European Committee for Electro-technical Standardization (CENELEC), Association of Radio Industries and Businesses (ARIB), Electric Power Research Institute (EPRI), and Federal Communications Commission (FCC).

There are many NB-PLC technologies. The most important ones are G3-PLC, PoweRline Intelligent Metering Evolution (PRIME), and IEEE 1901.2. These technologies use Orthogonal Frequency Division Multiplexing (OFDM) modulation technique and a frequency band that is below 500 KHz.

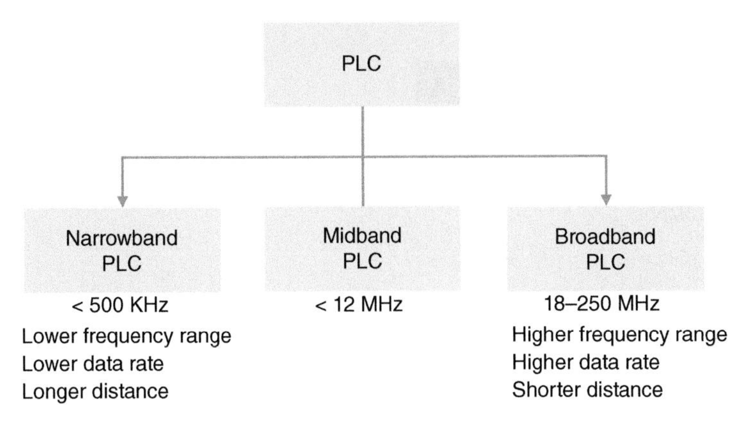

Figure 5.7 PLC technologies.

Table 5.3 The most important standardization bodies for narrowband PLC and their frequency bands.

Standardization body	Region	Frequency band
European Committee for Electro-technical Standardization (CENELEC)	Europe	3–148.5 KHz
Association of Radio Industries & Businesses (ARIB)	Japan	10–450 KHz
Electric Power Research Institute (EPRI)	China	3–500 KHz
Federal Communications Commission (FCC).	USA	10–490 KHz

There are some NB-PLC technologies that utilize simpler modulation techniques such as Binary Phase Shift Keying (BPSK) and Frequency Shift Keying (FSK). These technologies usually have lower data rates, and they are not in use by the industry today. An NB-PLC technology may operate in different frequency bands based on the geographical location that has been used. For example, G3-PLC can operate on CENELEC A band in Europe, on ARIB band in Japan, and on FCC band in the United States and other parts of the world.

With the intent to create a standard for communication among devices installed in a smart home with the Internet or among each other, the HomePlug power line alliance was established in 2000 [6]. Examples of the smart-home devices are plug-in Electric Vehicles (EVs), home appliances, smart lock systems, and smart meters. In 2001, the alliance published the HomePlug 1.0 specifications. Four years later, HomePlug AV was published that supported substantially higher data rates as compared to HomePlug 1.0. HomePlug Green PHY was published in 2011. This standard was designed to optimize for the power consumption, and it reduced the power consumption up to 75% as compared to HomePlug AV at the cost of reducing the data rate. HomePlug Av2 was published in 2012 to support data rates of 1 Gbps and above. Table 5.4 lists the HomePlug standards. Also, ITU-T G.hnn is another BB-PLC technology for communication of smart home devices that works over phone cables, coaxial cables, fiber optical cables, and power lines.

Since PLC signals can travel through very long transmission lines, they might be attenuated or become noisy. In these situations, the PLC signals might need to be refreshed. This can be done by filtering out the signal from the transmission line, demodulating the signal, modulating it again, and reinjecting the signal to the transmission line. These operations are performed in PLC carrier repeating stations.

Table 5.4 HomePlug standards.

BB-PLC HomePlug	Maximum data rate	Published
HomePlug 1.0	14Mbps	2001
HomePlug AV	200Mbps	2005
HomePlug Green PHY	10Mbps	2011
HomePlug AV2	1Gbps	2012

5.4.1 PLC for Smart Grid

Traditional electricity grids were built a long time ago, when an isolated local generator could provide the electricity needs of a community. At that time, each house in the community usually demanded only a small amount of electricity to turn on some light bulbs or small appliances. The utility companies delivered the electricity to their customers and the billing would be performed on a monthly basis. It is clear that the traditional grid cannot cope with the increasing demands of customers as well as the new methods of power generation. Even though utility companies have made tremendous efforts to enhance the capabilities and efficiencies of the traditional grid, those efforts have not been enough to be able to manage the needs of today's energy generation and consumption in an optimized manner. Today, customers have many devices in their homes that work with electricity, and they may need electricity for their electric cars and bicycles. Using solar technology, houses can generate electricity that can be sold to the utility companies. That is why the electricity grid needs to become smart to be able to manage the needs of utility companies and their customers. Smart homes must be able to communicate with the smart grid in order to manage electricity usage. Utilities must be able to measure electricity usage more frequently by smart meters and provide their customers with better information to manage electricity bills. Smart appliances must be able to adjust their schedule to decrease electricity demand during peak hours. The smart grid can provide detailed information about the electricity consumption in real time, which enables the utility companies to see and manage the electricity consumption, and generate power in such a way that matches the real-time demand. Smart grids can also reduce outages by automatically identifying the problems in order to restore power to customers in an efficient manner.

Generally speaking, a power grid can be divided into three sections: power generation, power transmission, and power distribution. These three sections are connected to each other through substations. Each substation consists of transformers, switchgears, grid protection, automation systems, and other infrastructure. Therefore, there is a need for a reliable real-time data transmission to ensure efficient operation of substations. The PLC technology has been very popular with many utility companies, since connectivity is provided by the infrastructure that they control. The utility companies need communications between their substations and also between the substations and the control systems.

Also, there is a need for communication between the distribution system and smart meters. The customers may need to use the electricity wiring inside the house for the communication among the smart devices or between these devices and the smart meter. One may think that when the next generation of substitutions start to roll out, they will have fiber optic cables and optical networks, and therefore, there is no need for PLC technologies within utility companies. However, it will take a very long time before all substations will be upgraded, due to the extensive cost of implementation. When this happens, PLC can still play the role of a backup solution in case of a fiber break or a fault in the optical network.

A smart meter is an important device in a smart grid system, and it is part of Advanced Metering Infrastructure (AMI). Generally speaking, an AMI system requires a two-way communication between the utility company and the meter in order to enable data collection in real time or in various intervals by the utility company, as well as sending information such as pricing to the customers. AMI can provide necessary information to improve the efficiency of the services provided by a utility company, detect malfunctions, and help the utility company and its customers to manage the cost in an efficient manner. The connection between the smart meter and the utility company can be provided by PLC technology. There are many smart meters that contain a PLC module for this purpose. It should be noted that Automatic Meter Reading (AMR) devices are another type of IoT metering devices that are usually read by the utility personal in the proximity of the meter. The AMR devices are usually walk-by or drive-by devices that their data is collected remotely by the utility personal and will be uploaded to the cloud. Also, some AMR devices may establish a one-way communication with the utility company in order to send their data. However, an AMR meter cannot make a two-way communication in order to receive information from the utility company.

Example 5.1 Investigating the application of Ethernet for power line distribution

Let us model the power line distribution system and investigate whether Ethernet can be used instead of NB-PLC technologies (i.e. PRIME, G3-PLC) for reading the residential smart meters.

The power line distribution system, also called access network, is a network from the medium/low voltage transformer to the houses that are connected to this transformer. Generally speaking, a power line distribution system consists of several trunks that build a tree-like network as shown in Figure 5.8a. As you can see from this figure, each house is connected to the power line distribution network using short cables. The wires inside each house also establish a network with a tree topology. Logically, we can think that Figure 5.8a can be simplified to a network with a bus topology as shown in Figure 5.8b. Even though the simplified figure shows a similar distance between each house and the shared medium, in practice this distance differs from house to house.

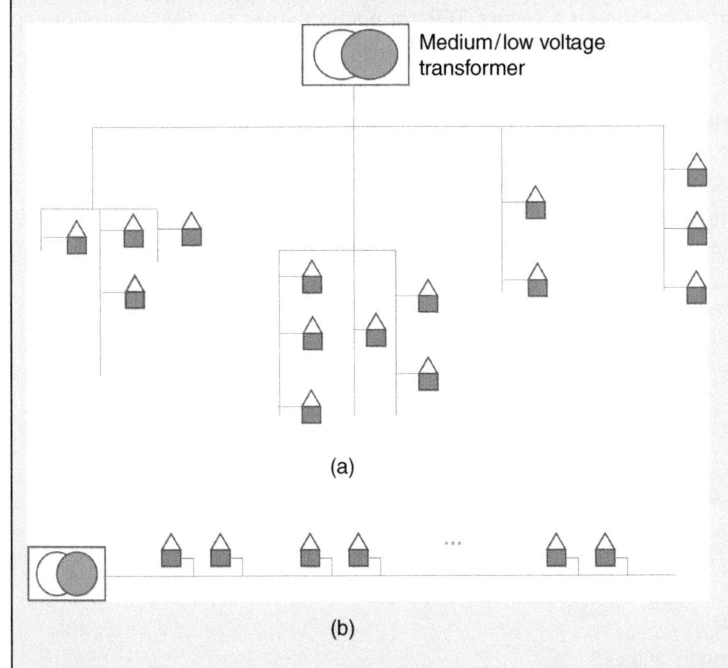

Figure 5.8 Power line distribution: (a) tree topology and (b) simplified bus topology.

(Continued)

Example 5.1 (Continued)

However, Ethernet technology is not a suitable communication method for the power line distribution network for the following reasons.

- The power lines have low impedance and strongly attenuate the high frequency component of a signal. Legacy Ethernet is designed to work on a network with a bus topology. However, it uses twisted pairs or coaxial cables that have better performance and less attenuation at higher frequencies.
- Power line system introduces a high amount of noise and interference caused by turning on and off appliances. Ethernet is designed to work with lower amounts of noise.
- The characteristic impedance of an Ethernet cable is usually constant, while the characteristic impedance changes dynamically in power line distribution cables.

Example 5.2 Communication between an electrical car and its charging station

Let us discuss the IoT connectivity schemes between an electric car and an EV charging station.

There is a need for a reliable communication between an Electric Vehicle (EV) and an Electric Vehicle Service Equipment (EVSE) such as an Electric Charging Station (ECS) as shown in Figure 5.9. A charging station may be installed in an indoor or outdoor location. The connection between a charging station and an EV can be implemented using PLC technology. In this case, after plugging in the charging cable, the vehicle uses PLC technology to make a connection with the charging station and obtains an IP address from the charging station over Dynamic Host Configuration Protocol (DHCP). To start the charging process, the charging station needs to authenticate the EV over a secure TCP connection. After the EV is authenticated, it starts to exchange information with the ECS. Examples of the exchanged information include the charging profile, rates, service information, and payment options. Then the charging process starts. The charging cable should not be polled out during this charging process. To accomplish this task, the charging cable will be physically locked to prevent power theft. The exchange of data between EV and ECS does not stop during the charging process, and they continue to exchange data regarding the status of the charge. When the charging is complete, the data regarding the process is sent via the Internet to the energy provider for billing.

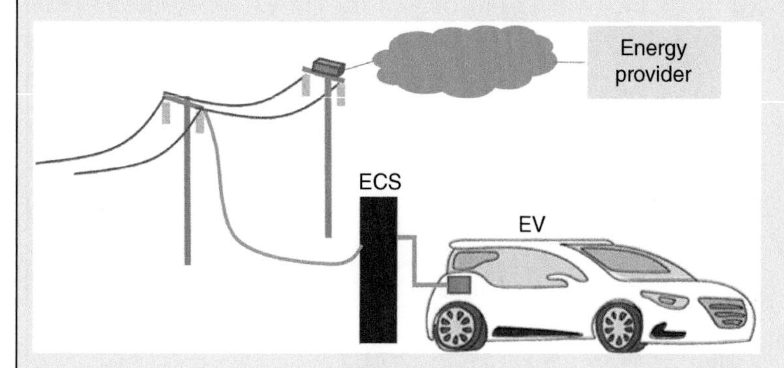

Figure 5.9 Connection between EV and ECS.

EVs often have large storage capacity and can draw large currents of around 100 A during the charging process. This certainly can put pressure, especially on low voltage transformers. It should also be mentioned that EVs can supply power back to the grid during peak hours, if needed.

Wireless connection can be used for connecting an EV to an EVSE. However, PLC can provide a better performance in this scenario. One of the PLC technologies that is suitable for this example is G3-PLC, which can be considered as a global solution for this application. As mentioned earlier, G3-PLC is used in different frequency bands (10–490 KHz) in various parts of the world. BB-PLC technologies can also be used for this application. As the number of EVs grows in the future, the need for PLC technologies such as G3-PLC becomes clearer.

5.5 Summary

In this chapter, wired IoT technologies were discussed. IoT devices can use Ethernet in a local area setting to be connected to an IoT gateway. Even though Ethernet is very popular, easily available, and can satisfy the requirements of many IoT applications, especially in terms of data rate, it is not suitable for most IIoT applications that have tight requirements in terms of latency, bandwidth, packet loss, and jitter.

Industrial Ethernet is a common name for many solutions offered over the standard Ethernet to solve the deficiencies of using Ethernet in industrial environments. However, these solutions neither are generalized nor vendor independent. Therefore, there is a need for a vendor-independent, generalized wired connectivity solution that can be used in industrial settings.

Ethernet TSN added several capabilities to the standard Ethernet including time synchronization, time scheduling, bandwidth reservation, redundant retransmission, and pre-emption to become a suitable candidate for IoT connectivity in industrial environments. TSN is also backward compatible with the standard Ethernet. In this chapter, we briefly discussed key features of TSN including the time synchronization mechanism and different traffic shaping and queuing approaches such as TAS, pre-emption, and CBS.

Power line communication, also known as Power Line Carrier, is another method used by IoT applications for data transmission. PLC can be divided into three main groups of NB-PLC, MB-PLC, and BB-PLC. NB-PLC technologies operate at lower frequencies; they support lower data rates and shorter distances as compared to other PLC technologies. Even though we used the term midband PLC technology, we can consider it as a medium frequency BB-PLC technology. A brief explanation of the use of PLC technologies in smart grid applications is also presented at the end of this chapter.

References

1 Vitturi, S., Zunino, C., and Sauter, T. (2019). Industrial communication systems and their future challenges: next-generation ethernet, IIoT, and 5G. *Proceedings of the IEEE* 107 (6): 944–961. https://doi.org/10.1109/JPROC.2019.2913443.

2 Intel (2022). Time-sensitive networking: from theory to implementation in industrial automation. https://www.intel.com/content/dam/www/programmable/us/en/pdfs/literature/wp/wp-01279-time-sensitive-networking-from-theory-to-implementation-in-industrial-automation.pdf (accessed 30 June 2022).

3 Cisco (2017). Time-sensitive networking: a technical introduction. https://www.cisco.com/c/dam/en/us/solutions/collateral/industry-solutions/white-paper-c11-738950.pdf (accessed 30 June 2022).
4 IEEE (2022). 802.1 Time-sensitive networking task group. www.ieee802.org (accessed 30 June 2022).
5 IEEE (2015). 802.1: 802.1Qbu - frame pre-emption. https://www.ieee802.org/1/pages/802.1bu.html (accessed 30 June 2022).
6 HomePlug Powerline Alliance (2010). Home Plug Green PHY - The Standard For In-Home Smart Grid Powerline Communications-Version 1.00. https://content.codico.com/fileadmin/media/download/datasheets/powerline-communication/plc-homeplug-green-phy/homeplug_green_phy_whitepaper.pdf (accessed 30 June 2022).

Exercises

5.1 How an IoT device that is connected to an Ethernet switch should be configured to be able to send its data to the Internet? Assume the Ethernet switch is connected to a router with an Ethernet connection.

5.2 What is a proxy ARP? Where a proxy ARP can be located on a network?

5.3 What does VLAN priority mean? How many VLAN priority classes does the Ethernet standard support?

5.4 What does it mean to say that Ethernet TSN provides deterministic communications?

5.5 Is it a good idea to put IoT devices on a separate VLAN in a local area network? Why?

5.6 Assume a router does not have any Ethernet interface.
- **a.** Can you connect this router to an Ethernet switch?
- **b.** Do you think this router can be used in any network at all?

5.7 Compare PROFINET and Ethernet in terms of the layers of the OSI model that each one supports?

5.8 What is industrial Ethernet?

5.9 Conduct research and identify the main difference between the following two Cisco switch products: Cisco IE-4000 series and Cisco Catalyst 2970

5.10 In a network:
- **a.** What is jitter?
- **b.** What is the reason for the jitter?
- **c.** How can an average jitter be calculated?

5.11 How bandwidth is reserved in a TSN network consisting of several switches?

5.12 In a TSN network, there are two TSN switches that both like to act as a grandmaster. TSN switch 1 is configured as (priority 1 = 64, clock class = 6, priority 2 = 128, and its MAC address is 62:61:60:25:12:13). The second TSN switch (switch 2) is configured as (priority 1 = A, clock class = B, priority 2 = C, and its MAC address is 62:61:60:25:12:10). Based on the following values for A, B, and C, which TSN switch becomes grandmaster?

a. A = 64, B = 7, C = 130
b. A = 128, B = 5, C = 128
c. A = 64, B = 6, C = 64
d. A = 64, B = 6, C = 128

5.13 For what types of applications, Ethernet TSN may use redundant transmission?

5.14 Why is PLC popular with the utility companies?

5.15 What is the difference between NB-PLC and BB-PLC in terms of frequency, data rate, and distance?

5.16 What is the most widely deployed PLC standard for home automation?

5.17 What is an Ethernet jumbo frame? What is the effect of jumbo frames on TAS guard bands in a TSN network?

5.18 Among all the HomePlug standards:

a. Which one provides the highest data rate?
b. Which one has the lowest power consumption?
c. Which one was published first?

Advanced Exercises

5.19 If TAS or pre-emption method can ensure that the data is delivered to a destination node within its required time frame, why is there a need for CBS?

5.20 Future substations will have fiber optic communications and utilize optical networks. Do you think PLC technology will be useful in the future?

5.21 A substation may have wave traps. Conduct research and find the answer to the following questions:

a. What is a wave trap and why a substation may need it?
b. How can we send data on a PLC system if the transmission line has a wave trap?
c. Do you think all substations have wave traps?

5.22 Conduct research and find out what is a PLC repeating station?

5.23 In a TSN network, the network is configured to provide the requirements of time-critical IoT applications. Assume best-effort traffic suddenly increases substantially on this network. How does this affect the time-critical IoT applications?

5.24 An Ethernet cable has four twisted pairs. Ethernet 10 Mbps uses two pairs, while faster Ethernet standards require four pairs for data transmission and reception. Do you think that we can use PoE technology on a 1 Gbps Ethernet switch?

5.25 Can an Ethernet switch work with non-IP packets?

5.26 G3-PLC alliance has introduced G3-PLC-RF as a new connectivity scheme that combines G3-PLC technology with Radio Frequency (RF) wireless technology. In this new technology, each node can use G3-PLC or RF for its communication. Each node selects the most suitable technology for its communication needs. In other words, a node should determine whether G3-PLC has better performance at a given point in time or the wireless technology. Explain how this new technology can open up new use cases for the PLC technology.

5.27 Single Pair Ethernet (SPE) is a new Ethernet standard that operates over a single twisted pair instead of the four-pair cabling used in the traditional Ethernet and can deliver up to 10 Mbps speeds at distances of up to 1 km. Conduct research on this Ethernet standard and answer the following questions:

a. What IEEE standard discusses SPE?
b. Why SPE can reduce the cost of IoT implementation?
c. Can power be transmitted over SPE cables?
d. Why SPE is important for industrial IoT and for providing connectivity for field devices in an industrial application?

Chapter 6 Unlicensed-band Wireless IoT

6.1 Introduction

One technology alone cannot provide the connectivity requirements of diverse Internet of Things (IoT) applications. That is why many technologies have been designed, developed, and utilized, where each technology satisfies the requirements for a subset of IoT applications. In the previous chapter, we discussed the most important wired technologies used for IoT use cases. In this chapter, we discuss unlicensed-band wireless technologies that can be used for this purpose.

A device cannot transmit data in a specific frequency band unless the device manufacturer has obtained the required license for using the spectrum. For example, telecommunication companies spend a huge amount of money to acquire a license that allows them to build their system on a specific radio spectrum. In other words, transmitting data in a spectrum without acquiring the appropriate licenses is illegal. Some government-oriented organizations may get a license to use a specific spectrum for their operations for free. Examples of these organizations are military, aviation, or companies that provide civilian emergency services. There are some spectrums that are assigned for everyone to use at no cost and without acquiring any license. These spectrums are called unlicensed bands. Even though these spectrums do not require any license, there are some rules that need to be followed when a device wants to use them. Transmission in unlicensed frequency bands should be performed with a limited power and a limited range in a personal or local area network. These limitations are not the same in different regions of the world. For example, the maximum transmit power is 1 Watt in the Americas, while the maximum allowed power is substantially less in Europe. The Industrial, Scientific and Medical (ISM) frequency bands are unlicensed-band spectrum that is defined by the International Telecommunication Union (ITU) regulations. Even though originally these bands were designated for industrial, scientific, and medical purposes, many short-range and/or low power systems operate in ISM bands. The list of ISM bands is shown in Table 6.1.

We divide unlicensed-band wireless technologies into two categories. The first category is the one that uses short-range wireless technology to communicate with an IoT gateway. The most important short-range wireless technologies are Zigbee, Bluetooth Low Energy (BLE), and WiFi. The second category are unlicensed-band wide area network technologies. The most important technologies of this group are Long Range Wide Area Network (LoRaWAN) and Sigfox.

There exists also licensed-band cellular technologies such as LTE for Machine-Type Communications (LTE-MTC or LTE-M), Narrowband Internet of Things (NB-IoT), and 5G. These technologies are provided by telecommunication companies and service providers and can provide the coverage, capacity, security, interoperability, and availability that is needed for many

Fundamentals of Internet of Things: For Students and Professionals, First Edition. F. John Dian.

Table 6.1 List of ISM bands.

Frequency band	Center frequency (MHz)	Bandwidth (MHz)
6.765–6.975 MHz	6.78	0.03
13.553–13.567 MHz	13.56	0.014
26.957–27.283 MHz	27.18	0.326
40.66–40.7 MHz	40.68	0.04
433.05–434.79 MHz	433.92	1.74
902–928 MHz	915	26
2400–2500 MHz	2450	100
5725–5875 MHz	5800	150
24–24.250 GHz	24125	250
61–61.5 GHz	61250	500
122–123 GHz	122500	1000
244–246 GHz	245000	2000

Source: Based on ITU Radio Regulations [1].

IoT applications. Due to the importance of cellular IoT technologies, we will discuss LTE-IoT technologies in Chapter 7, some of its key features in Chapter 8, and IoT in the 5G era in Chapter 9.

6.2 Zigbee Wireless Network

Zigbee provides a suite of short-range communication protocols and a radio architecture that is designed based on IEEE 802.15.4 specifications [2]. Zigbee creates a low-power, low-bandwidth radio with a low data rate of up to 250 Kbps, which is perfect for many IoT applications such as home automation or medical data collection systems. The Zigbee standard was published by the Zigbee Alliance, which has been rebranded itself to Connectivity Standards Alliance (CSA) recently.

Zigbee defines three devices in its network: coordinators, routers, and endpoints. Each network needs to have a coordinator that is responsible for setting up the network and allowing routers and endpoints to join the network. The coordinator is the root of a Zigbee network, and each network must have exactly one coordinator. Routers are responsible for routing the data traffic sent by endpoints. Each endpoint is usually connected to one or several sensors and sends its data to a router. Endpoints are not connected directly to each other, and each endpoint must be connected to a router. The endpoints do not route traffic and need to send their data to their routers. In other words, an endpoint is like a child and should have a router as its parents. Endpoints have the ability to sleep in order to save energy. When an endpoint wakes up, it can send a message to its router and request to receive its buffered messages. Therefore, routers have memory and when endpoints are asleep, they keep messages for their child endpoints. To provide short-range connectivity among the IoT endpoints, or between IoT endpoints and IoT gateways using Zigbee technology, each IoT device as well as the IoT gateway need to be equipped with a Zigbee module.

The Zigbee's devices can be categorized into two different types in terms of their functionalities. The first type is called Full Function Devices (FFD), and the second type is called

Reduced Function Devices (RFD) based on the Zigbee terminology. Coordinators and routers are FFDs, and endpoints are usually RFDs. Endpoints have reduced functionalities, are less complex, require less memory, and are low cost. As mentioned earlier, a coordinator is an important part of the Zigbee network. An example of a coordinator in a home automation network is a smart home controller. Examples of routers in home automation applications are light bulbs, thermostats, or fans. A router is usually powered, while endpoints are usually battery powered. Examples of endpoints in a home automation application are portable light switches, door/window sensing devices, smoke detectors, or motion sensing devices.

The topology of a Zigbee network is based on a mesh topology, which can also be represented by a tree or star topology in special cases. Therefore, we can also say that Zigbee supports star, tree, and mesh topologies.

Figure 6.1a shows a Zigbee mesh network. This network has one coordinator shown as C; four routers expressed as R_1, R_2, R_3, and R_4; and eight endpoints shown as E_i $(i = 1 \dots 8)$. The network is also connected to a Zigbee gateway, expressed as G, which is connected to the Internet. A Zigbee gateway plays the role of a router within the Zigbee network. The beauty of a mesh network is that each FFD device becomes a relay point to the rest of the network, and any endpoint may have several different paths to a specific destination endpoint. This makes the Zigbee network very robust and reliable. If a router fails to operate, traffic between a source endpoint and a destination endpoint can continue by going through other routers. The dynamic routing in Zigbee's mesh network, in which the link between a source endpoint and a destination endpoint may dynamically change due to a failure in the network, is called self-healing. For example, the traffic between E_8 and E_2 can go through different paths such as $R_4 - R_2$, $R_4 - C - R_1 - R_2$, $R_4 - C - R_3 - R_2$, $R_4 - C - R_3 - R_1 - R_2$, $R_4 - G - R_1 - R_2$, $R_4 - G - R_1 - R_3 - R_2$, and $R_4 - G - R_1 - C - R_3 - R_2$. By looking at Figure 6.1a, you can find more routes that connect E_8 and E_2. Some wireless technologies only support star topology. In a network with star topology, traffic between two endpoints should go through a central unit that is called principle, master, or access point. Star topology makes the routing simple, but makes the central node a bottleneck for the traffic. Also, the central node is a point of failure for

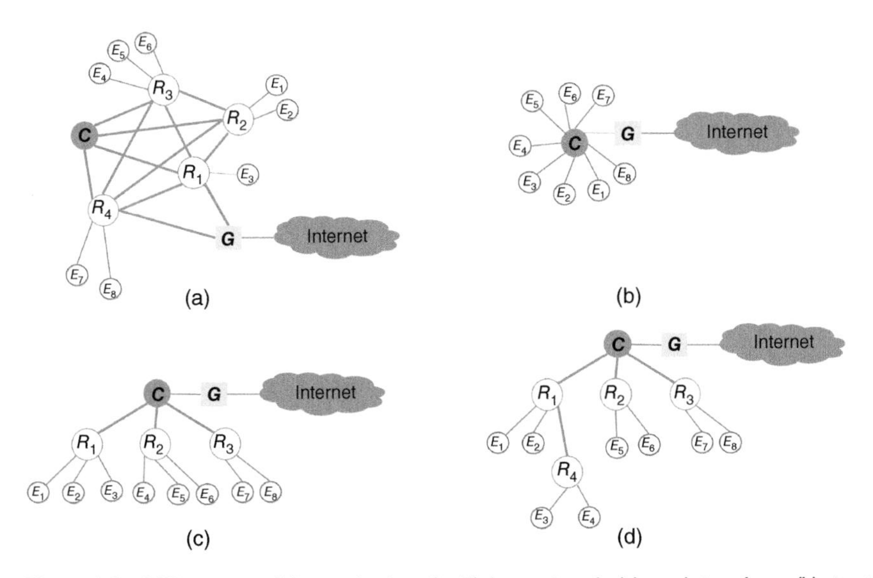

Figure 6.1 Different possible topologies of a Zigbee network: (a) mesh topology; (b) star topology; (c) tree topology; (d) clustered tree topology.

this network in terms of reliability. If a central unit fails, then the entire network stops operating. By supporting mesh topology, the Zigbee network is very scalable and the range of its network can also change dynamically. While there is a need for an extender to increase the range of a network in some wireless networks, there is no need for an extender in the Zigbee network.

A star topology is a special case of a mesh topology where no router is used, and all the endpoints are connected to a coordinator. This is shown in Figure 6.1b. Also, a mesh network can become a tree or clustered tree as shown in Figure 6.1c, d, respectively. In a tree topology, routers are not connected to each other, while in a clustered-tree topology the routers might be connected to each other in order to extend the network reach.

Zigbee operates in an unlicensed band of 2.4 GHz to 2.4835 GHz worldwide, where sixteen channels of 2 MHz are allocated for Zigbee's operation. These channels are 5 MHz apart. Even though 2.4 GHz is the main frequency band for Zigbee's operation, there are other unlicensed bands used in different parts of the world for Zigbee. For example, 902–928 MHz has been used in North America and Australia, and 868–868.6 MHz is used in Europe. In a Zigbee network, the coordinator only has one channel for communication, and therefore, all devices share this single channel to communicate. Zigbee defines two methods for channel access as shown in Figure 6.2. The first method is called the contention free method or beacon mode. In this method, the coordinator sends a beacon periodically for time synchronization, which provides the scheduled time that can be used by each endpoint in order to start its transmission. This time period is called Guaranteed Time Slot (GTS). The second method is called a contention-based method or non-beacon mode, in which each endpoint competes to access the channel. In this method, the coordinator and routers should not sleep, since any endpoints can wake up at any time and start transmission. In the contention-free method, the coordinator and routers can sleep, if there is no traffic in the network.

Zigbee is one of the enablers of IoT technology. For example, many smart meters installed in the United States and around the world are equipped with Zigbee modules. These meters send pricing and usage information to In Home Displays (IHDs). The connection from a smart meter to the Internet might be provided through Power Line Communication (PLC) technology, as explained in Chapter 5. Also, the communication inside a smart home can be established using BB-PLC technologies such as one of the HomePlug standards. One may think that using PLC for smart meters must be the most suitable option. It should be noted, however, that many smart home devices are battery powered and not connected to the electrical grid. Examples of these devices include wireless thermostats, smoke detectors, and light switches. Therefore, Zigbee has been used in many smart meters to provide the required connectivity for the communication between a smart home device and the smart meter.

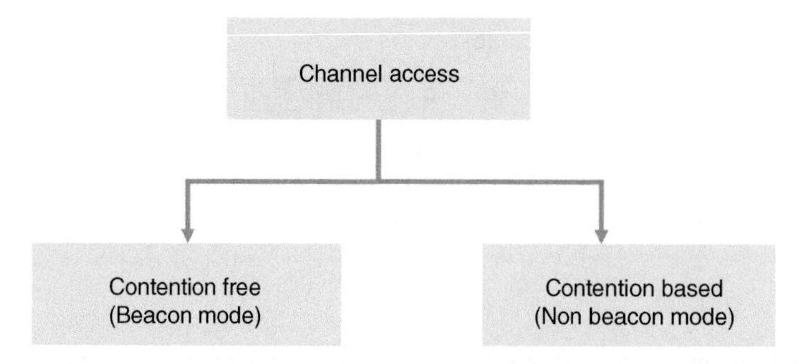

Figure 6.2 Channel access methods in a Zigbee network.

6.3 BLE Wireless Network

Similar to Zigbee, Bluetooth low energy is one of the enablers of the IoT technology, which can provide short-range connectivity between IoT devices and an IoT gateway. For this purpose, both the IoT device and the IoT gateway should be equipped with BLE modules. BLE was published by the Bluetooth Special Interest Group (SIG) in 2010 and uses unlicensed band for its operations. Due to the efficient architecture of BLE, it has become one of the most popular short-range communication technologies and it has been widely used in IoT applications. BLE modules are used in most smartphones today, and therefore, this technology is a suitable candidate for those IoT applications that want to connect to the Internet using smartphones.

A BLE module can be configured to play one of the three following roles: an advertiser, a scanner, or an initiator. An advertiser sends advertising messages periodically to discover scanners or initiators in its surrounding area. A scanner scans to receive advertisement messages from any nearby advertisers. An initiator has all the functionalities of a scanner, and it can also request for more information from an advertiser. BLE is used in one of the two following modes: discovery mode or connection mode. In discovery mode, the scanner or initiator discovers BLE advertisers and receives their advertising messages without making a connection with them. In the connection mode, an advertiser stops sending advertising messages and communicates with the scanners per agreed schedule [3].

The discovery process in BLE, where a scanner or an initiator can discover an advertiser, has many applications in IoT. Let us explain this with two simple examples.

Assume people who walk on a trail are wearing a BLE-enabled IoT device. Each IoT device has a BLE module that is configured as an advertiser. The wearable device advertises the name of the person who is wearing the device. There are other BLE devices, configured as scanners that are installed in certain locations along the trail. As a person comes close to one of these scanners, the scanner can discover the advertisement messages. Since the advertisement messages contain the name of the person who wears the IoT device, the scanner learns the names of the people who pass through that specific location on the trail at a specific time. If the scanner is directly connected to the Internet, or it is connected to the Internet through a gateway, it can also send this information to the cloud. Now, assume that the advertiser has also been configured to have other information, such as the occupation of the people who wear the wearable device. However, the advertiser does not transmit information about the occupation as part of its advertisement packet. Let us tweak this scenario slightly and use initiators instead of scanners to better understand the distinction between the roles of a scanner and an initiator. An initiator performs the same function as a scanner, but it can additionally request extra information from the advertiser after receiving an advertisement. In our example, if a doctor is needed in a certain area of the trail, the initiator can request the advertiser to provide the occupation data as well.

In the second example, we will look at how BLE's discovery mode can be used in a shopping scenario. Assume that consumers who shop at a store have downloaded a store-provided app to their smartphone. This app configures the smartphones' BLE module in scanning mode. On each shelf, there are some BLE devices that advertise the new items on the shelf or the items that are on sale. When a consumer approaches a shelf, these devices may send an advertising message to the customer. Additionally, the app may be customized to only display the advertisements for products that a consumer is interested in.

To make a BLE node consume less energy, a BLE in scanning mode can go to sleep mode and wake up periodically to discover advertisements from devices within range. By the same token, a BLE in advertising mode can go to sleep mode and wake up periodically to send advertisements. The discovery process in BLE has found lots of attention from IoT developers. For this reason, Bluetooth 5, published in 2016, has improved the capabilities of the discovery process.

After an advertisement is received by a scanner or an initiator, these devices can initiate a connection with the advertiser in order to establish a reliable communication in which resources are allocated in advance. Now that we have understood the difference between an initiator and a scanner, we may use them interchangeably during the rest of this chapter. For establishing a connection, a scanner sends a request to an advertiser. If the advertiser accepts this request, the scanner becomes a central node (master node), and the advertiser becomes a peripheral node (slave node). During connection establishment, the central and peripheral nodes can set up a Connection Interval (CI) that indicates time schedules at which a central node periodically allocates resources to a peripheral node for data transmission.

Let us explain this with an example: A medical IoT device needs to send the collected medical data from a patient to a central unit every 200 ms. To design this system, the medical IoT device becomes a peripheral node, and a device that is also equipped with a BLE module, and might be connected to the Internet, becomes a central node. These two BLE devices can establish a connection with a CI equal to 200 ms. Therefore, every 200 ms, the central unit asks the peripheral IoT medical device for its data.

Now that you are familiar with both the discovery process and the connection establishment process in BLE, we can explain these two modes of operation in more detail. Let us start with the discovery process. The scanner can be configured to wake up periodically to check for advertising messages. If the scanner is connected to electricity (not a battery-powered device), an IoT system designer may want the scanner to scan for advertising messages at all times. Otherwise, the scanner can be configured to wake up periodically to scan for advertising messages. To configure a BLE for this purpose, the BLE defines a parameter called the scan interval (T) as shown in Figure 6.3. The scan interval is the length of time between when a scanner wakes up to receive advertisements and when it should wake up the following time for this purpose. When a scanner wakes up, it starts to listen to advertising messages for a period represented by T_S. It is clear that T_S is a positive value and is always smaller than or equal to the scan interval. The maximum value of the scan interval for BLE is 10.24 seconds, meaning that at least every 10.24 seconds, the scanner should wake up to listen to advertisements. BLE uses three channels (37, 38, and 39) for advertising. During the scanning process, a BLE module scans these three advertising channels one by one.

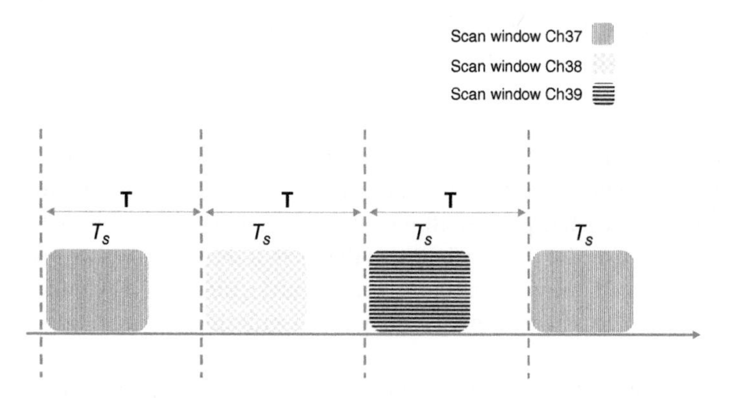

Figure 6.3 Scanning process in BLE.

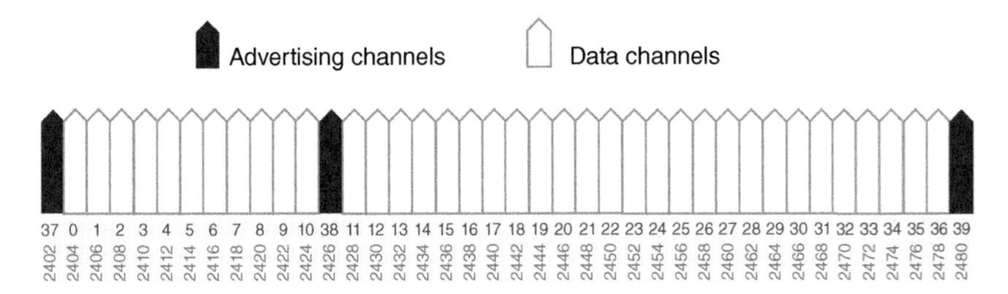

Figure 6.4 BLE advertising and data channels.

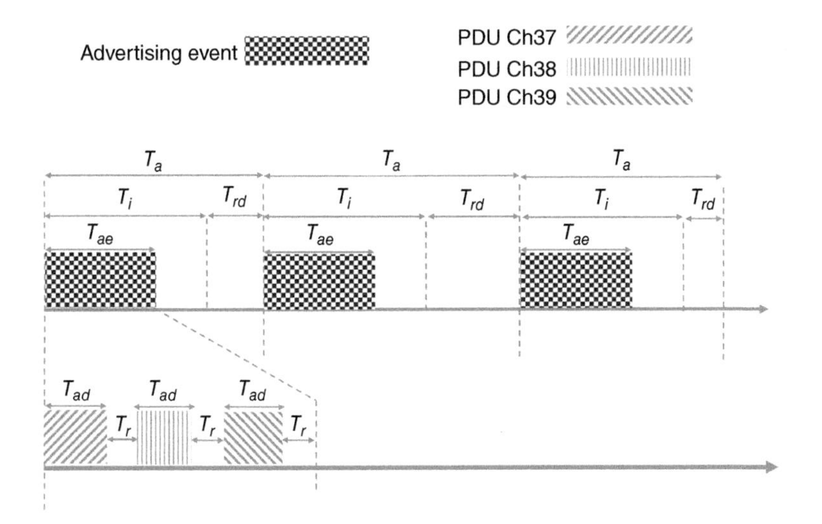

Figure 6.5 Advertising process in BLE.

BLE divides the unlicensed ISM band into 40 channels, where each channel is 2 MHz. The frequency range of 2.4 GHz unlicensed band is 2400–2483.5 MHz in most countries. As mentioned earlier, three channels of 37, 38, and 39 are used for advertising, while the rest of the channels are for data transfer. The channels of BLE are shown in Figure 6.4.

Advertisers are usually battery powered and should wake up regularly to generate advertising events in order to send their advertising messages. The timing used in an advertising process is shown in Figure 6.5.

As can be seen from this figure, the advertising process is based on six parameters:T_{ad}, T_r, T_a, T_i, T_{rd}, and T_{ae}. The explanation of these parameters is as follows:

T_{ad}: The duration of advertising messages on a predefined advertising channel during each advertising event.
T_r: Waiting period to receive a reply from a scanner.
T_a: Interval between two advertising events (contains T_i and T_{rd})
T_i: A fixed interval that is part of T_a.
T_{rd}: A variable pseudo-random delay interval.
T_{ae}: The duration of an advertising event.

BLE specifies that 20 ms $<T_i < 10.24$ ms, $0 < T_{rd} < 10$ ms, and T_i should be multiple of 0.625 ms.

Now, let us explain the connection establishment process. When a scanner intends to initiate a connection with an advertiser, it sends a request to the advertiser. If the advertiser is configured to accept connections, and if there is no whitelist, or if the scanners' address is on the whitelist, then the advertiser accepts the connection request. After accepting the request, the advertiser stops sending advertising messages and follows the parameters sent by the scanner as part of the connection request packet. At this point, the advertiser becomes a peripheral or slave node, and the scanner becomes a central or master node. In connection mode, the topology of the BLE network is a star topoloagy. One of the most important parameters of a BLE connection is called the connection interval or CI. The master node sends a message on every CI, and the slave node sends data and acknowledgment on every CI. To configure CI, there is a parameter called interval that needs to be set up. Each CI is equal to the interval parameter multiplied by 1.25 ms. The interval parameter can accept any value between 7.5 ms and 4 s, according to the BLE specifications. There are many parameters that are used to establish a connection. Besides CI, two other important parameters of a connection are latency and time out. Latency shows the number of connection events that a slave node is allowed to skip transmission. Time out shows the number of connection events that a central node continues to keep the connection open without receiving any data from a slave node.

Interested readers are encouraged to study [4, 5], and complete Assignment #5 that is a hands-on project on this topic and uses the BLE programming in [6].

6.3.1 Bluetooth 5

Bluetooth 5, published in 2016, supports a data rate of 2 Mbps, while BLE supports 1 Mbps physical layer. The higher data rates have been achieved at the cost of a slight reduction in the communication range. To improve the range of communication, Bluetooth uses two types of coded physical layers called S2 and S8. The coded physical layers use a technique called Forward Error Correction (FEC), in which each bit is replaced by 2, or 8 bits in S2, or S8 settings, respectively. The coded physical layer increases the range at the cost of lower data rates. Table 6.2 shows different physical layers in Bluetooth 5 in terms of data rate and range. If the range of communication using 1 Mbps physical layer is represented by x, the approximate range of communication and the data rate using other supported physical layers of Bluetooth 5 are shown in Table 6.2.

Bluetooth 5 improved the discovery process significantly as compared to BLE using a technology that is called advertising extension, in which in addition to the three channels (37, 38, and 39), all possible data channels can also be used for advertising. Bluetooth 5 uses the term primary channels for these three channels (37, 38, and 39), while it uses the term secondary channels for all other channels. Bluetooth 5 advertising messages can be longer as compared to BLE. A BLE advertising message can be up to 32 bytes, while for Bluetooth 5, the size of advertising messages can be up to 255 bytes. Also, in Bluetooth 5, the advertisement can be chained to increase the length of advertisements even further.

Table 6.2 The specifications of different physical layers in Bluetooth 5.

Bluetooth 5–PHY	Relative data rate	Range
PHY 2 M	2 Mbps	$0.8x$
PHY 1 M	1 Mbps	x
PHY coded S = 2	500 Kbps	$2x$
PHY coded S = 8	125 Kbps	$4x$

6.3.2 Bluetooth Mesh

Generally speaking, a BLE device can cycle between being a master and a slave node. This can be used to extend the range of a BLE network. For this purpose, a BLE device can become a master node for some slave nodes in the network for a certain period of time. During this time, the master node collects the messages sent by its slave nodes. Then it changes its role and plays the role of a slave node for a while. During this time, the BLE node would send the collected messages to a central node. This central node should be located somewhere in the network to extend the reach of the network. The process needs to be repeated, and the BLE device should toggle between being a master and a slave. Even though a developer can use this method to extend the reach of a BLE network, it is not an efficient strategy.

Bluetooth Mesh, published in 2017, introduces mesh topology for Bluetooth networks. Bluetooth mesh is a Bluetooth network that uses mesh topology. It is based on the concept of flooding to transmit data to a destination node. Flooding means that every message sent out by a source node is relayed by other nodes and is sent to all nodes in the range. This makes routing simple, but it is not an efficient routing strategy like the one used by the Zigbee network.

The details of Bluetooth mesh operations are outside the scope of this book. We encourage the interested readers to study [7] and [8].

6.4 WiFi Wireless Network

WiFi is one of those technologies that does not need any introduction. Everyone is familiar with WiFi-enabled devices, WiFi access points, and WiFi extenders. However, many people might not know the principle of data communications in a WiFi network.

In more than two decades, WiFi has introduced many standards such as 802.11, 802.11b, 802.11a, 802.11g, 802.11n, and 802.11ac. Each new standard has tried to increase the data rate as compared to the previous ones. The increase in data rate has occurred as a result of using higher spectrum for data transmission (also called channel binding), using more antennas (Multiple Input Multiple Output (MIMO technology)), using simultaneous connection between multiple streams and multiple devices (MU/MIMO), and utilizing more complex modulation schemes (higher Quadrature Amplitude Modulation (QAM)). By implementation of all these techniques, a higher data rate has become achievable. For example, the channel in 802.11g is 20 MHz, while 802.11n allows channel binding, and therefore, the use of 80 MHz channel is also possible. 802.11ac allows for a higher channel binding scheme, and 160 MHz channels are available in this standard. A detailed explanation of techniques such as channel binding, MIMO, MU/MIMO, or QAM is outside the scope of this book. However, readers must know that all these techniques made 802.11 standards evolve to provide higher data rates.

Even though having a higher data rate is indeed interesting, it is not what is required in most IoT applications, especially if the increase in the data rate is only achievable when the number of devices in the network is limited, or when devices need to consume a huge amount of energy to communicate in this fast network. Two important requirements for most IoT applications are: the IoT technology should work with a large number of IoT devices and it should consume less energy. However, these were not the important factors that have been considered in WiFi standards such as 802.11g, 802.11n, 802.11ac, or even 802.11ad, where the main focus has been on higher data rates.

There are two WiFi standards that address the needs of IoT. These two standards are 802.11ax, also called WiFi 6, and 802.11ah that is called WiFi Halow. They are designed with IoT in mind, and we explain these WiFi standards in this section.

6.4.1 WiFi 6

Even though 802.11ac has a superior performance as compared to 802.11n, or 802.11n has a better performance than 802.11g, the performance might not be much better when the number of devices or users in a WiFi network increases. For example, using a higher QAM modulation technique might be very effective when the noise level is not high. As the noise level increases in an environment, using a higher QAM modulation does not make a huge difference in increasing the data rate of a system. It is clear that when the number of users or devices increases in an environment, the level of noise also increases. Therefore, in an IoT network with a large number of devices, 802.11ac does not have substantially a better performance as compared to 802.11n. Similarly, channel binding is a method to increase the data rate. But this method might not be possible in a dense environment. For instance, we know that there are 25 non-overlapping channels of 20 MHz in a 5 GHz band. When the channel size is changed to 40 MHz, 80 MHz, or even 160 MHz, then the number of channels will decrease to 12, 6, and 2, respectively. In a dense environment, we use many access points in order to provide the capacity in smaller cells, and therefore, having large channels is not possible. Another method to increase the data rate is by using MIMO antennas, which is also not possible due to the fact that the cost of IoT devices becomes expensive. Using MU/MIMO is efficient where there is a large physical distance between the nodes of the network. As the number of IoT devices increases, the distance between them decreases, which makes using this technique also ineffective. In other words, all techniques that have been used to increase the performance of a WiFi network so far are not effective when there are a huge number of low-cost nodes in a WiFi network.

We are surrounded by access points, and as the number of devices and users in a WiFi network such as 802.11n or 802.11ac increases, it causes the network latency to increase and the data rate to decrease. However, WiFi 6 can generate consistent throughput and its latency does not decrease as the number of devices on the network increases.

WiFi 6 introduces several important techniques to increase its performance in situations where there are many IoT devices on the network, in addition to techniques that are used in other WiFi standards to increase the data rate. In this section, we briefly explain two important techniques that are used for this purpose.

First, WiFi 6 introduces air time efficiency. On a 20 MHz WiFi channel, one frame with the size of up to 2300 bytes can be sent. However, the average packet size is usually substantially smaller than 2300 bytes. Therefore, most of the airtime is actually wasted. WiFi 6 uses Orthogonal Frequency Division Multiple Access (OFDMA) in which each transmission is optimized by the application. This allows multiple streams with lower data sizes to be sent in parallel in order to occupy airtime for the transmission of 2300 bytes. In other words, instead of sending a lightly loaded frame, there can be information from many IoT devices that fill the data payload. This significantly improves the performance and works quite well when there are many devices on the network. The second technique, which is used in WiFi 6 to deal with performance improvement in densely populated networks, is cell coloring. A cell in a WiFi network is called Basic Service Set (BSS). For this reason, cell coloring is also called BSS coloring. In high density environments, there is a need for more access points in order to create smaller cells. Even in smaller cells, users and devices are usually transmitting at high power, which makes the channel look busy. When a WiFi-enabled device wants to transmit, it listens to the channel, and if the channel is not free, the device waits and does not transmit. The device will check the channel again at a later time, and if it finds that the channel is free, it starts sending its data. There are many situations where a device believes that the channel is not free, since other devices on the network are sending their data with high transmit power.

To solve this problem, WiFi 6 introduces cell coloring, in which an access point asks its associate devices to use a lower power in their small cell. This reduces noise and eliminates situations where devices on the network detect that the channel in their cell is busy, when actually it is not.

6.4.2 WiFi HaLow

The rapid growth of IoT has led the IEEE 802.11ah task group and the WiFi Alliance to design a standard that satisfies the requirements of ultra-low-power IoT devices and the ones that need long-range connectivity. In other words, the WiFi community found out that there is a need to optimize WiFi for IoT in order to connect to a larger number of IoT devices at much longer distances and at lower power levels as compared to the traditional WiFi standards.

WiFi HaLow operates in the unlicensed sub-1-GHz spectrum, where traditional WiFi standards mostly operate in the 2.4 and 5 GHz spectrum. The lower the frequency, the longer the signal can travel in the air, and the better its penetration through walls and barriers. Therefore, by operating at lower frequencies, its signals can pass through walls more easily as compared to the traditional WiFi signals. The unlicensed sub-1-GHz spectrum of WiFi HaLow varies between 750 and 950 MHz in different parts of the world. For example, this spectrum is 902 to 928 MHz for the Americas and 863 to 868 MHz in Europe.

WiFi HaLow offers data rates varying from hundreds of Kbps for larger coverage and tens of Mbps for shorter distances. WiFi-HaLow uses channels with substantially lower bandwidth as compared to the traditional WiFi standards. For example, 802.11ac allows for 20, 40, 80, and 160 MHz channel bandwidth, while the channel bandwidth for WiFi Halow is 1, 2, 8, and 16 MHz. Using a narrower bandwidth, a better modulation scheme, and a lower frequency, the communication range of WiFi HaLow is increased. WiFi HaLow provides larger address space for connecting devices per access point. While the maximum number of accessible devices in the traditional WiFi standards and even WiFi 6 is around 2000, it is four times larger for WiFi HaLow.

WiFi HaLow has substantially lower energy consumption as compared to other WiFi standards. This is achieved by implementing many techniques. Some of the most important techniques for this purpose are:

1. Cell coloring: This is similar to what we explained in WiFi 6. Each access point, which represents a cell, uses a different unique color (a 6-bit identifier) as part of its packet. Each device learns the color of its cell upon association with its access point. Signals with the same color use a low power threshold for deferral, while signals with different colors use a higher power threshold for deferral.
2. Target Wake Time (TWT): This allows an IoT device and an access point to pre-arrange a time for the IoT device to wake up from sleep and listen to beacons.
3. Restricted Access Window (RAW): This allows for an access point to give a green light to a subset of IoT devices to transfer their data, while others are forced to sleep.

Let us discuss the design of a WiFi network connecting 8000 IoT devices that are operating in a 1 km^2 square-shaped area. To simplify the situation, let us assume that these IoT devices are spread evenly across an outdoor area. We are interested in the number of access points required for this network if 802.11ah devices are used, as well as when all nodes are 802.11ac devices.

Generally speaking, there is no straight answer to this question, since the answer depends on many factors that are not specified in this example. However, we try to provide a rough estimate that can give readers some understanding about the design of WiFi networks as well as a differentiation between IEEE 802.11ac and 802.11ah. In general, a WiFi network should be designed to satisfy coverage and capacity requirements. Network coverage ensures network connectivity at all desired locations, while network capacity ensures that sufficient bandwidth has been allocated in order to satisfy the needs of IoT devices.

WiFi coverage can be determined based on many factors such as the location of access points, the power settings of WiFi devices as well as access points, the physical environment (indoor or outdoor), and the type of antenna that is used by access points and IoT devices. In an indoor environment, the number of walls, floors between the access point and the WiFi-enabled device, and the building material also need to be considered. Therefore, there are many parameters that should be considered when designing a network for coverage. However, network designers usually perform a rough estimate of the number of the required access points and the location of their installations. Then they perform a site survey to find any potential coverage issues. If necessary, they may add more access points or relocate access points to address coverage issues.

Planning the WiFi network for coverage might be simple but does not guarantee that the WiFi network provides a suitable connectivity scheme, if the capacity is not factored into the design. When designing a WiFi network for capacity, a designer should know the number of devices that are using the network simultaneously, the required bandwidth per device, the type of data, the quality of service required by the IoT devices, and the throughput of the access points.

The maximum number of the devices that an access point can support is another factor in designing a WiFi network. While an 802.11ac access point can connect up to 2007 devices, an 802.11ah can handle more than 8000 IoT devices.

To design a WiFi network for this example using IEEE 802.11ac, there is a need for four access points (8000 devices/2007 devices per an access point). Let us also assume that the range for an IEEE 802.11ac access point is 300 m. Since the area is 1 km by 1 km, the area can also be covered using four access points. In practice, an 802.11ac access point might not be able to give access to more than 100 or 200 devices due to their processing and storage limitations. Since some of the access points act as a Dynamic Host Configuration Protocol (DHCP) server as well, many of them provide only 255 IP addresses. Considering 200 devices per access point indicates the need for 40 access points (8000/200) for our example. IEEE 802.11ah access points can be connected to more than 8000 devices, and since they use lower frequency bands, one access point can cover the area in this example. Designing a WiFi network for capacity requires some knowledge about the required bandwidth for IoT devices as well as their transmission pattern.

6.5 LoRaWAN Wireless Wide Area Network

As explained earlier, besides short-range wireless technologies that use the unlicensed band, there are several long range wide area networks that also operate in the unlicensed band. In this section, LoRaWAN, one of the most popular Low Power Wide Area Network (LPWAN) technologies, is discussed briefly.

LoRaWAN defines four entities in its network architecture as shown in Figure 6.6. These entities are end nodes, gateways, network servers, and application servers [9]. There is not a unique association between an end node and a specific gateway, and therefore, transmitted data by an end node can be received by several gateways. An LoRaWAN gateway forwards the data that is received from an end node to a network server. The network server can be cloud based or can be privately owned. The connectivity scheme between the gateway and the network server can use any type of wired or wireless technology such as Ethernet, WiFi, cellular, or even satellite. The gateway is responsible for forwarding the data toward the network server, that is an intelligent device and

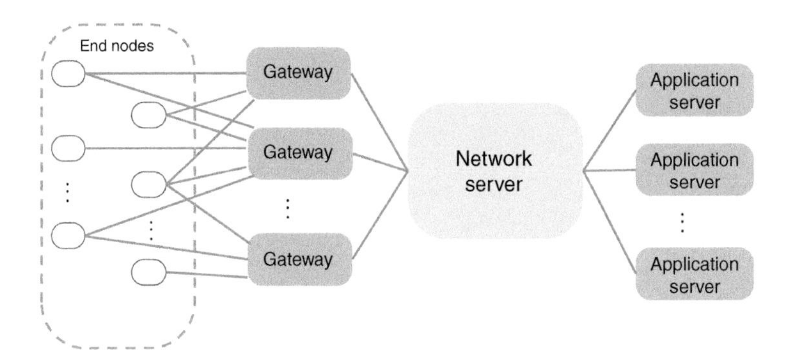

Figure 6.6 LoRaWAN architecture.

manages the LoRaWAN network. As mentioned earlier, data from one end node can be received by several gateways. Since each gateway forwards the data to the network server, this server might receive duplicated data from an end node. It is the responsibility of the network server to filter the duplicated data, ensure security of the transmission, and determine to what gateway the acknowledgments need to be transmitted. The LoRaWAN architecture can handle the mobility of the end nodes to some extent. LoRaWAN does not support a handover mechanism, but when an end node moves from a gateway to another gateway, the network server can detect the movement.

In the LoRaWAN architecture, end nodes communicate asynchronously with the gateway. This way, end nodes do not need to synchronize with the network. In general, time synchronization requires a substantial amount of signaling and results in the consumption of a significant amount of energy. This makes LoRaWAN suitable for many battery-powered IoT applications [10].

As can be seen from Figure 6.6, LoRaWAN uses a star topology. In order to ensure this network architecture has the capacity and the capability to handle a massive number of end nodes, and can provide a long range of coverage, the gateway must be capable of handling multiple channels in parallel.

One of the important features used by LoRaWAN is its adaptive data rate transmission. This indicates that if an end node and a gateway have a good connection, the end node can transfer its data at a greater data rate, reducing air time, as opposed to when the end node and the gateway do not have a good connection. LoRaWAN is scalable thanks to this feature. If a gateway is added to the network, the data rate of end nodes in proximity to the newly added gateway can be increased.

LoRaWAN defines the communication protocols and the architecture of a low power wide area network including the access control and application layer. LoRaWAN is over a physical layer called LoRa, which is based on spread spectrum modulation technology and operates in unlicensed bands. The specification of LoRaWAN in terms of frequency bands, number of available channels, channel bandwidth, transmit power, and data rate varies slightly in different regions of the world. For example, the frequency band for LoRaWAN in North America is between 902 and 928 MHz, while in Europe the spectrum of 867–869 MHz is allocated to LoRaWAN.

Let us revisit the same example that we discussed in the previous section, and find out how many gateways are needed to connect to 8000 IoT devices spread across a 1 km^2 square-shaped outdoor area. A gateway acts as a switch, and forwards the data packets to the network server. Communication between a gateway and a network server can be established via WiFi, Ethernet, or a cellular connectivity scheme. Generally speaking, there are two types of LoRaWAN gateways: single-channel and multi-channel gateways. Single-channel gateways are low-cost devices that can be used in small projects, which often do not support adaptive data rate transmission, and

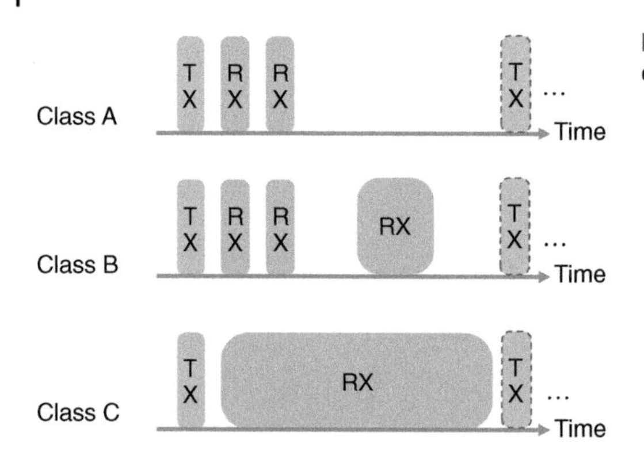

Figure 6.7 The operations of different classes of end nodes in LoRaWAN.

have limitations in their functionalities. Multi-channel gateways are the ones that are used in most applications, and they usually have 8 to 10 channels. It is interesting to note that many types of antennas can be used with LoRaWAN gateways, and these antennas have various radiation patterns. LoRa uses the spread spectrum modulation technique in its physical layer, and uses random access for its second layer. Therefore, the possibility of collision exists in LoRaWAN network. Calculation of the number of end nodes that a multi-channel gateway can support is outside the scope of this book. However, readers should understand that this number depends on the amount of transmitted data and the transmission pattern of the end nodes. These criteria can be used to determine the number of supported end nodes and the duration of time that a gateway is busy receiving data. Typically, an LoRaWAN gateway can support several hundreds to a few thousand end nodes.

The end nodes in LoRaWAN can be put in three different classes: class A, class B, and class C. Class A allows for bidirectional data transfer in which an end nodes' uplink data transmission is followed by two short periods of data reception in the downlink direction. This class is an excellent choice for IoT end nodes that only need the downlink transmission from the network server shortly after the IoT device has transmitted an uplink message. Therefore, this class is suitable for the most energy-efficient end nodes. Class B opens additional reception windows at scheduled time in addition to the ones existing in class A. To be able for class B end nodes to open their reception windows at scheduled times, they should receive time synchronization messages from the gateway. Class C end nodes have their reception window open at all times except the time that they are transmitting. The operations of different classes of end nodes in LoRaWAN are shown in Figure 6.7.

Figure 6.8 compares the three classes of end nodes in LoRaWAN in terms of latency and power consumption. As can be seen, class A devices consume lower amounts of power, but have the largest amount of latency, while class C devices consume the highest amount of energy, but have the lowest amount of latency among all three classes.

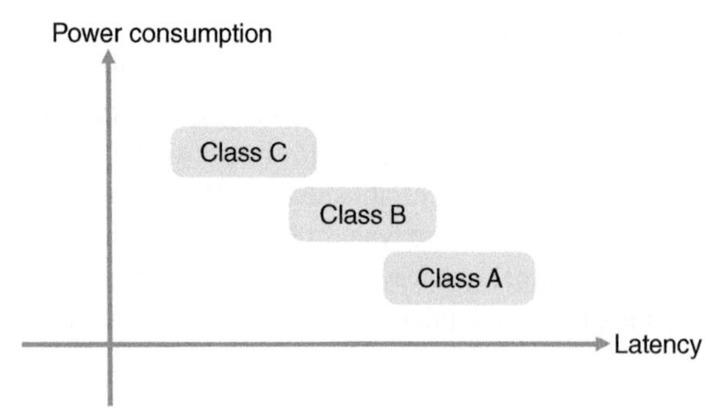

Figure 6.8 The comparison among different classes of end nodes in terms of latency and power consumption in LoRaWAN.

Interested readers are encouraged to take a look at Assignment #8, which is a hands-on project on this topic.

6.6 Summary

In this chapter, we discussed unlicensed-band wireless IoT technologies. Unlicensed bands are spectrums for everyone to use at no cost and without acquiring any licenses. Even though these spectrums do not require any license, there are some rules that need to be followed when a technology wants to use them.

We divide unlicensed-band wireless technologies into two categories. The first category is the one that uses short-range wireless technology in order to communicate with an IoT gateway. The most important short-range wireless technologies are Zigbee, BLE, and WiFi. The second category is unlicensed-band wide area network technologies. The most important technology of this group is LoRaWAN.

Zigbee uses a mesh topology and a dynamic routing algorithm. It provides a very reliable network in which IoT devices can join the network rapidly. A Zigbee network supports data rates below 250 kbps.

BLE uses a star topology, can provide a higher data rate than Zigbee, and has been widely used in many smart devices. Bluetooth 5 and Bluetooth mesh are also available that makes Bluetooth a suitable solution for various IoT applications.

WiFi technologies such as 802.11b/g/n/ac provide a very high data rate, and these technologies are available almost in all homes and offices. However, WiFi consumes lots of power and might not be suitable when a large number of devices become connected to WiFi networks. There are two WiFi technologies that suit many IoT applications. These two WiFi technologies are WiFi Halow and WiFi 6. These two WiFi technologies have been designed with IoT in mind and can handle a large number of devices.

One of the most important technologies that provides wide area connectivity in unlicensed bands is LoRaWAN. This technology belongs to low power wide area networks. A brief discussion on the LoRaWAN architecture and different classes supported by the IoT nodes in an LoRaWAN network is provided in Section 6.5.

References

1 ITU Radio Regulations. CHAPTER II – Frequencies, ARTICLE 5 Frequency allocations, Section IV – Table of Frequency Allocations.

2 ZigBee Alliance (2004). ZigBee Specification, Version 1.0. ZigBee Document 05-3474-21. https://zigbeealliance.org (accessed 30 June 2022).

3 Heydon, R. (2012). *Bluetooth Low Energy*. Prentice Hall.

4 Dian, F.J. and Vahidnia, R. (2020). Formulation of BLE throughput based on node and link parameters. *IEEE Canadian Journal of Electrical and Computer Engineering.* 43 (4): 261–272. https://doi.org/10.1109/CJECE.2020.2968546.

5 Dian, F.J., Yousefi, A., and Lim, S. (2018). A practical study on Bluetooth Low energy (BLE) throughput. In: *2018 IEEE 9th Annual Information Technology, Electronics and Mobile Communication Conference (IEMCON)*, 768–771. Vancouver, Canada: IEEE IEMCON (Nov. 2018). doi: https://doi.org/10.1109/IEMCON.2018.8614763.

6 Bluegiga (2015). BGscript scripting language-developer guide-version 4.1. https://device.report/m/ff2e4a62e769742e782cf8b8cae2ed87eda4492b792984be92767b92ff502ae4.pdf (accessed 30 June 2022).

7 Darroudi, S.M., Gomez, C., and Crowcroft, J. (2020). Bluetooth low energy mesh networks: a standards perspective. *IEEE Communications Magazine.* 58 (4): 95–101. https://doi.org/10.1109/MCOM.001.1900523.

8 Hernandez-Solana, A., Valdovino Bardaji, A., Perez-Diaz-De-Cerio, D. et al. (2020). Bluetooth mesh analysis, issues, and challenges. *IEEE Access.* 8: 53784–53800. https://doi.org/10.1109/ACCESS.2020.2980795.

9 Semtech (2019). LoRa® and LoRaWAN®:A Technical Overview. https://lora-developers.semtech.com/uploads/documents/files/LoRa_and_LoRaWAN-A_Tech_Overview-Downloadable.pdf (accessed 30 June 2022).

10 LoRa Alliance (2016). LoRa Alliance Wide area networks for IoT. https://www.lora-alliance.org/ (accessed 30 June 2022).

Exercises

6.1 An IoT device operates in the ISM bands and it is sold in the United States.
- **a.** Does the device manufacturer need to obtain a license to transmit in this spectrum?
- **b.** Can the device output power be 2 W?
- **c.** Can the device use a sub-GHz (below 1 GHz) frequency?

6.2 An IoT device operates at the 2.4 GHz ISM bands. The available spectrum for an IoT device is in the 2.4–2.483 GHz range.
- **a.** What is the bandwidth of the 2.4 GHz ISM bands (based on the information provided in Table 6.1)? What spectrum is available to the IoT device?
- **b.** If we divide the spectrum into 20 MHz channels with a guard band of 5 MHz, how many non-overlapping channels are possible?
- **c.** If we divide the spectrum into 1 MHz channels with a guard band of 1 MHz, how many non-overlapping channels are possible?

6.3 An IoT application that uses the Zigbee network is shown in Figure 6.E3.
- **a.** What is the topology of this network?
- **b.** An endpoint sends a message to E_2, but E_2 is in sleep mode. What happens in this situation?
- **c.** What devices in this network are FFDs, and which ones are RFDs?
- **d.** Endpoint E_1 sends a message to the gateway, and the gateway sends this message to the cloud. Assume the link between R_1 and the coordinator gets broken. Explain what happens in this situation?

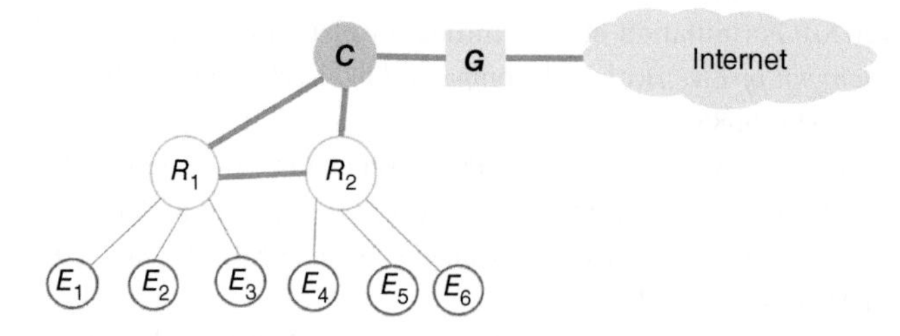

Figure 6.E3 Network in Exercise 6.3.

6.4 In a home automation application that uses a Zigbee network, we have the following devices: smart home controller, light bulbs, thermostat, fans, portable light switches, door/window sensing devices, smoke detectors, and motion security devices. Which of these devices can better play the role of coordinator, router, and endpoints?

6.5 The Zigbee network in Figure 6.E3 operates at 2.4 GHz.
a. Generally speaking, if a Zigbee network uses a 2 MHz channel, and channels in the spectrum are 5 MHz apart, how many channels are defined in this network?
b. A 2 MHz channel (let us say 2.4–2.402 GHz) is used between R_1 and E_1, what channel has been used between R_1 and E_2?
c. A 2 MHz channel (let us say 2.4–2.402 GHz) is used between R_1 and E_1, what channel has been used between R_2 and E_4?

6.6 What is the difference between two different types of channel access in a Zigbee network?

6.7 A Zigbee network can extend its reach by adding more routers. Can we add so many routers to make a wide area Zigbee network?

6.8 The architecture of Zigbee is based on mesh topology. Bluetooth mesh also uses mesh topology. Can we say that Bluetooth mesh and Zigbee networks are the same?

6.9 A BLE-based IoT device is connected to an IoT gateway that plays the role of a central BLE. The connection interval time should be 15 ms. What value should be considered at the time of connection establishment for interval?

6.10 A Bluetooth-based IoT device is connected to an IoT gateway that plays the role of a central node. Both the IoT device and IoT gateway use Bluetooth 5 technology. If both the IoT device and gateway use a 1 Mbps physical layer, the maximum distance between the IoT device and gateway is 150 m. If the Bluetooth module is set to coded physical layer $S = 8$, what is the maximum range of communication in this case?

6.11 A BLE-based IoT device is connected to an IoT gateway that plays the role of a central BLE. The connection interval is set to 30 ms, and the latency is set to 2. The IoT device generates data every 100 ms. Does this configuration work? If not, choose a suitable configuration for this system?

6.12 Why is WiFi 802.11g/n/ac not suitable for most IoT applications?

6.13 Why can WiFi HaLow penetrate through walls more easily than WiFi 802.11g/n/ac?

6.14 What is BSS coloring? Why is it used in WiFi networks?

6.15 What is the difference between LoRa and LoRaWAN?

6.16 The end nodes in LoRaWAN can be put in three different classes: class A, class B, and class C. What are the differences among these classes?

6.17 How has the discovery process been improved in Bluetooth 5 as compared to BLE?

Advanced Exercises

6.18 The manager of a museum would like to find out which paintings are more popular and which ones are less attractive. Visitors are asked to wear an IoT wristband that uses a BLE module. There is also a BLE device next to each painting, and a gateway that is connected to the Internet and also uses BLE technology. The BLE devices need to be configured by system designers.

- **a.** Explain the high-level design of a system that can determine how many people in a group of visitors have visited a particular painting.
- **b.** Explain the high-level design of a system that can determine how much time (on average) a visitor has spent to view a particular painting.
- **c.** If you design this system using BLE technology, and configure BLE nodes in connection mode, how many visitors (BLE slave nodes) can communicate with a painting (a BLE master node) at the same time?
- **d.** If a visitor stays for a short amount of time next to a painting, is it possible that the BLE node (the one next to the painting) will not be able to detect the visitor?

6.19 WiFi uses three 20 MHz non-overlapping channels in the 2.4 GHz band. Let these channels be 2.402–2.422 GHz, 2.427–2.447 GHz, and 2.452–2.472 GHz.

- **a.** What is the guard band between WiFi non-overlapping channels?
- **b.** If BLE and WiFi coexist, which BLE advertising channels do not overlap with these WiFi channels?

6.20 BLE has three advertising channels and 37 data channels. It uses a frequency hopping algorithm to cycle through the 37 data channels. If f_n is the frequency of the channel used in a connection event, f_{n+1} is the channel that needs to be used on the next connection event, and hop represents the predefined hop values (which are always between 5 and 16), then

$$f_{n+1} = (f_n + \text{hop}) \bmod 37$$

Assume a BLE starts from channel 4, and the hop sequence is 10, 6, 12, and 8. What channels are used by BLE for transferring data?

6.21 How can we have high power and low energy cellular technology? Explain this concept by differentiation between energy and power.

Chapter 7 Cellular IoT Technologies

7.1 Introduction

To develop new technology for the third generation of cellular networks, a global technical body called Third Generation Partnership Project (3GPP) was formed in 1998. 3GPP continued its work to develop the technical specifications for the fourth generation (4G) and the fifth generation (5G) cellular networks later on. 3GPP represents seven organizational partners that are large regional telecommunication companies in Europe, America, and Asia. These partners prepare the standards based on the technical specifications that are written by 3GPP. The organization of 3GPP's work is represented by a number that is called the release number. For example, 4G was first introduced in 3GPP Release 8, while Machine-Type Communication (MTC) and IoT were first introduced in 3GPP Release 12 and 13, respectively.

One may think that 3GPP Release 8 is allocated to the 4G network, but that is not true. A specific release number does not belong to a specific cellular network technology. For example, even though 4G was first introduced in 3GPP Release 8, it continued to be part of higher releases in parallel with other technologies. In a similar way, 3GPP Release 15 has introduced the 5G technology in parallel with 4G. In addition to broadband communications for providing voice and data communications, 3GPP considers MTC in 3GPP Release 12+. For providing cellular connectivity for IoT applications, 3GPP Release 13 introduced three technologies. As it is shown in Figure 7.1, these technologies are Extended Coverage GSM Internet of Things (EC-GSM-IoT), LTE for Machine-Type Communications (LTE-MTC or LTE-M), and Narrowband Internet of Things (NB-IoT). The enhancements of the LTE-M and NB-IoT technologies have been continued in 3GPP Release13+. The 3GPP documentations for these three technologies can be found in [1–8], which are also used in our introductory discussion of these technologies in this chapter. Interested readers can also study [9, 10] for some more detailed information about these technologies.

7.2 EC-GSM-IoT

The Global System for Mobile Communications (GSM) is a cellular network technology developed in Europe. This cellular technology has been widely used around the world, especially by European countries. Unlike the first generation of cellular networks, which used analog technology, GSM (as one of the technologies in the second generation of cellular networks) was based on digital technology. This means that GSM uses digital voice, as well as digital modulation and coding techniques for data transmission. There are several cellular technologies that belong to the second and third

Fundamentals of Internet of Things: For Students and Professionals, First Edition. F. John Dian.

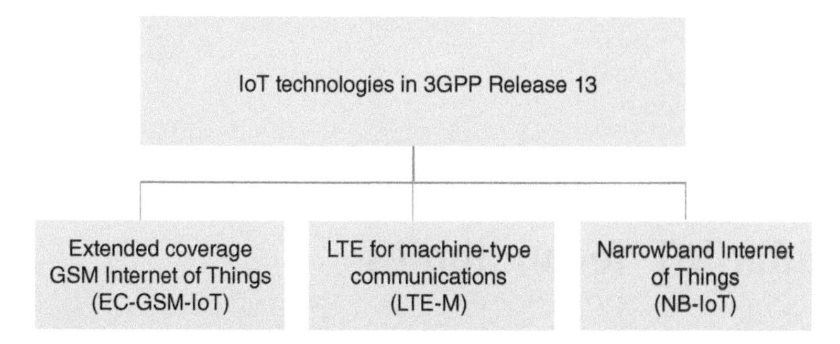

Figure 7.1 Three IoT technologies introduced in 3GPP Release 13.

generations of cellular networks and are based on GSM. Examples of these technologies are General Packet Radio Service (GPRS) and Enhanced Data rates for GSM Evolution (EDGE). It is clear that the newer generations of cellular networks can provide higher data rates as compared to GSM. However, GSM continues its presence as an important part of the cellular technology to provide voice communications and to be a fallback solution for data services in case of failure of the higher generations of cellular networks, or when these networks experience a network congestion. GSM technology has been deployed in many regions around the world, including Europe, Africa, and Middle East. Even though the Code Division Multiple Access (CDMA)-one technology was the second generation of cellular networks deployed in North America, some operators in this region also provided GSM access to their customers. GSM uses different spectrums depending on the region of deployment. Table 7.1 lists the frequency bands and the uplink and downlink spectrum for various deployments of GSM technology in different regions of the world. The rate of data transfer in the second generation of cellular networks is extremely low, and therefore one might think that network operators should decommission this technology. But it must be noted that GSM can provide a communication scheme that satisfies the requirements needed for many industrial equipment, and therefore, operators will continue to provide GSM technology to their customers. On the other hand, GSM can be an excellent choice for machine-to-machine-type communication as well as IoT applications that require a small rate of data transfer. GSM can operate in the 850MHz and 900MHz frequency bands, as shown in Table 7.1, and many operators throughout the world use these two frequency bands for their operations. In any wireless technology, utilizing lower frequencies results in lower attenuation, and therefore, it provides a higher range of communication. Since GSM can operate in lower frequency bands as compared to other cellular technologies, it becomes a good candidate to provide connectivity to IoT devices. Also, the second generation of cellular networks

Table 7.1 Some of the GSM spectrum depending on the region of deployment.

GSM band	Frequency (MHz)	Downlink	Uplink	Regions
GSM–850	850	869 – 893	824 – 848	America (North, Caribbean, Latin)
E-GSM–900	900	925 – 960	880 – 915	Asia pacific, Europe, Middle East, Africa
R-GSM–900	900	921 – 960	876 – 915	Asia pacific, Europe, Middle East, Africa
DCS–1800	1800	1805 – 1879	1710 – 1784	Asia pacific, Europe, Middle East, Africa
PCS–1900	1900	1930 – 1989	1850 – 1909	America (North, Caribbean, Latin)

Source: Based on [11].

used simpler technology than newer generations. The use of a simpler technology can be translated to the use of lower-cost devices in the GSM network. This also makes GSM an excellent choice for IoT. Even though in this book, our focus is on the use of higher generations of cellular networks for the IoT ecosystem, readers should understand that GSM has several good features in terms of simplicity and coverage that makes it an attractive choice for IoT applications. That was the reason that GSM was considered in 3GPP Release 13 as one of the cellular technologies that can be used to provide connectivity for IoT devices.

7.3 LTE-based Cellular IoT Technologies

Long-Term Evolution (LTE)-M and NB-IoT were first introduced in 3GPP Release 13 by extending and modifying the LTE technology in order to create technologies that are suitable for MTCs and IoT applications. The enhancements to LTE-M and NB-IoT are provided in higher 3GPP releases. We explain these two technologies and their performance in terms of channel bandwidth, duplexing, data rate, coverage, and mobility in this section.

7.3.1 LTE-M

LTE-M is an LTE-based cellular technology that is suitable for many IoT applications in terms of performance and functionality. LTE-M provides lower power consumption, deeper coverage, larger density per cell, and lower device cost as compared to LTE.

7.3.1.1 Channel Bandwidth

LTE-M technology uses part of the existing bandwidth of legacy LTE for its operation. LTE technology uses a channel bandwidth of 1.4, 3, 5, 10, 15, or 20 MHz. This means that operators have deployed LTE with one of these channel bandwidths. In 3GPP Release 13, the channel bandwidth for LTE-M was limited to 1.4 MHz. For example, if an operator deploys LTE with the channel bandwidth of 1.4 MHz and intends to support LTE-M, the whole bandwidth is needed to be used for LTE-M. But if an operator deployed LTE with the channel bandwidth of 20, 1.4 MHz would be used for LTE-M and the rest of the bandwidth could be used for LTE voice and data transmission. It should be noted that the entire spectrum of 1.4 MHz is not used for data transfer or signaling. In LTE-M, this spectrum consists of 1.08 MHz for data and signaling information and 0.42 MHz is used as guard band. Figure 7.2 shows the coexistence of LTE and LTE-M, where the operator has deployed LTE with a channel bandwidth of 20 MHz. The operator in this scenario is using 1.4 MHz of the bandwidth for LTE-M and IoT applications.

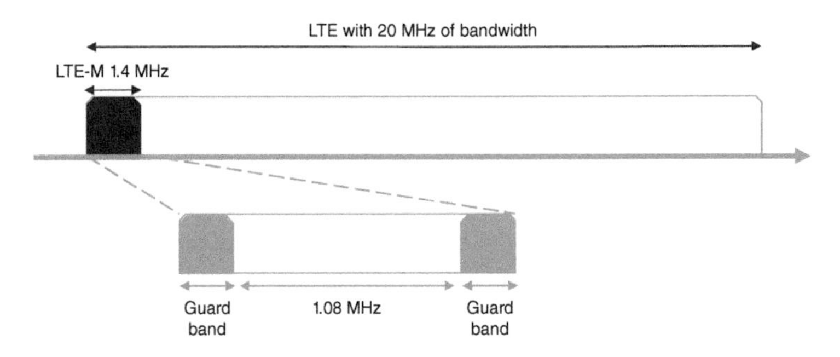

Figure 7.2 An example of coexistence of LTE and LTE-M.

Table 7.2 Maximum spectrum for LTE-M in 3GPP Release 13 and 14.

	LTE-M Cat-M1 Release 13	LTE-M Cat-M1 Release 14	LTE-M Cat-M2 Release 14	LTE-M Release 15
Maximum DL bandwidth (MHz)	1.4	1.4	5	5
Maximum UL bandwidth (MHz)	1.4	1.4	5	5

In 3GPP Release 14 and higher, the LTE-M bandwidth has been enhanced to enable a maximum channel bandwidth of 5 MHz in both the downlink and uplink directions, as indicated in Table 7.2. It is worth noting that 3GPP Release 14 specifies two types of devices: Cat-M1 and Cat-M2.

In general, the amount of data that a radio technology can carry depends on the channel bandwidth and the way that the data is modulated and encoded. A specific modulation and encoding scheme represents the average spectral efficiency of the technology that is used by the operator. Spectral efficiency is defined as the throughput of a system per channel bandwidth, and it is measured as bits per second per Hz.

$$\text{Spectral efficiency} = \frac{\text{Throughput (bps)}}{\text{Channel bandwidth (Hz)}} \tag{7.1}$$

For example, consider an operator that deploys LTE with 10 MHz channel spectrum in both downlink and uplink and uses spectral efficiency of 1.6 bps/Hz and 1 bps/Hz in the downlink and uplink directions, respectively. In this example, the bandwidth of the channel in the uplink and downlink would be 10 MHz. Therefore, the downlink data rate is $10\ \text{MHz} \times 1.6\frac{\text{bps}}{\text{Hz}} = 16\ \text{Mbps}$, while the uplink data rate is $10\ \text{MHz} \times 1\frac{\text{bps}}{\text{Hz}} = 10\ \text{Mbps}$. By using a more advanced modulation and encoding technique, we can increase spectral efficiency. However, this also increases the cost of manufacturing for mobile devices. Since the cost of an IoT device is an important factor, implementing sophisticated modulation and encoding schemes that require large processing power are not usually recommended for IoT devices.

Similar to LTE, LTE-M technology can adapt to changing radio conditions. One method to achieve this is to vary modulation and coding schemes, which changes the value of the spectral efficiency. When the radio condition is poor, the radio technology uses a less-efficient modulation in terms of bandwidth, as well as a stronger error coding scheme with more error correction bits. When the radio condition is somehow better (fair), the system may use the same modulation technique but with fewer error correction bits. When the radio condition improves further (good), the radio technology uses a more efficient modulation scheme but uses a higher number of bits for error correction. In situations where the radio condition is excellent, the radio technology uses a higher spectral efficiency modulation technique, and a lower number of error coding bits. Therefore, under excellent radio conditions, the radio produces the highest spectral efficiency. Figure 7.3 shows the use of different modulation and encoding schemes under four different radio conditions. LTE may use advanced modulation and encoding techniques that require complex processing, while LTE-M has some restrictions in using those techniques. Low power consumption and affordability are two important requirements for most Cellular IoT (CIoT) devices.

7.3.1.2 Duplexing

The channel bandwidth deployed by an operator needs to be used for the uplink and downlink transmission. The method utilized by both the base station and IoT devices to send and receive information inside one channel is called duplexing. The most common ways for providing full duplex operation are Time Division Duplexing (TDD) and Frequency Division Duplexing (FDD).

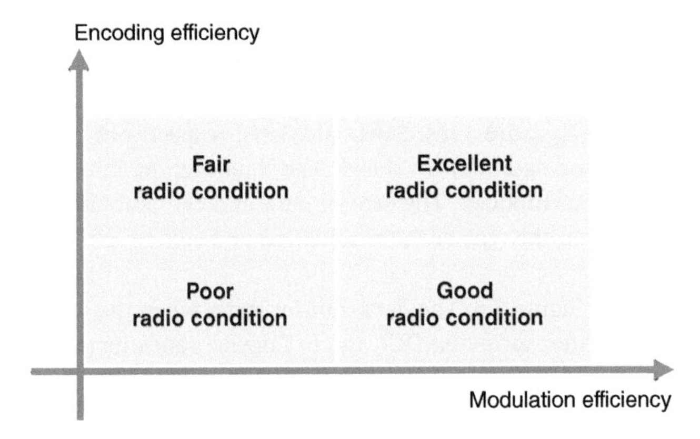

Figure 7.3 Selecting a suitable modulation and encoding scheme based on different radio conditions.

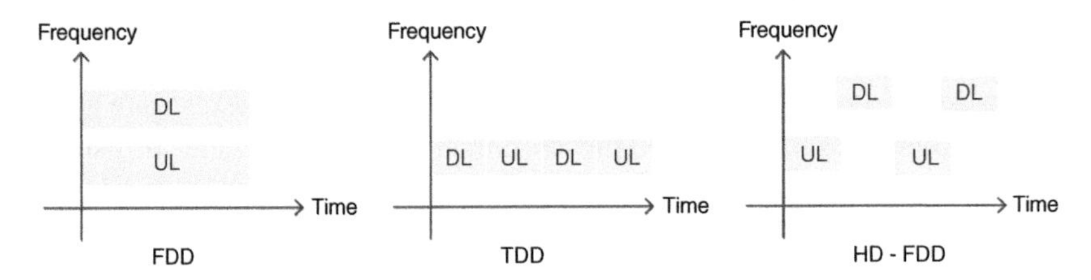

Figure 7.4 Comparison among FDD, TDD, and HD-FDD.

In the TDD duplexing method, both the UpLink (UL) and DownLink (DL) traffic use the same frequency band, but the transmission occurs at different times. In full duplex FDD, the data transmission and reception can happen simultaneously, but part of the channel bandwidth is used for UL and the other part is used for DL.

LTE-M in 3GPP Release 13, 14, and 15 support Half-Duplex FDD (HD-FDD) in addition to TDD and FDD. In HD-FDD, every node alternates between the transmission and reception modes as shown in Figure 7.4. The maximum achievable data rate using the HD-FDD scheme is lower than both FDD and TDD. However, this duplexing scheme is simpler and can be used in the design of IoT devices in order to reduce their cost. A comparison of these duplexing methods is also shown in Figure 7.4.

7.3.1.3 Data Rate and Latency

Figure 7.5 illustrates a simple example of the downlink data transfer in LTE-M. It starts with the base station transmitting a Downlink Control Information (DCI) signal to the IoT device.

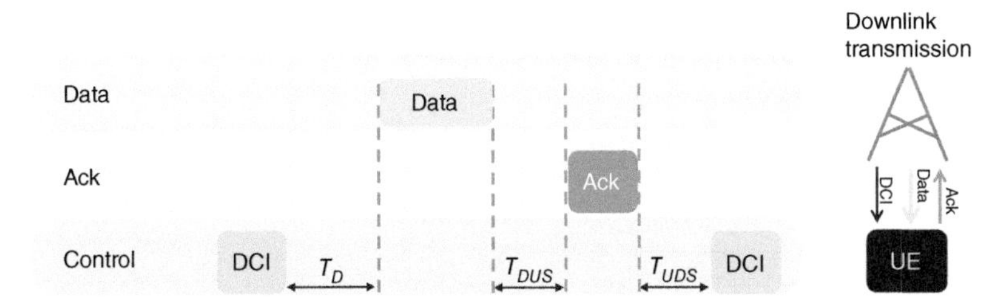

Figure 7.5 DL transmission in LTE-M.

Then the base station transmits its data. When the IoT device receives the data, it sends an acknowledgment to the base station. In Figure 7.5, T_D represents the cross subframe delay, T_{DUS} represents the downlink to uplink switching time, and T_{UDS} shows the uplink to downlink switching time. The values of T_D, T_{DUS}, and T_{UDS} are 1 ms, 3 ms, and 3 ms, respectively. The data section in Figure 7.5 represents the sensor data and overhead. The data section can use different forms of modulations and encoding techniques. The size of data and the amount of time that it takes to transmit the data depends on the type of the modulation and coding scheme that is used.

The LTE-M uplink data transfer is shown in Figure 7.6. The data transfer starts when the base station sends a DCI signal to the IoT device. After receiving DCI, the IoT device sends its data. The base station receives the data and sends an acknowledgment to the IoT device. Similar to the downlink transmission, the downlink to uplink switching time, uplink to downlink switching time, and cross subframe delay are part of the uplink data transmission.

To increase the data rate, legacy LTE uses Hybrid Automatic Repeat Request (HARQ) process. In ARQ, when a corrupted packet is received, the receiver discards this packet and sends a Negative Acknowledgment (NAK) packet to the transmitter in order to request for the packet to be re-sent. If the newly transmitted packet becomes also corrupted, the receiver discards the packet again and asks for the transmission of a new packet. The HARQ process enhances the ARQ scheme by using an error correction technique called soft combining, which no longer discards the corrupted data. In soft combining, the receiver buffers the corrupted packets and tries to perform error correction.

The number of HARQ processes for downlink in LTE-M is similar to legacy LTE, but LTE-M allows for the repetition of each transmission. The number of HARQ processes for uplink is different from legacy LTE.

We can increase the data rate by increasing the number of concurrent HARQ processes. Figure 7.7 shows a situation where two HARQ processes are used in the downlink direction. Let us represent T_{DCI} as the duration of time for the transmission of a DCI packet. Let us also represent T_{Data1}

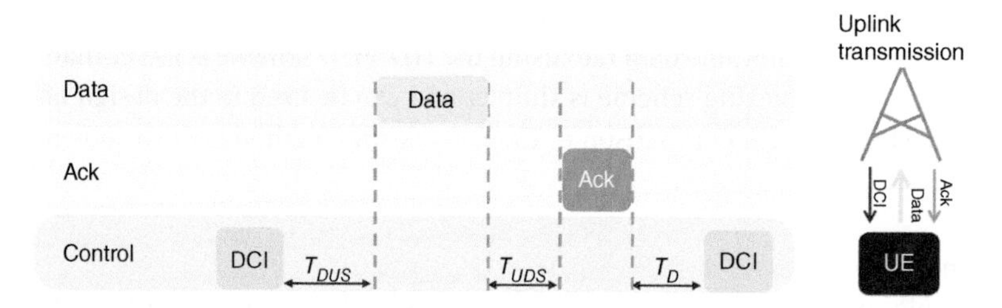

Figure 7.6 UL transmission in LTE-M.

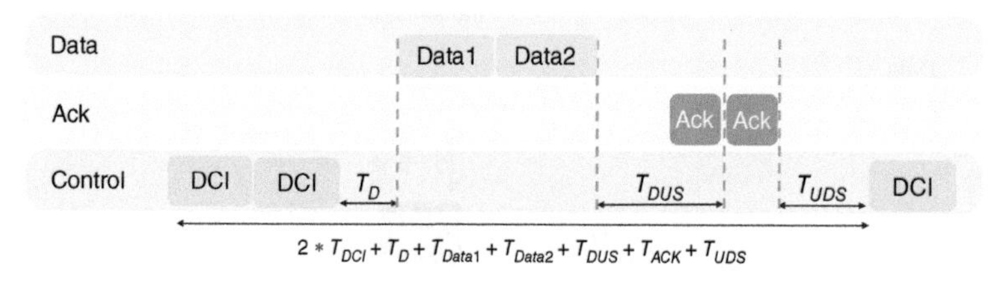

Figure 7.7 DL transmission with HARQ in LTE-M.

and T_{Data2} as the duration of time for the transmission of Data1 and Data2 packets, respectively. The total transmission time for sending these two data packets can be calculated as:

$$\text{Total Time} = 2 * T_{\text{DCI}} + T_D + T_{\text{Data1}} + T_{\text{Data2}} + T_{\text{DUS}} + T_{\text{ACK}} + T_{\text{UDS}} \tag{7.2}$$

The peak achievable data rate for LTE-M depends on the channel bandwidth, the maximum number of HARQ processes, and the maximum Transport Block Size (TBS). Due to the timing requirement in sending the data, we can usually say that by increasing TBS, the data rate will be increased. Increasing the number of HARQ processes can also increase the data rate and is a feature that is mostly useful in good radio conditions.

In LTE-M Release 13, the theoretical peak data rate is 1Mbps using 1000 bytes of TBS and when eight HARQ processes in both uplink and downlink are used. In 3GPP Release 14, the LTE-M spectrum is 5 MHz and devices can support larger TBS of up to 6968 bytes in the uplink direction and 4008 bytes in the downlink direction. It also supports a maximum of ten HARQ processes in the downlink and eight HARQ processes in the uplink. As a result of these modifications, the packet data rate in 3GPP Release 14 for LTE-M has been increased to 4Mbps and 7Mbps in the downlink and uplink directions, respectively. A new spectrum called non-balanced (non-BL) is also introduced in 3GPP Release 14, which supports 20 MHz and 5 MHz spectrum in the downlink and uplink directions, respectively.

A summary of the supported TBS, achievable data rates, and the maximum number of HARQ processes for LTE-M in different 3GPP releases is depicted in Table 7.3. It should be mentioned that the numbers in this table are representing the situations that LTE-M uses the FDD duplexing scheme. For example, in 3GPP Release 13, the peak data rate of LTE-M is 1Mbps when FDD is used. For half-duplex FDD, the resulting peak data rate is 300 kbps in the downlink and 375 kbps in the uplink. 3GPP Release 14 defines two types of IoT devices: Cat M1 devices and Cat M2 devices.

7.3.1.4 Power Class

The maximum power that an IoT device can transmit is limited and is defined as the devices' power class. In LTE-M Release 13, IoT devices support two power classes of 23 and 20 dBm. It is clear that an IoT device that uses the 23 dBm (200 mw) power class has 3 dB more power than an IoT device that transmits in the 20 dBm (100 mw) power class. Using higher transmit power increases the maximum transmit range. It might sound interesting to increase the transmit power above 23 dBm to increase the range of transmission. However, this causes many technical issues in addition to increasing the cost. The technical issues are safety issues related to maximum allowed radiation limits, the increased peak current of an IoT device, and the increased inter-cell interference. Actually, IoT devices tend to use lower transmit levels to be able to integrate the power amplifier into the IoT device and reduce the cost of the device. For this reason, 14 dBm as a new power class is added

Table 7.3 Summary of LTE-M specifications in various 3GPP releases.

	LTE-M Cat-M1 Release 13	LTE-M Cat-M1 Release 14	LTE-M Cat-M2 Release 14	LTE-M Release 15
Maximum DL TBS (bits)	1000	2984	4008	4008
Maximum DL data rate (Mbps)	1	1	4	>4
Maximum UL TBS (bits)	1000	2984	6968	4008
Maximum UL data rate (Mbps)	1	3	7	7
Maximum HARQ processes	8	8	10	10

to LTE-M. This class can especially be useful for wearable IoT technology. It is also interesting that the current draw in this power class is substantially less than other classes and that the IoT device can easily use coin-size batteries.

7.3.1.5 Coverage

The coverage of a cellular network indicates the areas that its radio signal can reach. Since IoT devices might be located in any location, cellular coverage is an important factor for IoT applications. There are many factors that affect the maximum coverage area. Examples of these factors are distance from transmitters, transmitter power, physical obstructions on the signal path, noise, and interference.

LTE networks are deployed based on a network planning process that determines the location and size of each cell. LTE IoT technologies are deployed on existing LTE networks. Therefore, the size and location of a cell in both LTE networks and LTE IoT networks are the same. However, due to unique requirements of IoT, the required coverage area for IoT devices might be different than what LTE provides for smartphones. For example, IoT devices might be designed to transmit data with a lower maximum power, which consequently reduces the range of communications. Also, for IoT devices located in challenging locations, coverage extension methods have been considered where these techniques do not exist for the LTE network.

3GPP has used Maximum Coupling Loss (MCL) as the metric to determine the coverage of a radio access technology, and it is the metric that we use in this book. MCL is the maximum power loss that a system can tolerate and still be operational. Simply put, MCL is calculated based on the difference between the power level at the transmitting antenna and the one at the receiving antenna port. The directional gain of the antenna does not have any effect on MCL calculations. MCL can be expressed as:

$$\text{MCL} = P_T - (\text{SINR} + \text{Noise figure} + \text{Noise floor}) \tag{7.3}$$

where P_T is the transmit power of an IoT device or a base station. It is evident that a base station has a higher transmit power (P_T) as compared to the IoT device. Signal to Interference and Noise Ratio (SINR) is the signal to noise ratio. The value of SINR depends on the communication link between the transmitter and the receiver. Noise figure is based on the receivers' frontend Low Noise Amplifier (LNA). The noise floor is impacted by a number of parameters, including signal bandwidth. Noise floor can be calculated based on Eq. (7.4). The sum of SINR and noise floor is called receiver sensitivity.

$$\text{Noise floor} = -174 + 10Log_{10}(\text{bandwidth}) \tag{7.4}$$

Another metric to evaluate the coverage of a radio access technology is Maximum Path Loss (MPL), which calculates the difference in radiated power levels at the transmitting and receiving antennas. Considering G_T and G_R as the directional antenna gain of the transmitter and receiver, respectively, MPL can be calculated as:

$$\text{MPL} = P_T + G_T + G_R - (\text{SINR} + \text{Noise figure} + \text{Noise floor}) \tag{7.5}$$

Figure 7.8 shows the difference between MCL and MPL for a smart meter application where a smart meter sends its data to the Internet. Smart meters and other IoT devices may require more extensive network coverage, and they should operate at a higher MCL, or MPL, than regular cellphones. For example, the cellular technology selected for a smart home application should be capable of providing sufficient coverage for smart meters throughout the house, including basements.

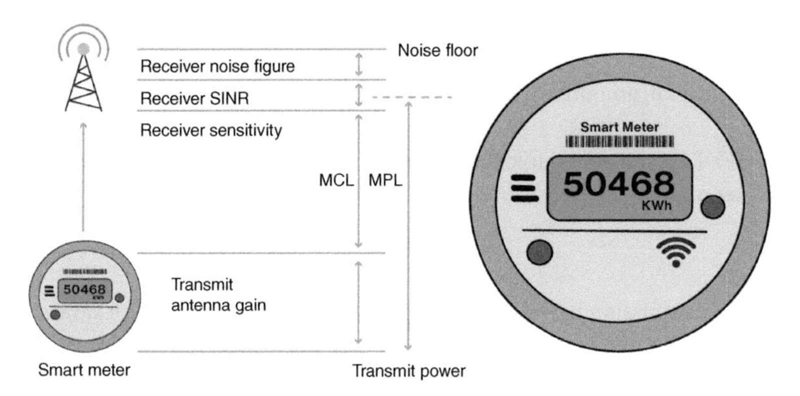

Figure 7.8 Cellular-based smart meter.

LTE-M defines two Coverage Enhancement (CE) modes, called CE mode A and CE mode B, in order to increase coverage of the LTE technology for IoT applications. The basic idea behind these two CE modes is the repetition technique. In this technique, both data and control signals are repeated multiple times over a period of time. For instance, CE mode A supports the repetition of data up to 32 times, and CE mode B supports the repetition of data up to 2048 times. The use of repetitions increases coverage but decreases the data rate. The default mode for operation of IoT devices using LTE-M is CE mode A. CE mode B provides better coverage but at the expense of throughput and latency. Therefore, CE mode B is more suited to stationary or very low speed applications that require low data rates and very low data transmission. One example of using CE mode B is for a smart meter that is installed in a basement.

In 3GPP Release 13, LTE can normally operate at an MCL of approximately 142 dB. The maximum MCL for LTE-M with CE modes A and B is 155 dB and 164 dB, respectively.

Another technique for extending coverage is called Power Spectral Density (PSD) boosting. While the serving base station can increase its transmit power in order to extend its coverage in the downlink direction, an IoT device can also put all its power on a smaller bandwidth to substantially increase the transmit power density.

3GPP Release 14 and 15 adds small features for CEs, but no major modifications are added.

7.3.1.6 Mobility

In many IoT applications, such as asset tracking, mobility is an important feature. After an IoT device connects to a cell, it also monitors its neighboring cells. If an IoT device finds a stronger signal from an adjacent cell, the reselection mechanism kicks in, and the IoT device connects to the base station that has the stronger signal.

Because LTE-M supports handover, if an IoT device is in the middle of a data transaction while moving, it will maintain its connection. In CE mode B, however, LTE-M does not support handover. For CE mode A, 3GPP Release 14 and 15 support use cases associated with higher velocity than those in 3GPP Release 13. A detailed discussion of the mobility of CIoT technologies is provided in Subchapter 8.4.

7.3.2 NB-IoT

Since there are many IoT applications that require sending small amounts of data (lower than what can be transferred by LTE-M), a special category, namely NB-IoT, has been incorporated into 3GPP Release 13 specifications in order to support these IoT applications.

7.3.2.1 Channel Bandwidth and Duplexing

NB-IoT reduces the channel bandwidth to 200 KHz (180 KHz plus a guardband). Since NB-IoT is used for IoT devices that transmit small amounts of data infrequently, its bandwidth did not change in later 3GPP releases.

NB-IoT defined half-duplex FDD as the only duplexing method in 3GPP Release 13 and 14. Support for TDD is introduced later on in 3GPP Release 15.

7.3.2.2 Data Rate and Latency

The downlink and uplink data transmission in NB-IoT is very similar to LTE-M. Figure 7.9 shows the downlink transmission. The values of T_D, T_{DUS}, and T_{UDS} for NB-IoT are different from the ones for LTE-M. These values are 4 ms, 12 ms, and 3 ms for the downlink transmission and 4 ms, 8 ms, and 3 ms for the uplink transmission, respectively.

In 3GPP Release 13, the maximum TBS value in the downlink and uplink directions is 680 bits and 1000 bits, respectively. NB-IoT in 3GPP Release 13 allows only one HARQ process. 3GPP Release 14 supports 2536 bits as the maximum TBS value in both the uplink and downlink directions using one HARQ. It also introduces optional support for two HARQ processes with 1352 bytes of data in the uplink direction and 1800 bytes in the downlink direction. For 3GPP Release 14, the downlink peak data rate for NB-IoT devices is 80 kbps and the supported uplink peak data rate is 105 kbps. A summary of NB-IoT specifications in various 3GPP releases is shown in Table 7.4.

In general, NB-IoT is designed for applications that are latency tolerant. However, the technology has defined 10 s as its maximum possible latency.

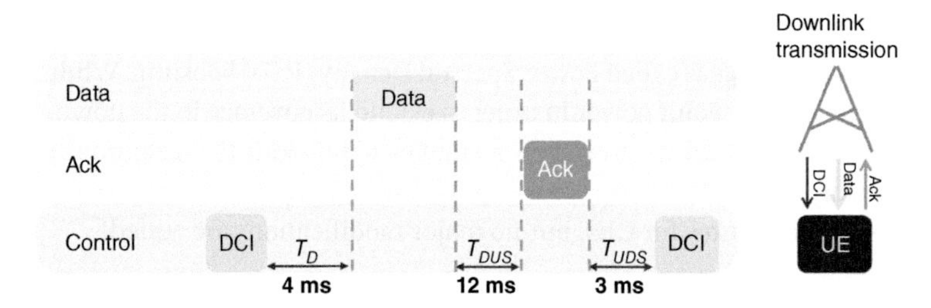

Figure 7.9 DL transmission in NB-IoT.

Table 7.4 Summary of NB-IoT specifications in various 3GPP releases.

	Cat-NB1 Release 13	Cat-NB2 Release 14	NB-IoT Release 15
Maximum DL TBS (bits)	680	2536	2536
Maximum DL data rate (Mbps)	0.025	0.127	>0.127
Maximum UL TBS (bits)	1000	2536	2536
Maximum UL data rate (Mbps)	0.062	0.159	>0.159
Maximum HARQ processes	1	1 or 2	2

Source: Based on [12].

7.3.2.3 Power Classes

A power class is the indication of the maximum power that an IoT device can transmit. In 3GPP Release 13, an NB-IoT device supports the 20 dBm power class. The 14 dBm power class was added in 3GPP Release 14.

7.3.2.4 Coverage

For applications where the IoT device is located in challenging and hard-to-reach areas, NB-IoT can enhance coverage using the repetition technique. NB-IoT can provide 164 dB of MCL, which is 20 dB more than what legacy LTE can provide. The maximum allowed repetitions for NB-IoT are 128 and 2048 repetitions for uplink and downlink, respectively.

3GPP Release 14 and 15 adds small features for CEs, but no major modifications are added. For example, in 3GPP Release 14, the network can make some restrictions for using the CE modes by an IoT device. The CE modes use repetitions and demand a substantial amount of network resources to be used for sending the same data. The base station can restrict the IoT device to use a CE mode. This can be very useful in high traffic situations where the resources used by an IoT device in a CE mode can be used by many other IoT devices that do not need any repetition.

7.3.2.5 Mobility

NB-IoT is suitable for IoT applications that use stationary IoT devices or the ones that have limited mobility inside a cell. Generally speaking, NB-IoT does not support handover between cells. When an IoT device moves out of a cell and enters into a new cell, the device loses its connection and experiences Radio Link Failure (RLF). It is possible for an NB-IoT device to try to connect to the new cell in order to continue its data transfer. However, for NB-IoT, full mobility and handover is not defined.

7.4 Practical Use Cases

In this section, we bring three scenarios to discuss the use of LTE-M and NB-IoT in different applications.

Scenario 7.1 Investigating cellular coverage

A forestry company intends to install 1500 generic sensors to read the vibration information of its equipment in one of its remote pulp mills. There is no cellular coverage in this remote location. However, if signals from a cellular network could reach this remote location, the maximum conducted loss would be around 160 dB. The sensors are configured to transmit 100 bytes of data every 15 minutes. Let us discuss the use of CIoT technology for this application.

In general, the coverage enhancement techniques that are available in both LTE-M and NB-IoT are not primarily intended for achieving coverage in remote forests where cellular coverage is not possible but for achieving coverage in challenging areas such as basements. This means that there must be decent cellular macro coverage available in the area but perhaps not exactly where the device is located. Therefore, none of the existing technologies can be considered a viable solution. In this application, first the data from the equipment needs to be transmitted to an IoT gateway where a decent cellular macro coverage is available. Then the IoT gateway can transmit the data to the cloud using a cellular technology.

Scenario 7.2 Investigating maximum data rate

A factory intends to install 1000 sensors to read the vibration and temperature of its equipment (e.g. oil pumps and motors). The oil pumps are installed in the basement in which MCL may reach up to 164 dB. The sensors are configured to transmit 15 000 bytes of data every 5 minutes. Due to the small size of the sensors, the IoT device that is connected to the oil pump is using a coin-size battery. In order to reduce the replacement cost for changing the battery, the sensors should transmit their data with minimum power so that the battery can last longer. Let us discuss which technology and which 3GPP release number works better for this application.

In this scenario, the uplink data rate is calculated as $\frac{15000 \times 8}{5 \times 60} = 400$ bps. Figure 7.10 shows the LTE-M Release 13 data rate in terms of the MCL in extended coverage regions. From this figure, one can see that the maximum possible data rate at MCL = 164 is around 200 bps, which is lower than the required uplink data rate. And therefore, it does not satisfy the required data rate within the coverage region. LTE-M in 3GPP Release 14 can use 5 MHz bandwidth instead of 1.4 MHz and achieve higher data rates and would be a good choice for this application. NB-IoT is not a good choice for this application, since its supported data rate is lower than the LTE-M network.

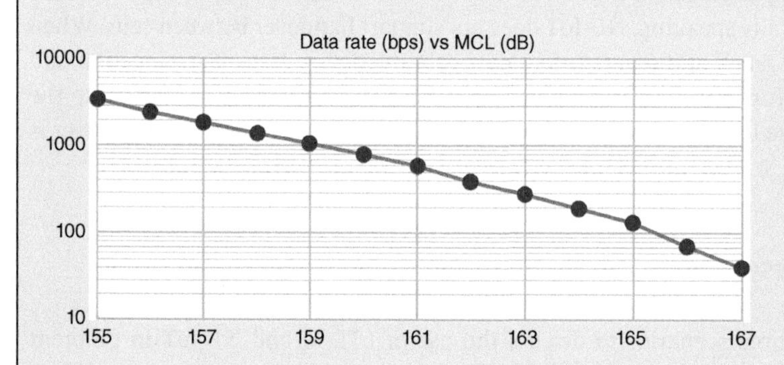

Figure 7.10 Data rate versus MCL for the LTE-M network. Source: Modified from Sierra Wireless [13].

Scenario 7.3 Investigating cellular radiation and safety

A mining company located in an area with poor cellular coverage is looking to increase its worker safety by implementing a suitable technology. The company provides its employees with wearable devices that are equipped with sensors to monitor the environment (e.g. CO_2) as well as vital health signs such as heart rate and body temperature. The wearable IoT device in this application is a wristband that uses LTE-M technology to transmit data that is collected from health and environment sensors. Let us discuss whether there is any safety concern on using these IoT devices or not.

As mentioned earlier, LTE-M defines three different power classes: 23, 20, and 14 dBm. 3GPP Release 13 supports 23 and 20 dB power classes, and the higher releases continue to support these two power classes. The 14 dBm power class is supported by 3GPP Release 14 and higher.

In areas with poor signal coverage (MCL of greater than 144 dB), an IoT device may use a large number of repetitions in enhanced coverage modes. As a result, one may think that the radiation (power density) absorbed by a user may go higher than the threshold levels defined by the International Commission on Non-Ionizing Radiation Protection (ICNIRP) standard. In other words, the increase in the coverage of LTE-M comes at the cost of increasing the duration of the Radio Frequency (RF) radiations at the transmitter antennas. However, from the far-field perspective, the radiation levels are similar to a smartphone, and therefore, it should not create any concern. As for the near-field, there will not be any concerns either, especially if the wearable IoT is a wristband that is not close to the brain. Therefore, it is completely safe to use LTE-M wearable devices for this application.

7.5 CIoT Frequency Bands

There are many chip manufacturers that make CIoT modules. To find the best module for an IoT project, there are several factors that need to be considered such as supported frequency bands; standard release number that demonstrates supported data rate, latency, coverage, and other features; certification; supported protocols; and supported physical Input-Output (I/O) by the chip.

A CIoT module may operate in a specific frequency band or a set of frequency bands. If a CIoT device supports only one fixed frequency band that is not supported by any service provider in that region, then the CIoT device cannot operate in that region. The frequency bands used by the telecommunication companies for CIoT in different countries or regions are quite different from each other. Two service providers in the same region may use different frequency bands. These frequency bands are often represented by Bxx, where xx shows the band number. Each frequency band supports a specific spectrum for the uplink and downlink transmission as well as a specific duplexing method. The bandwidth supported by different frequency bands is not the same. Table 7.5 lists some information about some of the existing frequency bands used by CIoT modules. The information includes the uplink and downlink frequency range, the available bandwidth, the duplexing mode, supported 3GPP release number, supported technology (NB-IoT or LTE-M), and the region that the frequency band is used. For instance, B2 uses a frequency range of 1850–1910 MHz for its uplink transmission, while 1930–1990 MHz is used for the downlink transmission. The available bandwidth of B2 in the uplink direction is 60 MHz. The bandwidth for this frequency band in the downlink direction is also 60 MHz. Therefore, the total bandwidth is 120 MHz. In North America, this band is used by service providers for both LTE-M and NB-IoT for 3GPP Release 13 and higher. From Table 7.5, it can be found that LTE-M in 3GPP Release 14 utilizes three new frequency bands of B25, B40, and B70. LTE-M also allocates B70, B71, B72, B73, B74, and B85 to 3GPP Release 15. NB-IoT in 3GPP Release 15 has also added six new bands (B5, B14, B71, B72, B73, B74, and B85). Also, NB-IoT adds four frequency bands to 3GPP Release 14. These bands are B11, B25, B31, and B70. An NB-IoT module that only supports 3GPP Release 13 does not operate in the frequency bands that are added for 3GPP Release 14, such as B11.

Service providers usually use multiple frequency bands. The frequencies used by LTE-IoT technologies exist in a spectrum from frequencies 400 up to 2600 MHz as shown in Table 7.5. Service providers use higher frequencies for densely populated areas. By using these frequencies, service providers deliver large amounts of data at high speeds across short distances. Due to the high attenuation of higher frequencies, this range of frequencies does not perform well over long distances. By the same logic, service providers use lower frequencies for rural areas, countryside, or any sparsely populated region. Low-frequency bands do not provide the data capacity of higher

Table 7.5 Existing frequency bands used by the CIoT modules.

Frequency band	Frequency (MHz)	Uplink (MHz)	Downlink (MHz)	BW (MHz)	Duplexing mode	Release LTE – M	Release NB – IoT	Regions
B1	1920/2100	1920/1980	2110 – 2170	60	FDD	R13	R13	Global
B2	1900	1850 – 1910	1930 – 1990	60	FDD	R13	R13	North America
B3	1800	1710 – 1785	1805 – 1880	75	FDD	R13	R13	Global
B4	1700/2100	1710 – 1755	2110 – 2155	45	FDD	R13	R15	North America
B5	850	824 – 849	869 – 894	25	FDD	R13	R15	North America
B7	2500/2600	2500 – 2570	2660 – 2690	70	FDD	R13	–	Europe, Middle East, Africa
B8	900	880 – 915	925 – 690	25	FDD	R13	R13	Global
B11	1400	1427.9 – 1447.9	1475.9 – 1495.9	20	FDD	R13	R14	Japan
B12	700	699 – 716	729 – 746	17	FDD	R13	R13	North America
B13	700	777 – 787	746 – 756	10	FDD	R13	R13	North America
B14	700	788 – 798	758 to 768	10	FDD	R14	R15	North America
B17	700	704 – 716	734 – 746	12	FDD	R15	R13	North America
B18	850	815 – 830	860 – 875	15	FDD	R13	R13	Japan
B19	800	830 – 845	875 – 890	15	FDD	R13	R13	Japan
B20	800	832 – 862	791 – 821	30	FDD	R13	R13	Europe, Middle East, Africa
B25	1900	1850 – 1915	1930 – 1995	65	FDD	R14	R14	North America
B26	800	814 – 849	859 – 894	35	FDD	R13	R13	North America
B27	800	807 – 824	852 – 869	17		R13	–	North America
B28	700	703 – 748	758 – 803	45	FDD	R13	R13	Europe, Asia
B31	400	452.5 – 457.5	462.5 – 467.5	5	FDD	R13	R14	Global
B39	1900	1880 – 1920	1880 – 1920	40	TDD	R13	–	China
B40	2300	2300 – 2400	2300 – 2400	100	TDD	R14	–	Asia, Europe
B41	2600	2496 – 2900	2496 – 2600	194	TDD	R13	–	Global
B66	1700/2100	1710 – 1780	2110 – 2200	90	FDD	R13	R13	North America
B70	1700/2000	1695 – 1710	1995 – 2020	25	FDD	R14	R14	North America
B71	600	633 – 698	617 – 783	65	FDD	R15	R15	North America
B72	400	451 – 456	461 – 466	5	FDD	R15	R15	Europe, Middle East, Africa
B73	400	450 – 455	461 – 466	5	FDD	R15	R15	North America
B74	1400	1427 – 1470	1475 – 1518	43	FDD	R15	R15	North America
B85	700	698 – 716	728 – 746	10	FDD	R15	R15	North America

Table 7.6 Some of the frequency bands used for CIoT by three major service providers in Canada.

Service provider	LTE frequencies (MHz)	Bands
Telus, Bell, Rogers	700	12, 13, 17
	850	5
	1700/2100	4
	1900	2
	2600	7

frequency bands, but perform substantially better when it comes to traveling long distances or penetrating into walls and buildings. When using middle range frequencies between 400 and 2600 MHz, such as 1800 MHz, the service providers try to find a balance in factors such as speed, data capacity, and distance. Service providers usually offer a select range of low, medium, and high-frequency services to suit the varying needs of their subscribers. Since different service providers tend to service different areas, with some focusing more on densely populated cities and others servicing suburbs or even rural areas, they offer various LTE frequencies and bands. For example, some of the LTE frequency bands deployed by the three major Canadian service providers are shown in Table 7.6.

To connect to the Internet, each CIoT module supports one or several communication protocols such as HyperText Transfer Protocol (HTTP), HyperText Transfer Protocol Secure (HTTPS), Message Queue Telemetry Transport (MQTT), and Constrained Application Protocol (CoAP). These protocols will be discussed in Chapter 9.

Each CIoT module also provides several I/O interfaces. Besides interfaces such as reset and antenna connection that exist in any CIoT module, the module may have other interfaces such as Universal Asynchronous Receiver/Transmitter (UART), General Purpose Input/Output (GPIO), and Universal Serial Bus (USB).

CIoT modules should have an interface for connection to a physical Subscriber Identification Module (SIM) card or interfaces for connection to an embedded SIM (eSIM) or integrated SIM (iSIM). While both eSIM and iSIM replace a physical SIM card, the main difference between them is in their implementation. An eSIM uses a dedicated chip, while an iSIM is embedded in one of the chips on the module and can provide a better security. To provide secure identification, SIM cards are used by service providers to store profiles that authenticate a device. Physical SIM cards may not be the best solution for CIoT applications and using eSIM or iSIM technologies might be considered as better choices. As the number of CIoT devices grows, the cost of production, maintenance, and management of physical SIM cards grows. A physical SIM card is relatively large in size, and therefore, it occupies a large area of the physical device. It can easily be damaged or stolen. eSIM and iSIM have similar functionalities as physical SIM cards, but they actually will be mounted on the Printed Circuit Board of the physical device itself and they cannot be removed. eSIM is a chip that is soldered to the PCB of a CIoT device. iSIM is an integral part of a chip that already is on the IoT device, and therefore, it occupies only a small fraction of the area of that chip. The size of eSIM and especially iSIM would be substantially smaller than a physical SIM card, and they do not need to be installed at the edge of a CIoT device to be accessible. Both eSIM and iSIM can be configured remotely, and therefore, SIM data can be updated without changing the physical card. Both eSIM and iSIM can be reprogramed easily when there is a need for modifying the carrier, changing the

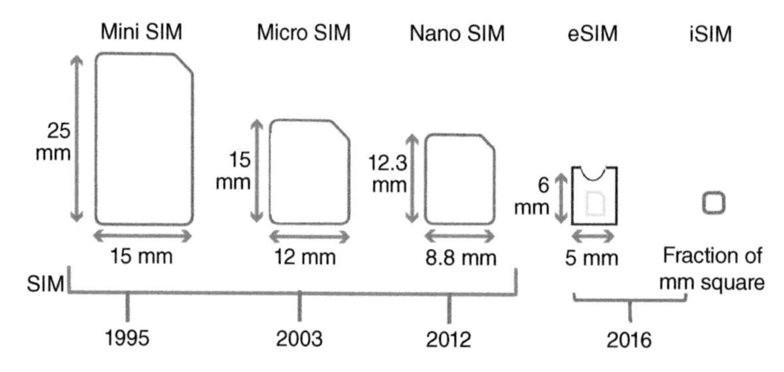

Figure 7.11 Size comparison of various physical SIM cards with eSIM and iSIM.

configurations and settings, or making some changes to the data plan. In asset tracking applications, using eSIM and iSIM, a single device can register with multiple operators. This way there is no need for international roaming that is costly and sometimes not available. A comparison of the physical size of a SIM card with eSIM and iSIM is shown in Figure 7.11.

7.6 Certification

Certification is another factor in choosing a CIoT module. Many CIoT chip manufacturers are trying to partner with local or global service providers, with the objective of increasing the sales of their CIoT modules and to strengthen their position in the CIoT chip market. As a result, a CIoT module might be in the list of certified modules supported by a service provider. In general, CIoT modules should not have an adverse effect on other wireless devices in their proximity. They must also be rigorously tested to work as intended when connected to the service provider's network. In situations where the selected CIoT module does not work properly on the network, the network operator and their customers may need to spend a lot of time to address and resolve possible technical problems and issues.

Since service providers usually provide SIM cards to their customers, a CIoT module is required to be certified by service provider. It should be noted that due to different configurations of cellular networks from one region to another, the service providers require additional testing to ensure that the CIoT module works correctly with their specific network configuration. In general, some CIoT deployments might be configured to provide specific features, and therefore, a CIoT module may need to be customized for that specific network. For this reason, developers would typically use CIoT modules that have been certified.

Generally speaking, there are three types of certifications that can be acquired by a CIoT manufacturer as shown in Figure 7.12. Regulatory certification such as Federal Communications Commission (FCC) or Network Access License (NAL), telecommunications-specific industry

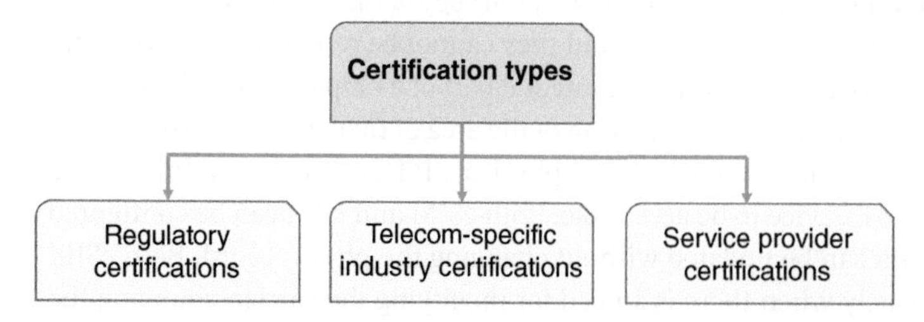

Figure 7.12 Various certification types.

certification such as Global Certification Forum (GCF), and PCS Type Certification Review Board (PTCRB) and service provider certifications such as the ones issued by AT&T or Verizon.

Regulatory certifications are usually mandatory for any types of electronic products that are sold or used in any specific market. Depending on the market, IoT module manufacturers might require one or several regulatory certifications. Each of these regulatory certifications may have a different approval process. FCC and NAL are examples of regulatory certifications. NAL certification is mandatory for telecommunication equipment that is exported to or sold in China, while FCC certification is a regulatory agency that regulates wired and wireless telecommunications across the United States.

Telecommunications-specific industry certifications ensure that the equipment functionalities conform to telecommunications industry standards. This can help to ensure interoperability among various equipment. GCF and PTCRB are two telecommunications-specific industry certifications. PTCRB is used in North America, while GCF is used worldwide.

Network operators may need to perform more testing specific to their network configurations in order to approve a CIoT module. Network operators usually have a white list of IoT modules that have been certified with their network.

7.7 CIoT Modules

CIoT modules are manufactured by more than 40 companies. These companies invest a significant amount of money to bring these modules to the market. Their investment includes the cost of the design, development, testing, initial market trials, and certification of the module. The fact that there exists a large number of companies competing in this market indicates that there is a huge market for these modules. The main players in CIoT module design are Fibocom, Gemalto M2M, Murata, Quectel, Sequans, Sercomm, Sierra wireless, SIMCom, Telit, U-BLOX, Wistron Neweb, Qualcomm, Samsung, Altair (Sony), MediaTek, HiSilicon, Nordic Semiconductor, Cheerzing, Digi, and Skyworks.

Let us consider a transportation company that has equipped all its buses to be connected to the cloud using CIoT technology. For this purpose, a system has been installed on each bus. According to GSMA TS.34 [14] terminology, and as it is shown in Figure 7.13, the bus in this example is

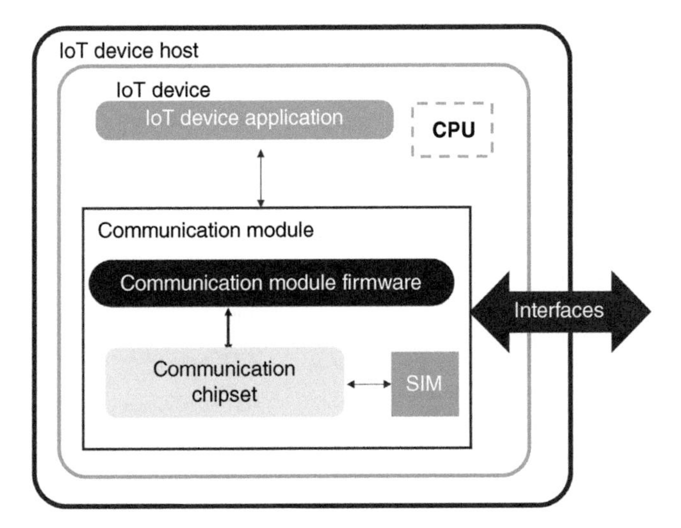

Figure 7.13 IoT device host and IoT device components. Source: Modified from GSMA [14].

considered an IoT device host. An IoT device host is an application-specific environment such as a connected bus. The IoT system installed on the bus is called an IoT device. This IoT device contains an IoT device application and a communication module. The IoT device application is a software program that controls the communications module and uses this module to interconnect to an IoT service platform. The communication module by itself consists of several components. These components are communication modem (communication chipset), SIM, and communication module firmware. Communications module firmware provides an Application Programming Interface (API) to the IoT device application and controls the communication modem. Communication modem provides connectivity to the mobile network.

Let us think about the requirements of an IoT device for a connected bus example. Certainly, a CIoT connectivity scheme that can be offered by a communication chipset is required. There might be a need for a cloud device agent to access cloud resources. There would also be a requirement for an application processor, which would allow the IoT application firmware and communication module firmware to perform their processing independently.

A CIoT module must have a communication module and one or several interfaces. Some CIoT module manufacturers also add a processor and an IoT cloud agent to their modules. If there is no IoT device application or processor onboard, the IoT device needs an external processor in addition to the CIoT module. However, if the CIoT module has a processor for application processing, then the CIoT module can be considered as an IoT device. Table 7.7 shows a list of some of the most certified CIoT modules from various manufacturers.

To develop their CIoT module, many CIoT module manufacturers take a module designed by another manufacturer and add some additional features, interfaces, or sensors to it. For example, Qualcomm MDM9205 has a dual LTE-M and NB-IoT modem, an ARM Cortex A7 processor, a GPS, many interfaces, a cloud agent, and an IoT Software Development Kit (SDK). This module has been used as the main component in the design of several other CIoT modules by other manufacturers.

Table 7.7 Some of the most certified CIoT modules.

Manufacturer	Country	Module	Type	Frequency bands	Certifications
Qualcomm	USA	MDM9205	Dual	1, 2, 3, 4, 5, 8, 12, 13, 18, 19, 20, 26, 28, 39	GCF, CE, CCC, FCC, PTCRB, Verizon, AT&T, RCM, Telstra, IFETEL, IC
Quectel	China	BG96	Dual	1, 2, 3, 4, 5, 8, 12, 13, 18, 19, 20, 26, 28, 39	GCF, CE, CCC, FCC, PTCRB, Verizon, AT&T, RCM, Telstra, IFETEL, IC
Sierra wireless	Canada	HL7802 (Tri-mode)	Dual	1, 2, 3, 4, 5, 8, 9, 10, 12, 13, 14, 17, 18, 19, 20, 25, 26, 27, 28, 65, 66	GCF, FCC, PTCRB, IC, AT&T, Verizon, Vodafone, Telstra
Nordic	Norway	nRF9160 SiP	Dual	1, 2, 3, 4, 5, 8, 12, 13, 14, 17, 18, 19, 20, 25, 26, 28, 66	Verizon, GCF, PTCRB, FCC, CE, ISED, ACMA RCM, NCC, IMDA, MIC, MSIP
Sercomm	Taiwan	TPB41	NB-IoT	3, 5, 8, 20	CCC, SRRC, NAL
Telit	England/ Italy	ME866A1-NA	LTE-M	2, 4, 12	AT&T, FCC, PTCRB
Huawei	China	ME309-562	LTE-M	2, 4, 12, 13	FCC, IC, PTCRB, AT&T, Verizon
Fibocom	China	M910-GL	Dual	1, 2, 3, 4, 5, 8, 12, 13, 19, 20, 26, 28, 39	CE-RED, GCF, FCC, PTCRB, VZW, AT&T, SRRC, NAL, CCC, RCM
Gemalto	Netherland	EMS31-X	LTE-M	2, 4, 12, 13	Verizon, AT&T, FCC, GCF, IC, UL, PTCRB, California RoHS
U-blox	Switzerland	SARA-R410M-02B	Dual	1, 2, 3, 4, 5, 8, 12, 13, 17, 18, 19, 20, 25, 26, 28, 39	FCC, ISED, IFETEL, GCF, RCM, CCC, SRRC, NCC, RED, PTCRB, Verizon, AT&T, Telstra, GMA
Altair (Sony)	Japan	ALT1250	Dual	1-5, 8–10, 12-14, 17–20, 25–28, 66, 71, 85	AT&T, KDDI, Verizon
Sequans	France	Monarch GM01Q	LTE-M	1, 2, 3, 4, 5, 8, 12, 13, 14, 17, 18, 19, 20, 25, 26, 28, 66	AT&T, KDDI, NTT Docomo, Orange, Telstra, GCF, PTCRB, FCC, IC, ACMA, JATE, TELEC

Table 7.8 Qualcomm MDM9205-based CIoT modules.

Manufacturer	IoT module
Telit	ME910x1 & xL865
Quectel	BG95 & BG77
Fibocom	MA510
Gemalto	EXS82
Gosuncn	GM100
MeiG	SLM156
Neoway	N27 and N28
SIMCom	SIM7070G & SIM7080G
Thundercomm	TurboX T95

Table 7.8 shows some of the manufacturers that are using Qualcomm MDM9205 as part of their CIoT module.

Interested readers are encouraged to take a look at Assignment #3 and Assignment #4 that are hands-on projects on this topic.

7.8 AT Commands

The communication chipset of an LTE-M or NB-IoT module can be considered a modem device. AT commands can be used to configure and control a modem. These commands are text-based instructions that start with "AT" in order to inform a modem about the start of a command. The AT commands can be combined to control or monitor the operation of a modem. These commands can be classified into two types: basic commands and extended commands. Extended commands are the ones that start with "+."

In general, a modem has two types of connections as shown in Figure 7.14. These two connections are the connection between a host and the modem and the connection between the modem and the network. Host-to-modem connection is established through an interface, which is usually a serial or USB connection. To control the modem using AT commands, a connection between the host and modem needs to be established. The connection between the modem and the network can be set to either as always-on or as on-demand connection. Always-on connection means that the connection will be established whenever a service is available, and therefore, if the service is available, then the connection exists. On the other hand, the on-demand connection is established only when the host-to-modem connection is requesting it. Using AT commands, we can configure and control LTE-M and NB-IoT modems.

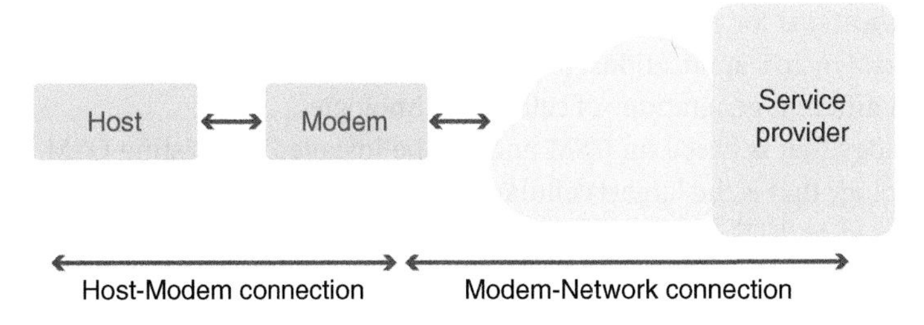

Figure 7.14 Modem connections.

In general, there are three parameters in determining the status of a modem. The first one is the modem condition. The condition of a modem can be online or offline. The second one is the state of a modem, which can be either data or commands. In the command state, the modem exchanges AT commands, while in the data state, the modem exchanges data with the host. The third parameter is called modem mode. A modem can be set up in a circuit switched or packet switched connection mode. These operating modes demonstrate how the network connection is established.

The communication between the host and the modem is buffered. When the modem is in command state, the modem buffers the input from the host until a Carriage Return (CR) control character is received. Most AT commands have one or more parameters. If an AT command does not have any parameters, it acts like a ping to the modem.

The AT command set has grown from the original set introduced by Dennis Hayes in 1981 and is now being used as part of the 3GPP standard. The original AT command set was used for phone line operations such as dialing or connection tear down. Due to the absence of a standard for AT commands at the time, many modem manufacturers designed their modems or changed the format of the AT commands in such a way that made the modems incompatible when receiving the same AT commands. For instance, some modems require spaces between commands or a small delay after receiving the reset command.

AT commands can be classified into four categories based on their functionalities. These four categories are testing, reading, setting, and executing. The *set* AT commands are used for making modifications in the settings, the *read* AT commands are for reading the setting, The *test* AT commands are used for testing the compatibility, and the *execute* AT commands are used for executing specific operations. Some of the existing standards for AT commands are V.25, which includes commands for any standard modem, GSM 07.05 for Short Messaging System (SMS), GSM 07.07 for GSM and GPRS modems.

Interested readers are encouraged to take a look at Assignment #3, which is a hands-on project on this topic.

7.9 Summary

Cellular IoT provides an excellent connectivity scheme and an effective adaptation method that suits many IoT applications. Even though CIoT can use the existing 2G and 3G cellular networks, larger bandwidth, lower latency, and greater density are possible using the 4G technology. CIoT can be enhanced substantially in terms of capacity, coverage, density per cell, latency, and power consumption by using the 5G network, which also enables a suitable connectivity scheme for critical applications.

Third Generation Partnership Project (3GPP) is a global technical entity that intends to develop globally accepted technical specifications for mobile communication systems. 3GPP organizes its work in terms of release numbers. For IoT applications, 3GPP has developed technical specifications in its releases 13+ based on different generations of cellular technologies.

EC-GSM-IoT is an IoT technology that is based on GSM and can be installed on existing GSM deployments. GSM is a 2G technology that is the largest cellular technology in the world. LTE-M is another IoT technology that is based on LTE. This technology uses flexible channel bandwidth of 1.4MHz or more. Therefore, LTE-M is more suitable for applications that need higher throughput and lower latency as compared to what other IoT technologies can support. NB-IoT is another IoT technology that is also based on LTE. NB-IoT as the name suggests is a narrowband technology that operates in a narrow spectrum of 200KHz. Even though we briefly discussed EC-GSM technology, but the main focus of this chapter was on LTE-M and NB-IoT technologies.

There are many chip manufacturers that make CIoT modules. To find the best module for an IoT project, there are several factors that need to be considered such as supported frequency bands; standard release number that demonstrates supported data rate, latency, coverage, and other features; supported protocols; certifications obtained by manufacturers; and supported physical Input-Output (I/O) interfaces. We also explained the names and features of most popular CIoT modules and their manufacturers in Section 7.7.

The use of physical SIM cards, eSIM, and iSIM were described in this chapter. To provide secure identification, SIM cards are used by service providers to store profiles that authenticate a device. Physical SIM cards may not be the best solution for CIoT applications, and the use of eSIM or iSIM technologies might be considered as better choices. A physical SIM card is relatively large in size, and therefore, it occupies a large area of the physical device. It can easily be damaged or stolen. As the number of CIoT devices grows, the production, maintenance, and management of physical SIM cards becomes extremely costly. eSIM and iSIM have similar functionalities as a physical SIM card, but they actually will be mounted on PCB of the physical device itself, and they cannot be removed. eSIM is a chip that is soldered to the PCB of an IoT device. iSIM is an integral part of a chip that already exists on the IoT device, and therefore, it occupies only a small fraction of the area of that chip.

References

1 Third Generation Partnership Project (2016). Cellular System Support for Ultralow Complexity and Low Throughput Internet of Things. *Technical Report 45.820 v13.0.0.*

2 Third Generation Partnership Project (2016). General Packet Radio Service (GPRS); Service Description. *Technical Specification 23.060 v14.0.0-stage 2.*

3 Third Generation Partnership Project (2016). Mobile Radio Interface Layer 3 Specification; Core Network Protocols. *Technical Specification 24.008 v14.0.0-stage 3.*

4 Third Generation Partnership Project (2016). Evolved Universal Terrestrial Radio Access (E-UTRA); Physical Channels and Modulation, *Technical Specification 36.211 v14.0.0.*

5 Third Generation Partnership Project (2016). Evolved Universal Terrestrial Radio Access (E-UTRA); User Equipment (UE) Radio Access Capabilities. *Technical Specification 36.306 v14.0.0.*

6 Third Generation Partnership Project (2017), Technical Specification Group Radio Access Network; Study on Scenarios and Requirements for Next Generation Access Technologies; (Release 14). *Technical Report 38.913, v14.2.0.*

7 Third Generation Partnership Project (2016). Evolved Universal Terrestrial Radio Access (E-UTRA) and Evolved Universal Terrestrial Radio Access Network (E-UTRAN); Radio Resource Control (RRC); Protocol Specification. *Technical Specifications 36.331 v13.3.0.*

8 Third Generation Partnership Project (2016). Cellular system support for ultra-low complexity and low throughput Internet of Things (CIoT). *TR 45.820.*

9 Liberg, O., Sundberg, M., Wang, E. et al. (2017). *Cellular Internet of Things: Technologies, Standards, and Performance.* Academic Press ISBN: 978-0-12-812458.

10 Kanj, M., Savaux, V., and Guen, M.L. (2020). A Tutorial on NB-IoT Physical Layer Design. *Communications Surveys and Tutorials, IEEE Communications Society* 22 (4): 2408–2446.

11 GSM frequency bands. https://en.wikipedia.org/wiki/GSM_frequency_bands (accessed 30 June 2022).

12 Dian, F.J. and Vahidnia, R. (2020). LTE IoT technology enhancements and case studies. *IEEE Consumer Electronics Magazine.* https://doi.org/10.1109/MCE.2020.2986834.

13 Sierra Wireless (2017). Coverage Analysis of LTE-M Category-M1-version 1.0. https://www.altair-semi.com/wp-content/uploads/2017/02/Coverage-Analysis-of-LTE-CAT-M1-White-Paper.pdf (accessed 30 June 2022).

14 GSMA (2021). IoT Device Connection Efficiency Guidelines-version 7.1. Official Document TS.34. https://www.gsma.com/newsroom/wp-content/uploads//TS.34_v7.1.pdf (accessed 30 June 2022).

Exercises

7.1 Can we say that 3GPP is a global standardization body?

7.2 Is 3GPP Release 15 just about 5G technology? Does 3GPP Release 15 cover anything in regard to the 4G cellular networks?

7.3 Why did 3GPP Release 13 consider EC-GSM-IoT as a technology for IoT?

7.4 Why is 14 dBm added as a power class to LTE-M in 3GPP Release 14 and higher?

7.5 An operator deploys LTE with 20 MHz channel spectrum in both downlink and uplink. If the operator uses the spectral efficiency of 1.5 bps/Hz in downlink, what is the maximum downlink data rate?

7.6 Why the use of the operator-certified IoT modules are always recommended? Remember, you can request your operator for the list of their approved modules.

7.7 Why do B39-B41 frequency bands have the same uplink and downlink frequencies?

7.8 A mobile vacuum cleaner sends its data to the Internet during the period that it vacuums a house. Since the vacuum cleaner is not stationary, do you think that NB-IoT can be used in this application?

7.9 What is the main advantage of NB-IoT over LTE-M? What is the main advantage of LTE-M over NB-IoT?

7.10 The common 3V lithium batteries (such as 2032) are 230 mAh. This means that if a constant current of 0.2 mA is drawn, the battery works for 1150 hours. Can we say that the battery works for 2.3 hours, if the current is 100 mA? Do you think a 20 dBm NB-IoT device can use coin-cell batteries? Why?

7.11 In which cellular IoT technology (LTE-M or NB-IoT) the number of possible HARQ processes is more? Why?

7.12 Does NB-IoT support full duplex FDD as part of its duplexing scheme?

7.13 NB-IoT is used as a connectivity technology for an IoT device that needs software updates on a regular basis. The software is in the range of 200–300 Kbytes. The device needs to perform software updates in less than one minute. Can this be done with NB-IoT in 3GPP Release 13? What about 3GPP Release 14?

7.14 In Figure 7.10, assume that data is sent in a 680-byte TBS, and its duration is 3ms. What is the data rate during the transmission of TBS? Can we say that the data rate during the transmission of TBS is the data rate that NB-IoT can support?

7.15 Do unlicensed LPWAN technologies work under the same regulatory bodies around the world?

Advanced Exercises

7.16 What is the difference between LTE-M Cat M1 in 3GPP Release 13 and LTE-M CAT M1 in 3GPP Release 14?

7.17 Graph the frequency band of the existing CIoT modules as tabled in Table 7.5 versus frequency.

7.18 LTE-M CE mode A supports up to 32 repetitions of data, and CE mode B supports up to 2048 repetitions of data. Under certain situations (accurate channel estimation and frequency tracking), we can assume that there is a linear relationship between the number of repetitions and coverage gain. By doubling the repetitions, approximately 3 dB coverage gain can be achieved. LTE-M mode A has an MCL of 155, and Mode B can achieve an MCL of 164 dB. Do you think that it is possible to do very accurate channel estimation and tracking for LTE-M?

7.19 How many new frequency bands have been added for NB-IoT in 3GPP Release 14?

7.20 An IoT module only supports B27 frequency band. Can this be an NB-IoT module? Can it be used in Europe?

7.21 If an IoT module has telecommunications-specific industry certifications, why is it possible to need an operator certification as well?

7.22 For every dB the TX power is decreased, there is a one dB decrease in MCL. If the supported MCL for a 23 dBm LTE-M IoT device is 164 dB, what is the MCL value for a 14 dBm LTE-M device?

7.23 What is the percentage area reduction of eSIM as compared to MiniSIM?

7.24 ALT1250 module from Altair (Sony) supports the following frequency bands B1-B5, B8-B10, B12-B14, B17-B20, B25-B28, B66, B71, and B85. Do you think that this module can support 3GPP Release 14?

7.25 IoT modules usually support several frequency bands and cellular technology.

a. Why do most IoT modules support several frequency bands?
b. Why do many IoT modules support dual LTE IoT modes (LTE-M and NB-IoT)?
c. Why do some IoT modules support GSM in addition to LTE-M and NB-IoT?

7.26 Research the names of seven organizational partners of 3GPP.

7.27 Explain how LTE-M uses different modulation and coding schemes (MCS) to adapt to different radio conditions. In which radio condition the spectral efficiency is maximum?

7.28 What method of duplexing is shown in Figure 7.E28? What are the advantages of using this duplexing scheme?

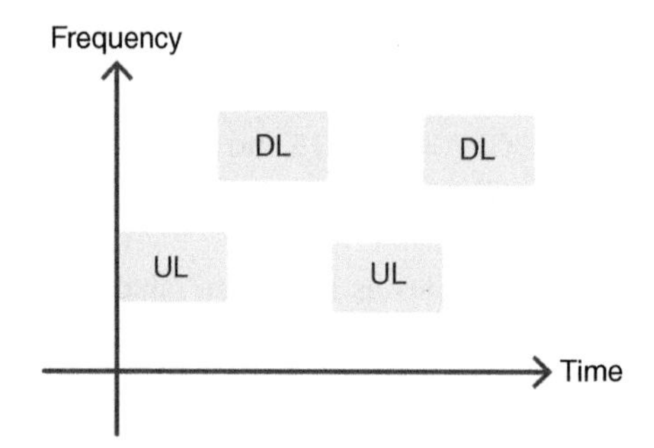

Figure 7.E28 Duplexing method in E2.28

7.29 Explain which coverage enhancement mode (mode A or mode B) is better for the LTE-M networks in each of the following situations:

a. A smart meter installed in a basement.
b. A fall detection device that, after detecting a fall, starts a voice communication channel with the person wearing the fall detection device.

7.30 Do all LTE-M modules support coverage enhancement mode A and mode B?

7.31 For downlink data transmission in NB-IoT (3GPP Release 13 and 14), the duration of the data packet is determined not only by the size of the data but also by the selected modulation and coding scheme, as well as TBS. The duration of data and maximum value of TBS for 12 different types of MCS are shown in Table 7.E31 for NB-IoT (3GPP Release 13 and 14). For example, if MCS = 4 and TBS = 680 bits, then the duration of the packet is 10 ms in both 3GPP Release 13 and 14. Based on this table, answer the following questions:

a. What is the maximum length of TBS in NB-IoT (3GPP Release 13) in downlink direction? What about 3GPP Release 14 with one HARQ process? What about Release 14 with two HARQ processes?
b. What happens if the packet size is smaller than TBS?
c. A packet with TBS = 680 bits needs to be sent in the downlink direction using NB-IoT (3GPP Release 13). Considering different MCS values, how many options for transmitting this packet exist?

Table 7.E31 Duration of data based on TBS and MCS in the downlink direction for NB-IoT in 3GPP Release 13 and 14.

	Duration (ms)							
MCS	1 ms	2 ms	3 ms	4 ms	5 ms	6 ms	8 ms	10 ms
0	16	32	56	88	120	152	208	256
1	24	56	88	144	176	208	256	344
2	32	72	144	176	208	256	328	424
3	40	104	176	208	256	328	440	568
4	56	120	208	256	328	408	552	680
5	72	144	224	328	424	504	680	872
6	88	176	256	392	504	600	808	1032
7	104	224	328	472	584	680	968	1224
8	120	256	392	536	680	808	1096	1352
9	136	296	456	616	776	936	1256	1544
10	144	328	504	680	872	1032	1384	1736
11	176	376	584	776	1000	1192	1608	2024
12	208	440	680	904	1128	1352	1800	2280
13	224	488	744	1128	1256	1544	2024	2536

Release 13   Release 14 Two HARQ processes   Release 14 one HARQ process

d. What is the maximum possible data rate for the downlink data transmission in NB-IoT (3GPP Release 13) based on Figure 7.9? Assume MCS = 4 and TBS is 680 bits. The duration of Ack is 1 ms.
e. Does your calculation in part (d) match the information provided in Table 7.5?
f. What is the maximum possible data rate for the downlink data transmission in NB-IoT (3GPP Release 14) with one HARQ process (based on Figure 7.9)? Assume MCS = 12 and TBS is 680 bits. The duration of Ack is 1 ms.
g. Does your calculation in part (f) match the information provided in Table 7.5?
h. What is the maximum possible data rate for the downlink data transmission in NB-IoT (3GPP Release 14) with two HARQ processes (similar to Figure 7.7), if MCS = 4 and TBS is 680 bits. Assume the duration of Ack is 1ms.

Chapter 8 CIoT Features

This chapter has been divided into four subchapters. Each of these subchapters discusses an important concept of Cellular Internet of Things (CIoT) in detail. These concepts include power saving schemes, uplink access methods, positioning capabilities, and mobility features of the CIoT technology.

In the first subchapter, we discuss power saving schemes implemented in LTE-IoT. Reducing power consumption of an IoT device is vital in order to ensure that CIoT can be utilized as a suitable connectivity scheme for many IoT applications. Most CIoT applications transmit small amounts of data, especially in the uplink direction, and they are more delay-tolerant compared to other IoT applications. However, CIoT devices are usually battery powered, and a long battery lifetime is an important design requirement. In order to reduce power consumption, an IoT device can turn off its radio and go to sleep mode, when it does not need to transmit or does not expect to receive any packet. Clearly, during this time, the device is unreachable. In this case, the IoT device receives the command or data from the base station when it activates its radio again. The trade-off between IoT device reachability and power consumption in IoT applications is usually in favor of reducing power consumption. However, different IoT applications have different requirements in terms of power consumption and delay.

In the second subchapter, we discuss uplink network access in CIoT. When a large number of devices need to communicate via a shared channel, there must be a process in place to ensure that communication among these nodes can be performed efficiently. By the same token, when there are many cellular nodes that want to connect to the same cell on the network, there must be a procedure to ensure efficient communications between each node and the cell. For example, there will be a collision if two or more IoT devices attempt to send data to a base station at the same time and on the same frequency. Access control is a process that controls communication and gives the right to each node in the network to transmit at a specific time or certain frequency in order to avoid collision. This is quite similar to when multiple people at a meeting try to speak at the same time. Access control acts as a moderator, ensuring that no one interrupts other participants during the meeting. There are many multiple access control protocols that can be used for controlling communication among multiple nodes in a network. In the second sub-chapter, we discuss these access control protocols, and explain how CIoT devices access a cellular network.

In the third sub-chapter, we discuss positioning methods in CIoT. Positioning is the process of determining the geographical location of an IoT device. Besides providing location information, positioning also enables us to determine the velocity of a mobile IoT device. The positioning process uses radio signals to estimate the location or velocity of a CIoT device. The radio signal can be a satellite signal, mobile radio signal, or a combination of both. There are many IoT use cases that need to provide Location Based Services (LBS), and therefore, require positioning estimation.

Fundamentals of Internet of Things: For Students and Professionals, First Edition. F. John Dian.

The IoT devices that are used for emergency services also usually need location information. It is clear that different IoT applications require different positioning accuracy. While some applications may need a positioning accuracy of a few meters or less, accuracy in the range of hundreds of meters or larger would be good for many applications.

The mobility process in CIoT is covered in the final subchapter. Because CIoT uses existing cellular networks, the signaling and features used for mobility in a CIoT device should be identical to those used in a smartphone system. However, CIoT mobility management differs from that of a cellular phone network. For example, CIoT devices are often battery powered, and in order to extend their battery life, they go into sleep mode and turn off their transmission and reception circuits. This means sending less mobility signals uplink and not always receiving downlink signaling information, such as mobility-related signaling messages. Due to specific requirements of IoT devices in terms of power consumption and device complexity, the mobility in LTE-IoT technologies is not exactly similar to the mobility provided by the LTE technology for smartphones.

Chapter 8.1 Low-power Consumption Schemes

8.1.1 Introduction

When a CIoT device is turned on, it needs to negotiate with the base station to join the network. The negotiations involve transmission of signaling information between the CIoT device and the base station. Clearly, the CIoT device needs to consume power to transmit signaling information. Generally speaking, we like CIoT devices to go to sleep mode when they do not need to transmit and do not expect to receive data. Assume that an IoT device becomes completely off in such a way that when it turns its radio on again, it needs to reconnect to the network. This may defeat the purpose of reducing power consumption. Because the amount of energy that is needed for transmitting signaling information to rejoin the network might be more than the amount of energy that is saved as a result of turning off the radio. In other words, turning off the IoT device for a short time may actually increase the power consumption, while turning off the IoT device for a long period of time reduces the power consumption at the cost of increasing the network latency. Therefore, it might be more beneficial to put the IoT device in an idle state instead of completely turning off its radio. By doing this, the reconnection to the network can be performed with less signaling.

A CIoT device in the idle state can activate and inactivate its reception from time to time. The frequency at which the IoT device inactivates reception has a direct relationship with latency. Therefore, it is a good idea to allow the CIoT device to choose this frequency based on the maximum amount of latency that it can tolerate.

In many applications, when an IoT device connects to the network and sends its data to the base station, it might be beneficial to stay connected and frequently listen to possible downlink transmissions from the base station. In this situation, the frequency of inactivity usually is lower than when the device is in the idle state.

There is always a trade-off between the battery lifetime and device reachability. This always needs to be taken into account in IoT solution developments. Due to the existence of this trade-off, IoT developers need to optimize power consumption for each individual application.

3GPP publications explain the specifications of the power-saving methods employed in LTE-IoT. The details of these power-saving strategies are covered in numerous 3GPP documents. Examples of these documents are [1–5].

8.1.2 Power Saving Techniques in 3GPP Release 13

When a CIoT device, also called a User Equipment (UE), is powered on, it enters what is called a Radio Resource Control (RRC) idle mode. The UE at this point does not have an established

Fundamentals of Internet of Things: For Students and Professionals, First Edition. F. John Dian.

physical connection to the base station and needs to set up a connection. After the network connection is established, UE enters what is called an RRC connected mode. In this mode, UE can access the network and request communication resources. For saving power, the base station can release the RRC connection that causes UE to move to the RRC idle mode again. By doing that, UE may later resume to the RRC connected mode and avoid the connection setup. In other words, UE does not need to perform network entry procedures such as scanning, authentication, security key exchange, or obtaining an IP address from the network. This saves a considerable amount of signaling overhead, and UE can also avoid renegotiating security. The RRC modes of operation, and transition between the RRC idle and connected modes are shown in Figure 8.1.1.

In 3GPP Release 13, when the RRC connection is released, an LTE-M or NB-IoT device can use the extended/enhanced Discontinuous Reception (eDRX) or the Power Saving Mode (PSM) method in order to reduce its power consumption. These two methods are shown in Figure 8.1.2. It should be mentioned that a CIoT device may also use a combination of both methods.

The objective of both eDRX and PSM is to reduce the energy consumption of UE. eDRX achieves this objective by putting UE in the RRC idle mode and periodically checking for paging information and downlink messages. The process used by the network to send signaling information in order to find the location of a device is called paging. We will discuss paging in more detail in subchapter 8.4. Once UE has connected to a base station, it finds out about the paging cycle of the cell. During every paging cycle, the base station transmits a broadcast message in the downlink direction, which is called a paging event or a paging occasion. UE can wake up during the paging occasion in order to listen to paging messages.

In eDRX, no resources are allocated to UE, but UE continues periodically listening to broadcast information sent by the base station. If UE notices any data is available to be downloaded, it triggers

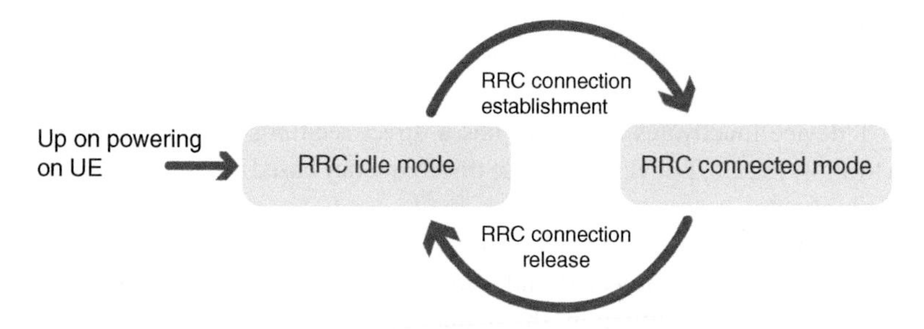

Figure 8.1.1 RRC connection establishment and release.

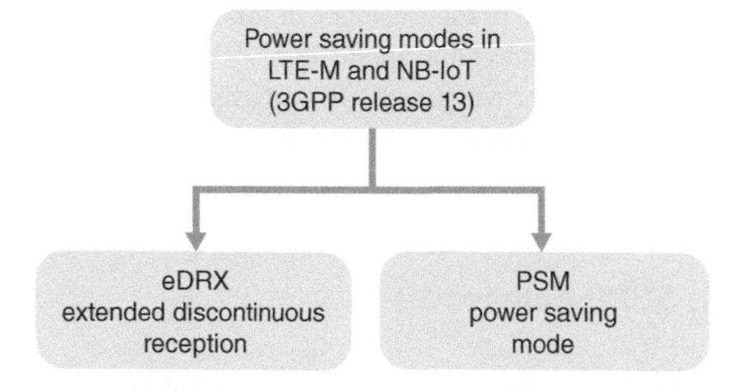

Figure 8.1.2 Power saving modes in LTE-M or NB-IoT (3GPP Release 13).

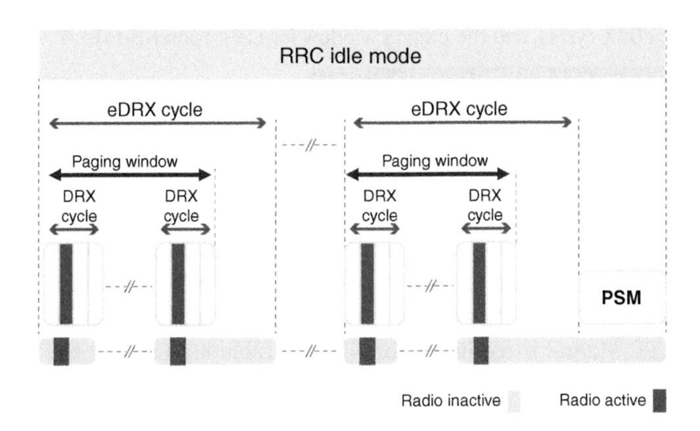

Figure 8.1.3 eDRX cycle and PSM mode in the RRC idle mode.

an RRC resume. This process involves substantially less signaling information than the signaling required for establishing a new connection. The eDRX method consists of several cycles that are called eDRX cycles as shown in Figure 8.1.3. During each eDRX cycle, there are several DRX cycles. It is clear that in the RRC idle mode, new resources cannot be requested by UE, but UE tracks the synchronization signals. When eDRX expires, UE starts the PSM process in which it turns off its radio completely and goes to deep sleep. Therefore, UE cannot be accessed by the base station anymore.

An eDRX cycle consists of two parts as shown in Figure 8.1.3. The first part is called the paging window, in which UE listens to paging messages from time to time. In legacy LTE, this part ranges from 2.56 to 40.96 seconds. In the second part, UE is not active. The duration of this part is from the end of the paging window to the end of the eDRX cycle.

During each DRX cycle, UE needs to monitor the downlink control channel in order to check whether there are any paging messages or not. If UE identifies any message, it needs to decode the Downlink Control Indicator (DCI) that carries the resource scheduling information.

Let us consider a situation where only one DRX cycle is used in a paging window as shown in Figure 8.1.4. In this case, the DRX cycle and the paging window are the same. It should be noted that the concept of the DRX power saving technique was introduced in 3GPP Release 12.

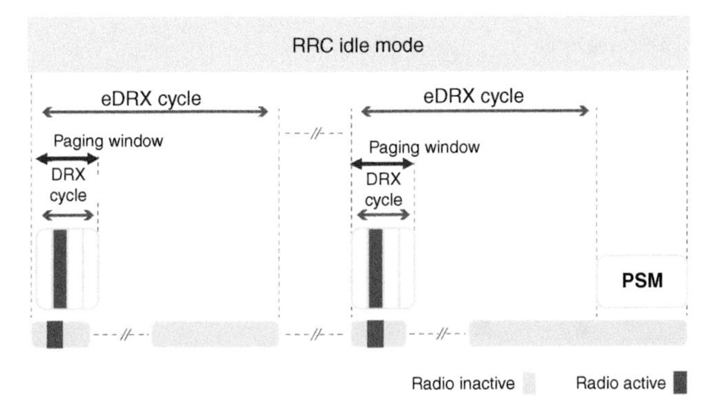

Figure 8.1.4 An example of an RRC idle mode with only one DRX cycle in its paging window.

Table 8.1.1 Possible values of DRX cycles, eDRX cycles, and the paging window for LTE-M and NB-IoT.

CIOT technology	DRX cycle (s)	eDRX cycle (s)	Paging window (s)
LTE-M	0.32, 0.64, ... , 2.56	5.12, 10.24, ... , 2621.44	1.28, 2.56, ..., 20.48
NB-IoT	1.28, 2.56, ... , 10.24	20.48, 40.96, ... , 10485	2.56, 5.12, ..., 40.96

The duration of a DRX cycle can vary from 0.32 to 2.56 seconds for LTE-M and from 1.28 to 10.24 seconds for NB-IoT. The duration of eDRX for LTE-M can be set to be from 5.12 to 2621.44 seconds and for NB-IoT can have a range from 20.48 to 10 486.76 seconds. Therefore, the maximum period for the eDRX cycle is 44 minutes for LTE-M, and three hours for NB-IoT. Table 8.1.1 shows the possible values of DRX, eDRX, and the paging window for both LTE-M and NB-IoT.

As mentioned earlier, long battery lifetime is one of the requirements of most CIoT applications. Therefore, the eDRX approach is very effective. Putting UE in the PSM mode further reduces the power consumption, but in the PSM mode, the radio is completely off and UE does not listen to the paging information.

In addition to the DRX cycle in the RRC idle mode as explained in this section, the connected DRX that is used in the RRC connected mode is another power saving strategy taken into account to reduce power consumption of a UE. Figure 8.1.5 shows DRX in both RRC idle and connected modes.

In the RRC connected mode, after an IoT device transmits its data, it goes to reception mode. In other words, when there is no data transmission, UE keeps its reception on for a while. It then enters the RRC connected mode DRX cycle (if this option is enabled). In LTE-M and NB-IoT, the maximum DRX cycles in the connected mode are 10.24 and 9.216 seconds, respectively. So, the duration of the maximum DRX in the RRC connected mode is substantially lower than the eDRX in the RRC idle mode. The DRX in connected mode is called Connected DRX (C-DRX), while the DRX in idle mode is called Idle DRX (I-DRX).

The eDRX process is controlled by a set of timers as shown in Figure 8.1.6. There are three timers that are controlled by UE. The first one is the activity timer (T3324). This timer becomes active when a UE moves from the RRC connected mode to the RRC idle mode. When the timer expires, UE goes to PSM mode. The timer value shows the amount of time that UE remains reachable and can receive downlink messages in the RRC idle mode. The value of T3324 can be in the range of 0 to 11,160 seconds, or roughly 3.1 hours. The second timer is called Paging Time Window (PTW). The value of this timer shows the duration of the paging window, and it consists of several DRX

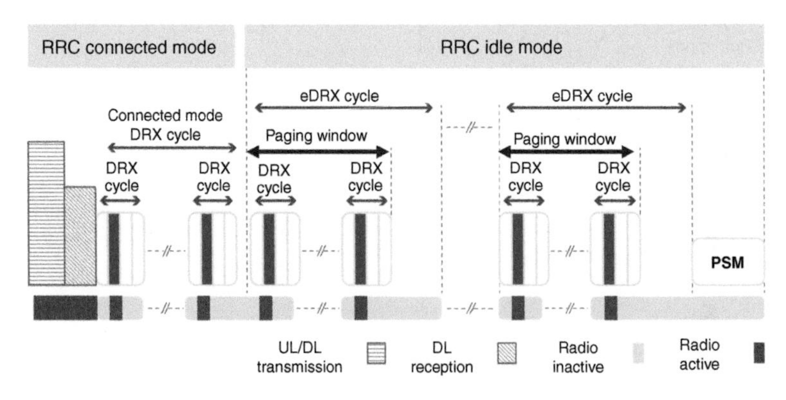

Figure 8.1.5 DRX in both the RRC idle and connected mode.

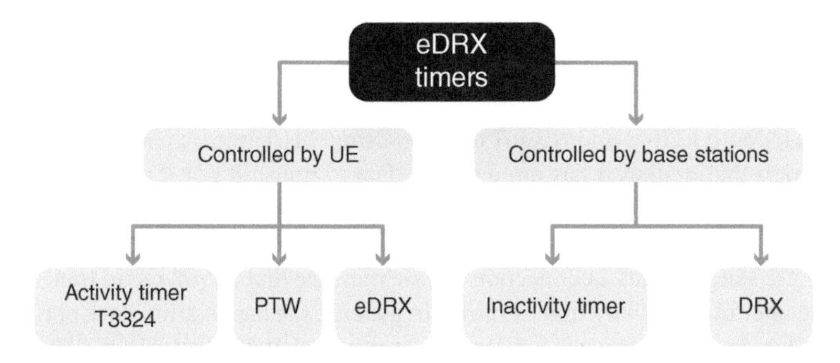

Figure 8.1.6 eDRX timers.

cycles. The third timer is the eDRX cycle timer that determines the eDRX cycle. The number of eDRX cycles can be determined by the value of the activity timer and PTW.

For the eDRX process, there are two timers that are set by the base station, and therefore, UE cannot modify the timers. The first one is the inactivity timer. When this timer expires, RRC will change its mode of operation from the RRC connected to the RRC idle mode. The second timer that is controlled by the base station is the DRX cycle timer that shows the duration of the DRX cycle. The DRX cycle consists of several paging events. Each paging event is 1280 ms. In a DRX cycle, UE listens for one paging event and sleeps during the other paging events within the DRX cycle.

In the PSM mode, the radio is off, and therefore, UE does not receive any notification. To be able to control the duration of PSM, a timer, referred to as Tracking Area Update (TAU) or timer T3412, is dedicated for this purpose. When this timer expires, UE wakes up to perform a TAU in which it updates the network about its availability. The TAU timer can be configured for a long period of up to 413 days (more than a year) for NB-IoT and LTE-M. These timers are shown in Figure 8.1.7. We explain TAU in detail in subchapter 8.4. The device exits the PSM mode once it needs to transmit data in the uplink direction or when it is mandated to send TAU messages in order to give some information about UE to the network. Interested readers are encouraged to study [6] which explains the trade off between energy and latency in NB-IoT network, and optimizes energy consumption based on the PSM and eDRX parameters.

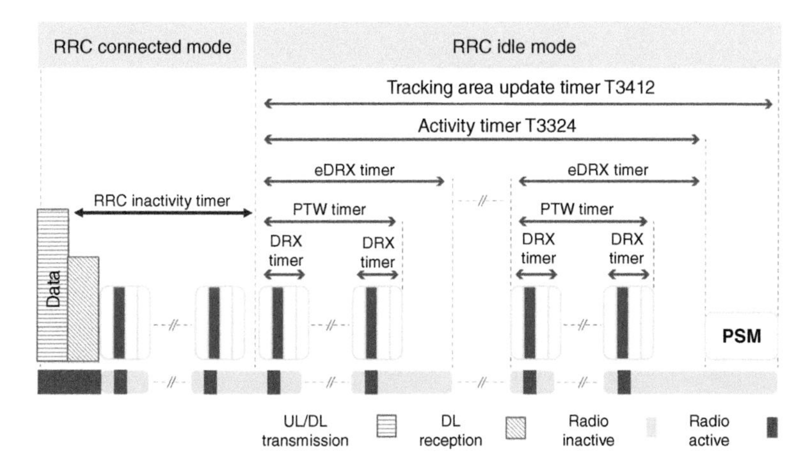

Figure 8.1.7 TAU and active timers.

8.1.3 Power Saving Techniques in 3GPP Release 14

To further reduce power consumption, 3GPP Release 14 adds a new feature called Release Assistance Indication (RAI) for both LTE-M and NB-IoT devices. Using this feature, a CIoT device can indicate to the base station that neither it has more uplink data to transmit nor it expects to receive any more downlink data. This causes an early transition from the RRC connected mode to the RRC idle mode.

In 3GPP Release 13, the base station sends a connection release message that allows UE to transition to the RRC idle mode after the inactivity timer expires. In general, the base station does not know the status of the UE's data buffer or the expected downlink traffic by UE. For this reason, this new feature has been added to 3GPP Release 14, in which UE does not need to wait for the base station to send the appropriate signaling messages for transition from the RRC connected mode to the RRC idle mode. Figure 8.1.8a shows the transition from the RRC connected mode to idle mode when the RAI feature is disabled. In Figure 8.1.8b, the use of the RAI feature is shown. One may think that UE can set the inactivity timer to zero in order to transition to the RRC idle mode. But we should remember that the inactivity timer is set by the base station and not by UE.

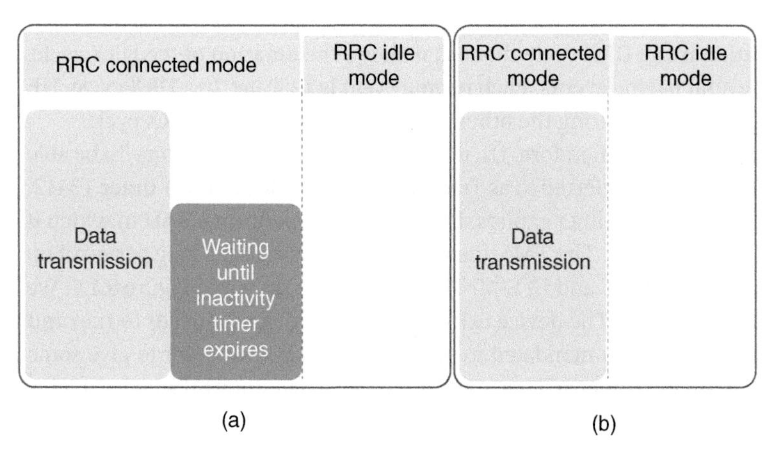

Figure 8.1.8 Transition from the RRC connected mode to idle mode without and with RAI feature.

8.1.4 Power Saving Techniques in 3GPP Release 15

To further reduce power consumption, 3GPP Release 15 adds several new features. The most important one is Wake Up Signal (WUS). It also introduces a new feature called Early Data Transmission (EDT), which is actually a multiple access method. We will explain EDT in detail in subchapter 8.2. EDT can save power when an IoT device intends to get network access. We explain the WUS technique in this section.

8.1.4.1 Wake Up Signal

WUS was introduced in 3GPP Release 15. The benefit of WUS is that it reduces the unnecessary power consumption related to monitoring of the control information [7]. Without WUS, UE needs

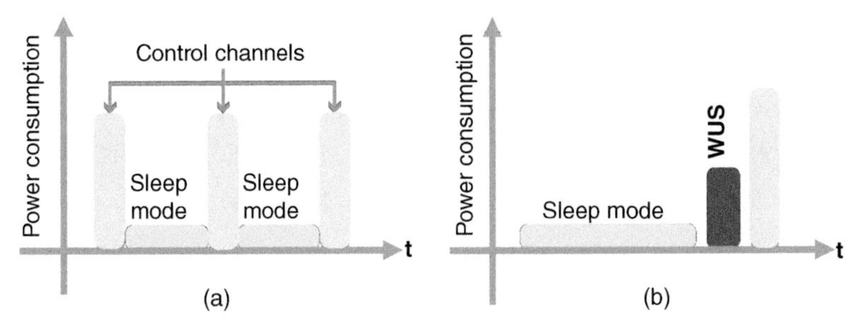

Figure 8.1.9 WUS power saving introduced in 3GPP Release 15: (a) without WUS and (b) with WUS.

to monitor the control information at each paging event. In the WUS approach, UE only needs to decode this channel when WUS is detected. If UE does not detect WUS, it stays in the sleep mode. The WUS signal can be detected with a simple method using a matched filter that is a type of low power receiver. Therefore, it does not need to consume much energy for performing operations such as decoding or demodulation. Figure 8.1.9a and b show the eDRX operation without and with WUS, respectively.

8.1.5 Power Consumption for Various Use Cases

Some applications have very relaxed requirements on reachability and can accept long periods of being unreachable. In these types of applications, the most energy-efficient strategy for reducing power consumption is PSM. For example, a smart water meter may send its counter data weekly or monthly. In these use cases, it is possible to reach a 10-year battery lifetime using both LTE-M and NB-IoT devices. For more uplink-driven data applications, with no or small amounts of downlink data transmission, and non-real-time requirements, utilizing PSM might be a better choice. During PSM, a CIoT device draws current which is at the microampere level when the device is in the deep sleep mode.

eDRX is a suitable strategy for applications such as asset tracking, and smart grids, in which an IoT application should often listen to the network for incoming messages. With choosing a long eDRX cycle, eDRX reduces power consumption, yet still maintains a relatively fast response as compared to PSM. It should be noted that the eDRX current drain is in the range of several milliamperes or microamperes depending on the applied paging cycle length. Generally speaking, utilizing DRX is more beneficial as compared to eDRX for CIoT devices that prefer a lower interval of reachability.

Let us consider the example of a CIoT-based alarm system that only gets triggered once per year on average. In a rare situation, when the system gets triggered, the alarm needs to send a packet in near real time (with minimal delay). In this example, it is important to ensure that the alarm system is working properly. To verify the operability of the alarm system, the network can send a page request message and receive a page response. The network can then send an application-layer message to a server that triggers the reporting of any existing alarm. In this application, eDRX with a shorter paging cycle is preferred to reduce the delay. Certainly, eDRX can be used in conjunction with PSM to find a balance between battery lifetime and service responsiveness for a specific IoT application.

Example 8.1.1 Evaluation of the CIoT power saving parameters

A CIoT device transmits data every 2 hours and 40 minutes to a server. The server never sends any commands or data to the IoT device. The following timer values are set for this IoT device.

$$T_{PTW} = 1.28\,sec,\ T_{eDRX} = 5.12\,sec,\ T_{3324} = 512\,sec,\ \text{and}\ T_{3412} = 10\,000\,sec.$$

The base station uses $T_{DRX} = 0.32\ sec$. During each DRX, the IoT device consumes 100 mw when it monitors the paging events, and 1 mw otherwise. In the PSM mode, the IoT device consumes 0.015 mw.

a. Does the CIoT device transition to the RRC connected mode from the RRC idle mode before timer T3412 expires?
b. What is the number of paging events during each paging window? What is the number of eDRX cycles? How much power does the IoT device consume in each RRC idle mode?
c. In this scenario, why the timer values, set by the IoT device, do not minimize the power consumption in the RRC idle mode?
d. What timer parameters minimize the power consumption of the IoT device in the RRC idle mode?
e. Using the parameters in (d), what is the amount of power consumption in the RRC idle mode?
f. If the parameter in section (d) is used, what is the amount of power reduction in the RRC idle mode as compared to the settings used in this example?

Solution

a. The IoT device sends its data to the Internet every 2 hours and 40 minutes or 9600 seconds. Timer T_{3412} expires after 10 000 seconds, which is more than three hours. Therefore, after 9600 seconds when there are 400 seconds (10 000-9600) left on timer T_{3412}, the IoT device needs to transmit its data and leaves the RRC idle mode and transitions to the RRC connected mode.
b. The number of paging events during each paging window is $\frac{T_{PTW}}{T_{DRX}} = \frac{1.28}{0.32} = 4$. The number of eDRX cycles can be calculated as $\frac{T_{3324}}{T_{eDRX}} = \frac{512}{5.12} = 100$. Therefore, there are 4×100 paging events in each RRC idle mode.

The power consumption in each DRX can be calculated as shown in Figure 8.1.10.

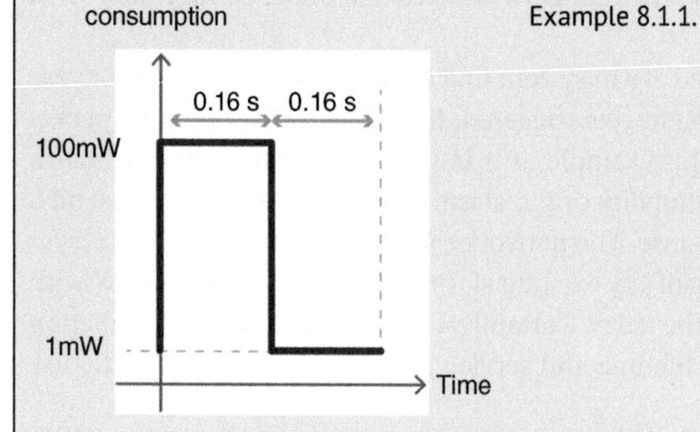

Figure 8.1.10 Power consumption in each DRX for Example 8.1.1.

The power consumption is 100 mw $\times 0.16 + 1$ $mw \times 0.16 = 16.16$ mw during 0.32 seconds. Therefore, the average power consumption would be $\frac{16.16}{0.32} = 50.5$ mw.

The total energy consumption during all eDRX cycles is 50.5 mw $\times 4 \times 100 = 20.200$ w during 512 seconds. Therefore, the average power consumption during the eDRX cycle would be $\frac{20200}{512} = 39.453$ mw.

The power consumption during PSM is 0.015 mw for 9088 (9600-512) seconds. Therefore, the average power consumption during each RRC idle mode can be expressed as:

$$\frac{39.453 \times 512 + 0.015 \times 9088}{9600} = 2.11mw$$

c. Since in this application, the IoT node does not expect any downlink message from the server, it is better to stay in PSM mode all the time during the RRC idle mode.

d. $T_{3412} = 9600\ sec\ and\ T_{3324} = 0.$

e. The average power consumption would be 0.015mw.

f. The amount of average power reduction would be 2.11 mw – 0.015 mw = 2.103 mw.

Example 8.1.2 Effect of ramp-up and down in the DRX cycle

Practically, each DRX cycle in Example 8.1.1 includes a ramp-up and ramp-down section as shown in Figure 8.1.11. If the duration of ramp-up is 15 ms and the duration of ramp-down section is also 15 ms, calculate the power consumption during the RRC idle mode for Example 8.1.1. Considering the effect of these sections, what is the percentage of change in power consumption?

Solution

The power consumption in each DRX cycle can be calculated as:

$$\frac{100-1}{2} \times 15\ ms + (100 \times 0.16\ sec) + \frac{100-1}{2} \times 15\ ms + (1 \times 0.157\ sec) = 16.175\ mw$$

Therefore, the average power consumption would be $\frac{16.175}{0.32} = 50.546$ mw. The average power consumption without considering these sections was 50.5 mW. This shows an increase of $\frac{50.546-50.5}{50.5} \times 100 \approx 0.5\%$.

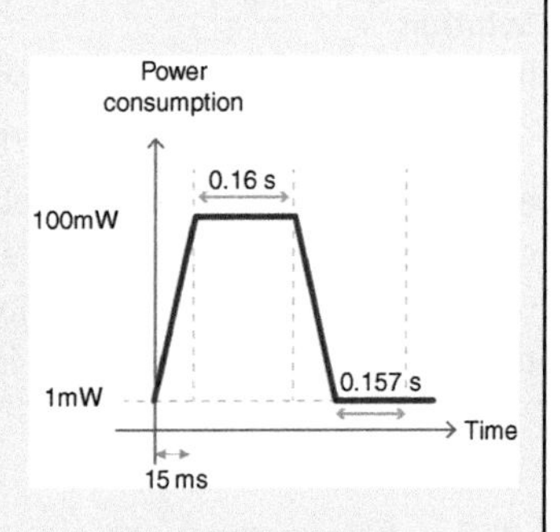

Figure 8.1.11 Power consumption in each DRX for Example 8.1.1 considering a ramp-up and a ramp-down section.

Example 8.1.3 Calculating the power consumption

An IoT device is consuming 450 mw to transmit a data packet and expects to receive data from the base station for five seconds. The duration of the data transfer is also five seconds. After that the IoT device is not sending any data and does not expect to receive any data for 24 hours. The DRX inactivity timer is set to five seconds. The power consumption for listening to the downlink channel is 250 mW. In this scenario, assume that eDRX is disabled. After the inactivity timer expires, the IoT device transitions to the RRC idle mode.

a. Calculate the amount of average power consumption in the RRC connected mode.
b. If the DRX inactivity timer was set to zero, what was the average power consumption?
c. The IoT device supports NB-IoT in 3GPP Release 13. Can, after five seconds, the IoT device transition to the RRC idle mode?
d. If the NB-IoT device and network was based on 3GPP Release 14, can, after five seconds, the IoT device transition to the RRC idle mode?

Solution

a. The IoT device is transmitting 450 mw for five seconds and consuming 250 mw for the next five seconds. So, the average power consumption would be:

$$\frac{450 \times 5 + 250 \times 5}{10} = 350\,mw$$

b. In this case, the average power consumption is 450 mw in five seconds.
c. No, since 3GPP Release 13 does not support RAI feature. In general, the base station does not know the status of the buffer of an IoT device. That is why the inactivity timer is not set to zero.
d. Yes, this is possible for 3GPP Release 14 that supports RAI feature.

Example 8.1.4 Latency evaluation

An IoT device may receive downlink data from a base station at any random time during the RRC idle mode as shown in Figure 8.1.3. The probability distribution function of this random signal is uniform. What is the minimum average delay that UE experiences in the downlink direction in the RRC idle mode?

Solution

If the downlink information comes sometime within the paging window, then the average delay would be $\frac{T_{DRX-Cycle}}{2}$. If the arrival of the downlink information occurs sometime within the eDRX cycle, then the average delay can roughly be calculated as $\frac{T_{DRX-Cycle}}{2}$. If we consider that the downlink information can happen randomly within the RRC idle mode, then the average delay would be almost $\frac{T_{DRX-idle}}{2}$ especially if the duration of PSM is substantially large as compared to the duration that the device is in the RRC idle mode.

8.1.6 Summary

In many IoT applications, low power consumption is an important issue. For battery-powered CIoT devices, the amount of power consumption determines the lifetime of the battery. Therefore, a lot of attention has been given to lower the power consumption in CIoT applications. When a CIoT device is powered on, it enters what is called a Radio Resource Control (RRC) idle mode and sets up a connection. After the network connection is established, the device enters what is called an RRC connected mode. For power saving, the base station can release the RRC connection that causes the device to move into the RRC idle mode again, where it can sleep and reduce power consumption.

Cellular IoT devices use techniques such as DRX, eDRX, PSM, RAI, and WUS to reduce power consumption. The main principle methods in 3GPP Release 13 are eDRX and PSM. RAI and WUS are techniques that were added in 3GPP Release 14 and 15, respectively.

The idea behind the DRX and eDRX is to put a CIoT device in the RRC idle mode and periodically check for the downlink messages. eDRX consists of several DRX cycles. The eDRX process is controlled by a set of timers. In PSM mode, the CIoT radio is off, and therefore, the device does not receive any downlink message. The duration of PSM is also controlled by a timer.

Using the RAI feature, a CIoT device can indicate to the base station that it neither has more uplink data to transmit nor it expects to receive any more downlink data. This causes an early transition from the RRC connected mode into the RRC idle mode, and results in further reduction of power consumption. WUS allows a CIoT device to stay in sleep mode until it detects a wake up signal. WUS signal can be detected using a matched filter that is a type of low power receiver. Therefore, the CIoT device does not need to wake up periodically to monitor downlink messages.

References

1 Third Generation Partnership Project (2016). Evolved Universal Terrestrial Radio Access (E-UTRA) and Evolved Universal Terrestrial Radio Access Network (E-UTRAN); Radio Resource Control (RRC); Protocol Specification. Technical Specifications 36.331 v13.3.0.

2 ETSI (2017). LTE – Evolved Universal Terrestrial Radio Access (E-UTRA); User Equipment (UE) radio transmission and reception. 3GPP TS 36.101 version 13.6.1 Release 13.

3 ETSI (2017). LTE – Evolved Universal Terrestrial Radio Access (E-UTRA); User Equipment (UE) radio transmission and reception. 3GPP TS 36.101 version 14.5.0 Release 14.

4 ETSI (2017). LTE – Evolved Universal Terrestrial Radio Access (E-UTRA); User Equipment (UE) radio transmission and reception. 3GPP TS 36.101 version 15.3.0 Release 15.

5 ETSI (2016). Digital cellular telecommunications system (phase 2+); Universal Mobile Telecommunications System (UMTS); LTE. TS 123682.

6 Sultania, A.K., Blondia, C., Famaey, J. et al. (2021). Optimizing the energy-latency tradeoff in NB-IoT with PSM and eDRX. *IEEE Internet of Things Journal* 8 (15): 12436–12454. https://doi.org/10.1109/JIOT.2021.3063435.

7 Liberg, O., Sundberg, M., Wang, E. et al. (2017). *Cellular Internet of Things: Technologies, Standards, and Performance*. Academic Press. ISBN: ISBN: 978-0-12-812458.

Exercises

8.1.1 Why completely turning off the radio of a CIoT device for a short period of time might increase its average power consumption?

8.1.2 CIoT devices go to the RRC idle mode instead of going to deep sleep.

a. What is the reason behind putting a CIoT device in the RRC idle mode?

b. In the RRC idle mode, a CIoT device activates and deactivates its reception periodically. For which applications longer deactivation is possible?

8.1.3 What is the difference between transitioning to the RRC idle mode after the device is powered on and when it transitions from the RRC connected mode at a later time?

8.1.4 What happens when the inactivity timer expires? Which entity controls the inactivity timer?

8.1.5 When an IoT device transitions from PSM to the RRC connected mode, does the IoT device need to reconnect itself to the network again and perform security operations?

8.1.6 Can a CIoT device receive messages when it is in the PSM cycle?

8.1.7 Can eDRX be used without PSM in the RRC idle mode? How does this relate to T_{3412} and T_{3324} timers?

8.1.8 Give examples of several operations that an IoT device does not need to perform, when it transitions from the RRC idle mode to the RRC connected mode for gaining access to the network?

8.1.9 For which IoT applications the use of RAI power saving feature is efficient?

8.1.10 WUS is a power saving feature introduced by 3GPP.
- a. Which 3GPP release number supports WUS?
- b. How is a WUS signal detected?
- c. How does WUS reduce power consumption?

Advanced Exercises

8.1.11 In this chapter, we only discussed one kind of DRX in the RRC connected mode. But LTE supports the concept of short DRX and long DRX cycles. Short DRX (as the name suggests) permits a CIoT device to have a shorter DRX cycle. The number of such short DRX cycles is also very limited. If there was no data transfer during the ON period of the short DRX cycles, then the CIoT device transitions to the long DRX cycles. This is shown in Figure 8.1.E11. What is the reason for short and long DRX cycles?

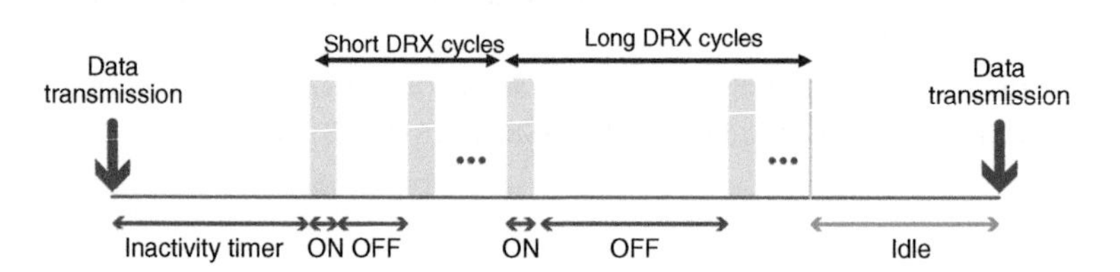

Figure 8.1.E11 Short and long DRX cycles.

8.1.12 An IoT device is consuming 375 mW to transmit a data packet for four seconds and expects to receive some data from the base station for four seconds. After that, the IoT device is not sending any data and does not expect to receive any data for 24 hours. The DRX inactivity timer is set to four seconds. The power consumption for listening to the DownLink (DL) channel is 300 mW. In this scenario, assume that eDRX is disabled. After the inactivity timer expires, the IoT device transitions to the RRC idle mode. In general, the base station does not know the status of the IoT device data buffer or the expected downlink traffic by the IoT device. That is why the inactivity timer, set by the base station is not zero.

a. Calculate the amount of average power consumption in the RRC connected mode.
b. If the DRX inactivity timer was set to zero, what was the average power consumption?
c. The IoT device supports NB-IoT in 3GPP Release 13. Can, after five seconds, the IoT device transition to the RRC idle mode?
d. If the IoT device and network was based on NB-IoT in 3GPP Release 14, can the IoT device transition to the RRC idle mode after five seconds?

Chapter 8.2 Uplink Access

8.2.1 Introduction

Choosing the best multiple access protocol for a network depends on many factors. For example, if the number of the nodes are limited and they have frequent traffic, the channelization-based multiple access protocols such as Frequency Division Multiple Access (FDMA), Time Division Multiple Access (TDMA), or Code Division Multiple Access (CDMA) might be a good choice. In FDMA, the available bandwidth is divided into several non-overlapping frequency bands, and each frequency band is allocated to one of the nodes for its data transmission. In TDMA, the available bandwidth is shared among network nodes, but each node is given a specific time slot to send its data. In CDMA, a unique code is allocated to each node in order to send its data at all times and in all available frequency bands. Therefore, each node in CDMA occupies the entire available bandwidth and all nodes can send their data simultaneously. When there is a massive number of IoT devices in a network with a need for small and infrequent data transmission, the use of channelization-based multiple access protocols would not be efficient.

Another category of the multiple access protocols is polling, in which a node in the network is designated as a primary node (like a base station) and the other nodes are called secondary nodes. The data transmission should always go through the primary node that controls the communication and gives the right of access to the secondary nodes. The primary node controls the communication link, while the secondary nodes transmit their data based on the instructions of the primary node. The primary node goes around and asks other nodes one by one, whether they have any data to send or not. This multiple access method cannot be effective in situations where there are a large number of IoT devices in a network.

There are other multiple access protocols that have been widely used in local area networks, but they are not suitable for cellular IoT applications. Token passing is an example of these protocols. In this type of multiple access protocol, the nodes must be arranged in a logical order of predecessor and successor. Each node that has a token can send its data. Then, this node needs to transmit the token to the next node as specified in a predefined logical order. This type of multiple access control does not provide any efficiency for IoT applications. Due to the infrequent traffic generated by IoT nodes and the massive number of them in a network, token passing results in huge amounts of traffic generated for passing the token to the nodes that mostly do not have any data to transfer.

Random Access (RA) is another class of multiple access technologies that has been widely used in data communication and networks. It defines a multiple access method in which each node has the right to send its data without being controlled by any other node on the network. However, if two or more nodes try to transmit at the same time, there will be collisions that result in corruption

Fundamentals of Internet of Things: For Students and Professionals, First Edition. F. John Dian.

of data. Ethernet used an RA protocol for its operation in early versions of its standards, before becoming a collision-free standard using full-switched Ethernet technologies. The RA process is a good choice for a massive number of IoT devices that send small amounts of data infrequently. In this chapter, we discuss cellular IoT uplink data transmission.

8.2.2 Random Access Process

First, let us make this clear that the RA process does not intend to send the actual data, and it only tries to access the network, and request for uplink resources. After a successful RA process, an IoT node has the time and frequency information for its data transmission [1]. The RA process is completed when four messages are exchanged between the IoT device and the base station. These four messages are called Msg1, Msg2, Msg3, and Msg4 as shown in Figure 8.2.1.

The first message is initiated by the IoT node, and it is called Random Access Preamble (RAP) or simply Msg1. Whenever an IoT device intends to send its data to the Internet, it listens to the broadcasted information from the base station in order to find the first available contention period to send Msg1. These available contention periods are called Random Access Opportunities (RAO) and are allocated by the base station to IoT devices that need to transmit their data. During an RAO, CIoT nodes compete to obtain the required resources for their data transmission. During the first available contention period, an IoT node transmits Msg1 that is actually a preamble. An IoT node should randomly choose one preamble from a set of available preambles to send as Msg1. In theory this set should contain 64 preambles, but it is possible that all preambles are not available in a cell. The IoT device can obtain the list of available preambles from the information that is broadcasted by the base station. These 64 preambles are orthogonal, which means multiple IoT devices can

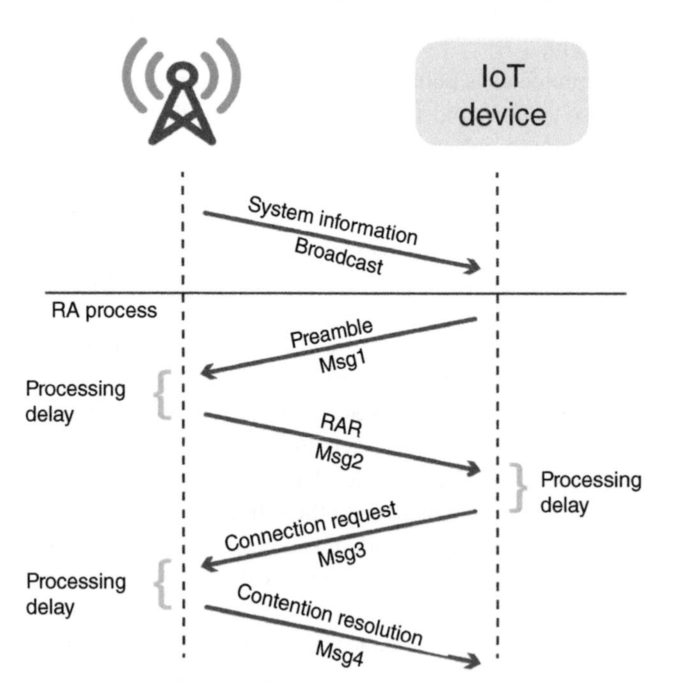

Figure 8.2.1 Random access messages.

access the network at the same time, if they use different preambles. If two or more IoT devices choose the same preamble at the same time, the transmission of these preambles would result in a collision.

In its first attempt to transmit Msg1, a CIoT device adjusts its transmit power according to the strength of the reference signals received from the base station. If the transmission of Msg1 fails, the IoT device performs power ramping that increases the transmit power after each failure in the RA process in order to reduce the probability of subsequent failure in the transmission of Msg1.

The second message is initiated by the base station, and it is called Random Access Response (RAR) or Msg2. After receiving a preamble without error, the base station sends a packet back to the IoT device. This packet provides information regarding the uplink resources that the IoT device needs for sending Msg3. The base station gives the IoT device a temporary identity. It also sends a Time Alignment (TA) value for the uplink synchronization. TA is calculated based on the distance between the IoT device and the base station to ensure that the downlink and uplink subframes are synchronized at the base station.

Different preambles are associated with different coverage levels. Therefore, the IoT device should also choose a preamble based on its coverage mode. After the IoT device sends a preamble based on its coverage level, it sets up a timer that is called a response window timer. If the timer expires and the IoT device still has not received any response (Msg2), then the RA process has not been successful and needs to be repeated.

If two or more IoT devices choose the same preamble, and if preambles get corrupted due to the collision, then no Msg2 will be sent in response to collided preambles. If preambles do not become corrupted, all IoT devices receive the same Msg2. Since all these IoT devices use the same resources in terms of time and frequency in order to send Msg3, the transmission of their Msg3 results in collisions. It should also be noted that if the base station does not have sufficient resources for a successfully decoded preamble, then it does not send any Msg2.

The third message is initiated by the IoT device, and it is called connection request or Msg3. By sending Msg3, the IoT device requests for a connection. The IoT device sends Msg3 using the resources (time and frequency) granted by the base station in Msg2. The IoT device also activates a timer called a contention resolution timer. If this timer times out and the IoT device has not received Msg4 yet, then the IoT device has failed the contention resolution process. In this case, the IoT device should start the RA process again by sending Msg1.

The fourth and the last message is initiated by the base station, and it is called contention resolution or Msg4. The base station performs contention resolution, and by sending Msg4 to the IoT device identifies the time that the IoT device can start its data transmission. Msg4 contains the unique contention resolution ID received from Msg3.

After the RA process is complete, the security of the data bearers needs to be established or resumed. As we discussed in subchapter 8.1, when an IoT device is first powered on, it enters the RRC idle mode and at this point does not have an established physical connection to the base station. After the IoT device establishes its connection to the network, it enters the RRC connected mode. In the RRC connected mode, the IoT device is able to access the network and request for network resources. For reducing power consumption, the base station releases the RRC connection that causes the IoT device to go back to the RRC idle mode. However, this time it stores the current context in order to avoid the connection setup process and procedures such as scanning, authentication, key exchange, or IP address assignment. Therefore, after the RA process is complete, the IoT device moves to the RRC connected mode and starts transferring data in Msg5. If the data size is larger than what Msg5 can handle, more messages need to be sent.

8.2.2.1 Random Access Dependency to the Coverage Level

To send Msg1, a CIoT device selects a preamble based on its coverage level. Therefore, before sending the preamble signal, the IoT device needs to estimate its coverage level. For this purpose, the IoT device measures the received power of reference signals that are broadcasted by the base station.

In an LTE-M network, the base station can configure up to three threshold values for the Reference Signal Received Power (RSRP). These RSRP threshold values are used by IoT devices to determine four coverage levels. The base station broadcasts these threshold values to IoT devices. When an IoT device fails to transmit Msg1, it continues further attempts until the reception of Msg2 or until it reaches the maximum allowed number of attempts. As we mentioned earlier, the IoT device also applies power ramping, to increase the power and reduce the possibility of collision. In LTE-M, the IoT device also performs coverage level ramping and moves to higher coverage levels after several failed transmissions, which increases the number of repetitions in its next attempts. If the base station supports CE mode B, then the base station can define up to four coverage levels. For base stations that only support CE mode A, the base station can define up to two coverage levels. Figure 8.2.2 shows a scenario where the base station has defined the maximum number of coverage threshold values for the LTE-M network [2].

In NB-IoT network, IoT devices are also allowed to perform coverage level ramping. IoT devices have to know the maximum number of preamble transmission attempts, at a coverage level. If the number of attempts reached the maximum value, then the IoT device considers moving to the next higher coverage level, if one exists. NB-IoT defines three coverage levels of normal, extended, and extreme, where each coverage level is associated with an Maximum Coupling Loss (MCL) number. The normal coverage mode is associated with an MCL of 144 dB, while extended and extreme coverage modes are associated with MCL values of 154 and 164 dB, respectively.

8.2.2.2 Access Barring (AB)

When a large number of IoT devices access the network at the same time, then it is possible that a large number of collisions occur, and an IoT device may experience multiple unsuccessful attempts before obtaining access. Obviously, these situations increase latency, which is not suitable for many IoT applications. To reduce the number of simultaneous accesses in each RAO, the concept of access barring is introduced to randomly delay the initiation of access requests by IoT devices. This delay redistributes the RA process initiation over time. The base station instructs IoT devices to randomly delay the beginning of their RA process according to barring parameters. In other words, some IoT devices are barred from accessing the network in a certain time frame. Therefore, the possibility of collision becomes less, and the possibility of gaining access to the network becomes higher. Two types of widely used barring methods are Access Class Barring (ACB) and Extended Access Barring (EAB).

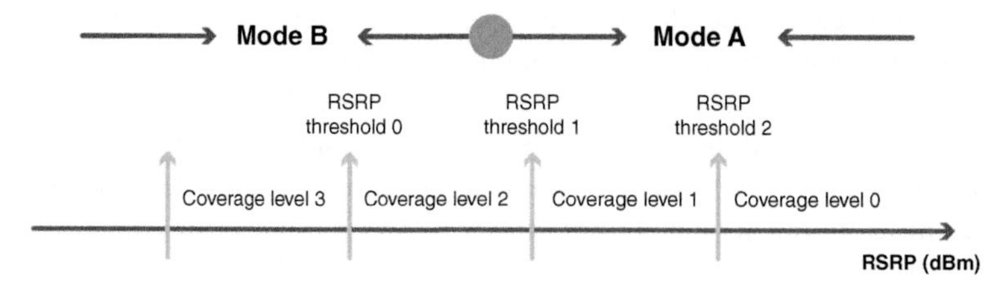

Figure 8.2.2 LTE-M coverage levels.

In ACB, the base station broadcasts an ACB factor $p \in [0, 1]$ as well as a barring time at each RAO. Each IoT device that has experienced failure in the RA process generates a random number based on the Bernoulli trials. The device is allowed to transmit only if the generated random number is smaller than the ACB factor. Otherwise, it waits sometime according to a barring time and a random number. In this way, at every RAO, some IoT devices are barred from sending Msg1 and initiating the RA process. An ACB factor is a parameter that plays an important role in controlling ACB, and therefore, it is crucial for the base station to choose a good value as the ACB factor. If the ACB factor is too small, then not many IoT devices are allowed to initiate the RA process and many resources might be wasted. If the base station selects a large ACB factor, most IoT devices are allowed to transmit, which makes the barring method inefficient.

In EAB, each IoT device is a member of a barring class. The classes are numbered from 0 to 9, and this information is stored in Subscriber Identity Module (SIM). At each RAO, IoT devices belonging to one specific class are permitted to access the network, while devices belonging to other classes are barred from initiating the RA process. Therefore, IoT devices need to wait for their turn in terms of class number in order to access the network. EAB mainly focuses on reducing the number of collisions in the network rather than reducing latency. The base station broadcasts a 10-bit pattern in which one of its bits is one, and all other nine bits are zero. Each bit in this pattern represents a barring class where one means unbarred and zero represents barred. The index of each bit in the pattern, which is one, indicates the EAB class that is not barred from initiating the RA process. For example, the bitmap pattern of 0100000000 shows that only IoT devices with class 1 are unbarred and all the other IoT devices are barred. The base station broadcasts the bitmap pattern, and IoT devices accordingly understand whether they can initiate an RA process or not. IoT devices check the broadcasted information from the base station and start the RA process when they are allowed. EAB also defines classes from 11 to 15 for very high priority devices allocated for emergency cases. If the bitmap pattern is not broadcasted, then no IoT device is barred from accessing the network. EAB is very effective in high traffic situations.

LTE-M supports both ACB and EAB, while NB-IoT supports only EAB. The classification of AB control schemes for IoT devices is shown in Figure 8.2.3.

8.2.2.3 Preamble Formats

The preamble is composed of four symbol groups that are transmitted at different frequencies. Frequency hopping is applied to these symbol groups. Each sample group is composed of a Cyclic Prefix (CP) and five Orthogonal Frequency-Division Multiplexing (OFDM) symbols. There is also a guard period that starts after the end of the OFDM symbols and continues until the end of the LTE subframe. Figure 8.2.4 shows one symbol group in which the duration of CP is represented by

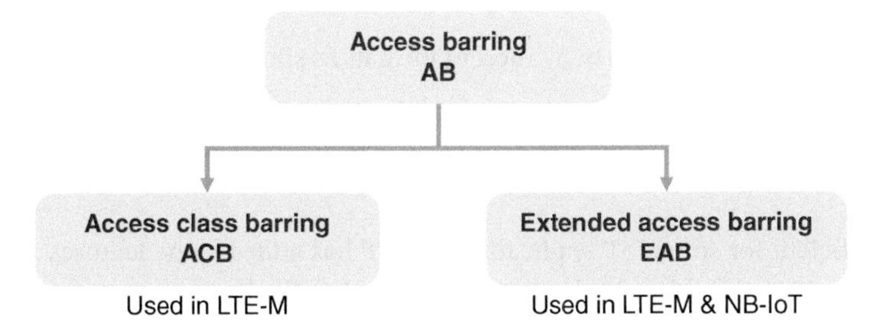

Figure 8.2.3 Access barring schemes used in IoT devices.

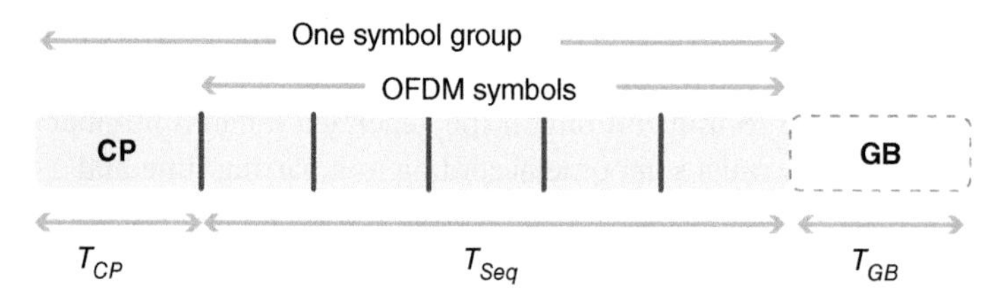

Figure 8.2.4 Preamble fields.

Table 8.2.1 LTE-M preamble formats.

Preamble format	T_{CP} (μs)	T_{sea} (μs)	T_{GB} (μs)	Cell range (Km)
Format 0	103	800	97	15
Format 1	684	800	516	78
Format 2	203	1600	197	30
Format 3	684	1600	716	108

Table 8.2.2 NB-IoT preamble formats.

Preamble format	T_{CP} (μs)	T_{seq} (ms)	Cell size (Km)
Format 0	66.7	1.333	10
Format 1	266.67	1.333	40

T_{cp}, the duration of the sequence of five OFDM symbols is shown as T_{Seq}, and the duration of the guard band is expressed as T_{GB} [3].

The length of preambles varies based on the preamble format. In LTE-M, a preamble can accept four different formats as shown in Table 8.2.1. Larger T_{cp} values would provide better tolerance in a fading environment. In other words, a larger T_{Seq} value is better for decoding the preambles in situations where the value of Signal to Noise Ratio (SNR) is higher. Comparing Format 1 with Format 0 or Format 3 with Format 2 shows that choosing a larger T_{GB} has been considered for larger cell sizes. The cell size can be roughly calculated using Eq. (8.2.1), where c is the speed of light (300 km/ms).

$$\text{Cell range} = \frac{C \times T_{GB}}{2} \tag{8.2.1}$$

In NB-IoT, the preamble usually accepts one of the two different formats as shown in Table 8.2.2.

8.2.3 RA Advancements

To make the RA process more efficient for some IoT applications, 3GPP has added some features to the RA process in 3GPP Release 15 and Release 16. In this section, we briefly discuss two main features called EDT, introduced in 3GPP Release 15 and Preconfigured Uplink Resources (PUR) that is part of 3GPP Release 16.

EDT enables an IoT device to send small-size data as part of its Msg3 or a base station to send small-size data as part of its Msg4. PUR enables IoT devices with deterministic traffic patterns to skip the transmission of Msg1, as well as the reception of Msg2. Enabling both PUR and EDT makes an RA process that does not need to transmit Msg1 and receive Msg2, and can transmit its data inside Msg3, or receive information inside Msg4.

8.2.3.1 Early Data Transmission

EDT intends to improve the performance of a network in terms of latency and power consumption for those IoT applications that have just a small amount of data to transmit [4]. Generally speaking, the size of data that an IoT device transmits depends on the application. The IoT device may intend to send small data or have a multi-packet transmission of large amounts of data. In 3GPP Release 13, the network does not know the amount of data that the IoT device would like to send prior to sending Msg3. To be able to use data transmission as part of the RA process and send data inside Msg3, the IoT device must inform the base station in advance regarding the size of its data. For this purpose, the IoT device can show its interest in using EDT feature during the transmission of Msg1 by choosing from the list of preambles that are assigned to EDT. The base station broadcasts the maximum Transport Block Size (TBS) that can be used for the Msg3 transmission as part of Msg2. If the TBS value is smaller than the data size, using the EDT is not practically possible, and data transfer should be performed after the reception of Msg4. If the TBS value is substantially greater than the data size, the IoT device can use EDT, but it needs to add padding that reduces the efficiency of EDT.

Assume the IoT device is in the RRC idle mode, and the size of user data is smaller than TBS. In this situation, the RA process with the support of EDT can be summarized as:

1. The IoT device reads the system information that is broadcasted by the base station and retrieves information such as available preambles and the maximum TBS from the system information.
2. The IoT device indicates that it would like to use EDT by randomly selecting a preamble from a set of available preambles that are assigned for EDT.
3. The base station receives the preamble and sends Msg2 including an uplink Msg3 grant to an appropriate TBS.
4. The IoT device transmits the necessary information (the IoT device's identity, resume identity, RRC message) and also includes the user data in Msg3.
5. The base station receives Msg3. Since the IoT device has sent all its data in Msg3, the IoT device stays in RRC idle mode.

It should be noted that for EDT, somehow security needs to be established before the transmission of Msg3 [5].

8.2.3.2 Preconfigured Uplink Resources

Different types of RA process established before 3GPP Release16 consist of four steps (Msg1, Msg2, Msg3, and Msg4) that require two round-trip transmission of messages between the IoT device and the base station. PUR is introduced in 3GPP Release 16 in order to reduce signaling overhead by implementing a two-step RA process instead of a four-step process. It eliminates the transmission of the Msg1 and Msg2, which results in reducing the latency, and lowering the power consumption of IoT devices when they access the network. Both the IoT device and base station can enable PUR.

An IoT device can request the network to enable PUR by sending a request message, while the network can enable the PUR feature based on the IoT device's traffic pattern.

If the IoT device knows about the available resources and the required configurations for transmission of its Msg3, the transmission of Msg1 and Msg2 is not needed. For instance, an IoT device needs to know the time-frequency resources, Modulation and Coding Scheme (MCS), TBS, number of repetitions, power control parameters, and the number of repetitions for the coverage enhancement modes. The IoT device should also know about the value of TA, TA validation criteria, and the number of allowable skips. It is clear that TA is calculated based on the distance between the IoT device and the base station to ensure that the downlink and uplink subframes are synchronized at the base station, and therefore, it is needed before the transmission of Msg3.

The IoT device needs to validate TA to ensure that the serving cell has not changed. PUR can be efficient if the IoT device has not been moved to another cell and if it has a deterministic traffic pattern. It is clear that if the serving cell and the IoT device want to transmit at previously scheduled times, PUR may cause collisions and reduce performance. An IoT device should ensure that the RSRP has not changed substantially and the Time Alignment Timer (TAT) has not expired. It should be noted that the base station continuously measures the timing of uplink signals and adjusts the uplink transmission timing by sending the TA command to the IoT device. If the IoT device has not received a TA command until the expiry of TAT, it might have lost the uplink synchronization. If the validity of TA is not satisfied, PUR cannot be implemented and it should use the normal RA procedure (defined by 3GPP Release 13).

When PUR is enabled, the allocated resources are available to IoT devices based on a preconfigured pattern. The assumption is that the IoT device has a periodic or deterministic traffic pattern. However, it can be configured in such a way that the IoT device can be allowed to skip transmitting a certain number (maximum eight) of consecutive allocated occasions [5].

In general, we can say that using PUR provides higher efficiency when the frequency of the data transmission is higher. By using PUR, the IoT device does not need to transmit Msg1 and wait to receive Msg2. Due to the possibility of collisions, PUR becomes more effective as the number of nodes in the network increases. More collisions mean more signaling and more latency for transmitting data. The frequency of transmission has a direct relationship with latency and signaling. The more often an IoT device needs to transmit its data, the network gains more efficiency in activating PUR. As mentioned earlier, the traffic pattern and the frequency that an IoT device generates traffic differs from one IoT application to another.

PUR is more effective when used together with EDT for small-sized data. In a poor coverage area with large numbers of repetitions, PUR becomes less effective as compared to situations where the CIoT device is located in a good coverage area where it does not need any repetitions or it needs only a small number of repetitions.

8.2.4 Summary

Uplink access in CIoT is performed using the RA process. In this method, the nodes can access the network at any time. As defined by 3GPP Release 13, the basic RA process for CIoT devices consists of four messages. The first message, Msg1, is a preamble that is sent by the IoT device. The second message is initiated by the base station, and it is called RA response or Msg2. After receiving a preamble without error, the base station sends Msg2 to the IoT device. This packet provides information regarding the uplink resources that the IoT device can use for sending Msg3. The third message is initiated by the IoT device, and it is called a connection request or Msg3. By sending Msg3, the IoT device requests for a connection. Msg3 will be sent using the resources (time and frequency) granted by the base station in Msg2. The fourth and the last message is initiated by the base station, and it is called contention resolution or Msg4. The base station performs contention resolution and by sending Msg4 identifies the time and the spectrum that the IoT device can use to start data transmission.

The RA method using these four messages may result in collisions. Due to the limited number of preambles, there is the possibility of collision in Msg1. Also, several nodes that have chosen the same preamble may start sending Msg3 at the same time. If a collision happens, the RA process should start again. But there are a number of methods available to reduce the possibility of collisions. Power ramping increases transmission power in order to increase the possibility of a successful transmission after consecutive failures. Coverage level ramping uses a higher coverage level after several failed attempts, which increases the number of repetitions in those attempts.

Access barring randomly delays the initiation of access requests by IoT devices. This delay redistributes the RA process initiation over time. The base station instructs IoT devices to randomly delay the beginning of their RA process according to barring parameters. Two important types of barring methods are Access Class Barring (ACB) and Extended Access Barring (EAB).

Early data transmission and preconfigured uplink resources are two methods introduced in 3GPP Release 15 and 16 in order to increase the performance of the RA process. EDT is effective when an IoT device or base station has a small amount of data to send. In this case, the IoT device can send its data during Msg3, or the base station can send its data during Msg4. This reduces the latency, number of signaling messages, and power consumption of CIoT devices. PUR eliminates the transmission of Msg1 and Msg2, which results in reducing the latency and lowering the power consumption of an IoT device. PUR can be used for applications that have deterministic traffic pattern. Applications that not only have a periodic or deterministic traffic patterns but also transmit small amounts of data can take advantage of both EDT and PUR.

References

1 Dian, F.J. and Vahidnia, R. (2021). Random access process enhancements for cellular internet of things (CIoT). *IEEE IEMCON* https://doi.org/10.1109/IEMCON53756.2021.9623252.

2 Liberg, O., Sundberg, M., Wang, E. et al. (2017). *Cellular Internet of Things: Technologies, Standards, and Performance.* Academic Press ISBN: 978-0-12-812458.

3 Dian, F.J. and Vahidnia, R. (2021). NB-IoT preamble signal: a survey. *WF-IoT* 107–112. https://doi.org/10.1109/WF-IoT51360.2021.9596011.

4 Höglund, A., Van, D.P., Tirronen, T. et al. (2018). 3GPP Release 15 early data transmission. *IEEE Communications Standards Magazine* 2 (2): 90–96.

5 Dian, F.J. and Vahidnia, R. (2020). A simplistic view on latency of random access in cellular internet of things. *IEEE IEMCON* 391–395. https://doi.org/10.1109/IEMCON51383.2020.9284948.

Exercises

8.2.1 Give at least four possible reasons for the occurrence of an unsuccessful random access process.

8.2.2 Three NB-IoT devices are connected to a cell that transmits the bitmap pattern of 0010000000 for EAB. Device #1 has a class barring of 2, device #2 has a class barring of 8, and device #3 has a class barring of 3. Which one of these devices is barred? Why?

8.2.3 Are ACB and EAB supported by both LTE-M and NB-IoT technologies?

8.2.4 In an NB-IoT network, there are three IoT devices. One is in the normal coverage level, one is located in the extended coverage level, and the last one is located in the extreme coverage level. The maximum number of preamble transmission attempts in each coverage level is 1 for the device in the normal coverage level, 2 for the one in the extended coverage level, and 3 for the IoT device in the extreme coverage level. Assume that all RA initiations result in failure. Show the coverage level ramping for these devices as shown in Table 8.2.E4.

Table 8.2.E4 Exercise 8.2.4 data.

Coverage Level	First attempt	Second attempt	Third attempt
Extreme	Device #3		
Extended	Device #2		
Normal	Device #1		

8.2.5 What is power ramping?

8.2.6 What is the duration of each LTE-M preamble format? If each subframe is 1 ms, what is the number of subframe(s) used for the transmission of a preamble?

8.2.7 What are the names of the RA advancements made by 3GPP in Release 15 and 16?

Advanced Exercises

8.2.8 Does an IoT device activate the PUR feature, or is it the duty of a base station to activate this feature?

8.2.9 To use PUR, the IoT device needs to validate some conditions before using this feature. What are some of these conditions? What happens if the conditions are not satisfied?

8.2.10 If PUR is enabled for a CIoT device, the resources allocated for this device are available periodically or based on its traffic pattern. The assumption is that the IoT device has a periodic or deterministic traffic pattern. What happens if the CIoT device does not have data to send in some of the allocated times?

Chapter 8.3 Positioning

8.3.1 Introduction

Positioning is the process of determining the geographical location of an IoT device. Besides location, it can also determine the velocity of mobile IoT devices. There are many IoT use cases that need to provide location-based services, and therefore, require positioning estimation.

Positioning can become very challenging in dynamically changing environments and for IoT devices that are mobile. In general, IoT devices that require positioning information have to accurately estimate the location and/or velocity in all environments including urban, rural, indoor, or outdoor areas. Low latency in position estimation is another important requirement for many IoT applications. Latency in positioning defines the time from when an IoT device starts to estimate the position until the time that it obtains the location information. It is clear that different IoT applications may have different requirements in terms of the resolution of location information or the amount of latency that they can tolerate.

When it comes to positioning, people usually associate positioning with the Global Positioning System (GPS) developed by the United States in 1995. But GPS is just one of the systems that exist to provide positioning information. Besides GPS, there are other positioning systems, called Global Navigation Satellite Systems (GNSS), which use satellite radio signals for position estimation. For example, GLObal Navigation Satellite System (GLONASS) of Russia, Galileo of European countries, or even Compass of China are all part of the GNSS system.

To estimate the position using satellite radio signals, there is a need for receiving unobstructed line-of-sight signals from at least four satellites. It is not always easy to receive four satellite signals in all environments. Therefore, GNSS is not suitable for environments where no satellite signal is available, or the areas where the number of received satellite signals is less than four. Since the power consumption of a satellite receiver is usually high, the use of satellite technology is also not attractive in battery-powered IoT applications. While GNSS provides reliable coverage for some IoT applications, other solutions are needed to address those IoT applications whose requirements cannot be fulfilled by using a GNSS system. For example, GNSS does not work in indoor areas, while there are many indoor IoT applications.

One of the problems with the GNSS system is its large acquisition time in situations where the GNSS receiver does not receive the satellite signal or when the receiver is off for a long time. Time To First Fix (TTFF) is a metric that indicates the duration of time that is needed for a GNSS receiver to acquire satellite signals and consequently estimate its position. Assisted GNSS (AGNSS) is a system that can significantly improve the start-up acquisition time of a satellite system. In an AGNSS system, the cellular network assists the GNSS receiver to receive satellite broadcast data. In general,

Fundamentals of Internet of Things: For Students and Professionals, First Edition. F. John Dian.

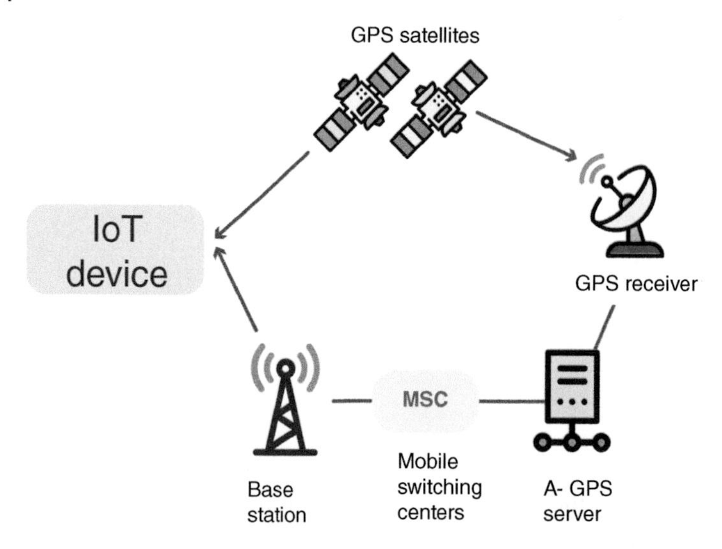

Figure 8.3.1 An AGNSS system.

satellites broadcast two types of data: almanac and ephemeris. Almanac data provides information on the status of all satellites in the GPS network, while ephemeris data is highly detailed orbital and clock-correction data for each satellite. This information is required for determining positioning. Each satellite sends only its own ephemeris data. Assistance data usually contains the almanac and/or ephemeris data that is received as part of the satellite navigation message. In an AGNSS system, the base station transmits the contents of the navigation message. This enables the GNSS receiver to know which satellites are in sight, which consequently makes the search region to become smaller. This consequently reduces the start time to a few seconds as compared to tens of seconds or even minutes. Figure 8.3.1 shows an AGNSS system in which the satellite sends almanac and ephemeris data to a stationary receiver that is connected to a GNSS server and mobile services. The satellite's navigation messages are then transmitted through the base station toward an IoT device that requires location information.

Assisted GNSS cannot solve the problem with indoor positioning. Since many IoT applications are located indoors, GNSS and AGNSS are not suitable for these IoT applications. Cellular technologies can provide a solution for indoors and urban areas. In this case, position estimation can be obtained by using existing cellular infrastructure and technologies. Positioning has been part of LTE cellular technology since 3GPP Release 9. Regulations for emergency calls were the primary motivator for considering positioning in LTE. As IoT is introduced in 3GPP Release 13, some of the existing LTE positioning solutions are also utilized as LTE-IoT positioning schemes in 3GPP Release 13. The advantages that a reliable positioning scheme could provide for IoT applications, such as asset tracking systems, were the main reasons behind including and advancing the positioning capability in LTE-IoT. 3GPP technical specifications [1–3] discuss details about positioning in LTE-IoT.

8.3.2 LTE Positioning

In 3GPP Release 13, the only positioning method available for LTE-M and NB-IoT was Cell IDentity (CID). As shown in Table 8.3.1, 3GPP Release 14 introduced two basic positioning

Table 8.3.1 3GPP positioning schemes.

Technology	Release 13	Release 14
LTE-M	CID	ECID, OTDOA
NB-IoT	CID	ECID, OTDOA

solutions for IoT applications. The first one was Enhanced Cell Identity (ECID) that was the same as ECID introduced in 3GPP Release 9 for LTE. The second one is Observed Time Difference Of Arrival (OTDOA) that is also a 3GPP Release 9 positioning method in LTE. In this section we explain these positioning methods.

8.3.2.1 CID

CID estimates the position of a device with the knowledge of its serving base station. The method reports the identity or geographical information of a cell to which the device is connected. As far as cellular coverage exists in a region, CID is easily accessible, and it is part of the connection information. Therefore, the device can access the CID information very fast. However, the location information only shows the cell that the device resides in. Location accuracy is dependent on cell size. This method does not meet the requirements that are needed in most IoT applications. Figure 8.3.2 shows a situation where the location of the device is determined by a cell number. Knowing the border coordinates or the size of a cell can give us a better idea about the position accuracy of the CID method.

8.3.2.2 ECID

There have been various attempts to enhance the accuracy of the CID positioning scheme, resulting in a set of ECID methods. Therefore, ECID can be considered as an enhanced version of CID that uses additional information in order to measure the position more accurately as compared to CID. In ECID, an IoT device performs and collects necessary measurements and information from the serving cell and neighboring cells and reports a message back to the base station. Some typical

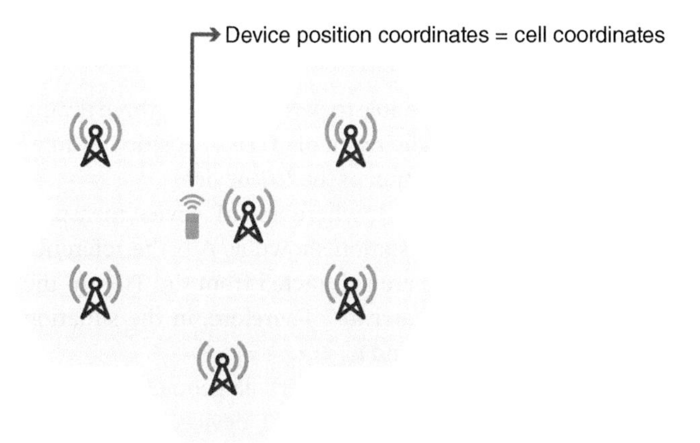

Figure 8.3.2 CID positioning scheme.

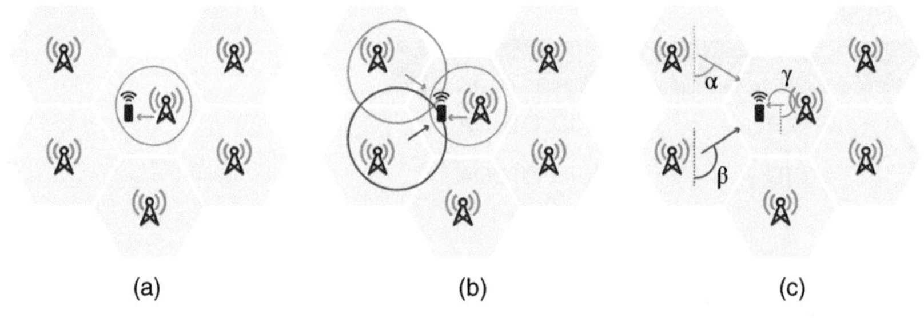

Figure 8.3.3 Different ECID schemes: (a) using distance to serving cell; (b) using distance to serving cell as well as neighboring cells; (c) using AoA from different base stations.

examples of ECID schemes are shown in Figure 8.3.3. For example, in Figure 8.3.3a, the location of a device is determined based on the strength of the received signal inside the cell. The IoT device measures the power of the received signal from the serving cell in order to determine the distance. In Figure 8.3.3b, the position is determined using the power received from two neighboring base stations in addition to the power received from the serving cell. This method can determine the position that is the intersection of three circles, each circle indicating the distance between the IoT device and one of the base stations. In Figure 8.3.3c, the IoT device measures the Angle of Arrival (AoA) from different base stations and uses this information in order to estimate the position.

An IoT device usually performs different types of measurements. For example, an IoT device might measure RSRP, Reference Signal Received Quality (RSRQ), and Rx-Tx time difference along with the serving cell ID. RSRP is the average power received from a single reference signal, while RSRQ indicates a measure of the quality of the received reference signal. RSRP is measured in dBm, while RSRQ is measured in dB. The time difference between the reception and transmission time of a serving cell is represented by the Rx-Tx time. In other words, the Rx-Tx time is the time difference between the start of a downlink message and the transmission of the corresponding uplink message. From this information, ECID estimates the position of an IoT device.

8.3.2.3 Observed Time Difference of Arrival (OTDOA)

OTDOA is a multi-lateration method in which the IoT device measures the Time of Arrival (ToA) of signals received from multiple base stations. Since base stations broadcast the downlink Positioning Reference Signal (PRS) periodically, the IoT device measures the PRS time difference of arrivals of several neighboring cells in comparison to a reference cell. The measurements performed by IoT devices are called Reference Signal Time Difference (RSTD).

For position estimation using OTDOA, an IoT device must be able to perform at least three timing measurements from geographically dispersed base stations. The result of OTDOA position estimation would be two coordinates (latitude/longitude) of the location of the IoT device.

The OTDOA positioning method is illustrated in Figure 8.3.4, where an IoT device measures three TOAs of PRSs as τ_1, τ_2, and τ_3. In this figure, the base station shown as A is the reference base station. The TOAs from several neighbor base stations are subtracted from the TOA of the reference base station to form "observed time difference of arrival." Therefore, in the situation shown in Figure 8.3.4, two OTDOAs are formed: $t_{21} = \tau_2 - \tau_1$ and $t_{31} = \tau_3 - \tau_1$.

It can be proven that each time difference equation represents a hyperbola geometrically. The intersection point of these hyperbolas determines the location of the CIoT device. For example, since there are two time difference equations in Figure 8.3.4, we have two hyperbolas. By calculating the point at which these two hyperbolas intersect, the location of the device can be determined.

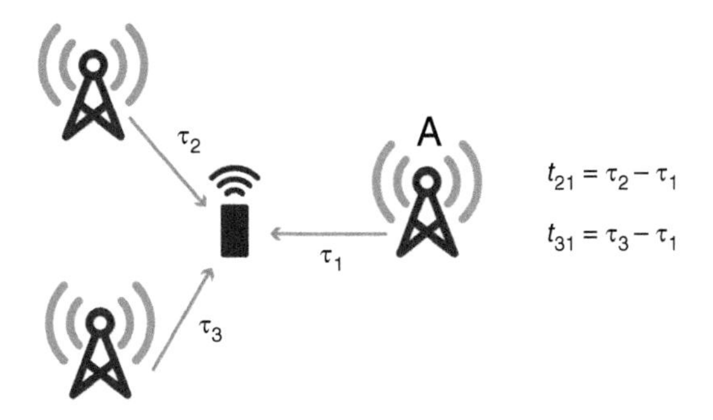

Figure 8.3.4 TOAs from two neighboring base stations and a reference base station.

8.3.2.3.1 Basic OTDOA Navigation Equations

TOA measurements performed by an IoT device have a direct relationship with the geometric distance between the IoT device and the base station. Let us say that the known coordinates of a base station are $X_i = [x_i, y_i]$ and the unknown coordinates of an IoT device are $X_t = [x_t, y_t]$. The RSTD measurements are defined as the time difference between the arrival of the PRS from two neighboring base stations and a reference base station. Let us assume that all base stations start sending the PRS at the same time, and then RSTD between the base station i and the reference base station can be expressed as:

$$RSTD_{i,R} = \frac{\sqrt{(x_t - x_i)^2 + (y_t - y_i)^2}}{c} - \frac{\sqrt{(x_t - x_R)^2 + (y_t - y_R)^2}}{c} \tag{8.3.1}$$

In Eq. (8.3.1), c is the speed of light, and $X_R = [x_R, y_R]$ represents the coordinates of the reference base station. This equation represents a hyperbola. If an IoT device can measure at least two RSTD values from two neighboring base stations, then it can estimate the location of the IoT device $X_t = [x_t, y_t]$. In other words, by having two neighboring base stations ($i = 2$) in addition to the reference base station, the location of the IoT device $X_t = [x_t, y_t]$ can be determined. In this situation, we have two parabolas and the intersection of these two hyperbolas is the location of the IoT device. Usually, measurements from more than two neighboring base stations ($i > 2$) are desired to perform position estimation more accurately. The hyperbolas resulted from two neighboring cells ($i = 2$) are shown in Figure 8.3.5.

There is no reason that base stations should be synchronized and send PRSs at the same time. If T_i and T_R represent the start time of transmitting the PRSs by the base station i and the reference base station, respectively, then $T_i - T_R$ is the time offset between the two base stations. This time offset is referred to as Real Time Differences (RTDs). It is clear that the value of $T_i - T_R$ should be

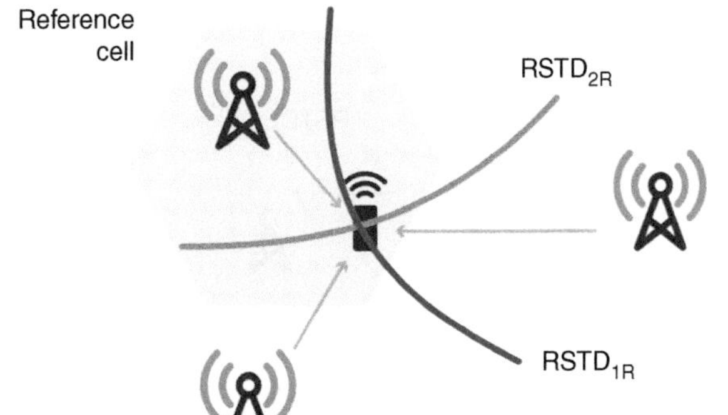

Figure 8.3.5 OTDOA uses the intersection of at least two hyperbolas for the position estimation.

zero in a synchronized network. To consider base stations that are not completely synchronized, Eq. (8.3.1) can be written as:

$$RSTD_{i,R} = \frac{\sqrt{(x_t - x_i)^2 + (y_t - y_i)^2}}{c} - \frac{\sqrt{(x_t - x_R)^2 + (y_t - y_R)^2}}{c} + T_i - T_R \tag{8.3.2}$$

An IoT device needs to measure the ToA for the reference base station as well as all other base stations that are part of OTDOA calculations. If $RSTD_{i,R}$ measurements were noise and interference free, the i hyperbolas would intersect at one point, which was the exact location of the IoT device. For several different reasons that we will discuss later in Section 8.3.6, there might be errors in measuring ToAs. If N_i and N_R are the errors in the estimation of the ToA for the received signals from the base station and the reference base station, respectively, then Eq. (8.3.2) can be modified to show the effect of measurement errors.

$$RSTD_{i,R} = \frac{\sqrt{(x_t - x_i)^2 + (y_t - y_i)^2}}{c} - \frac{\sqrt{(x_t - x_R)^2 + (y_t - y_R)^2}}{c} + (T_i - T_R) + (N_i - N_R) \tag{8.3.3}$$

To be able to estimate the location of IoT devices precisely, the coordinates of each neighboring base station $X_i = [x_i, y_i]$, the coordinates of the reference base station $X_R = [x_R, y_R]$, the transmit time offsets $T_i - T_R$, and the values of $RSTD_{i,R}$ that is measured by each IoT device are needed. Any uncertainty in these parameters has an impact on the accuracy of the position estimation process.

To show the effect of uncertainties in ToA measurements, each hyperbola can be represented by two marginal hyperbolas that are separated by a constant width. The location of the IoT device is the intersection area formed by the intersection of these marginal hyperbolas as shown in Figure 8.3.6.

OTDOA positioning method can be combined with the GNSS data. Clearly, using the GNSS data can improve the OTDOA estimation accuracy, while using OTDOA without GNSS can reduce the cost, and the additional space that is needed to integrate the GNSS receiver.

8.3.2.3.2 Positioning Reference Signals (PRSs)

We discussed that OTDOA measurements are based on the arrival time of PRSs. One may think that these measurements can be performed on any downlink synchronization signals. Examples of these signals are the synchronization signals (primary and secondary synchronization signals) as well as reference signals. However, these signals are not suitable in situations where signals from multiple neighboring base stations have to be detected by an IoT device. Therefore, PRSs have been introduced to allow proper timing measurements of an IoT device from the base stations' signals in order to improve the OTDOA positioning performance.

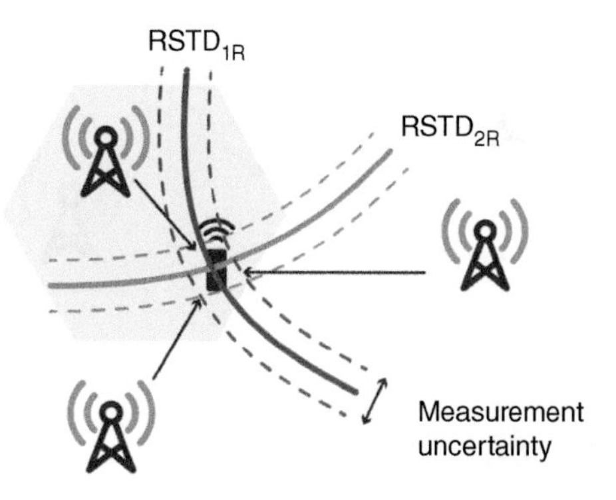

Figure 8.3.6 Effect of OTDOA measurement error on the position accuracy.

8.3.3 Positioning Architecture for LTE-IoT

A typical architecture of the positioning system for LTE-IoT is illustrated in Figure 8.3.7. The architecture consists of several entities, such as Mobility Management Entity (MME), Gateway Mobile Location Center (GMLC), and Evolved Serving Mobile Location Center (E-SMLC), which is also known as location server, in addition to base stations and IoT devices.

There are two types of position estimation processes as it relates to IoT devices. One is assisted positioning in which the IoT device provides RSTD measurement values to the location server. Then the location server performs the computation and sends back the estimated position coordinates to the IoT device. In the second type, the IoT device performs both position measurements and computations. A location server or E-SMLC is a logical or physical entity that is responsible for managing the position estimation of IoT devices. A location server may also compute (assisted positioning) or just verify the final location estimate. Both ECID and OTDOA methods are assisted positioning methods, and the computation is not done inside an IoT device. It is clear that cellular IoT devices need to be low cost, and for this purpose, the computation will be performed on the location server. However, in GNSS or AGNSS systems, both assisted positioning and device-based positioning are considered.

Location requests can be initiated by the IoT device or by the network. If the IoT device needs its location information, it can initiate the process. If the network initiates the process, it can be rejected by the IoT device using the device privacy settings. However, on some occasions, typically used for emergency services, the IoT device cannot reject the request.

Let us consider a rescuing service as an example in which we try to determine the location of a person who is wearing an IoT device. In this scenario, the GMLC is the first node of the network that receives a request from an external client that needs the location information of a particular IoT device. After successful registration and authorization, the request for positioning information is sent to MME, which then transmits a positioning request to a location server. This triggers a positioning process in which the location server sends a request to the IoT device. The IoT device performs the measurements and sends the measurement values to the location server. If the assisted positioning method is used, the location server calculates the location of the IoT device and sends the results to MME. Depending on the application, MME can send the location coordinates to the IoT device or to a GMLC interested readers may refer to [4] to study these protocols.

The signaling between the location server and the IoT device is carried out using the LTE Positioning Protocol (LPP). The signaling between the location server and the base station is based on the LTE Protocol Positioning annex (LPPa) protocol as shown in Figure 8.3.7 [5]. The explanation of these protocols is outside the scope of this book.

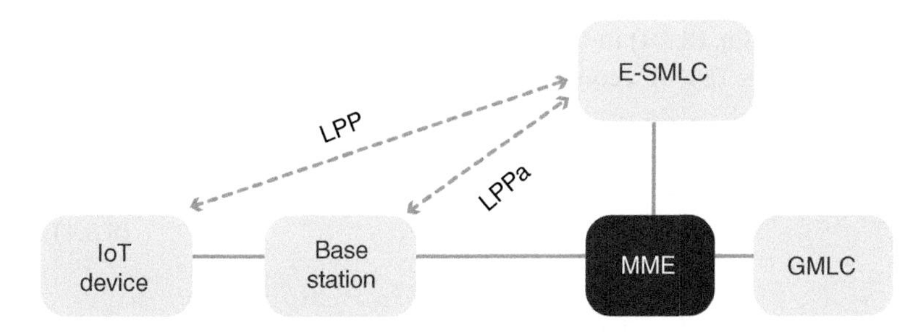

Figure 8.3.7 Positioning architecture for LTE-IoT.

The location server and the IoT device exchange many messages over the LPP protocol. For example, the IoT device can provide the location server with its capability parameters, such as supported frequency bands in response to a request from the location server. In addition, the location server can assist an IoT device by recommending a list of cells and their specific PRS configuration. The location server also provides the IoT device with a search window and calculates the RSTD value that the IoT device is supposed to measure in order to help the IoT device in its operation. For this purpose, the location server needs to have a rough idea about the location of the IoT device. This rough knowledge can be obtained using CID or ECID positioning schemes. The location server selects neighboring cells around the estimated location as the candidates for RSTD measurements. By having a rough idea about the location of the CIoT device, and the location of each selected neighboring cell, the location server can calculate the RSTD value that the IoT device is expected to measure. The main message between the location server and the IoT device occurs when the location server requests for location information and the IoT device sends the RSTD measurements or computed location coordinates to the location server.

In general, there are two different architectures for LTE-IoT. The first one is called Control Plane (CP) and the other is called User Plane (UP).

In a CP architecture, all the signaling messages that are used to initiate a positioning event and perform the position estimation process happen over control channels. What we explained in Figure 8.3.7 is an example of a CP architecture, in which an MME transmits a positioning request to the location server in order to trigger the positioning process. In this case, the location server returns the location coordinates to MME after performing position estimation.

In a UP architecture, all the signaling messages that are used to initiate a positioning event and perform the position estimation process happen over data channels. The UP architecture can be implemented with any types of cellular networks (2G, 3G, 4G, or 5G), since the signaling for position estimation is sent as user data of that particular cellular network. Therefore, UP architecture is not dependent on a particular cellular technology. The location server in this architecture is called Secure User Plane Location (SUPL). So, instead of having the E-SMLC entity as the location server in the CP architecture, the UP architecture introduces SUPL as the location server.

8.3.4 RSTD Measurement Performance

In LTE, the requirements for accurate RSTD measurements specify that mobile devices should report RSTD measurements for the reference cell as well as all neighboring cells under specific conditions. To satisfy these conditions, the signal to noise ratio of PRS for the reference cell should be greater than −6 dB and for neighboring cells should be greater than −13 dB. The accuracy requirement is a function of the PRS bandwidth. If the bandwidth is higher, the accuracy is higher as well. LTE position accuracy is expressed in Eq. (8.3.4) in terms of T_s. This value is the basic time unit defined in LTE and is equal to $1/(2048 \times 15000)$ seconds, which is around 32 ns.

$$\begin{array}{ll} \pm 15T_s \text{ and } \pm 21T_s & BW < 5\,\text{MHz} \\ \pm 6T_s \text{ and } \pm 10T_s & 10\,\text{MHz} < BW \leq 5\,\text{MHz} \\ \pm 5T_s \text{ and } \pm 9T_s & BW \geq 10\,\text{MHz} \end{array} \tag{8.3.4}$$

In Eq. (8.3.4), two numbers are expressed for each bandwidth range. The first number indicates the accuracy for intra-frequency measurements and the second number shows the accuracy for inter-frequency measurements. For example, for a bandwidth of lower than 5 MHz, the accuracy is $\pm15T_s$ for intra-frequency measurements and $\pm21T_s$ for inter-frequency measurements. Based on Eq. (8.3.4), we can find that the accuracy requirement for intra-frequency measurements is around ±150 m for low bandwidth and around ±50 m when the bandwidth is 10 MHz or higher. By the same token, the accuracy requirement for inter-frequency measurements is around ±210 m for low bandwidth and it is around ±90 m when the bandwidth is 10 MHz or higher.

8.3.5 PRS Signals

PRSs for both LTE-M and NB-IoT are based on and somewhat similar to PRSs in LTE. However, there are some differences between LTE-M and NB-IoT PRSs and the one used in LTE. We discuss PRSs in LTE, LTE-M, and NB-IoT in this section.

8.3.5.1 LTE PRS Signals

PRS is transmitted in the downlink direction to enhance positioning measurements of a mobile device. To achieve good positioning accuracy, PRSs need to be transmitted in such a way that they can be detected by an IoT device, and they can provide good measurement resolution for position estimation. PRS is distributed in time and frequency over a 1 ms period, that is, the duration of a radio subframe in LTE. A PRS may be sent over a number of consecutive subframes, which is called positioning occasions. In LTE, a positioning occasion may take 1, 2, 4, or 6 ms. Positioning occasions are transmitted periodically with a period of T_{PRS}. In LTE, T_{PRS} can be 160, 320, 640, or 1280 ms. The PRS frequency-time allocation for LTE and for positioning occasions of 2 ms is shown in Figure 8.3.8.

To better detect a weak PRS that is transmitted by a base station that shares the same frequency shift with a strong PRS transmitted by another base station, the base station with the stronger PRS may be muted in certain occasions. This is called PRS muting in LTE terminology. Each base station has a PRS muting pattern that shows whether the base station is allowed to transmit PRSs or not. The PRS muting helps to reduce interference when the signal to noise ratio of the PRS is low. However, PRS muting may increase Time To Fix (TTF) of positioning. This is a question for network administrators to decide how to configure the PRS muting to have a better reception of PRSs at the cost of increasing the TTF.

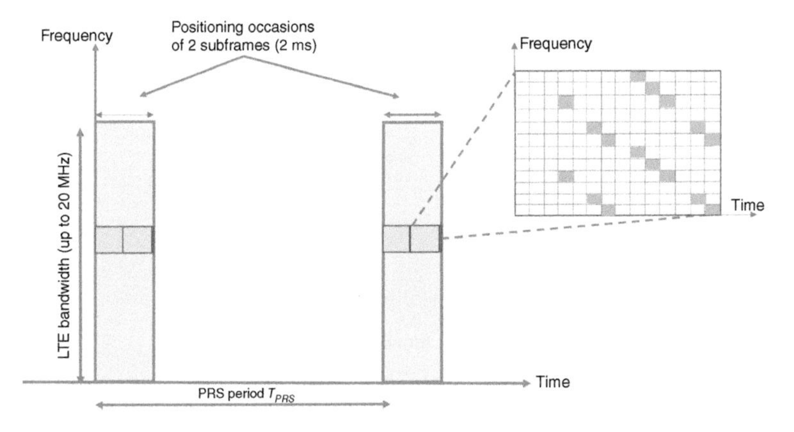

Figure 8.3.8 The PRS frequency-time allocation for LTE with the positioning occasion of 2 ms.

8.3.5.2 LTE-M PRS Signals

Since LTE-M has a lower spectrum (1.4 MHz in 3GPP Release 13, and up to 5 MHz in 3GPP Release 14), and also extends coverage for deployments in deep indoor and challenging environments, it introduces denser PRSs as compared to PRSs in LTE. To achieve denser PRSs, LTE-M introduces longer PRS positioning occasions of 10, 20, 40, 80, and 160 ms in addition to 1, 2, 4, and 6 ms that existed in LTE. LTE-M also introduces smaller positioning occasion intervals of 10, 20, 40, and 80 ms in addition to the positioning occasion intervals of LTE that are 160, 320, 640, and 1280 ms. Figure 8.3.9 compares PRSs for LTE and LTE-M. Figure 8.3.9a shows the positioning occasions of up to 6 ms in the LTE network as compared to Figure 8.3.9b, which is the positioning occasions of up to 160 ms in the LTE-M network. Since the bandwidth of LTE-M is smaller than LTE, LTE-M can use longer positioning occasions of up to 160 ms as shown in Figure 8.3.9b. It can also use small positioning occasions with smaller T_{PRS} values as 10 ms.

LTE-M uses PRS muting almost similar to LTE. Another feature added to PRS in LTE-M is frequency hopping. Figure 8.3.9 shows the location of PRSs in a subframe in terms of frequency and time. The location of a PRS is fixed within each subframe in LTE; however, LTE-M uses frequency hopping and the location of the PRS changes using known patterns.

8.3.5.3 NB-IoT PRS Signals

PRSs in NB-IoT are called NPRS. There are three different ways to configure NPRS in NB-IoT. The first method is exactly similar to what we explained for LTE or LTE-M in which the location of PRSs can be configured as the length of each positioning occasion and the frequency of occurrence of the positioning occasions. Since NB-IoT has a lower spectrum as compared to LTE and LTE-M, and also extends coverage for deployments in deep indoor and challenging environments, therefore NB-IoT introduces denser PRSs as compared to PRSs in LTE and LTE-M. To achieve that, NB-IoT introduces longer PRS positioning occasions of 10, 20, 40, 80, 160, 320, 640, and 1280 ms. The maximum length of PRS positioning occasions in LTE was 6 ms and in LTE-M was 160 ms. NB-IoT does not consider smaller than 10 ms as the length of its positioning occasion. The positioning occasion intervals defined for NB-IoT are the same as the ones defined for LTE.

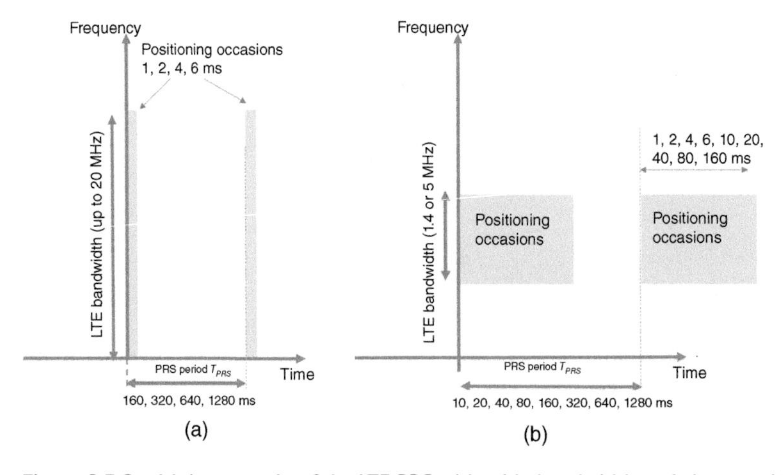

Figure 8.3.9 (a) An example of the LTE PRS with wide bandwidth and short positioning occasions. (b) An example of the LTE-M PRS with narrow bandwidth and wide positioning occasions.

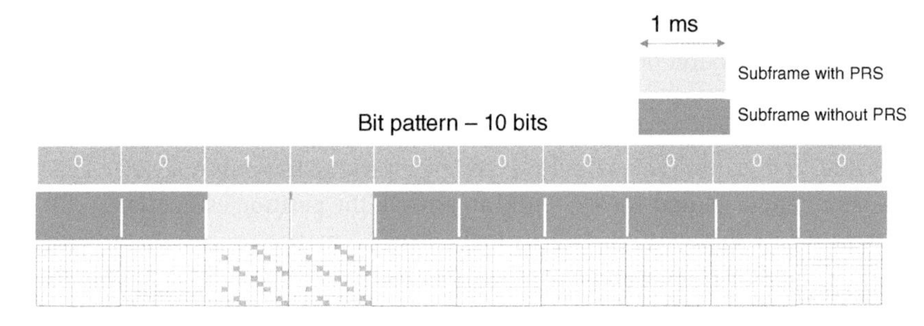

Figure 8.3.10 Bitmap pattern used in NB-IoT for location of PRSs.

In the second method, the base station transmits a bitmap of 10 bits or 40 bits every 10 or 40 ms, respectively. Each bit in this bitmap indicates the existence of an NPRS at each 1 ms. For example, a bitmap of 0011000000 shows the existence of NPRSs in the next 10 ms as shown in Figure 8.3.10. If NPRS exists during a 1 ms period, exact locations of NPRSs in terms of frequency and time within this period are fixed for the NB-IoT system.

The third method uses a combination of the first and second methods, in which NPRS exists when both the first and second methods indicate the existence of PRSs.

8.3.6 RSTD Error Sources

RSTD measurements are very sensitive to interference. On the other hand, the accuracy of RSTD measurements has a huge impact on the precision of the OTDOA position estimation. Therefore, to understand how accurately OTDOA can estimate the location of an IoT device, we should understand sources of interference in RSTD measurements.

The wireless signal often suffers from multipath propagation and fading. Therefore, RSTD values measured by IoT devices might be far from accurate in some scenarios.

The most important source of error in the OTDOA positioning method is multipath. Simply put, multipath occurs when the wireless signal arrives at the receiver through various paths due to reflection, diffraction, or scattering as illustrated in Figure 8.3.11. Obstacles such as buildings or trees on the signal path can cause transmitted signal to arrive at the receiver from different

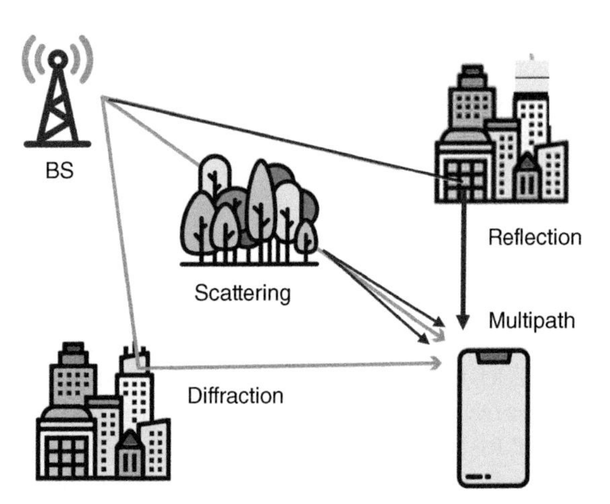

Figure 8.3.11 Multipath effect caused by reflection, diffraction, and scattering.

directions. Reflection involves a change in direction of the signal when it bounces off an object, diffraction involves a change in direction of the signal as it hits the edges or sharp corners of an object in its path, and scattering happens when a signal touches an object that has irregularities with dimensions smaller than or on the order of the wavelength of the signal.

These reflected, diffracted, and scattered signals also experience various amounts of delays. This results in an RSTD measurement error and consequently an error in the position estimation.

OTDOA needs at least three base stations to participate in the error estimation process. Increasing the number of base stations and RSTD measurements increases accuracy. Especially in multipath environments, the existence of many RSTD measurements might make it possible to remove or reduce errors in these measurements. If there are additional accurate measurements with less errors caused by multipath, it is possible to remove or reduce the weight of those measurements, which have experienced larger errors.

8.3.7 Summary

The process of determining the geographical location of a mobile IoT device by means of radio signals is called positioning. There are many IoT use cases that require location information, and therefore, positioning has a crucial role to play in many IoT applications. Besides using satellites for positioning that demand high power consumption and do not work efficiently in indoor scenarios, position estimation can be well obtained by using existing cellular infrastructure and technologies.

Positioning has been offered in the LTE cellular technology since 3GPP Release 9 and has evolved with IoT support in 3GPP Release 13. LTE-M and NB-IoT in 3GPP Release 13 use CID method. In 3GPP Release 14, ECID and OTDOA positioning solutions have been introduced in order to increase the performance of the position estimation process. ECID can be considered as an enhanced version of CID that uses additional information in order to measure the position more accurately as compared to CID. For example, ECID can use the AoA from different base stations in order to estimate the position. In OTDOA, RSTD measures the PRS and Time Difference Of Arrivals (TDOA) for several neighboring cells compared to a reference cell in order to perform OTDOA. We explained that RSTD measurements are performed by IoT devices. The more accurate the measurements, the more precise position estimation can be designed. The brief discussion on the RSTD performance and error sources discussed in Section 8.3.6 gives readers an understanding behind the uncertainty of a positioning process.

A high level description of the LTE-IoT architecture and its components were discussed in this chapter to give readers some idea about the names and the functionalities of the main entities that are involved in the LTE positioning process. Generally speaking, there are two different positioning architectures for LTE-IoT, which are called control plane and user plane architectures.

References

1 3GPP TS 36.305 (2016). Evolved Universal Terrestrial Radio Access Network (E-UTRAN); Stage 2 functional specification of User Equipment (UE) positioning in E-UTRAN. V14.0.0.

2 3GPP R1-167743 (2016). Requirements for NB IoT positioning enhancements evaluations. https://www.3gpp.org/ftp/tsg_ran/WG1_RL1/TSGR1_86/Docs/ (accessed 30 June 2022)

3 3GPP R1-167790 (2016). Requirements for NB-IoT/eMTC positioning. https://www.3gpp.org/ftp/tsg_ran/WG1_RL1/TSGR1_86/Docs/ (accessed 30 June 2022)

4 3GPP TS 36.355 (2017). Evolved Universal Terrestrial Radio Access (E-UTRA); LTE Positioning Protocol (LPP). V14.2.0.

5 Lin, X., Bergman, J., Gunnarsson, F. et al. (2017). Positioning for the Internet of things: A 3GPP perspective. *IEEE Communications Magazine.* 55 (12): 179–185. https://doi.org/10.1109/MCOM.2017.1700269.

Exercises

8.3.1 Does AGNSS work in an indoor environment since it gets assistance from the cellular networks?

8.3.2 A circular cell has a radius of 5 km. The cellular network uses CID positioning. What is the position accuracy of the method? Assume the cell ID shows the coordinates of the center of the cell.

8.3.3 Do you think the accuracy of CID is better in rural areas or urban areas? What about OTDOA or GNSS?

8.3.4 Is ECID introduced in 3GPP Release 9 or 14?

8.3.5 What is the difference between intra-frequency and inter-frequency measurements of RSTD?

8.3.6 What is the effect of the PRS bandwidth on the accuracy of OTDOA positioning?

8.3.7 Give at least three reasons for error in the RSTD measurement? Which one is the most important one?

8.3.8 What phenomenon causes a wireless signal to change direction and bend in the following scenarios?

a. The signal changes its direction over the tops of buildings or around street corners.
b. The signal encounters the surface of a large building wall or an asphalt road.
c. The signal encounters a chain link fence or the leaves of a tree.

Advanced Exercises

8.3.9 An RSTD measurement between a reference cell located at (0, 0) and a neighbor cell located at (2, 1 km) is measured and is $2.91\mu s$. Assuming both base stations transmit positioning reference signals at the same time, is it possible that the position of the device be at (0.5 and 0.5 km)?

8.3.10 Two RSTD measurements are performed by a device. $RSTD_{1,R}$ between a reference cell located at (0, 0) and a neighbor cell located at (4, 1 km) is measured and is $2.91\mu s$. $RSTD_{2,R}$ between a reference cell and another neighbor cell located at (1, 4 km) is also $2.91\mu s$.

Assuming all base stations transmit positioning reference signals at the same time, what are the coordinates of the device?

8.3.11 An RSTD measurement between a reference cell located at (0,1 km) and a neighbor cell located at (3, 1 km) is performed. If the PRS from the reference cell arrives $1\mu s$ earlier, what is the equation of hyperbola generated by this measurement?

8.3.12 Figure 8.3.9 compares PRS signals of LTE and LTE-M. Figure 8.3.9a is for LTE, which supports positioning occasions of up to 6ms as compared to Figure 8.3.9b, which is for LTE-M that supports positioning occasions of up to 160 ms. Figure 8.3.9b shows a situation where the network uses longer positioning occasions of up to 160 ms. LTE-M can also use small positioning occasions with smaller T_{PRS} value of 10 ms. Complete Figure 8.3.9 to show this situation.

Chapter 8.4 Mobility

8.4.1 Introduction

In this chapter, we discuss the concept of mobility in Cellular IoT (CIoT). One may think that CIoT devices that are based on technologies such as LTE-M or NB-IoT may have exactly the same mobility performance as legacy LTE. Even though it is true that these technologies are deployed on the existing LTE network, they are not completely identical in terms of mobility.

To be able to provide mobility to IoT devices, the network should be able to have an understanding of the location of these IoT devices. When an IoT device connects to a base station in order to establish communication and deliver its data uplink, the network learns about its location. In addition, when a device, which is connected to the network, moves from one cell to another, it must disconnect from one base station and connect with another. As a result, the network can determine the position of a mobile device depending on the strength and quality of the signal it receives. However, if the network needs to communicate with an IoT device to transfer data in the downlink direction, it must first determine the mobile device's location in order to send the data to the cell in which the mobile device is located. In this situation, it is possible that the device was in sleep mode when the previous connection was made and since then has moved away from that cell. And therefore, the network does not know the location of the device. The process used by the network to send signaling information in order to find the location of a device is called paging. Paging should be performed based on a paging strategy and should be fast and effective. Paging strategy should optimize the factors involved in the paging process. For example, paging all base stations in a network at the same time reduces the latency of the paging process at the cost of generating a huge amount of signaling traffic, which is not practical. When the network manages to find the cell in which the IoT device is located, the data can be directed to the base station that represents the cell. If an IoT device can notify the network of a cell change as soon as it occurs, or if the network can quickly locate the position through the paging process, the network will be able to transfer data to the correct base station in a timely fashion. CIoT devices are mostly battery powered and to prolong their battery lifetime, they try to consume as less power as possible by going to sleep mode and turning off the transmission and reception circuits. This means sending less mobility signals uplink, and not always receiving the downlink data packets or signaling information such as paging messages. Due to specific requirements of IoT devices in terms of their power consumption and device complexity, the mobility in LTE-IoT technologies might not be exactly similar to the mobility provided by the LTE technology for smartphones.

In this chapter, we discuss the fundamentals of mobility in LTE networks, and the changes that have been made to LTE in order to provide mobility for the LTE-IoT technologies. GPP publications

Fundamentals of Internet of Things: For Students and Professionals, First Edition. F. John Dian.

explain the specifications of the mobility for LTE and LTE-IoT networks. The details are covered in numerous 3GPP documents. Examples of these documents are [1–5].

8.4.2 Mobility

In subchapter 8.1, we discussed RRC idle and connected modes. When a device is first switched on, it goes to the RRC idle mode and attempts to select a cell to camp on. The process of finding a cell is called cell selection. When a mobile device selects a cell, it also needs to monitor several neighboring cells. If the mobile device detects that it can receive a stronger signal from one of the neighboring cells as compared to the serving cell, it should start the reselection process. Cell selection and cell reselection are two important processes used in the LTE and LTE-IoT technologies and are the basis of the mobility operation for these technologies. In this section, we discuss cell selection and cell reselection processes in more detail.

8.4.2.1 Cell Selection

The cell selection process determines a suitable cell for a mobile device to camp on and synchronize to. In an LTE network, base stations broadcast two synchronization signals that can be used by a mobile device to obtain the identity of a cell and the timing of the radio frames. These two downlink synchronization signals are Primary Synchronization Signal (PSS) and Secondary Synchronization Signal (SSS). To perform cell selection, an IoT device should search for PSS and synchronize to the SSS signal that are broadcasted by the network. The network also broadcasts information about system bandwidth and other information that is useful in the cell selection process. After synchronization to SSS, an IoT device can receive information from the network including the unique cell identity and tracking area.

8.4.2.2 Cell Reselection

After a mobile device selects a cell, it should continue to monitor some of the neighboring cells and perform continuous measurements of the signal's power and quality to ensure that the device is camped on the best cell. It is clear that continuous measurements enable the IoT device to perform quick handover between cells, which consequently results in having a good link for communications.

Repeated reselection processes among cells in a relatively short period of time are called the ping-pong effect. To avoid a ping-pong effect in the reselection process between two cells, the network provides the mobile device with a predetermined period called $T_{Reselect}$. After a selection or reselection process takes place, the mobile device is not allowed to start another reselection process for a period of $T_{Reselect}$. This period is for low-speed mobile devices, while high-speed devices can scale down this period using another parameter that is sent by the network and it is called Scaling Factor (SF). The amount of time after the reselection process that a high-speed CIoT device is not allowed to start another reselection process is $T_{Reselect} \times SF$.

It is obvious that performing continuous measurements of signals' power and quality ensures that the device is camped on the best cell. However, CIoT applications, where usually the devices have limited mobility, experience less frequent data transmission and can tolerate larger delays, while continuous measurements might not be needed. It is clear that constantly measuring the power of the signal and noise has an impact on the power consumption of a mobile device. It is a good idea

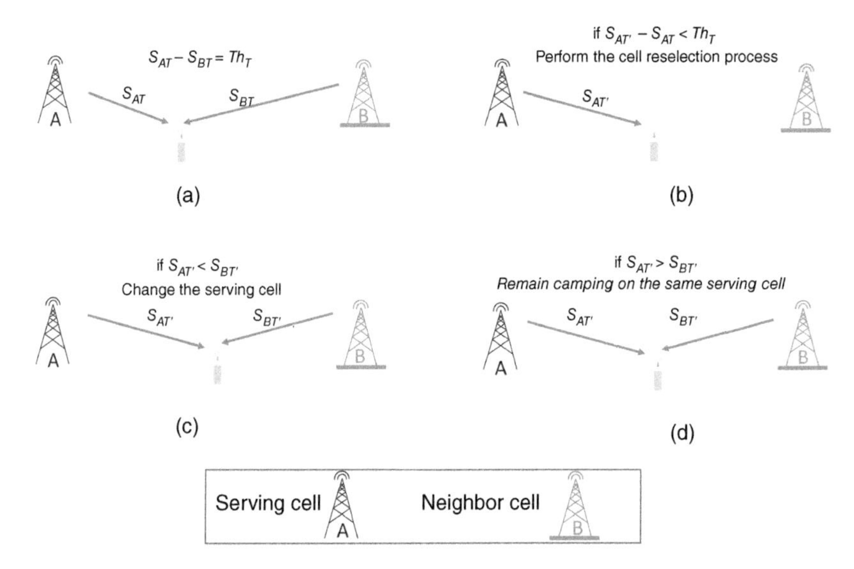

Figure 8.4.1 (a) Serving cell and the best suitable neighbor cell at time T. (b) Power measurement from serving cell at time T'. (c) Cell reselection process by measuring signals from the most suitable neighbors that resulted in changing the serving cell. (d) Cell reselection by measuring signals from best suitable neighbors, but device remains camping on the same serving cell.

to optimize the frequency of performing these measurements in order to increase the lifetime of the batteries used in mobile devices. There is a trade-off between finding the best suitable cell to camp on and reducing power consumption of a mobile device. For this reason, measurements can only be triggered for the cell reselection process when certain events take place. Signaling failure, a measured signal level of below the minimum allowed level; no cell reselection in the past 24 hours; and a drop in the level of signal, which is more than a predefined threshold value, are examples of the events that trigger the cell reselection process.

Figure 8.4.1 shows a scenario that a mobile device performs a reselection process due to changes in the signal power. Assume that at time T, a mobile device completes its reselection process and selects cell A as the most suitable cell to camp on. At time (T), let us assume that the best neighboring cell is cell B. We represent the difference in the signal strength between cell A and cell B as Th_T as shown in Figure 8.4.1a. At time (T'), the device wakes up to examine whether cell A still is the most suitable cell or not, as shown in Figure 8.4.1b. The device measures the signal strength of cell A ($S_{AT'}$) and calculates whether the signal has degraded by more than Th_T as compared to the previously measured signal level (S_{AT}), or not. This may initiate cell selection measurements. In this scenario, the mobile device might handover to cell B as shown in Figure 8.4.1c. The mobile device may also remain camping on cell A after the cell reselection process is completed as shown in Figure 8.4.1d.

If a mobile device receives a sufficiently strong signal and it is in good radio coverage, the device might prefer to stop performing any measurements in order to improve the life of its batteries.

8.4.2.3 Signal Measurements Used for Mobility

As mentioned earlier, there is a need for signal strength/quality measurements when a mobile device performs the cell selection, reselection, or handover process. Among other parameters, a mobile device in an LTE network measures two specific parameters on the reference signal. These

Table 8.4.1 The range of values for RSRP and RSRQ in different RF conditions.

RF conditions	RSRP (dBm)	RSRQ(dB)
Poor	< – 100	< – 20
Fair	–90 *to* –100	–15 *to* –20
Good	–80 *to* –90	–10 *to* –15
Excellent	> – 80	> – 10

two parameters are very important in understanding how mobility is performed in the LTE mobile network. The first one is RSRP, and the other one is RSRQ.

Received Signal Strength Indicator (RSSI) is one of the parameters that usually gets measured in wireless communications and indicates the signal strength for the whole band of transmitted signal. RSRP is the same as RSSI, but only indicates the strength of the reference signal, not the whole band. Similar to RSSI, RSRP is expressed in dBm. RSRP is specific to LTE systems and is used to make ranking among several different candidate cells based on the strength of their signals. RSRP measurement is used for both RRC idle and RRC connected modes.

Signal to Noise Ratio (SNR) is another parameter that usually gets measured in wireless communications and shows the strength of the signal as compared to the strength of the noise. RSRQ is the same as SNR, but only indicates the strength of the reference signal as compared to the strength of noise. In other words, RSRQ indicates the quality of the received signal. If RSRP shows excellent power, we only have information about the signal, but not about the noise. The RSRQ measurement is very useful when the information obtained from the RSRP is not sufficient to make a reliable decision regarding handover or cell reselection. The RSRQ measurement is usually applicable in the RRC connected mode and for handover process. Similar to SNR, RSRQ is expressed in dB.

Handover is usually initiated when the RSRP or RSRQ of a serving cell drops below a threshold value. A low RSRQ or a poor RSRP often indicates that the mobile device is at the edges of a cell. If the mobile device moves closer to a destination cell, handover to the destination cell usually enhances the radio conditions. Table 8.4.1 shows an approximate range of values for RSRP and RSRQ under different Radio Frequency (RF) conditions in LTE networks [6].

8.4.2.4 Idle Mode Versus Connected Mode Mobility

Mobility is handled differently when a device is in the RRC idle mode than when the device is in the RRC connected mode. In the RRC idle mode, the mobility is controlled by the mobile device, while in the RRC connected mode, the network manages the mobility, and the network determines when and to which cell the mobile device should be connected and how the handover needs to take place. In RRC idle mode, cell reselection is performed by the mobile device. The network usually provides the mobile device with a list of neighboring cells for performing measurements. In the RRC connected mode, the network is responsible for the handover process, and it is the entity that decides in which cell the mobile device needs to camp on. Typically, the mobile device reports measurements to the network, which decides to which cell the mobile device needs to handover. There is another type of handover that the network does not use the measurement report from the mobile device. This type of handover is called blind handover. Blind handover refers to a situation where the network directly orders a mobile device to switch or redirect to a certain cell without requiring the measurements.

8.4.2.5 Mobility Architecture

In a cellular network, the coverage area is divided into cells. Some of the cells can be grouped together to make a Tracking Area (TA), which is a logical grouping of cells in an LTE network. Each TA is represented by an identifier called Tracking Area Identifier (TAI). Each cell has only one TA, and each mobile device registers with one TA. However, a TA may be connected to more than one MME. Some TAs can also be grouped together to make a Tracking Area List (TAL) [7]. It would be useful if instead of assigning one TA to each mobile device, a mobile device can be assigned to a list of TAs. In this case, when a mobile device would receive a TA list from the network, it would keep the list, until it would move to a cell that is not included in its list. Network operators optimize the number of cells in each TA, the number of TAs in each TAL, and the allocation of TAs in a TAL when they perform network planning. Figure 8.4.2 shows a coverage area that has been divided into three TAs, each consisting of three cells.

The main component of the LTE network that is responsible for managing the mobility of devices is MME that controls and manages the mobility-related messages between a mobile device and the network. Figure 8.4.2 shows a situation where each TA is controlled by an MME. MME is also responsible for TAUs and paging. In other words, all messages related to updating TAs as well as paging messages are processed and managed by MME. The amount of MME signaling has a direct relationship with the number of cells that it monitors as well as the mobility pattern of mobile devices. The network should be designed in such a way that an MME can handle all mobility-related messages in a timely manner. Another component of the LTE mobility is Home Subscriber Server (HSS), which is a database maintaining the profile of mobile devices. This database stores information related to the current locations of mobile devices on MME level and the services granted to them.

After a mobile device registers with the network and selects a cell, a special TAL is allocated to the mobile device by MME. This TAL contains the list of TAs that are close to the location that the mobile device is residing in. When a mobile device moves out of the current TAL, it will be assigned a new TAL by a process called tracking area update, or TAU. TAU is initiated by the mobile device.

Figure 8.4.3 shows an example a connection between mobility components MME, HSS, and the coverage area structure of TALs, TAs, and cells. In this figure, the coverage area is divided into five TAs and three TALs, where TAL1 = TA1 + TA2, TAL2 = TA3 + TA4, TAL3 = TA4 + TA5. There

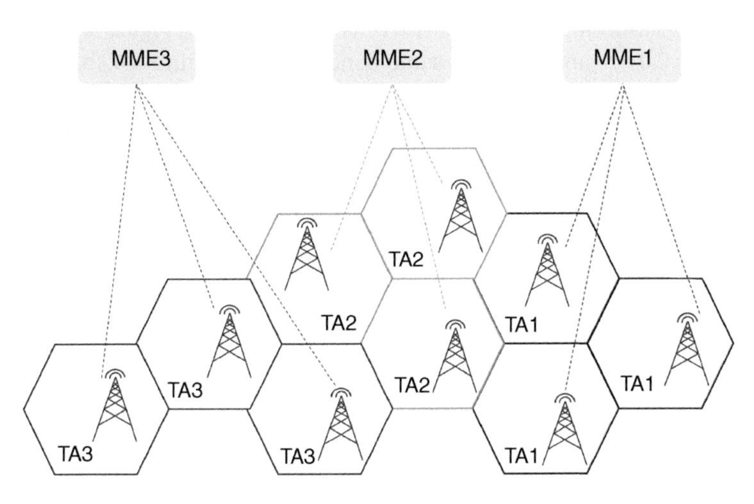

Figure 8.4.2 A coverage area divided into three TAs.

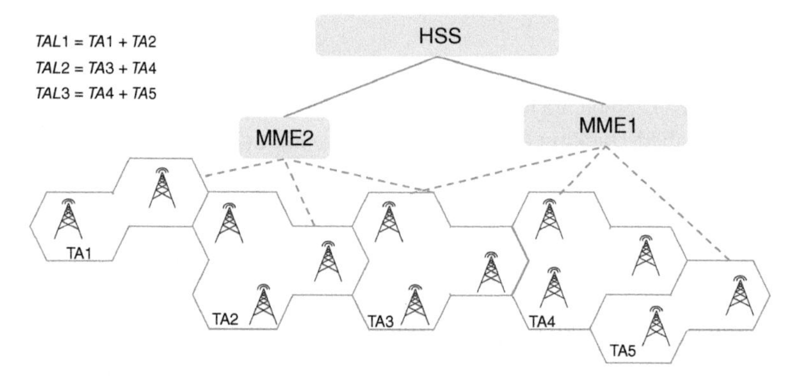

Figure 8.4.3 Example of a connection between mobility components MME, HSS, and the coverage area structure of TALs, TAs, and cells.

are two MMEs (MME1 and MME2), where MME1 manages TAL2 and TAL3 and MME2 manages TAL1 and TAL2.

MME usually records the current TAs of its mobile devices, and HSS usually should record the current MME for each mobile device. For the TAU process, MME needs to be updated only in situations where both TALs belong to the same MME (intra-MME), and HSS needs to be updated in situations where the two TALs belong to two different MMEs (inter-MME).

For the paging process, first, the current MME of a mobile device needs to be found by sending a query to HSS. Then MME should broadcast a paging message to all cells in TAL of the mobile device.

LTE networks allow high-speed mobile devices such as high-speed trains to maintain connectivity up to speeds of 500 km/h in rural areas or 350 km/h in urban areas.

8.4.2.6 Intra-Frequency vs. Inter-Frequency Mobility

If a mobile device performs mobility-related measurements in a situation where the center of the spectrum supported by the serving cell and the one supported by a neighboring cell are the same, then cell measurements are called intra-frequency cell measurements. Figure 8.4.4a–c depict situations in which intra-frequency cell measurements between two cells are required. In Figure 8.4.4a, both cells use the same frequency carrier and have the same bandwidth. In Figure 8.4.4b, the neighbor cell has a larger bandwidth, and Figure 8.4.4c shows a situation where the neighbor cell

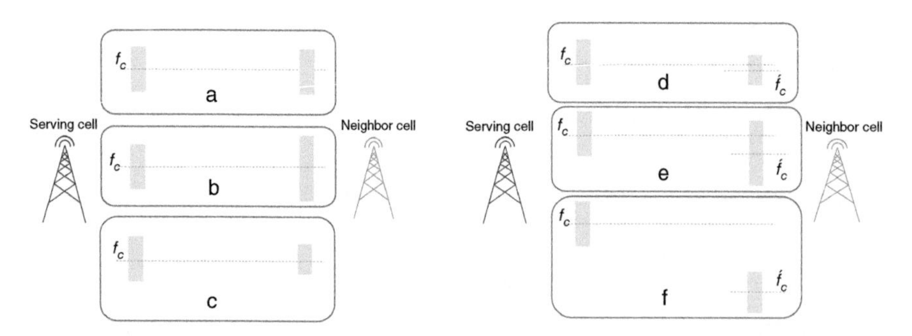

Figure 8.4.4 Different formats of the intra-frequency and inter-frequency mobility: (a, b, and c) different forms of intra-frequency measurement; (d, e, and f) different forms of inter-frequency measurement.

has a smaller bandwidth as compared to the serving cell. Therefore, an intra-frequency handover happens when two cells or TAs have the same center frequency, but these cells may have similar or different spectrums. Let us consider a scenario that one operator has a 10 MHz bandwidth deployment in B5 band of 850 MHz in Neighborhood A. In another neighborhood, let us say Neighborhood B, which is close to Neighborhood A, the operator has deployed a 5 MHz system in the same frequency band of B5 (850 MHz). When a mobile device moves from Neighborhood A to Neighborhood B, the mobile device has to perform intra-frequency measurements.

If a mobile device performs mobility-related measurements in a situation where the center of the spectrum supported by the serving cell and the one supported by a neighbor cell are not the same, then the cell measurements are called inter-frequency cell measurements. Figure 8.4.4d–f depict situations in which inter-frequency cell measurements between two cells are required. As we can see in these figures, the carrier frequencies of the two cells are different. Figure 8.4.4d shows a situation where the neighbor cell has a lower bandwidth. Figure 8.4.4e shows a situation where the neighbor cell has a larger bandwidth as compared to the serving cell. Figure 8.4.4f shows a situation of non-overlapping bandwidth.

There is an important difference between performing intra- and inter-frequency measurements. To perform intra-frequency measurements, the mobile device can perform the measurement at any time and the network does not need to allow a gap between when the mobile device measures the signal level of the serving cell and the one from the neighboring cell. However, for inter-frequency measurements the mobile must be given a time to stop listening to its serving cell and start listening to a neighbor cell. In other words, the mobile device should switch its carrier frequency in order to perform inter-frequency measurements. Therefore, the mobile device is not able to perform inter-frequency measurements without having measurement gaps. This makes inter-frequency measurements more complex. During a measurement gap, no downlink or uplink transmissions are allowed.

8.4.2.7 General Idea about TAU Strategies

One can consider many methods for a device to inform the network of its location. These methods can be divided into two main groups. The first group includes static schemes that are based on the network topology and are independent of the device and its mobility pattern. The second group of schemes includes dynamic approaches that take advantage of the mobility pattern of the device. Therefore, dynamic methods usually need a level of understanding of the mobility history of the device, and while they might have better performance, they generate more signaling messages and need more processing.

Two extreme situations for static location updating schemes happen when a device updates its location continuously when it crosses a cell and when the device does not update its location information at all. If the device updates its location information whenever it moves from one cell to another, then the network always knows the exact location of the device at the cell level and there is no need for paging in this situation. Overall, this can be a good strategy for devices with low mobility and frequent connections to the network. This is not a practical approach for LTE, and it would not be a good strategy for LTE-IoT. The other extreme happens when the device never updates its location information. This results in extensive paging and is especially inefficient for devices with high mobility or frequent data transmission. Another static method would be to ask a device to inform the network of its location only if the device has moved to a set of predefined cells identified as the reporting cells. Therefore, the device does not need to update its location whenever it moves into a new cell. Instead, it informs the network when it enters one of the reporting cells. Even though this method can be considered as a strategy for optimizing the transmission of the paging and the location-update signaling messages, it is very challenging to design the number and locations of reporting cells without having a good knowledge about the mobility of the devices in the network. The most important static method (which is also used in the LTE network) is based on updating the device location whenever the device moves to a new tracking location. In this scheme, as we explained earlier, the paging needs to be sent to all the cells inside TA.

Using dynamic methods for the location update, the location information should be updated based on the device history and mobility pattern. For example, a device can send its location to the network at a constant time interval. Each device may use a different time interval that suits its performance. Dynamic methods can also be established based on the profile of each device. In this approach, the network keeps a profile for each device. This profile has a list of the most probable locations that the device has been at different time periods in the past. The locations on the list are being paged one after another until the device is found. Skipping to update the location information when the device stays in a cell for a short time or providing location information when the device moves away a certain distance from the cell where its last location update has happened are other examples of dynamic methods. Even though dynamic methods might perform better for some specific applications, they also add to design complexity.

8.4.2.8 General Idea about Paging Strategies

As mentioned earlier, the paging process is used to find the cell that a mobile device resides in, doing so by polling the cells with the highest likelihood of hosting that mobile device. The network performs paging by transmitting paging messages to a set of possible cells in the network during a time period called the polling cycle. If during this attempt, the network does not find the mobile device, it selects another set of cells to be used in the next polling cycle. Therefore, the signaling overhead as it relates to paging has a direct relationship with the number of polled cells and the number of polling cycles until the device is found.

As we explained earlier, the network uses a paging process in order to determine the current serving cell of a CIoT device within the network. MME is in charge of keeping the data regarding the history of a CIoT device since an earlier session in the same cell. The maximum delay of the paging process is dependent on the maximum number of paging cycles. In order to reduce the paging overhead, all paging strategies try to have a good estimate of the location of the IoT device before starting the paging process.

8.4.2.9 TAU and Paging Optimization

There is always a trade-off between the paging and TAU signaling overhead. Designing small size TAs or TAL reduces the paging at the cost of increasing the TAU signaling overhead. For example, TAL of only one TA and TA of only one cell eliminate paging, but increase the TAU overhead substantially. Considering larger sizes TA or TAL has the opposite effect. Therefore, the main goal of network designers is to optimize TAU and the paging-related signaling overhead.

8.4.2.10 Doppler Effect

Change in the frequency of the radio wave due to mobility is called Doppler effect or Doppler shift. When a mobile device moves toward a base station, the wave compresses and the frequency increases. When a mobile device moves away from the base station, radio waves are spread further apart and the frequency decreases. The Doppler shift causes frequency deviation, and therefore, both the network and the mobile device should be able to tolerate the change in frequency due to mobility. The Doppler shift between a mobile device and a base station can be calculated as:

$$Doppler\ shift = v \times \frac{f}{c} \tag{8.4.1}$$

where c is the speed of light, f is the frequency of the radio wave, and v is the velocity of the mobile device. For example, a high-speed train with the speed of 486 km/h and a cellular device operating at 2.1 GHz induces a Doppler shift of 945 Hz.

8.4.3 NB-IoT Mobility

NB-IoT does not support full mobility and the handover process. If an NB-IoT device is mobile and goes from one cell to another cell, the IoT device becomes disconnected from its serving cell and should try to establish a new connection with the new cell. For this reason, NB-IoT devices are used in applications where the device either is stationary or has low mobility. When an NB-IoT device is transmitting data and leaves its serving cell, the communication link fails. In this situation, the device starts finding another suitable cell to resume data transfer. It is clear that efficiency can be increased if the IoT device is able to quickly recover the lost connection.

Since handover is not supported in NB-IoT, the main strategy to support idle mode mobility is cell reselection [8]. When a cell is selected, the device can monitor up to 16 intra-frequency neighboring cells and up to 16 inter-frequency neighboring cells. During the connected mode, if the received signal from the serving cell becomes weak, the communication link fails, which forces the mobile device to go back to the RRC idle mode. In the RRC idle mode, the mobile device uses the cell reselection mechanism to camp on a new serving cell and consequently initiates a random access process to move to the RRC connected mode in order to resume its data transfer.

8.4.4 LTE-M Mobility

In order to use the right signaling format for paging, MME needs to inform all the involved base stations in the paging process that the IoT device is an LTE-M device. MME may also provide a paging coverage enhancement level that indicates an estimate of the required number of repetitions.

LTE-M Cat-M1 devices support intra-frequency RSRP measurements and handover process in the RRC connected mode. LTE-M Cat-M2 devices support intra-frequency RSRQ measurements and inter-frequency RSRP/RSRQ measurements, and therefore, they have complete mobility support. LTE-M Cat-M2 can also support higher mobility as compared to LTE-M Cat-M1 devices.

3GPP Release 15 introduced support for high-speed mobility for LTE-M coverage enhancement mode A. It defines 200 Hz Doppler spread, which allows high-speed mobility of up to 240 km/h at 1 GHz and up to 120 km/h at 2 GHz [9].

8.4.5 Summary

When a CIoT device is switched on, it attempts to select a cell to camp on. This process is called cell selection. When a mobile device selects a cell, it needs to monitor several neighboring cells by measuring their signal strength and quality. If the mobile device detects that it can receive stronger signals from one of the neighboring cells as compared to the serving cell, it should start a process called the reselection process in order to change its cell. Two metrics that are used to determine the strength and quality of a signal are Reference Signal Received Power (RSRP) and Reference Signal Received Quality (RSRQ). Handover is usually initiated when the RSRP or RSRQ of a serving cell falls below a certain threshold value. Often, a low RSRQ or a poor RSRQ means that the mobile device is at the edges of a cell. If a mobile device performs mobility-related measurements in a situation where the center of the spectrum supported by the serving cell and the one supported by a neighboring cell are the same, then the cell measurement is called intra-frequency cell measurement; otherwise, it is called inter-frequency cell measurement.

Generally speaking, cells in a cellular network can be grouped together to form a TA, which is represented by an identifier called TAI. Each cell has only one TA and each mobile device registers with one TA. However, a TA may be connected to more than one MME, which controls and manages the mobility-related messages between a mobile device and the network. Some TAs can also be grouped together to make a TAL.

There are two main ways for the network to find the location of a device. Either the device sends this information to the network, or the network tries to page different cells to find the location of the device. There are many methods for a device to inform the network of its location. These methods can be divided into two main groups of static and dynamic schemes. Static schemes are independent of the device and its mobility pattern. For example, a CIoT device can send its location at a constant time interval. The dynamic schemes use the mobility pattern of the device. For example, a device may send its location information at a higher or lower interval based on its mobility information. The paging process is used to find the cell that a mobile device resides in by polling the most possible cells. The network performs paging by transmitting paging messages to a set of possible cells in the network during a time period called a polling cycle.

LTE-M supports full mobility and handover, while NB-IoT does not. In NB-IoT, if a CIoT device goes from one cell to another cell, it becomes disconnected from its serving cell and should try to establish a new connection with a new cell. For this reason, NB-IoT devices are used in applications where the device either is stationary or has low mobility within a specific cell.

References

1 3GPP TS 24.301 V13.5.0 (2016). Technical specification. Group core network and terminals; Non-Access-Stratum (NAS) protocol for Evolved Packet System (EPS); Stage 3 (Release 13).

2 3GPP TS 24.303 V13.0.0 (2015). Technical specification. Group core network and terminals; mobility management based on dual-stack mobile IPv6; Stage 3 (Release 13).

3 3GPP TS 24.303 V14.0.0 (2017). Technical Specification. Group core network and terminals; mobility management based on dual-stack mobile IPv6; Stage 3, (Release 14).

4 3GPP TS 24.303 V15.0.0 (2018). Technical specification. Group Core network and terminals; mobility management based on dual-stack mobile IPv6; Stage 3, (Release 15).

5 3GPP TS 24.303 V16.0.0 (2020), Technical specification. Group core network and terminals; mobility management based on dual-stack mobile IPv6; Stage 3, (Release 16).

6 Parikh, J. and Basu, A. (2016). Effect of mobility on SINR in long term evolution systems. ICTACT Journal of Communication. *Technology.* 7 (1): https://doi.org/10.21917/ijct.2016.0182.

7 Deng, T., Wang, X. et al. (2015). Modeling and performance analysis of tracking area list-based location management scheme in LTE networks. *IEEE Transactions on Vehicular Technology* 65: 1–1. https://doi.org/10.1109/TVT.2015.2473704.

8 Dian, F.J. and Vahidnia, R. (2020). LTE IoT technology enhancements and case studies. *IEEE Consumer Electronics Magazine.* 9 (6): 49–56. https://doi.org/10.1109/MCE.2020.2986834.

9 Liberg, O., Sundberg, M., and Wikström, G. (2020). *Cellular Internet of Things: From Massive Deployments to Critical 5G Applications*, 2ee. Academic Press https://doi.org/10.1016/C2018-0-01131-7.

Exercises

8.4.1 Do IoT devices need to perform power measurements at the time of the cell selection process, or only the cell reselection process requires power measurements?

8.4.2 What are the two main signals used for the cell selection process in an LTE network?

8.4.3 There are two CIoT devices that receive a $T_{Reselect}$ value of five seconds and an SF value of 0.2 from the network. After a cell reselection process, for how long are these CIoT devices not allowed to start another reselection process? Can this time be different for these two devices?

8.4.4 What is the Tracking Area Update (TAU) process? Which entity activates TAU?

8.4.5 In an area, the reference cell and one of the neighbor cells have non-overlapping bandwidth. Do you think that inter-frequency measurement or intra-frequency measurements need to be performed?

8.4.6 If a large number of CIoT devices simultaneously perform TAU updates, it may cause congestion.

a. Give a practical example of a situation that a large number of CIoT devices may simultaneously start performing TAU updates.

b. Give a solution to solve the congestion problem in (a).

8.4.7 In Figure 8.4.E7, TAL1 has TA1, TA2, TA3, and TA4 in its list. TAL2 consists of TA2, TA7, and TA9. Assume that the network gives TAL1 to UE#2 and TAL2 to UE#1.

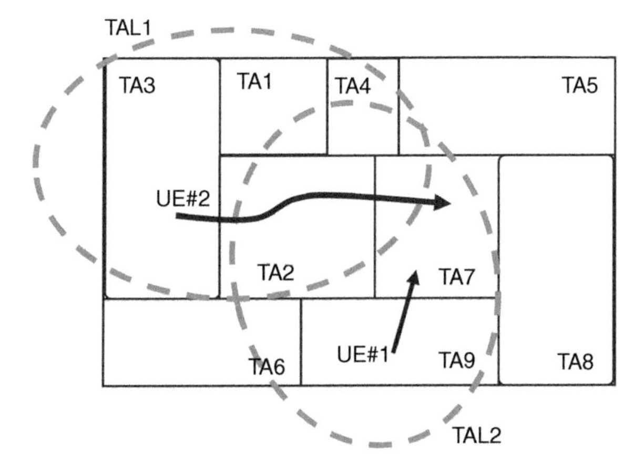

Figure 8.4.E7 Exercise 8.4.7.

a. Does UE#1 perform any TAU when passing from TA9 to TA7?
b. Does UE#2 perform any TAU when passing from TA2 to TA7?
c. If UE#1 or UE#2 needs to be paged, where will the paging message be sent?

8.4.8 Why are TAs grouped to form TAL?

Advanced Exercises

8.4.9 The RSRP and RSRQ of a serving cell during a test period is shown in Figure 8.4.E9. Explain what can be understood from this graph.

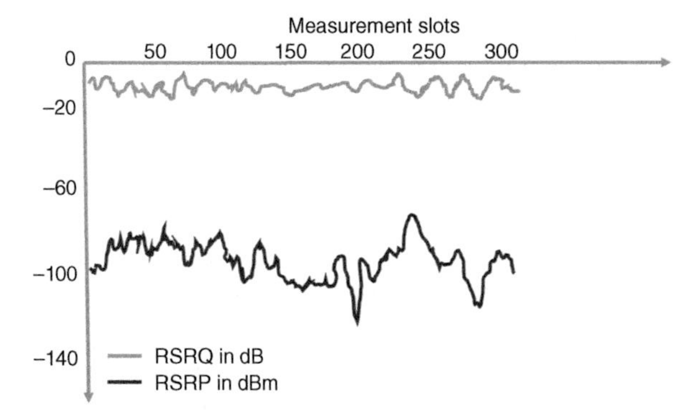

Figure 8.4.E9 Measurements obtained in Exercise 8.4.E9.

8.4.10 The RSRP and RSRQ measurement from two cells are recorded and shown in Figure 8.4.E10. Show the location that the handover needs to be performed.

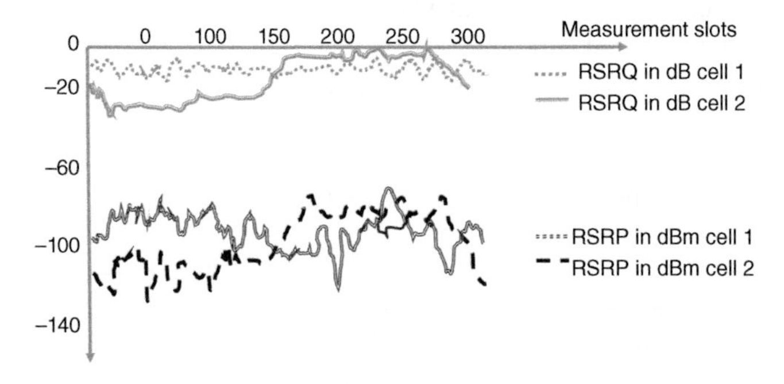

Figure 8.4.E10 Measurements obtained in Exercise 8.4.10.

8.4.11 What is the maximum Doppler shift induced to a mobile IoT device traveling with the velocity of 100 km/h and operating in frequency band B2?

8.4.12 What effect the size of TA or TAL has on the amount of signaling for paging and TAU?

8.4.13 What is a good paging strategy when the IoT device is:
a. a metering device that is installed in a basement?
b. a mobile IoT device that moves around frequently?

Chapter 9 IoT Data Communication Protocols

9.1 Introduction

IoT devices are connected to the Internet using Internet Protocol (IP). That is why they are part of the Internet of Things (IoT) ecosystem. In machine-to-machine communication, a machine may use a non-IP protocol and transmit non-IP packets. However, communication among IoT devices is based on the IP protocol. Since IoT devices use IP protocol as part of their Layer 3, it makes sense for these devices to use protocols such as User Datagram Protocol (UDP), Transmission Control Protocol (TCP), or a secure version of TCP as their Layer 4 protocol.

To satisfy various requirements of IoT applications, IoT devices use different types of application-layer protocols (Layer 7). Each of these protocols have their own advantages and disadvantages, and therefore, there is a need to choose a suitable application-layer protocol depending on the requirements of a specific IoT application. For example, a suitable application-layer protocol for IoT devices that are used in financial systems and require reliable transactions would be different from the ones used for IoT devices that perform simple operations in a constrained network.

One of the most important application-layer protocols, which is widely used in communication networks, is the HyperText Transfer Protocol (HTTP). Even though HTTP might be useful for some IoT applications, it is usually suitable for computers, smartphones, and other high capacity devices. HTTP usually requires a high bandwidth communication channel, and it is not suitable for IoT devices with limited resources.

The traditional application-layer protocols are based on a request-response approach, in which a client asks a server for a service. Many application-layer protocols that are used in the design of IoT applications do not follow this model and use another model that is called the publish-subscribe model. A publish-subscribe model separates an IoT client that sends a message, called a publisher, from another IoT client, which receives the message, called a subscriber. To be able to make this separation, there is no direct connection between the publisher and the subscriber, and the communication between these two entities is performed by an intermediate entity, called a broker, which is responsible for routing and distribution of messages. This establishes a loosely coupled connection among IoT devices. The publish-subscribe model is very flexible and scalable, which makes it a good model for IoT applications. In this model, the broker is required to efficiently relay a message to clients that have subscribed to receive that message. The broker can distribute a message based on the subject line of a message or based on the message context. In a protocol established based on the publish-subscribe model, a publisher does not know the status of a subscriber, does not have any information whether the subscriber has received the message or not, and does not know whether the message has arrived at the subscriber side safe and sound or not. Therefore, to establish reliable communication using protocols that are based on the

Fundamentals of Internet of Things: For Students and Professionals, First Edition. F. John Dian.

publish-subscribe model, additional flow of messages is required. Subscribers and publishers in a publish-subscribe model do not need to be online simultaneously. It should also be noted that even if there are no subscribers for a message, the publisher can still send its message to the broker.

Besides HTTP, other important application-layer protocols, which are used in the IoT ecosystem, are Message Queue Telemetry Transport (MQTT) protocol, Constrained Application Protocol (CoAP), and Advanced Message Queuing Protocol (AMQP). HTTP is based on a request-response model, while other protocols that we mentioned use a publish-subscribe model entirely or support this model as part of their architecture. In this chapter, we explain several application-layer protocols, which are used in the design of IoT systems in order to give readers insight into their strengths and limitations.

9.2 HyperText Transfer Protocol (HTTP)

HTTP is a protocol that is used mainly to access data on the World Wide Web (WWW) and is the foundation of data transmission over the web. It was an effort by both the Internet Engineering Task Force (IETF) and the World Wide Web Consortium (W3C). Today, there exists three versions of HTTP, known as HTTP 1, HTTP 2, and HTTP 3. As it is shown in Figure 9.1a, HTTP is based on a request-response model in which a client sends a request to a web server and the server responds to the client's request.

As shown in Figure 9.1b, an HTTP request contains a request line, zero or multiple header lines, a blank line, and an optional request body. The response contains a status line, zero or multiple header lines, a blank line, and an optional response body.

A request line is represented by "request-type URL HTTP-Version." Examples of important request types are shown in Figure 9.1c. These are commands such as GET, HEAD, POST, and PUT. Simply put, using a GET command, the client can ask to receive a resource such as a document. Using a HEAD command, the client can ask for information about a resource without receiving the resource itself. A POST command can be used by a client to send some information to a server. A PUT command can be used to send a resource to a server. Similar to the request line, a status line (which is part of HTTP response) is represented by "HTTP-version status-code status-phrase."

HTTP can be used as an application-layer protocol in the design of an IoT application. However, as mentioned earlier, due to the extensive power consumption of HTTP, it is not efficient for many IoT applications. It is filled with headers and rules that make the size of packets large, and therefore, it is not suitable for IoT devices that work in networks with constrained resources. It supports a one-to-one communication model in which there is one client, one server, and one request at a time.

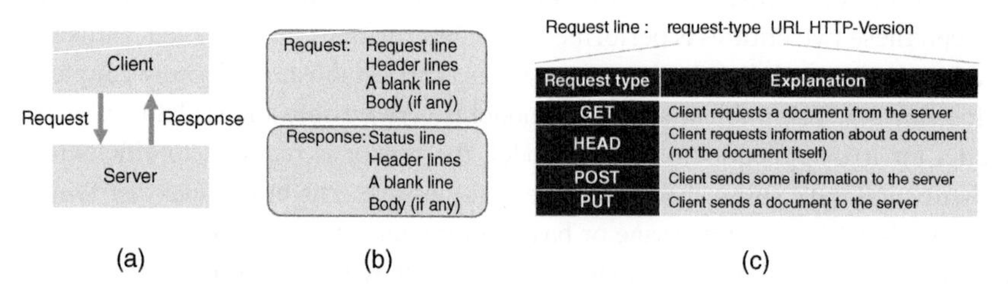

Figure 9.1 (a) HTTP model, (b) HTTP request and response format, and (c) some of the important request types.

HTTP uses TCP for reliability, and therefore, everything sent over HTTP is acknowledged. TCP port 80, which is a well-known TCP port number, is assigned to HTTP messages. Originally, HTTP would open a separate TCP connection for any object on a website. This was called non-persistent mode, which was not efficient. HTTP version 1.1 (and higher) operate in a mode known as persistent mode. HTTP in persistent mode, opens one connection, and gets all the objects from a website. The server then leaves the connection open for a while, in which the length of the time that the connection is open depends on the server. Figure 9.2 shows a situation where a client requests three separate files (F1, F2, and F3) from a server. In Figure 9.2a, TCP uses the non-persistent mode, and therefore, there is a need for a separate TCP connection for each file. Figure 9.2b shows the same situation in persistent mode in which, using only one TCP connection, all three files are transferred to the clients. In other words, the client opens a TCP connection and sends a request. The server then sends a response and leaves the connection open for more requests. The client reads the response and can send more requests, or it can request to close the connection. In the original design of HTTP, the server was stateless. In other words, servers did not remember the history of the requests. HTTP being stateless made the design of the HTTP server easy. But later on, they found out that having a history of old transactions between a client and a server is beneficial. So, the concept of cookies was added to HTTP. Cookies allow the server to remember previous information. HTTP is a text-based protocol and uses American Standard Code for Information Interchange (ASCII) characters. Most of the new protocols are in text format for human readability, and they are not in binary, which slightly saves the size of messages.

The details of HTTP are outside the scope of this book, but readers should understand that using request and response commands, a client can retrieve data such as objects inside a web page from an HTTP server which hosts the web page. A good explanation of HTTP can be found in [1]. Interested readers are encouraged to take a look at Assignment #6, which is a hands-on project on this topic.

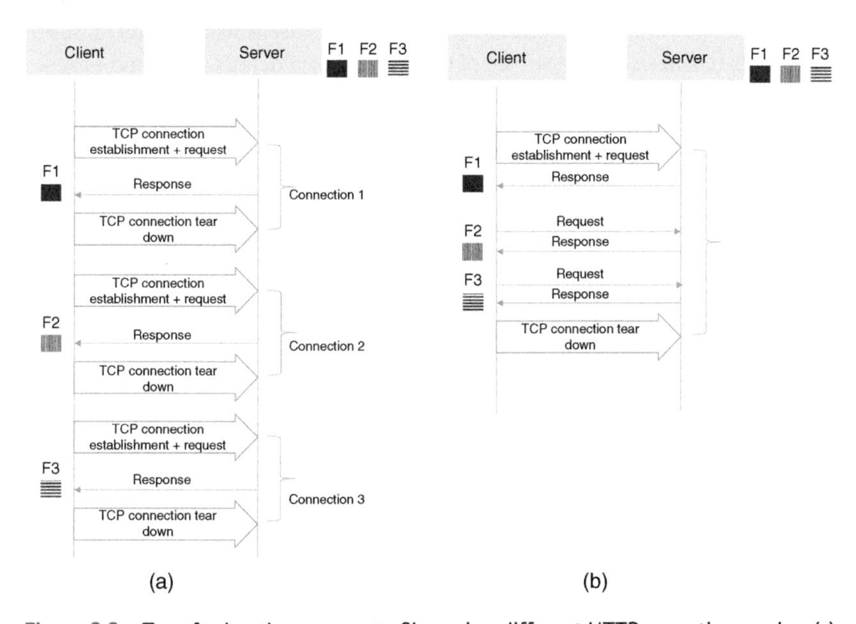

Figure 9.2 Transferring three separate files using different HTTP operation modes: (a) non-persistent mode and (b) persistent mode.

9.3 Message Queue Telemetry Transport (MQTT) Protocol

MQTT is an IoT messaging protocol, originally designed by IBM, and uses a publish-subscribe model. This application-layer protocol can relay data among IoT devices, also called MQTT clients, through a centralized server, called the MQTT broker. As explained earlier, and similar to any publish-subscribe-based protocol, IoT devices that use MQTT do not have a direct connection with each other, and MQTT enables these devices to communicate with one another via a broker. This architecture is very scalable and suits many IoT applications nicely. Later, IBM submitted MQTT to the Organization for the Advancement of Structured Information Standards (OASIS) specification body. Also, there is a variation of the MQTT protocol that is called MQTT for Sensor Networks (MQTT-SN). This protocol is developed for battery-powered IoT devices on non-TCP/IP networks. The specifications of MQTT protocol can be found in [2].

MQTT is an ideal application-layer protocol for data transfer among IoT devices that are battery powered and need a low bandwidth communication protocol. It is a lightweight and text-based protocol, which makes its messages simple and readable. A full MQTT client software can fit into microcontrollers with 32 KB of flash and 2 KB of RAM. Each MQTT message has a subject line, called a topic. An MQTT topic can be a simple string of text like PressureSensor#1, or it can have more hierarchy levels, which are separated by a slash character such as sensors/temperature. For example, assume that in a smart factory, temperature sensors are installed for all the devices located in different rooms. Each room might be located on different floors of different buildings. Therefore, a topic such as:

sensors/building_number/floor_number/room_number/temperature/device_name

might be used by each sensor, where topic "sensors/building2/floor3/room4/temperature/heater" represents a temperature sensor installed on a heater that is located at the fourth room in the third floor of the second building. It should be noted that MQTT topics are case sensitive. MQTT subscribers can use two wildcards to subscribe to a subset of topics. These wildcards are + and #. A + wildcard shows a single level of hierarchy, while # represents the remaining levels of a hierarchy. Let us take a look at the following four topics. The first topic (Topic 1) represents the temperature sensors of all the devices located in the fourth room on the third floor of the second building. The second topic (Topic 2) represents the temperature sensors of all the devices located in the fourth room of all the floors in the second building. The third topic (Topic 3) represents all the sensors on all the devices located in the fourth room on the third floor of the second building. The last topic (Topic 4) represents all the sensors on the second building.

Topic 1: "sensors/building2/floor3/room4/temperature/+"
Topic 2: "sensors/building2/+/room4/temperature/+"
Topic 3: "sensors/building2/floor3/room4/#"
Topic 4: "sensors/building2/#"

MQTT can handle one-to-one, one-to-many, many-to-one, and many-to-many communications. When an MQTT client sends a message, it uses a topic for its message that can be considered a unique identifier. The message is not addressed to any specific MQTT client, and therefore, any MQTT client that subscribes to that specific topic can receive the message. For example, if one MQTT client publishes a message with the topic "temperature" and only one MQTT client subscribes to receive it, then a one-to-one communication is established. If many MQTT clients subscribe to topic "temperature," then a one-to-many communication has been created. If many MQTT clients publish exactly the same topic, and there is one or multiple MQTT clients that

Figure 9.3 A bidirectional MQTT system for cellular IoT.

have subscribed to that topic, a many-to-one or many-to-many communication model has been established, respectively.

An MQTT client can act as both a publisher and a subscriber at the same time in which the MQTT client publishes some messages with unique topics and subscribes to one or several other topics. This creates bidirectional communication among IoT devices. Figure 9.3 shows the architecture of a bidirectional MQTT system in a cellular IoT network.

IoT devices that use the MQTT protocol (MQTT clients) do not need to be compatible with each other, since they communicate through a broker. Also, adding and deleting IoT devices to the MQTT architecture is easy and can be done without changing the infrastructure, since MQTT clients can subscribe to a topic or unsubscribe at any time.

Figure 9.4 shows an example of using the MQTT protocol for a smart gas station system. The gas station is equipped with a car identification unit consisting of an IP camera and a Radio Frequency IDentification (RFID) reader. The car identification unit is installed at the entrance of the gas station. The IP camera reads the license plate of any car entering the station. The RFID reader reads the Vehicle Identification Number (VIN). The car identification unit sends the license plate number and VIN number of any car entering the gas station under the topic "Vehicle Info" to an MQTT broker. Therefore, the car identification unit is an MQTT client. The gas station has several gas pumps. At each pump, a gas pump system sends a message to the MQTT broker with the VIN number, as well as the amount and type of gas purchased by a customer, with the topic "Vehicle Gas Info." So, each gas pump is also an MQTT client. The underground storage gas tank is also an MQTT client and publishes its gas usage to the broker with the topic of "Tank Level." In this scenario, the payment terminals are also MQTT clients and provide payment information to the broker using the topic "Payment Info." Let us consider that there is a central Processing Unit (PU) in this scenario that has subscribed to all the preceding topics. Upon reception of these topics, it analyzes the data and calculates whether the amount of gas leaving the tank is the same as the amount of gas coming out of all the gas pumps. If they are not the same, it might indicate a leak somewhere in the system. The central PU can send a message to the broker with the topic "Alarm."

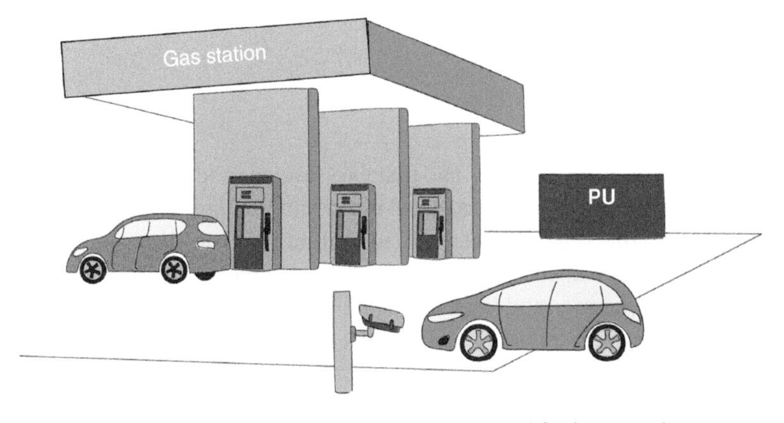

Figure 9.4 Smart gas station that uses MQTT protocol for its operation.

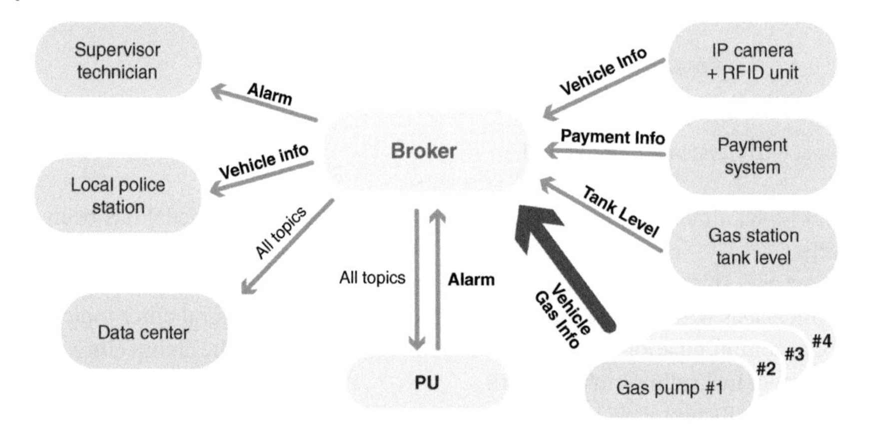

Figure 9.5 MQTT protocol used in a smart gas station.

A supervisor technician can subscribe to receive messages with the topic "Alarm." A local police station can subscribe to messages with the topic "Vehicle Info" in order to match the list of the cars that enter the gas station with the list of stolen cars. A data center can subscribe to all of these topics in order to receive all the information and save them to a database.

Figure 9.5 shows the connection between MQTT clients and the MQTT broker, as well as the topics of MQTT messages used in the smart gas station example.

There are many software programs that can be installed on a server to play the role of an MQTT broker. Examples of these programs are Mosquito, Mosca, and Emqttd. Many companies provide MQTT broker services. For example, Google supports the MQTT protocol by running a broker that listens to the port mqtt.googleapis.com:8883.

Since an MQTT broker is a centralized node in an MQTT architecture, it can be considered a single point of failure. If the MQTT broker is down, no MQTT client can publish a topic or subscribe to a topic. To solve this problem, most MQTT brokers support automatic switching to a backup broker, if the primary broker fails to operate. The broker software also performs load balancing to share the load among multiple brokers locally, in the cloud, or a combination of both.

MQTT can manage messages asynchronously. If the connection between an MQTT broker and an MQTT subscriber gets broken, the broker can hold these messages until the connection is re-established, and then forwards the messages to the MQTT subscribers. This is not similar to HTTP, which is a synchronous protocol and a client must wait for the server to send its response.

Interested readers are encouraged to take a look at Assignment #4, which is a hands-on project on this topic.

9.3.1 MQTT Connections

MQTT defines three levels of Quality of Service (QoS). The first QoS level is called level 0 or "fire and forget." When an MQTT client sends a message, it does not receive any acknowledgment from the broker at this level, and therefore, the MQTT client that had sent the message does not know whether the message is transferred or not and does not repeat the transmission in case of a failure. In other words, the MQTT client fires a message and forgets about it. This method is simple and has the lowest overhead for sending a message. It should be noted that QoS is managed by an MQTT broker, but the MQTT client can ask for a specific QoS.

In the second QoS level, called level 1, acknowledgment is required for any message. To ensure that a message has been transferred to a broker successfully, the broker sends an acknowledgment back to the sender. It is possible that the acknowledgment from the broker could get lost

Table 9.1 MQTT QoS levels.

QoS	Model	Description
0	Fire and forget	Messages are delivered without duplication; no acknowledgment is required
1	At least once	Messages are delivered at least once with possible duplication; acknowledgment is required
2	Exactly once	Messages are delivered without duplication

or corrupted. In this case, the MQTT client resends the message. In other words, the MQTT client sends duplicate messages until it receives an acknowledgment from the broker. This guarantees the transfer of the message. However, the message may reach the broker more than once.

In the third QoS level, called level 2, acknowledgment is required for any message, but the message is delivered only one time and without duplications. This level uses a sequence of four handshakes to ensure that the message has been received by the broker and the acknowledgment has been received by the MQTT client. When the handshake is complete, both the MQTT client and the broker know that the message has been transmitted exactly one time. So, it is not possible that the broker receives duplicates. Table 9.1 shows these three QoS levels.

MQTT messages are transmitted over TCP, and therefore, MQTT messages are encapsulated inside a TCP header. As we discussed in Chapter 1, TCP provides connection-oriented communication for a reliable and in-order delivery of packets. Since TCP is a connection-oriented protocol, it needs to establish a connection before data transmission, and needs to terminate the connection after the data is transmitted. In other words, the entire operation requires three phases: connection establishment, data transfer, and connection termination. TCP connection establishment requires a three-way handshake between the MQTT publisher and broker. The data transfer phase would be different depending on the MQTT QoS level. In QoS level 0, the MQTT publisher sends its message and then sends a disconnect request packet. In MQTT QoS level 1, the MQTT publisher sends its message and waits to receive an acknowledgment from the broker before sending a disconnect packet. QoS level 2 requires a four-way handshake to ensure the broker has only received the message once. Figure 9.6 shows the signaling involved in transmission of a message between an MQTT publisher and an MQTT broker using QoS level 0.

TCP is a reliable communication protocol that ensures MQTT messages published by an MQTT client are received by the broker correctly, and MQTT messages sent by a broker to MQTT clients have a reliable transport. However, MQTT sends its own acknowledgment in QoS levels 1 and 2 in order to ensure the validity of the message itself. For example, if a broker receives a message with different payload format than expected (due to some bugs on the MQTT client side), the broker does not acknowledge the correct reception of the message, and the MQTT client with QoS level 1 or 2 needs to publish the message again. In this scenario, TCP provides reliable communication, but the MQTT client has not sent the data in the correct format due to some problems [3].

9.3.2 Security of MQTT Protocol

Many IoT applications utilize low-cost devices and do not have the resources to establish a secure connection. In MQTT architecture, the security measures are configured on the MQTT broker and all the devices should authenticate with the broker on port 8883, which is the standard TCP port

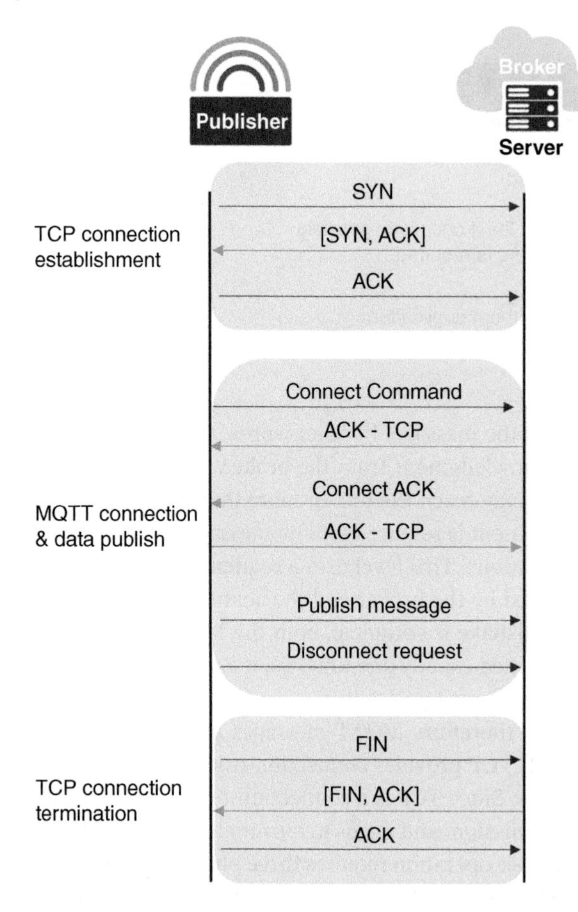

Figure 9.6 MQTT publisher-broker handshake in QoS level 0.

number reserved for secure MQTT connections. Therefore, when implementing security measures on the broker, the capabilities of MQTT clients should be considered. To be able to provide a secure MQTT session, MQTT requires that all MQTT clients have a client IDentification number (ID). The client ID links a topic to an MQTT client. Also, during authentication, the MQTT client should provide the broker with a username and password. However, the username and the password is not encrypted and would be sent in plaintext format. MQTT clients that require a more secure authentication approach are required to transmit an MQTT client certificate to the broker. A client certificate is basically a file, which contains information such as the client's name, the name of its Certificate Authority (CA), a digital signature, expiration date, and a serial number. Even though the MQTT client certificate is very secure, by combining two methods of authentication such as both the client certificate, as well as the username and password, a more secure authentication strategy is achievable.

9.3.3 MQTT Last Value Queue (LVQ)

MQTT has the ability to support Last-Value-Queues (LVQs) that are useful when a client connects to an MQTT broker and subscribes to a topic for the first time. In this situation, the MQTT client can get the latest state of the topic and then receive updates on it. This can assist a newly subscribed MQTT subscriber to get a status update after it subscribes to a topic. In other words, LVQ can eliminate the wait time for a new subscriber to receive its first update. A broker only keeps one

LVQ message for a topic. When the broker sends an LVQ message to a subscriber, it activates a flag, called the LVQ flag. Therefore, the MQTT subscriber knows that the message is an LVQ message and can decide how it wants to handle this message.

9.3.4 MQTT Last Will and Testament (LWT)

If an MQTT client intends to disconnect from a broker, it sends a disconnect packet to the broker. But if the connection between an MQTT publisher and an MQTT broker is broken due to a network failure or problems at the client's side, then the broker does not receive any disconnect packet. This is called an ungraceful disconnect. Each MQTT client should communicate with the broker within a defined period, which is called the keep-alive period. If the client fails to communicate with the broker within this period, then this is considered an ungraceful disconnect. It is clear that MQTT subscribers do not know about an ungraceful disconnect. To solve this problem, each MQTT publisher can define its Last Will and Testament (LWT), when first connected to a broker, and as part of a packet called the connect packet. The connect packet contains an LWT topic, an LWT message, and its QoS among many other information. When a broker detects an ungraceful disconnect, the broker sends the LWT message to all the clients that have subscribed to the LWT's topic. In other words, to ensure that the stale data is not delivered to the subscribing client when the publishing client is ungracefully disconnected, the MQTT broker provides subscribing clients with the LWT message.

Let us look at an example to demonstrate how LWT works. There are many machines in a smart factory. Each machine may have a large number of sensors. If a sensor is damaged, it has experienced power failure, or it has a connectivity problem with the broker; the MQTT subscriber is unaware of the ungraceful disconnect between that sensor and the broker. Assume that all sensors have defined their LWT with a topic represented by sensorID/offline, where sensorID is a unique number indicating a specific sensor. In the LWT message, the sensor has included some information about itself, the data that it measures, or the machine that it is connected to. Assume that an MQTT client has subscribed to the LWT topic for all of the sensors in this smart factory. When the broker experiences an ungraceful disconnect from a sensor, the broker sends the LWT message of the sensor to that MQTT client. This client can send an email to the technical team with the LWT message of that sensor in order to resolve the sensor's technical problem.

9.4 Constrained Application Protocol (CoAP)

Another application-layer protocol, widely used by IoT applications, is CoAP [4]. It is a very lightweight protocol that makes it suitable for IoT resource-constrained devices. This protocol was developed by IETF. CoAP uses the GET, PUT, POST, and DELETE commands similar to, but not exactly like, HTTP. CoAP and HTTP have many similar features with the difference that CoAP is optimized for IoT. CoAP has low overhead, has a good packet structure, and runs over UDP. Generally speaking, it is very simple to parse CoAP messages.

Instead of MQTT topic, CoAP uses Universal Resource Identifier (URI). You can find a kind of similarity between CoAP URIs and MQTT topics. For example, a temperature sensor that publishes its sensor information to a server, using either the CoAP or MQTT protocols, can use the following formats, respectively:

CoAP sensor publishing to CoAP server: URI: "coap://devices/sensors/temperature"

MQTT client publishing to an MQTT broker: topic: "/devices/sensors/temperature"

Figure 9.7 Smart home.

Let us look at the application of the CoAP in a smart home as shown in Figure 9.7, where there is a need to control smart devices such as light switches, light bulbs, door locks, thermostats, fans, or appliances. These devices can use the CoAP and play the role of a CoAP client. Each CoAP client registers with a CoAP server. A smart light switch, for example, can send a PUT command to a CoAP server to modify the state of a light bulb, such as turning it on or off or changing the intensity or color of a light bulb. Similarly, a fan may send its daily usage information to a CoAP server or a thermostat may send its status to a CoAP server. Also, a CoAP client can read some information from the CoAP server using the GET command.

One may think that all this could be done by MQTT as well. Even though this is true, we should consider that MQTT is a message-oriented protocol, which means it only focuses on the delivery of messages, while CoAP is all about operations with constrained resources. Therefore, CoAP is preferable in terms of implementation costs and for constrained devices that could not afford a TCP stack. Even though CoAP has many limitations as compared to MQTT, it is more cost-effective. CoAP introduces a lower latency in delivering messages, consumes lower power, and therefore, it is more suitable for battery-powered IoT devices. It is also easier to integrate CoAP with other web-like services.

9.4.1 CoAP Messages

CoAP supports two types of messages. The first type is called confirmable messages, also known as CON, and the second one is called non-confirmable messages, which are known as NON messages. A confirmable message needs a reliable communication, and therefore it needs to receive an acknowledgment from the server, while non-confirmable messages do not need any acknowledgment. Each of these messages (CON or NON) has a unique message ID. When a client sends a confirmable message with a unique ID, it should receive an acknowledgment from the server with the same message ID. The CoAP client keeps sending the message with a default timeout and based on an exponential back-off strategy between two retransmissions until an acknowledgment is received. The ACK messages can carry useful data, but they do not need separate ACKs. Figure 9.8a shows a client that sends a message with a unique ID of 0XAA62.

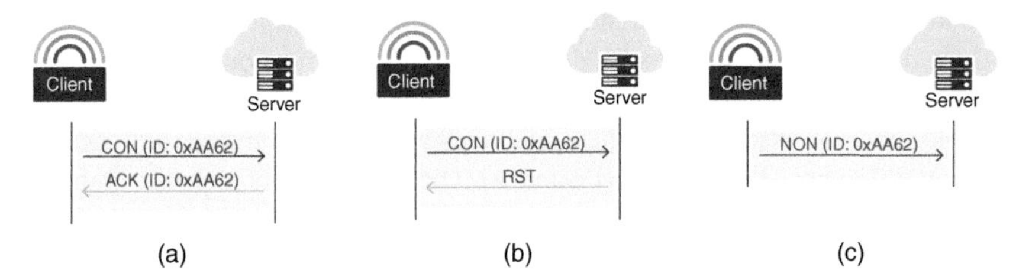

Figure 9.8 CoAP messages: (a) an acknowledged confirmable message; (b) a not processed confirmable message; (c) a non-confirmable message.

If a server cannot process this message, it sends a reset message (RST) instead of an acknowledgment packet as shown in Figure 9.8b. When a client sends a non-confirmable message, the server does not acknowledge the arrival of the message. Although non-confirmable messages are not acknowledged, they still have a message ID for duplicate detection. This is shown in Figure 9.8c.

When a CoAP client sends a request using a CON or NON message, the CoAP server sends an acknowledgment that may contain either a response or an error code. If the server is not able to provide a response immediately, it sends an acknowledgment packet whose response section is blank. Whenever the response is available, the server sends the client a new CON or NON message. This message contains the response, and then the client acknowledges receipt of the response. It should be noted that if the request sent by a client is an NON message, the server response would be an NON message as well.

To provide multicasting capability, CoAP supports group communications. In this case, a single request can be transmitted to several recipients at once. This enables CoAP to create a one-to-many communication model.

Similar to HTTP, CoAP is a plaintext protocol. This means CoAP needs an additional encryption protocol in order to secure communication.

9.4.2 CoAP Observers

Originally, CoAP was based on the request-response model. Later on, a capability called observer is added to CoAP in order to reduce the need for constant polling of a CoAP server by a CoAP client. With this new capability, we can say that CoAP supports both the request-response, as well as, publish-subscribe architecture. For example, a smart light switch can change the state of a light bulb (such as its color or on/off position) by sending a put command to the CoAP server. Since the change in status of the light bulb should happen momentarily, the light bulb can be configured in observer mode. In this case, the CoAP server sends a message to the light bulb in order to change its status. Even though the CoAP observer capability is very similar to publish/subscribe in MQTT, there is a difference between CoAP and MQTT in this regard. In MQTT, the goal is to send every message, while in CoAP, all registered observers will get to the latest resource state eventually.

9.5 Other IoT Protocols

In addition to HTTP, MQTT, and CoAP, there are other application-layer protocols that are used in IoT applications. Advanced Message Queuing Protocol (AMQP), Data Distribution Service (DDS), and Extensible Messaging and Presence Protocol (XMPP) are examples of these protocols. We briefly explain these three protocols. The details of each protocol is outside the scope of this book.

AMQP was established by a working group consisting of 23 organizations and later reorganized into an OASIS member section. AMQP is very similar to MQTT. It is based on the publish-subscribe model, runs over TCP, and uses a broker in its architecture. Similar to MQTT, it has a text-based format and defines three different types of message delivery including at-most-once where each message is not delivered at all or it is delivered only once, at-least-once where each message will be delivered one or multiple times, or exactly-once where each message will be delivered only once. However, there are many differences between MQTT and AMQP. AMQP offers a broader range of messaging circumstances as compared to MQTT, and it supports different types of acknowledgments and queuing techniques. AMQP was originally developed by the finance community and is customer driven, while the MQTT was vendor driven and was originally developed by IBM. AMQP has more advanced features, but it has more overhead than MQTT. The MQTT is a very lightweight protocol and works nicely in a low bandwidth network, while AMQP requires high bandwidth and more resources for its operations. In contrast to MQTT, AMQP does not support LVQs. MQTT can support different topics, in a simple hierarchy. An MQTT topic is global, equivalent to one queue that can cause some limitations in practical situations, while we can define many queues in AMQP and these messages are not usually sent directly from a client to another. The messages may be redirected, and clients might be interested in different subsets of messages. For example, a taxi company has a fleet of cars. The company may be interested to listen to different subsets of topics arranged by customers, car models, license numbers, drivers, or depots. The company may also want to change its subscribed subsets from time to time. AMQP separates the structure of a message from its delivery scheme. The AMQP broker is divided into two independent sections: exchange and queues. The exchange section receives the published messages and distributes them to queues based on predefined roles. AMQP clients connect to these queues, in order to obtain data with specific topics.

DDS is a messaging standard that uses a publish-subscribe architecture similar to MQTT. However, unlike MQTT, it is a broker-less model and does not need any centralized infrastructure such as a broker. In other words, DDS supports a decentralized architecture. Clients that generate data communicate directly with the applications and clients that need the data for processing in a peer-to-peer fashion. DDS can run over TCP or UDP and usually can handle a higher rate of messages as compared to MQTT. Due to differences in their architectures, MQTT and DDS are usually suited for different types of IoT applications.

XMPP is an application-layer protocol designed by the Jabber open source community for chat applications. It is based on a client-server architecture, runs over TCP, and it is based on eXtensible Markup Language (XML). While MQTT does not define any format for its messages, it is possible to define a message format in XMPP in order to receive structured data from clients. Unlike MQTT, XMPP does not support different levels of QoS.

9.6 Summary

There are many application-layer protocols that are used in the IoT ecosystem. There is not one specific application-layer protocol that satisfies the needs of all IoT applications, and therefore, it is a good idea to have some understandings about what each of these protocols can offer.

In this chapter, we discussed two main protocols that have been widely used by IoT applications, in more detail. These two protocols were MQTT and CoAP. We also briefly discussed HTTP and AMQP that can be used in IoT applications that can access higher bandwidth, tolerate more latency, and accept higher power consumption. Finally, we introduced DSS and XMPP as examples of other application-layer protocols that have been used in some IoT applications.

Application-layer protocols are usually designed based on two models. These two models are the request-response model and the publish-subscribe model. In the request-response model, a client sends a request to a server, and the server responds to the request. For example, an IoT system can send data to a server, and another IoT system can send a request to receive the data. In the request-response mode, a client should poll the server. In a publish-subscribe model, a client can publish to a server and the published data will be sent to MQTT clients that have subscribed to the data.

MQTT is a lightweight application-layer protocol that uses the publish-subscribe model. The server in MQTT is called a broker. MQTT is one of the most important IoT protocols, if not the most important one. Each MQTT message has a topic, and MQTT clients can subscribe to any topic that they are interested in. MQTT provides reliable communication and supports three different QoS levels that they have been defined for handling MQTT messages. MQTT runs over TCP.

CoAP is another lightweight application-layer protocol that originally used a request-response model. However, it later supported a new feature, called CoAP observer, which was based on the publish-subscribe model. Therefore, we can say that CoAP works based on both the request-response and publish-subscribe models. CoAP runs over UDP, and it is a better choice as compared to MQTT for IoT devices with very limited resources.

In Section 9.5, a brief explanation of several other application-layer protocols is provided. These protocols were AMQP, DDS, and XMPP. Even though these protocols are not as popular as MQTT, or CoAP, they offer several interesting features that make them suitable for some IoT applications.

References

1 Forouzan, B.A. (2021). *Data Communications and Networking with TCP/IP Protocol Suite.* New York, NY: McGraw Hill.

2 OASIS (2019). MQTT Version 5.0. OASIS Standard. https://docs.oasis-open.org/mqtt/mqtt/v5.0/mqtt-v5.0.pdf (accessed 30 June 2020)

3 Vahidnia, R. and Dian, F.J. (2021). *Cellular Internet of Things for Practitioners.* Vancouver: BCcampus. https://pressbooks.bccampus.ca/cellulariot/ (accessed 30 June 2022).

4 CoAP (2013). IETF RFC 7252. http://https://datatracker.ietf.org/doc/html/rfc7252 (accessed 30 June 2022)

Exercises

9.1 Which IoT protocol would you select to best address the requirements of the following applications:

a. In a telematics application where you need to transmit relatively high volumes of data with very high reliability. However, you are not worried about the authentication or security of the connection.

b. A surveillance camera that is installed on an agricultural land performs edge processing and transmits the analyzed data to a server. The data is used to detect the types and locations of insects in the area in order to send a drone with proper spray to destroy the insects. There should be a very little delay in transmitting the data.

9.2 NB-IoT devices are more resource constrained as compared to LTE-M devices. For data transmission in each of these two networks (LTE-M or NB-IoT), which protocol is preferred (MQTT or CoAP)?

9.3 Is HTTP suitable for IoT applications that require low power consumption?

9.4 In an IoT application, the speed of a fan is controlled according to the temperature that is measured by a sensor. The fan has four different speed settings. The IoT application chooses the speed based on the temperature data. How can we benefit from using the CoAP for this application to reduce the need for constantly polling the temperature sensor?

9.5 In MQTT protocol:
- **a.** What TCP port number is reserved for secure MQTT?
- **b.** Can multiple MQTT clients publish the same topic?
- **c.** Can MQTT clients find the identity of an MQTT client who has published a message?
- **d.** What does an MQTT broker do with messages that get published to topics that no one has subscribed to?
- **e.** Does an MQTT broker retain a message forever?

9.6 What are the transport protocols (L4) for MQTT and CoAP?

9.7 Give some examples of the security problems that can be created using an unsecure MQTT server in home automation applications.

9.8 Which protocol is more suitable for applications which require higher data rates, smaller latency, and lower power consumption?

9.9 In a smart building, we would like to open or close a set of smart blinds/ curtains based on the position of the Sun.
- **a.** Explain a solution for this IoT application based on a suitable IoT protocol.
- **b.** Do you use unicast communication or multicast communication for this application?

9.10 An IoT device uses MQTT protocol and publishes the data regarding a temperature sensor to a broker. The data that is sent by the IoT device represents the amount of change since the last reading. At which QoS level should the IoT device be configured to operate properly?

9.11 By comparing an MQTT broker to a post office, explain how MQTT messages are transferred.

9.12 Can an MQTT client authenticate with an MQTT broker using a client certificate in addition to the username and password?

9.13 Map CoAP confirmable and non-confirmable messages with MQTT QoS levels.

9.14 Where do you think the MQTT broker in Figure 9.3 is located?

9.15 How can MQTT establish a many-to-many communication model?

Advanced Exercises

9.16 Two IoT devices, which are using the MQTT protocol, send temperature information to an MQTT broker. Choose an appropriate topic for these two devices, if:

a. Device#1 sends indoor temperature and Device#2 sends data related to outdoor temperature.
b. Both devices are sending the indoor temperature of the same place.

9.17 How does CoAP perform multicasting?

9.18 What does it mean to say that MQTT is an asynchronous protocol? Is HTTP an asynchronous protocol as well?

9.19 How do CoAP observers work?

9.20 Which of the following sentences is true about the AMQP?

a. It is a lightweight protocol.
b. It is based on a request-response model.
c. It runs over UDP.
d. It can be used in low bandwidth networks.
e. The functionality of an AMQP broker is divided into two independent sections: exchange and queues. The exchange section receives the published messages and distributes the messages to queues based on the predefined roles.

9.21 In which of the following IoT applications, MQTT QoS level 1 can be used. Explain why.

a. In an IoT application, when we turn a switch on, a signal will be sent to turn on a boiler. The boiler is an MQTT subscriber and the switch is an MQTT publisher.
b. In an IoT application, when you turn a switch on, a signal will be sent to a robot to move 2 m forward.

9.22 An MQTT client is subscribing to a topic shown as "#." What does this mean?

9.23 In an IoT application which uses MQTT protocol, heavy machinery equipment is controlled by a controller. The controller is an MQTT publisher that sends appropriate commands to an MQTT broker. The broker pushes the commands to the equipment. Assume that the controller loses connection with the broker and the equipment is also subscribed to receive the LWT topic of the controller. Give examples of some of the practical messages that the LWT may contain.

9.24 Find the names and addresses of at least three public MQTT brokers.

9.25 Conduct research online, and find the IoT protocols that are supported by Microsoft Azure IoT Hub.

Chapter 10 IoT in 5G Era

10.1 Introduction

The future of Internet of Things (IoT) is brighter, now that 5G is around the corner. Some countries have implemented 5G, at least partially, and 5G deployments are underway by many telecommunications operators. These operators are making significant investments on 5G spectrum and infrastructure in anticipation of exponential growth of IoT applications that drive the increased demand for the availability of 5G cellular technology. 5G can bring many opportunities for IoT and can accommodate many different verticals, use cases, and applications.

IoT is expanding rapidly, and similarly, 5G is also expanding at a fast pace. It is interesting to witness how these two technologies evolve. 5G has been designed with IoT in mind, and IoT sees an opportunity to expand its use cases using 5G capabilities.

The higher demand from consumers, especially business communities, as well as the availability of low cost devices are the two main factors that drive the adoption of IoT and 5G. Even though earlier generations of cellular networks such as 2G, 3G, and especially 4G continue to provide the needs of IoT applications, 5G can provide many benefits that are not available by the 4G cellular network or earlier generations. 5G has been designed to support applications that require low latency, high reliability, or extremely fast data transfer. 5G can also support massive numbers of IoT devices (mobile or stationary) with various Quality of Service (QoS) requirements.

A combination of 5G and edge computing that distributes computation in different locations along the IoT connectivity path can support use cases, such as autonomous driving or industrial IoT, which can take advantage of distributed intelligence to meet their time-critical requirements.

The 5G implementation comes in phases. 3GPP Release 15 was established in 2018 as the first phase of 5G standardization. The 3GPP Release 16, the second phase of 5G, is published in 2020, which enhances the capabilities of the 3GPP Release 15 specifications, and supports ultra-reliable low-latency communication as well as non-public networks. 3GPP Release 17 will be published in 2022. The radio access technology for 5G is called 5G New Radio (NR), which is designed as a standard for the air interface of the 5G network. Initial 5G implementation will be based on existing 4G infrastructure. The implementation then moves to the 5G core network. The range of spectrum used in a 5G cellular network can be classified into two ranges. The first one is called Frequency Range 1 (FR1) that includes sub-6 GHz frequency bands. The second range is called Frequency Range 2 (FR2), which includes frequencies above 24 GHz and below 100 GHz. The FR2 range is also called millimeter wave, since the wavelength used in the FR2 spectrum is in the millimeter range.

In this chapter, we explain how IoT and 5G impact each other and discuss the new use cases that can be built by 5G IoT. We also discuss important features of 5G such as network slicing and

Fundamentals of Internet of Things: For Students and Professionals, First Edition. F. John Dian.

network exposure that can make a revolution on how IoT applications can use 5G to satisfy their diverse requirements.

10.2 5G Vision

The objective of 5G is to substantially enhance the capabilities of earlier generations of cellular networks in order to provide a suitable connectivity solution, not only for voice and data communications but also for IoT applications with massive numbers of devices, or those IoT applications with low-latency and ultra-high-reliability requirements.

It has usually taken a decade for a new generation of the cellular network to be deployed. If we look at the evolution of cellular technologies, we had the first generation of cellular networks emerge in the 80s, the second generation in the 90s, the third generation in 2000, the fourth generation around 2010, and the fifth generation around 2020. The main reason for the long period of time between these generations is that all local or global players providing cellular connectivity should come together, to discuss their needs, and define the necessary requirements of a new generation, and agree on the spectrum and the technology as a whole.

The vision of the 5G cellular network is shown in Figure 10.1. The vision is expressed in terms of multiple metrics including peak data rate, user experienced data rate, spectrum efficiency, mobility, latency, connection density, network energy efficiency, and area traffic capacity. In this section, we explain these metrics and provide a comparison between 5G metrics with the ones defined in 4G.

Peak data rate: This metric defines the maximum data rate of the network. 4G can use the spectrum of 40 or 100 MHz for data transfer. It is clear that the use of smaller spectrum results in a lower data rate. Also, the spectral efficiency is lower when a mobile device is at the cells' boundary as compared to the situation where the mobile device is close to the base station. Therefore, peak data rates can be achieved in the downlink direction, for mobile devices that are located close to

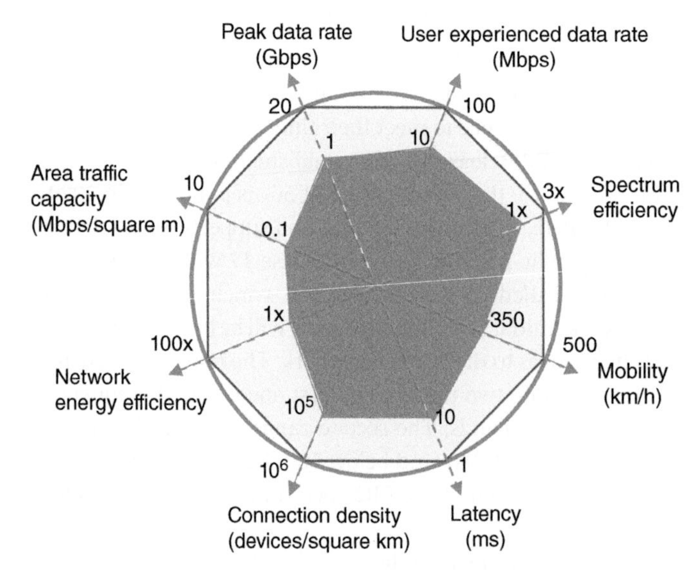

Figure 10.1 5G vision. Source: Modified from [1].

the base station, and when the largest possible spectrum is utilized. For example, 4G can use a 100 MHz spectrum to transmit data with the spectral efficiency of 15 bps/Hz in the downlink direction to a mobile device that is close to the base station. In this situation, a peak data rate of 1 Gbps is achievable. 5G's objective is to achieve a peak data rate of at least ten times more than the peak data rate of 4G.

User-experienced data rate: This metric defines the average data rate that a user can experience when transferring data. It is clear that resources such as time and frequency slots for data transfer are shared among all mobile devices in a cell. The data rate that a user can experience is dependent on many factors, such as the location of the mobile device within a cell, the number of active mobile devices, and the traffic pattern of these devices. The user-experienced data rate is a metric that provides an estimate of the achievable average data rate in practical situations. 5G's vision is for this metric to be ten times more than what was defined in 4G.

Spectrum efficiency: This metric defines the data throughput per unit of spectrum per cell (or per area). Therefore, the unit for this metric is bps/Hz/cell or bps/Hz/km^2. The term "spectrum efficiency" is also called spectral efficiency or bandwidth efficiency. 5G needs the spectrum efficiency to be three times more than the spectrum efficiency of the 4G network.

Mobility: This metric defines the maximum mobility of a mobile device. Generally speaking, there are many classes of mobility such as stationary, pedestrian, vehicular, and high speed. The mobility of the pedestrian class is below 10 km/h, while the mobility of the vehicular class is between 10 and 120 km/h. The mobility of high-speed class is above 120 km/h. 5G needs to accommodate the mobility of high-speed vehicles in order to provide service to people who travel by vehicles such as high-speed trains as well as the devices that exist in these trains.

Latency: This metric defines the duration of time from when the network allocates the resources to a cellular node for data transmission until the time that the destination node receives the data. The concept of latency is somehow confusing. There are many ways to define latency, and that is the main reason behind the ambiguity. First, when a cellular node wants to send a packet, it needs to access the network. We discussed access control in further detail in Subchapter 8.2. Some system designers like to market their system as a faster system by eliminating access control as part of the latency. Therefore, we should seek to understand whether access control is considered as part of the latency or not. It certainly takes time to access the network and there are many factors that determine access time. It should be noted that there are situations where accessing the network takes a long time. Second, the latency of sending a message might be different than the latency of transmitting a single packet. A message may contain several packets, and clearly the duration of time to send a message is larger than the time it takes to send only one packet. Finally, when a cellular node transmits a packet, it usually receives an acknowledgment that guarantees the reception of the packet. It is clear that defining latency as the period between the time that a packet is sent and the time that an acknowledgment is received would be different as compared to a situation in which receiving the acknowledgment is not needed. Based on the definition that we provided for latency, a 4G network can provide a latency below 10 ms. 5G wants to reduce latency to 1 ms. Latency is an integral metric for IoT applications. There are many IoT applications that need substantially lower than 10 ms of latency. These applications cannot use 4G cellular networks as a reliable connectivity scheme. By reducing latency to 1ms, these applications can take advantage of 5G for their connectivity needs.

Connection density: This metric defines the total number of devices per unit area, such as square km. This is an important metric as it relates to the number of IoT devices that can use the cellular technology in an area simultaneously. 5G wants to increase the connection density 100 times more than the connection density of the 4G network.

Network energy efficiency: This metric defines the number of bits that can be transmitted per 1 J of energy. This is an important factor that indicates whether the 5G network can be sustainable or not. We know that the 5G traffic would be substantially higher than the traffic on the 4G network. If 5G wants to work with the same network energy efficiency defined by 4G, then it consumes significantly higher energy. Since this cannot be sustainable for telecommunication companies, there have been enormous efforts to increase the number of bits that can be transmitted in 1 J of energy. Due to its importance, 5G intends to increase network efficiency 100-fold in comparison to 4G.

Area traffic capacity: This metric defines the total traffic throughput served per a geographic area. The unit of this metric is Mbps/m^2. 5G intends to increase this metric by 100 times as compared to the value defined as the area traffic capacity of the 4G network.

10.3 5G's Main Application Areas

The 5G cellular network targets three distinct clusters of application areas. These three areas are enhanced Mobile BroadBand (eMBB), massive Machine Type Communications (mMTC), and Ultra-Reliable and Low-Latency Communications (URLLC).

eMBB supports the delivery of voice and high-speed data. It provides a substantially larger system capacity, data throughput, and better spectral efficiency than the 4G network. For example, it supports the delivery of high-definition IoT cameras used for surveillance or inspection operations.

mMTC supports machine-to-machine type communications. It supports a large number of stationary and mobile IoT devices that have a diverse range of requirements in terms of bandwidth, data rate, and other QoS parameters.

URLLC is an important part of 5G and is vital for ultra-reliable and critical IoT systems. For example, it supports self-driving cars or enhanced factory automation applications that require a connectivity scheme that can provide reliable communications with low latency. Three key application areas of 5G are shown in Figure 10.2.

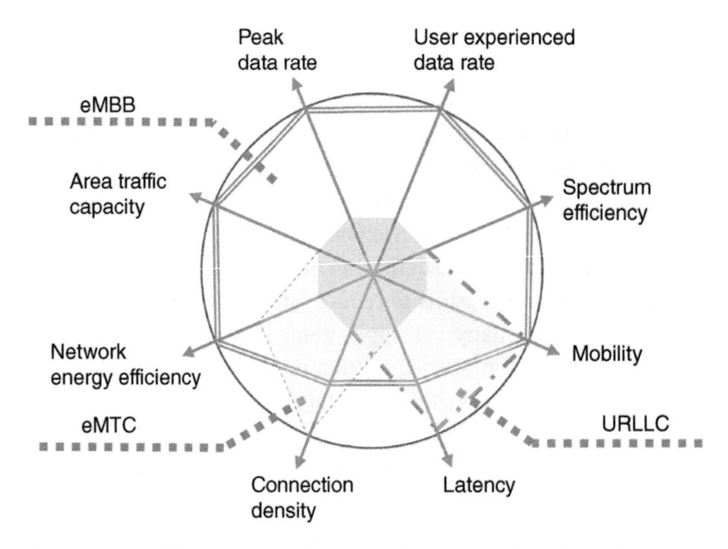

Figure 10.2 5G's key application areas. Source: Modified from [1].

10.4 5G Implementations and Features

In this section, we explain two different ways to implement 5G and some of the important features of 5G that did not exist in earlier generations of cellular networks.

10.4.1 Standalone and non-standalone 5G Network

The 5G network is a combination of a 5G Radio Access Network (RAN) and a 5G core. Simply put, a 5G RAN contains the radio access technologies and base stations, while the 5G core is responsible for operations such as authentication, signaling, paging, mobility, and security. At this time, most 5G deployments are based on the implementation of the 5G RAN for data communications, while the signaling and control plane uses the 4G core. This implementation is called Non-StandAlone (NSA) deployment. When the 5G core is deployed, both the data plane and signal plane pass through the 5G RAN. This is called StandAlone (SA) mode. A comparison of the 5G NSA and SA modes is shown in Figure 10.3.

10.4.2 5G Network Slicing

Network slicing is a form of network virtualization that aims to provide service differentiation. Generally speaking, virtualization can help to manage a resource in a more efficient way among many users. For example, a computer has many resources such as processing power, storage, operating system, or interfaces. By virtualization, we can change a computer to several Virtual Machines (VMs) in which each VM has its own required processing power, storage, operating system, or interfaces. Similarly, we can look at a 5G core as an entity with many resources and correspondingly perform virtualization on the 5G network. In other words, the 5G network can be divided into many slices. A slice is a logical network that serves a specific application, business partner, or customer. For example, at a very high level, a 5G network can be divided into three slices in terms of key application areas: eMBB, mMTC, and URLLC, as shown in Figure 10.4.

The 5G network can also be virtualized to give services to different business partners. If the network assigns a slice to a business partner, it allocates some of its resources to that specific partner. For example, a large company, factory, airport, university, or a small service provider might be considered a business partner. A 5G network operator may build part of its resources close to a base

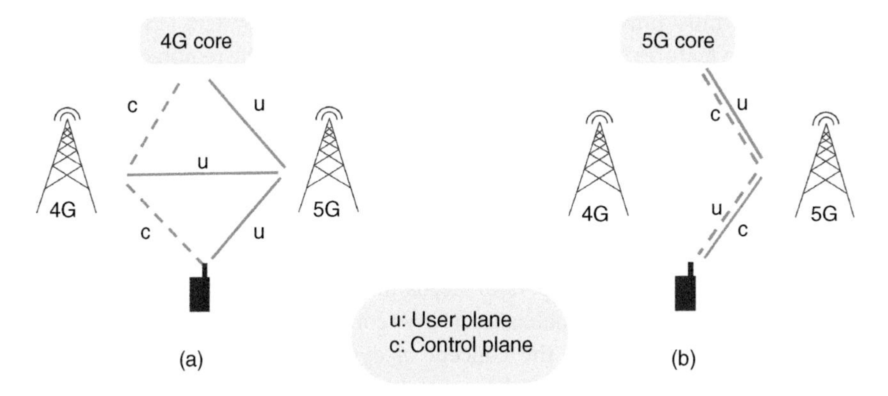

Figure 10.3 Two methods for the implementation of a 5G network: (a) NSA implementation and (b) SA implementation.

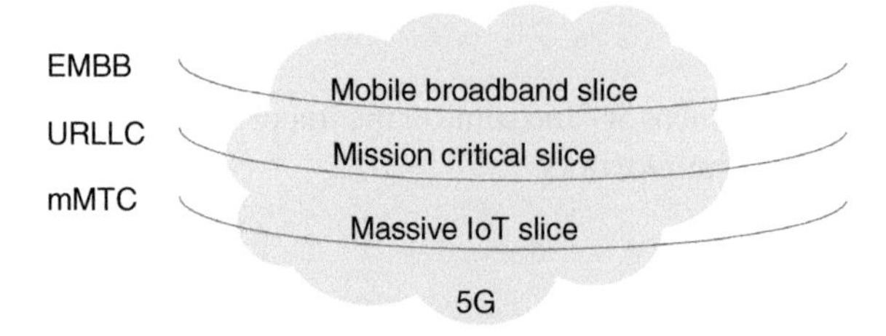

Figure 10.4 5G network slicing based on its main application areas.

station, in proximity to its business partner, or even at the location of the business partner. We will further discuss this concept, when we talk about private 5G networks. Each customer looks at a network slice as a separate network. Therefore, using network slicing, it is possible to create a virtual network that fits the requirements of a use case or business. In other words, the diversity of businesses that needs the 5G network can have an impact on how the network is built and managed.

Network slicing can be classified into two classes: single slicing and bundle slicing in which a bundle can be made of multiple single slices. For example, a smart car requires high-throughput data for its entertainment equipment, ultra-reliable and low-latency communications for autonomous or assisted driving, and mobility among different cells or even different operators networks. These diverse requirements can be assigned by an operator as a bundle of multiple single slices.

Let us explain the 5G network slicing using some examples:

- A logistics company needs wide area coverage, mobility, and low latency for its applications. In this case, network operators can provide a mobility-enabled, low-latency slice to this customer.
- As a pilot project, a city uses some stationary smart fire alarm systems in one of its neighborhoods. In this case, the operator can provide a mobility-disabled, high-availability slice to this customer.
- A large stadium needs high-throughput voice and data communication services for its visitors and can be given a throughput-optimized eMBB slice by a telecommunication company for this purpose.
- A large robotic manufacturing company needs a very reliable and low-latency connectivity scheme for its operations, and for this reason receives a critical, low-latency slice from a service provider.

As can be seen from these examples, network slicing can provide each use case with a specified set of functionalities. The operator must provide network slicing with a specific System Level Agreement (SLA), which specifies the features such as the requested data rates, or the required capacity in order to satisfy the needs of a customer or an application. Network slicing is an innovative part of the 5G architecture that allows telecommunication service providers to optimally allocate part of their network to a specific application, use case, or customer.

This is all possible since 5G network operators have the capability to allocate parts of their network based on the demanded specifications of their customers. IoT applications have diverse requirements, and 5G network slicing can ensure that their demands can efficiently be granted.

To be able to perform network slicing, the operator needs to make use of Software Defined Networking (SDN) and Network Function Virtualization (NFV) concepts.

SDN is a network architecture that brings the concept of dynamic network configuration to a network by separating the data plane from the control plane. The control plane uses a centralized controller or several centralized controllers. The controller(s) have a big picture of the network that they control and are able to program the data plane devices. There are open interfaces that are defined between the controllers and data plane devices. The SDN network is managed by a software program that runs on top of the controllers. Simply put, the hardware data plane infrastructure needs to efficiently forward traffic as fast as possible. Therefore, the operator must be able to dynamically configure this infrastructure using a software program, in order to optimize network operations. In other words, data plane devices do not perform signaling, and only transfer data in the most efficient way possible. The 5G network is customer centric and supports network slicing, while the earlier generations of cellular networks were operator-centric networks. In order to be customer centric, operators need to use SDN to centralize the control element of a network, and bring programmability to the network. SDN orchestration enables coordination among all networking hardware and software elements that perform various services. SDN needs Application Programming Interface (APIs) to communicate with hardware infrastructure. Standard APIs can be used to program different hardware infrastructures in a network. These infrastructures are designed by a wide variety of manufacturers.

NFV abstracts and virtualizes network functions such as routing, load balancing, and policy management, and also transfers these functions to virtual servers. With virtualization, the new operator's network looks more like a data center rather than the traditional telecommunication network. The network core becomes a data center and the network's radio access, or RAN, looks like distributed data centers. NFV brings efficiency and agility for operators to quickly deploy new services for their customers. The interested readers can study [2] for more detailed information about SDN, NFV, and their application in modern networks. Also, they are encouraged to study [3] which provides them with an introduction to Network slicing.

10.4.3 Private 5G Network

As part of a 5G network, we can have public 5G as well as private 5G networks. There are three methods to offer a private 5G network as shown in Figure 10.5. These three methods are known as standalone, virtual, and hybrid private networks.

Standalone private networks can be purchased from an operator and need to be managed by a customer. For example, an airport might be interested in buying a standalone 5G private network and deploying it physically at its airport vicinity. This network is isolated from the 5G public network and can provide a dedicated solution for the airport.

Virtual private 5G networks use virtualization on public 5G networks. In this case, the customer's network is logically private, but the network uses the 5G public network physically. For example, smart cars might use virtual private networks for their operations.

Hybrid private 5G networks use a combination of both standalone and virtual private networks. For example, a manufacturing plant may use both a standalone 5G private network and a virtual private network to have a better distribution of its functionalities.

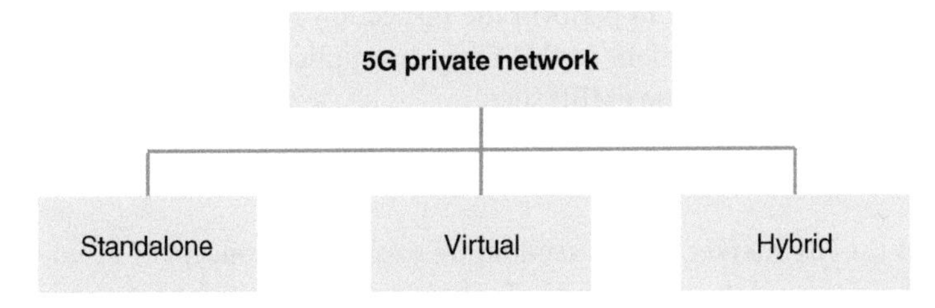

Figure 10.5 Private 5G network classification.

10.4.4 Network Exposure

Until now, customers of a cellular network did not have any visibility and access to the network capabilities on demand. The 5G network exposure enables customers and partners to have access to network services and resources. This would create endless opportunities for customers to build innovative applications that involve modifying some of the network capabilities. 5G operators should assign security and data integrity policies to their customers in order to give a customer a secure way to access network capabilities. Network exposure is a function of the 5G core network and can play an important role in the design of IoT applications.

In a 5G network, an operator can unlock a capability in such a way that it becomes accessible by a customer through network APIs. Several network APIs can be combined in order to make a service API that can be easily used to execute an operation on the network. It is extremely important to have access to a set of rich, easy-to-use service APIs that address the needs of various customers. Using these high-level service APIs, great opportunities can be created for the developers with a limited knowledge about telecommunication systems in order to develop new IoT applications.

The concept of network exposure had not been used by the telecommunication companies before 5G. Mostly due to security reasons, telecommunication companies have always tried to prevent their customers from having access to configure the network based on their needs. However, if these companies would allow their customers to configure the network, it would enable IoT system developers to use network resources based on the needs of their applications. A similar situation happened for smartphone manufacturers. At the beginning, smartphones offered only a limited number of apps that were designed by smartphone manufacturers. Later, when these manufacturers unlocked parts of their system for app developers, we witnessed that app developers designed a huge number of apps that revolutionized the smartphone industry. It should be noted that access for network configuration must be given in a secure manner. That is why network exposure must be performed under tight security.

Ericsson has performed an experiment to show the applicability of 5G network exposure for IoT applications [4]. In this experiment, a drone, which is used to perform tower inspections, is connected to a private 5G network. A user remotely controls the drone and inspects the tower using a management application. The application uses APIs to trigger an exposure server that resides on a 5G core network. To start the inspection, the user starts the mission that sends API#1 to the exposure server on the 5G core. API#1 triggers the exposure server to onboard the drone and authenticate the application to communicate with the drone. This establishes a low-throughput eMBB slice. The application sends a flight map to the drone, and the drone starts flying toward the tower. The inspection camera does not need to provide high-quality video, as the drone goes toward the tower. When the drone is close to the tower, the application sends API#2 to increase the bandwidth on the eMBB slice. When it is close to the tower, the application sends API#3 to the exposure server to add a URLLC slice for drone remote control and to increase the quality of the video on eMBB slice as shown in Figure 10.6. This enables the person who is performing the inspection to control the drone and have better video quality to perform the inspection remotely. After the inspection is complete, the inspector ends the mission. At this time, the application sends API#4 to erase the URLLC slice and lower the quality of the eMBB slice.

10.4.5 Fixed Wireless Access

It is expected that 5G will expand the IoT market. In this section, we explain the concept of Fixed Wireless Access (FWA) that can help expand the market for IoT. There are many locations in rural areas, and many places in underdeveloped and developing countries where no wired broadband

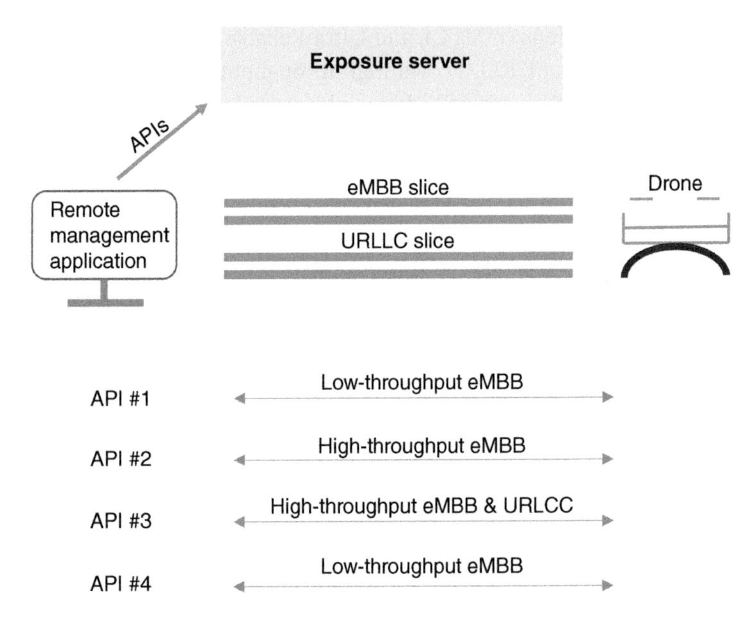

Figure 10.6 Ericson experiment to show the application of the network exposure.

Figure 10.7 (a) Wireline broadband connectivity. (b) Broadband connectivity through 5G FWA.

connection is available. Due to the high cost of implementation and low revenue expectations, the operators do not plan to install wireline infrastructure and are not interested in providing wired connections to these locations.

FWA can provide Internet access for these locations by taking advantage of wireless technology. 5G FWA can be utilized to provide Internet access in a quick and cost-effective manner. It can provide high-speed broadband access to homes, communities, and businesses that did not have Internet access. This means that by utilizing 5G FWA, many IoT devices can also be used in these locations. In other words, 5G enables an efficient connectivity scheme to these locations that subsequently enables the use of IoT in these places. Figure 10.7a shows a house that is connected to the Internet using wired Wide Area Network (WAN) technologies such as different types of Digital Subscriber Line (xDSL) including Asymmetric DSL (ADSL) or Very high-speed DSL (VDSL), cable TV, Passive Optical Network (PON), or Fiber-To-The-X (FTTX) technologies. Figure 10.7b shows a house that receives the Internet connection using 5G FWA.

10.5 Summary

In this chapter, we briefly discuss the 5G cellular network and its features. 5G targets three distinct clusters of use cases or application areas. These three areas are enhanced Mobile BroadBand (eMBB), massive Machine Type Communications (mMTC), and Ultra-Reliable and Low-Latency Communications (URLLC). eMBB, mMTC, and URLLC cover high-throughput IoT, massive IoT, and critical IoT applications, respectively. mMTC is specially designed to provide service to massive numbers of IoT devices in a network. URLLC is defined to address the needs of critical IoT devices that require ultra-reliable and low-latency communications. Therefore, 5G is designed with IoT in mind. 5G IoT is part of 3GPP Releases 16, 17 and interested readers are encouraged to study [5] for more information on the evolution of the 5G cellular networks.

5G has several revolutionary features that allow operators and customers to optimize the network and their applications, respectively. Network slicing is a form of network virtualization that aims to provide service differentiation. A 5G network can be virtualized by assigning a slice of its resources to a customer based on the customer's needs. A slice is a logical network that serves a particular application, business partner, or customer. Network exposure is another important feature of the 5G network that provides customers with visibility and access to the network capabilities and opens up new opportunities for customers to develop innovative applications.

The 5G architecture establishes a distributed data center scheme in which the edge data centers can be created in customer's locations as private 5G networks, on top of the base stations towers, or in telecommunication offices close to the base stations, as well as in the centralized cloud. This architecture can provide the latency, reliability, throughput, and privacy requirements of various customers and businesses.

In this chapter, the concept of FWA is briefly discussed. 5G FWA can provide broadband access to locations with no wired Internet connectivity. It can provide Internet access in a quick and cost-effective manner to locations where providing wired connectivity is not feasible. This enables the use of IoT in these places and increases the IoT market as a result.

References

1 ITU-R (2015). IMT vision – framework and overall objectives of the future development of IMT for 2020 and beyond. Recommendation M.2083-0. https://www.itu.int/dms_pubrec/itu-r/rec/m/R-REC-M.2083-0-201509-I!!PDF-E.pdf.

2 Stallings, W. (2015). *Foundations of Modern Networking*. Addison-Wesley.

3 GSMA (2017). An introduction to network slicing. https://www.gsma.com/futurenetworks/wp-content/uploads/2017/11/GSMA-An-Introduction-to-Network-Slicing.pdf (accessed 30 June 2022)

4 Nafees, S. (2020). Network exposure and the case for connected drones. https://www.ericsson.com/en/blog/2020/6/network-exposure-and-the-case-for-connected-drones (accessed 30 June 2022).

5 5GAmericas (2020). The 5G evolution: 3GPP releases 16-17. https://5gamericas.org/wp-content/uploads/2020/01/5G-Evolution-3GPP-R16-R17-FINAL.pdf#Free5GTraining (accessed 30 June 2022).

Exercises

10.1 5G supports 24–100 GHz frequency range, called FR2, as part of its spectrum. What is the range of wavelengths of FR2? Why is this range of spectrum called millimeter wave?

10.2 Network energy efficiency is an important metric in the design of the 5G network. Compare this metric in the 4G and 5G networks. Why is improving network energy efficiency important for the 5G network?

10.3 What is the difference between 5G and 5G new radio (5G NR)?

10.4 If the peak data rate of a 5G network is above 20Gbps, why do users only experience a data rate of 100Mbps?

10.5 Someone tells you that the latency of a network is less than 5 ms. Does this show the amount of time that it takes for a message to go through the network?

10.6 What are the three main application areas of the 5G network?

10.7 Does a 5G standalone private network need to be connected to a public 5G network? Who owns and manages a 5G private network?

10.8 SDN plays an important role in high-speed networking.
a. What does SDN stand for?
b. What is the main functionality that an SDN architecture offers?
c. What is the difference between a traditional router and an SDN router?
d. Is SDN part of the 5G network architecture?

10.9 What is the difference between 5G RAN and 5G core?

10.10 What is the 5G non-standalone deployment?

Advanced Exercises

10.11 Assume a new cellular technology uses a spectrum of 200 MHz and 300 MHz, and its maximum spectral efficiency (at the edge of the cell), minimum spectral efficiency (close to a base station), and the average spectral efficiency are summarized in Table 10.E11.

Table 10.E11 Data for Exercise 10.11

Spectral efficiency (bps/Hz)	Min.	Max.	Average
Downlink	0.12	30	4.4
Uplink	0.06	13.5	2.8

a. Find the peak data rate achievable by this cellular technology. In what spectrum is the peak data rate achievable?
b. What is the average downlink data transfer when using 200 MHz spectrum?
c. What is the uplink data rate using a 200 MHz spectrum near the edge of a cell?

10.12 5G NR is capable of providing downlink throughput of 2.3Gbps with certain configurations when it uses 100 MHz channel bandwidth? What is the spectrum efficiency of this network?

10.13 A taxi company requires a 5G network slice to manage its dispatch system and track its vehicles. Each taxi is equipped with an IoT device. The taxi company wants to trace the location of its vehicles and find the optimized routes to their destinations. This network also needs to be secure. Can an operator assign a network slice for the taxi industry?

10.14 A network slice has been defined for live broadcast of sports through Augmented Reality (AR) or Virtual Reality (VR) technology.

a. Why does an eMBB slice not work in this case?

b. What type of network slice do you think this use case needs?

10.15 A system is connected to the cloud using a 5G network. Can the network be sliced based on different users of the same system?

10.16 How a 5G network can be seen as a distributed data center?

Chapter 11 IoT and Analytics

11.1 Introduction

The combination of Internet of Things (IoT) and analytics is a perfect fit, a dream team, and a great match made in heaven! Each helps the other to advance in its own respective domain. In other words, they complement each other. There is no way to design IoT applications without analytics. The reasoning behind installing one or more sensors on a physical object and collecting its data is that we plan to perform data analysis, and then provide the physical object with the opportunity to make smart decisions. This is where analytics play a critical role in bringing intelligence to the decision-making process. In a similar fashion, one of the best opportunities for analytics is through IoT. There are three driving factors to advance analytics. The first is the advancements in processing power and parallel processing, the second is the advancements of new analytics algorithms, and the third is the existence of vast amounts of data. In many modern analytical methods, we use data to train an algorithm in order to behave in a particular way. And if more data is available to us, we can design more accurate algorithms. In other words, we need to feed these algorithms with more data, to better train them. Today, we are collecting a vast amount of data every day. However, many companies that collect data do not want to share their data. There are many organizations that may open source their algorithms, but they do not share their data with others. For example, Amazon does not share its customer purchase behavior data, Facebook does not share its social data, and Google does not share its search data. So, for the analytics community to advance their algorithms today, there is an opportunity to advance them by using IoT data. IoT can provide the analytics community with an abundance of real-time data that can be used to advance analytical algorithms designed for various applications.

Generally speaking, there are three types of analytics that are called descriptive analytics, predictive analytics, and prescriptive analytics, as shown in Figure 11.1.

Descriptive analytics is the simplest and the most basic form of analytics. An organization might use descriptive analytics to find out what is happening internally within the organization. For example, descriptive analytics can be used to make sense of the collected data from IoT devices in a smart factory. Assume an IoT device is installed in a smart factory and collects the current draw of one of its machinery as well as its temperature every five seconds. Using descriptive analytics, we can perform some types of statistical analysis on data in order to better understand the collected data. For example, descriptive analytics can be used to calculate the average current draw of a machine or its average temperature over a five minute period. Descriptive analytics can also be used to identify the root cause of a problem and the reason behind what has happened. For example, it can be used to determine if excessive current draw or temperature rise was the reason that a machine in a smart factory failed to function properly. While an organization can use

Fundamentals of Internet of Things: For Students and Professionals, First Edition. F. John Dian.

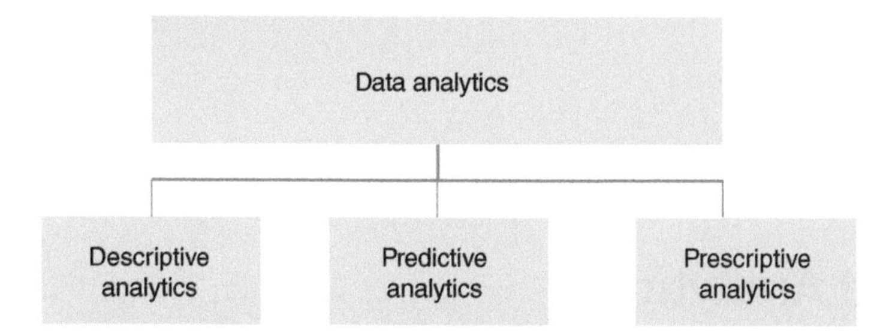

Figure 11.1 Different types of analytics.

descriptive analytics to better understand the collected data, descriptive analytics do not attempt to predict the behavior of a system in the future.

Predictive analytics models the behavior of a system and predicts future events. It answers the question of what will most likely happen. For example, with predictive analytics, we may be able to find the estimated time-to-failure for a machine that is showing a 10% increase in measured temperature.

Prescriptive analytics is the most complex form of analytics. After analyzing the data, it can generate recommended actions and strategies that need to be taken. An organization might use prescriptive analytics to answer the question of what an organization needs to do in order to maximize profit or optimize its production. For example, with prescriptive analytics, we may be able to determine the scheduling time for servicing a machine in a smart factory based on the temperature data, or determine an estimate of time-to-failure, and the tasks that should be performed to better monitor or control the temperature increase for those machines that have shown excessive temperature increases in the past.

The complexity and the value of each type of data analytics is shown in Figure 11.2. Descriptive analytics is the simplest form of analytics and adds less value to the organization as compared to other types of analytics. It does not provide enough intelligence to predict future events or model the behavior of a smart system. Predictive analytics are more complex and add more value to the organization. Predictive analytics uses collected data to model the behavior of the system, while prescriptive analytics goes even deeper into determining the potential actions that need to be taken.

In this chapter, we discuss different types of analytical methods and their corresponding techniques that are used to better represent the data, identify abnormalities, find patterns, predict events, and generate insights from vast amounts of data that can be generated by IoT devices. We start with the concept of the data pipeline and some of the processing operations that need to be performed in order to prepare the data before it is fed into analytical tools.

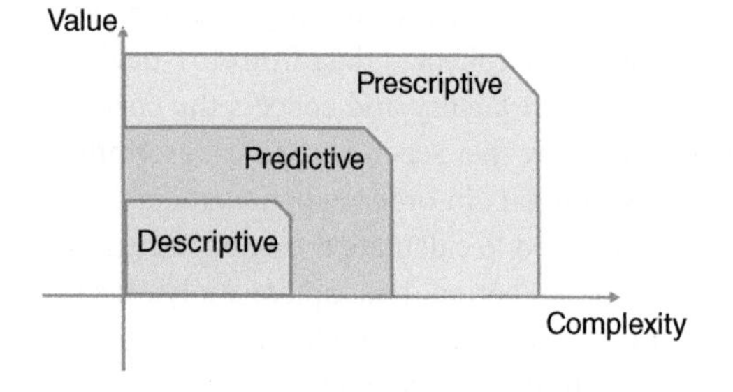

Figure 11.2 The complexity versus the value of different types of analytics.

11.2 Data Pipeline

A data pipeline represents the flow of data between two systems. In an IoT ecosystem, the data pipeline between a group of sensors and the cloud should be designed in such a way that data flows smoothly from one service to another. This data pipeline should be scalable to be able to handle the flow of data as the number of IoT systems grows. The cloud should also have the capacity to receive and process data, and therefore, we need to know the demand that is placed on a cloud backend.

Since data may be lost, corrupted, or arrive out of order to the cloud, the data must go through several data processing operations that prepare the data to be fed to analytical tools. First, the data may come from multiple sources and needs to be integrated. Second, the quality of data needs to be checked and the data needs to be cleaned. If needed, the format of the data also needs to be adjusted. Finally, the data needs to go to a data storage area such as a database, a data warehouse, or a data lake. After these operations, analytical tools and services can analyze the stored data. The result of analyzing the data is used to create an intelligence that is needed for making smart decisions, and it might also be sent to an IoT-based management system.

The transportation of data from one or more sources to a destination, where it can be stored and further analyzed, is called data ingestion. Three types of data ingestion include batch processing, stream processing, and micro-batch processing. Batch processing is scheduled processing and relies on stored data, while stream processing is based on real-time processing and usually relies on continuous data. In batch processing, the data is first stored in a storage area such as a data warehouse for a period of time and is processed at a scheduled time. For example, processing data hourly, daily, or weekly are examples of batch processing. In stream processing, the data is processed as it arrives. The result of stream processing might trigger some real-time actionable tasks or send the processed data to a real-time dashboard. Micro-batch processing is a type of real-time processing in which the data is stored for a very short amount of time before getting processed.

In an application, we may use both batch and stream processing together. Consider a scenario in which we are collecting data from a number of IoT devices. The data is transmitted to the cloud and pushed to a cloud gateway. Behind the gateway, there is an analytics service unit that cleans the data and sends a copy of the data to a storage location for batch processing. In this application, the data is analyzed daily using analytics tools. In addition, the data is sent to a stream processing analytics in which data is analyzed in real time. The result of this stream processing might be used to update a real-time dashboard.

Many IoT applications involve large data sets of sensor data that should be analyzed in near real-time in order to be used for decision making. These large data sets often have complex patterns, where each of these patterns is a representation of a unique event. The need for processing and analyzing complex patterns in real time in order to identify various events is called Complex Event Processing (CEP). [1]

11.3 AI

Intelligence is the ability to process information in such a way that it can be used to make smart decisions. The field of Artificial Intelligence (AI) is the science that focuses on building algorithms that process information and make future decisions. AI brings the cognition and intelligence of a human to its algorithms with faster decision-making capability, more processing power, and higher accuracy. Therefore, the objective of AI is to determine future trends and events from data without any human intervention.

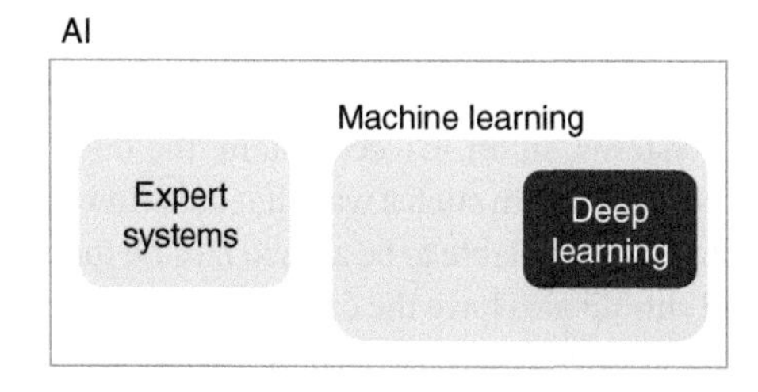

Figure 11.3 Relationship between AI, expert systems, machine learning, and deep learning.

An AI algorithm might be very simple, consisting of a series of if-then statements. On the other hand, an AI algorithm may use a very complex statistical model to process the data, predict future events, and make smart decisions. In developing an AI algorithm, the idea is not to instruct and teach the algorithm how to process the data in a step-by-step fashion. Instead, we want the AI algorithm to learn from the data in order to make its own decisions. The use of a series of if-then statements to program a machine is called rule engine, or expert system. The rule engine is a subset of AI in which the rule engine provides the instruction on how the algorithm should perform to be similar to the actions that an expert takes on a similar situation. The expert system gives the AI algorithm the cognition of a human expert. Therefore, the objective of an expert system is to transfer knowledge from an expert person into a series of rules that can be applied to data in a parallel fashion.

Machine learning is a subset of AI that focuses on teaching an algorithm to learn from experiences without explicitly being programmed. One of the main differences between machine learning and rule engine algorithms is that machine learning algorithms can modify themselves as they receive more data. We briefly explain machine learning algorithms in Section 11.4. A subset of machine learning algorithms is called deep learning algorithms, which have gained a lot of attention due to their incredible performance, and we will discuss them in Section 11.7. The relationship between AI, expert systems, machine learning, and deep learning is shown in Figure 11.3.

11.4 Machine Learning

Machine learning algorithms are a subset of AI algorithms that have the ability to automatically learn from data in order to improve themselves. Machine learning algorithms can be classified into five categories. These categories are supervised learning, unsupervised learning, semi-supervised learning, reinforcement learning, and deep learning as shown in Figure 11.4.

In supervised learning, a machine learning algorithm is trained using some training data sets, which include both inputs and their corresponding outputs (also called labels). Therefore, supervised learning refers to a group of algorithms that are trained using the labels associated with the input data. If the input data to a system is expressed as X and the output as Y, then Y is a dependent variable of X, and the relationship between input and output can be modeled as $Y = f(X)$.

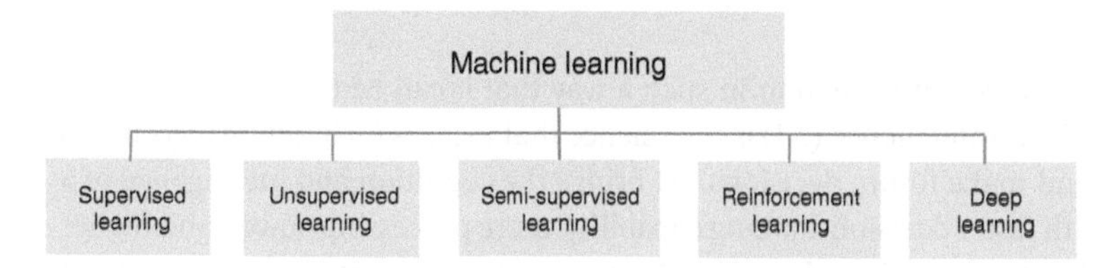

Figure 11.4 Different categories of machine learning algorithms.

A supervised learning algorithm tries to model the system and predict the output for any applied input data. During training, the supervised machine learning algorithm predicts the output associated with an input training data set and compares its prediction with the label provided in the training data. This comparison can help the algorithm to modify the model by minimizing the gap between its predicted output and the actual output (label). The method is called supervised learning, since when the algorithm makes a prediction during training, it is under supervision of a supervisor (label) that can help the trainee (algorithm) to improve its performance. There exists various techniques to implement supervised learning. The most important methods to develop supervised learning models are classification and regression. Regression techniques predict continuous outputs, while classification techniques predict discrete responses. Regression techniques try to predict the best function to relate input and output data, while classification techniques try to predict the best category that an input data belongs to. For example, based on the received data from a smart hockey helmet, a classification technique may predict whether a collision has been strong enough to cause a concussion or not. A regression technique, on the other hand, may predict the maximum speed that a hockey player can glide after being on the ice for a specific amount of time. We will discuss some of the important classification and regression techniques in Section 11.5.

Unsupervised learning refers to a group of machine learning algorithms that are trained with data sets that do not have any labels. In unsupervised learning, a machine learning algorithm attempts to learn from input in order to train itself. This can be done by identifying patterns inside data and understanding possible data features that can help in dividing input data into different clusters. For example, an insurance company may want to find similar groups of drivers based on their driving habits, through the data collected from their smart cars. An online music company might be interested in finding groups of people with similar interests in different types of music, in order to recommend the people in each group with songs that they might be interested in purchasing. The music company may analyze the collected data from their smart headsets for this purpose. Generally speaking, collecting labeled data is more expensive and in some cases, the label must be provided by experts, and therefore it needs some levels of human intervention. For this reason, there is an interest in developing unsupervised machine learning algorithms. The most important method for developing unsupervised machine learning models is clustering. We will discuss clustering and some of the important algorithms for performing clustering in Section 11.6.

Semi-supervised learning is a hybrid scheme that falls somewhere between supervised and unsupervised learning. In semi-supervised learning, the training data contains data sets that have labels, similar to supervised learning. It also contains data sets that are similar to unsupervised learning and do not have any labels. In many real-world applications, machine learning algorithms are designed based on semi-supervised learning schemes.

Reinforcement learning refers to a group of machine learning algorithms in which a physical system (called an agent) interacts with its surroundings (called an environment) to train itself. The agent is not provided with any training data. Instead, the agent learns from its own actions. The agent may choose an action based on its prior experiences (exploitation) or may choose a new action (exploration). The success of an action is measured as a numerical value (called a reward). The agent tries to choose its actions in such a way that maximizes its reward value. For example, let us consider an IoT-based smart race car that trains itself to move along a race track. During training, the smart race car (agent) chooses different actions (changing speed, turning steering wheel, etc.) on the race track (environment). Assume a camera sends a signal to the agent that can be used by the agent to measure the success of its action and determine a number as the award associated with the action. The smart race car can enhance its algorithm after going through the race track many times. Reinforcement machine learning can be implemented for IoT applications in which the IoT device can train itself to its environment without using any training data.

A deep learning algorithm focuses on using neural networks to automatically extract useful patterns or features from data, and learn from these features to train the algorithm. Therefore, deep learning means teaching algorithms how to learn a task directly from data using the concept of neural networks. We will discuss deep learning in Section 11.7.

One may think that machine learning algorithms are complex and need to be performed in the cloud. In other words, machine learning algorithms are part of cloud computing, and they cannot be implemented in the fog or as part of edge computing. It is true that most machine learning algorithms require computational resources which can be provided in a cloud environment, but machine learning algorithms can also be implemented in the edge. Tiny Machine Learning (TinyML) is a concept where machine learning algorithms are implemented on embedded systems. TinyML provides low latency, low bandwidth, and low power capability to edge and fog computing devices. There are some microcontrollers that are capable of performing TinyML. Therefore, neural networks and machine learning algorithms can even be implemented on IoT devices. This enables these devices to perform edge computing using AI.

11.5 Supervised Machine Learning Techniques

The most important methods to develop supervised learning models are classification and regression. In this section, we discuss the most important techniques belonging to these methods.

11.5.1 Classification

The process of predicting the class of an input data set is called classification. This process belongs to supervised machine learning algorithms where the training data set includes input data and outputs (labels). The objective of classification is to train the machine learning algorithm to correctly predict the output class, when it is provided with new data. We discuss several important classification techniques such as decision trees, random forests, and Support Vector Machine (SVM) in this section.

11.5.1.1 Decision Tree

Decision trees are one of the commonly used methods for supervised machine learning. Even though the decision tree can be used for both classification and regression, we explain the decision tree as a classifier in this section. The simple idea behind a decision tree classifier is to predict discrete output values by implementing decision rules obtained from the input data set, which are also called features. The decision tree uses a tree data structure that starts with a root node. In the root node, a feature is evaluated and one of the branches is selected. Therefore, each node in a tree acts as a decision rule. Then the splitting process divides the data into two or more subnodes based on other features. The process continues in order to further split the subnodes. A node that cannot be split any further is called a leaf node. A node, which can be divided into subnodes, is called a parent node, and each of its subnodes is called a child node. The process of making a decision tree from a training data set is a recursive operation. We will now go through an example that explains how to design and use a simple decision tree.

Assume a senior person is wearing a medical wearable IoT device that detects falls and provides an alarm if a fall occurs. Four signals are measured by the wearable IoT device. Table 11.1 depicts

Table 11.1 Fall detection based on various ranges of three signals.

Observations	Signal A	Signal B	Signal C	Signal D	Output
1	Low	Low	Low	High	No fall
2	Low	Low	High	High	No fall
3	High	Low	Low	High	Fall
4	Medium	Low	Low	Medium	Fall
5	Medium	High	Low	Low	Fall
6	Medium	High	High	Low	No fall
7	High	High	High	Low	Fall
8	Low	Low	Low	Medium	No fall
9	Low	High	Low	Low	Fall
10	Medium	High	Low	Medium	Fall
11	Low	High	High	Medium	Fall
12	High	Low	High	Medium	Fall
13	High	High	Low	High	Fall
14	Medium	Low	High	Medium	No fall

the various ranges of these four signals to show whether a fall has occurred or not. For example, the first signal (signal A) is classified into three ranges of low, medium, and high, while the second signal (signal B) and the third signal (signal C) are classified into two ranges of low and high. The fourth signal is signal D that has three ranges as low, medium, and high. The information in Table 11.1 contains 14 observations that were used as our training data set. These four signals are the inputs to the fall detection algorithm, and the output signal shows whether or not a fall has happened. For this scenario, we wish to create a classification scheme based on a decision tree. It may be difficult to understand how a fall is caused by these four input signals by glancing at the information in this table. It is obvious that as the number of input signals increases or each signal is classified into more ranges, understanding the input-output relationship becomes more challenging.

It should be noted that all combinations of the input signals are not included in the training data set. A decision tree is designed based on the training data set, but should be able to predict the output correctly for any new data. Assume after designing the decision tree, a new data (signal A = Medium, signal B = High, signal C = Low, signal D = High) occurs and the decision tree should detect whether these combinations indicate the occurrence of a fall or not. Since the training data set does not contain all possible combinations, the new data is not usually part of the training data set. We now explain a method to design a decision tree for this simple example that gives readers a basic understanding on how to design and use a decision tree classifier.

Designing a decision tree for this example involves finding the probability that a fall occurs according to each input signal. Decision trees should look at each input signal and split the training data set based on the various ranges of each input signal. Let us start building a decision tree by splitting the training data set based on the three ranges introduced for signal A. This is shown in Figure 11.5. For example, signal A is low in five observations. When signal A is low, observations 9 and 11 indicated a fall and observations 1, 2, and 8 did not show any fall. There are also five observations where signal A is medium. In this scenario, three observations indicate a fall, while two observations indicate no fall. There are four observations when signal A is high. These include observations 3, 7, 12, and 13. All these observations indicate a fall. We call this subset a pure subset. A pure subset occurs when the output for all the observations indicates the same output. A pure subset does not need to be split any further.

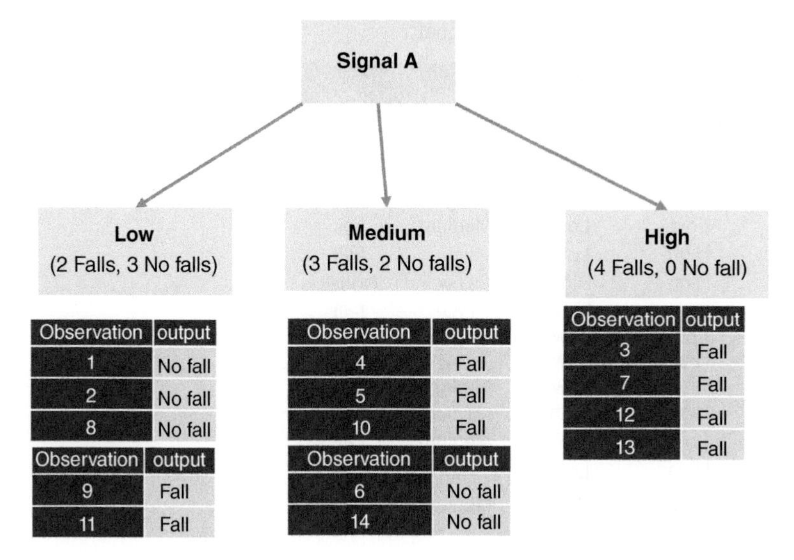

Figure 11.5 Building a decision tree by splitting the training data set based on three ranges of signal A.

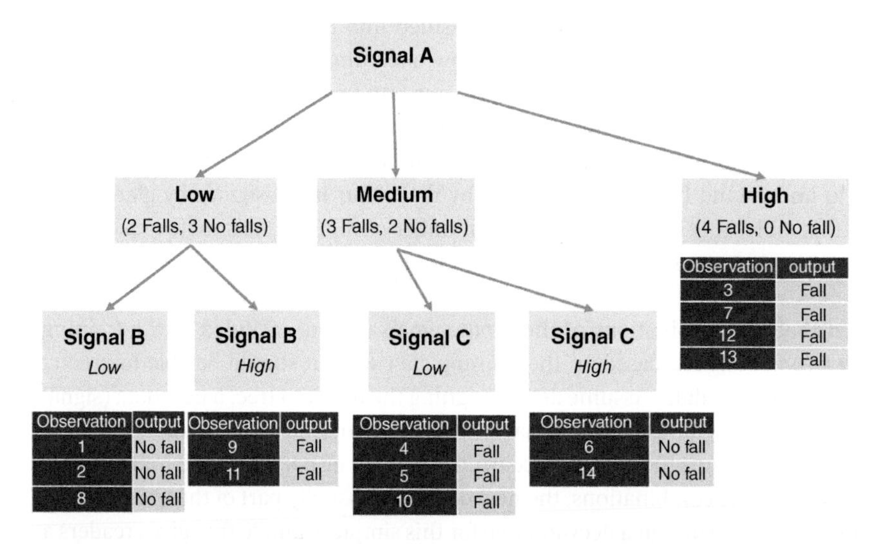

Figure 11.6 Continuation of the splitting process until all subsets are pure.

We can continue the splitting process to its depth by splitting the observations based on other signals to get to pure subsets. Let us perform the second round of the splitting process, and split the branch signal A-(low) based on signal B and the branch signal A-(medium) based on signal C. We do not need to split signal A-(high) any further, since it has provided a pure subset. As you can see in Figure 11.6, this splitting process results in pure subsets in all branches of this decision tree. Therefore, using the designed decision tree, we can determine that the new data (signal A = Medium, signal B = High, signal C = Low, and signal D = High) indicates a fall.

In our example, we did not explain how we chose the order of features used in the splitting process. For example, why did we start the splitting process with signal A, and not with signal B, signal C, or signal D? We now explain a method to systematically perform the splitting process by using a metric called entropy [2], which indicates the purity of a subset, expressed as:

$$H(s) = (-P_+ \, log_2 P_+) - (P_- \, log_2 P_-) \tag{11.1}$$

where P_+ and P_- are the probability of observations with positive and negative outputs in a subset, respectively. The entropy gives a number between 0 and 1, which indicates the purity of a subset or the certainty of an output prediction. For example, if all the observations are the same, the entropy is zero, while if 50% of the observations are positive and 50% of the observations are negative, the entropy becomes one. If we have 14 observations, with 9 showing a fall and 5 showing no fall, then the entropy can be calculated as:

$$P_+ = \frac{9}{14}, \quad P_- = \frac{5}{14}, \quad H(s) = -\frac{9}{14}\, log_2 \frac{9}{14} - \frac{5}{14}\, log_2 \frac{5}{14} = 0.94 \tag{11.2}$$

To determine which of the three signals to choose for splitting, we first need to calculate the splitting gain of each signal. Figure 11.7a–d shows the splitting process based on signal A, signal B, signal C, and signal D, respectively.

From these four possibilities, we choose the one that provides us with a higher splitting gain. The splitting gain is defined as:

$$\text{Splitting gain} = H(s) - \sum Pr_c H(S_c) \tag{11.3}$$

where $H(s)$ is the entropy of a parent branch, $H(S_c)$ is the entropy of each of its child branches, and Pr_c is the probability of each child branch.

The splitting gain for data shown in Figure 11.7a is:

$$H(s) = -\frac{9}{14}\, log_2 \frac{9}{14} - \frac{5}{14}\, log_2 \frac{5}{14} = 0.94 \tag{11.4}$$

$$\begin{cases} H\left(S_{\text{Signal A-Low}}\right) = -\frac{2}{5}\, log_2 \frac{2}{5} - \frac{3}{5}\, log_2 \frac{3}{5} = 0.971, Pr_{\text{Signal A-Low}} = \frac{5}{14} \\ H\left(S_{\text{Signal A-Medium}}\right) = -\frac{3}{5}\, log_2 \frac{3}{5} - \frac{2}{5}\, log_2 \frac{2}{5} = 0.971, Pr_{\text{Signal A-Medium}} = \frac{5}{14} \\ H\left(S_{\text{Signal A-High}}\right) = -\frac{0}{4}\, log_2 \frac{0}{4} - \frac{4}{4}\, log_2 \frac{4}{4} = 0, Pr_{\text{Signal A-High}} = \frac{4}{14} \\ \text{Splitting gain} = 0.94 - \left(\frac{5}{14} \times 0.971 + \frac{5}{14} \times 0.971 + \frac{4}{14} \times 0\right) = 0.246 \end{cases} \tag{11.5}$$

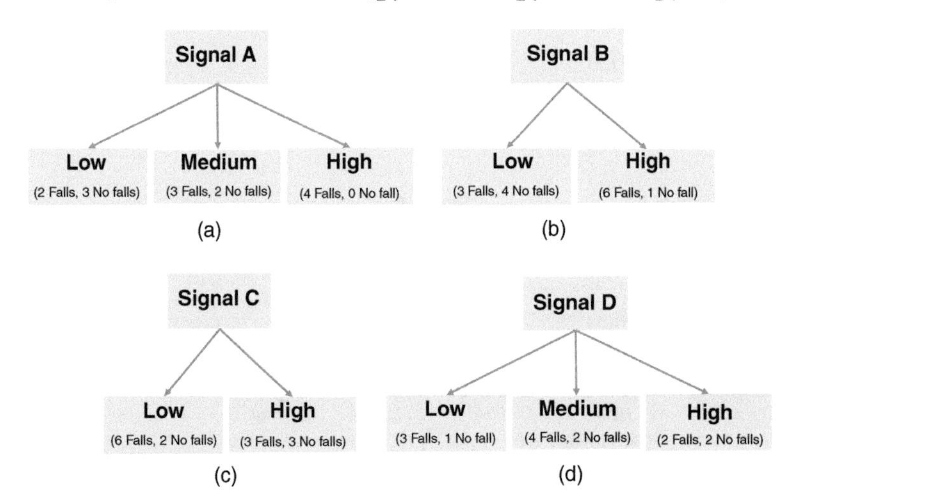

Figure 11.7 Four possible scenarios for splitting of the decision tree in our example.

We can perform the same calculations to find the splitting gain of Figure 11.7b. In this case:

$$H(s) = -\frac{9}{14}\log_2\frac{9}{14} - \frac{5}{14}\log_2\frac{5}{14} = 0.94 \tag{11.6}$$

$$\begin{cases} H\left(S_{\text{Signal B-Low}}\right) = -\frac{3}{7}\log_2\frac{3}{7} - \frac{4}{7}\log_2\frac{4}{7} = 0.985, Pr_{\text{Signal B-Low}} = \frac{7}{14} \\ H\left(S_{\text{Signal B-High}}\right) = -\frac{6}{7}\log_2\frac{6}{7} - \frac{1}{7}\log_2\frac{1}{7} = 0.591, Pr_{\text{Signal B-High}} = \frac{7}{14} \\ \text{Splitting gain} = 0.94 - \left(\frac{7}{14}\times 0.985 + \frac{7}{14}\times 0.591\right) = 0.1518 \end{cases} \tag{11.7}$$

The splitting gain for Figure 11.7c, d can be calculated in a similar way, which results in 0.048 and 0.029, respectively. Therefore, since the splitting gain of signal A is the largest (0.246 > 0.1518 > 0.048 > 0.029), we start with signal A. At each stage of the splitting process, we perform the same calculations by finding the largest splitting gain in order to determine the best feature to be used for the splitting process.

Let us consider the signal A-low subnode and decide if we should split this subnode based on signal B, signal C, or signal D as shown in Figure 11.8.

The splitting gain based on signal B is

$$H(s) = -\frac{2}{5}\log_2\frac{2}{5} - \frac{3}{5}\log_2\frac{3}{5} = 0.971 \tag{11.8}$$

$$\begin{cases} H\left(S_{\text{Signal A-Low Signal B-Low}}\right) = -\frac{0}{3}\log_2\frac{0}{3} - \frac{3}{3}\log_2\frac{3}{3} = 0, Pr_{\text{A-Low,B-Low}} = \frac{3}{5} \\ H\left(S_{\text{Signal A-Low Signal B-High}}\right) = -\frac{2}{2}\log_2\frac{2}{2} - \frac{0}{2}\log_2\frac{0}{2} = 0, Pr_{\text{A-Low,B-High}} = \frac{2}{5} \\ \text{Splitting gain} = 0.971 - \left(\frac{3}{5}\times 0 + \frac{2}{5}\times 0\right) = 0.971 \end{cases} \tag{11.9}$$

The splitting gain based on signal C is:

$$H(s) = -\frac{2}{5}\log_2\frac{2}{5} - \frac{3}{5}\log_2\frac{3}{5} = 0.971 \tag{11.10}$$

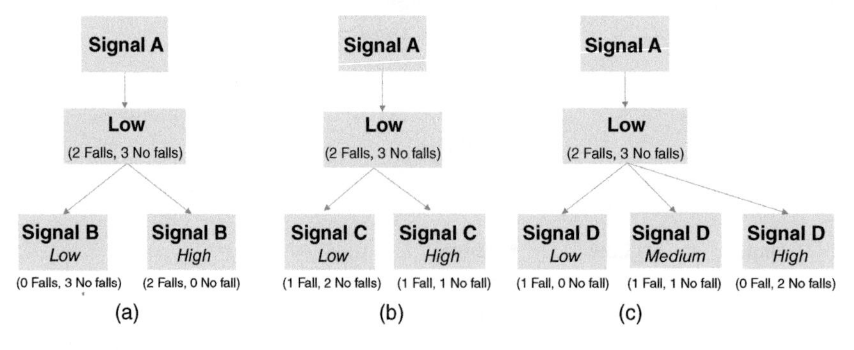

Figure 11.8 Splitting process: (a) based on signal B; (b) based on signal C; (c) based on signal D.

$$\begin{cases} H\left(S_{\text{Signal A-Low Signal C-Low}}\right) = -\frac{1}{3}\,log_2\frac{1}{3} - \frac{2}{3}\,log_2\frac{2}{3} = 0.918, Pr_{\text{A-Low,C-Low}} = \frac{3}{5} \\ H\left(S_{\text{Signal A-Low Signal C-High}}\right) = -\frac{1}{2}\,log_2\frac{1}{2} - \frac{1}{2}\,log_2\frac{1}{2} = 1, Pr_{\text{A-Low,C-High}} = \frac{2}{5} \\ \text{Splitting gain} = 0.971 - \left(\frac{3}{5} \times 0.918 + \frac{2}{5} \times 1\right) = 0.02 \end{cases} \tag{11.11}$$

The splitting gain based on signal D is:

$$H(s) = -\frac{2}{5}\,log_2\frac{2}{5} - \frac{3}{5}\,log_2\frac{3}{5} = 0.971 \tag{11.12}$$

$$\begin{cases} H\left(S_{\text{Signal A-Low Signal D-Low}}\right) = -\frac{1}{1}\,log_2\frac{1}{1} - \frac{0}{1}\,log_2\frac{0}{1} = 0, Pr_{\text{A-Low,D-Low}} = \frac{1}{5} \\ H\left(S_{\text{Signal A-Low Signal D-Medium}}\right) = -\frac{1}{2}\,log_2\frac{1}{2} - \frac{1}{2}\,log_2\frac{1}{2} = 1, Pr_{\text{A-Low,D-Medium}} = \frac{2}{5} \\ H\left(S_{\text{Signal A-Low Signal D-High}}\right) = -\frac{0}{2}\,log_2\frac{0}{2} - \frac{2}{2}\,log_2\frac{2}{2} = 0, Pr_{\text{A-Low,D-High}} = \frac{2}{5} \\ \text{Splitting gain} = 0.971 - \left(\frac{1}{5} \times 0 + \frac{2}{5} \times 1\right) = 0.571 \end{cases} \tag{11.13}$$

Since the splitting gain of signal B is the largest, we choose signal B as the feature for splitting at this stage.

This simple example shows how decision trees can be used to analyze the collected data from a wearable medical device. Interested readers can study [3] which classifies analytical techniques that are used in wearable IoT.

11.5.1.2 Random Forest

A decision tree can handle multi-dimensional data sets and can produce very good predictions for training data sets. This is especially true if we design the decision tree to its depth. However, this may result in a problem called overfitting, in which even though a model predicts the outputs for the training data sets perfectly (with a small amount of error), it produces a large error when the model is used for predicting new data.

Generally speaking, if a model produces large errors for training data sets, the problem is called underfitting. We are interested in models that have low bias (having low error when applied to the training data set) and low variance (having low error when applied to new data sets). Random forest solves the overfitting problem of a decision tree by using multiple decision trees in parallel.

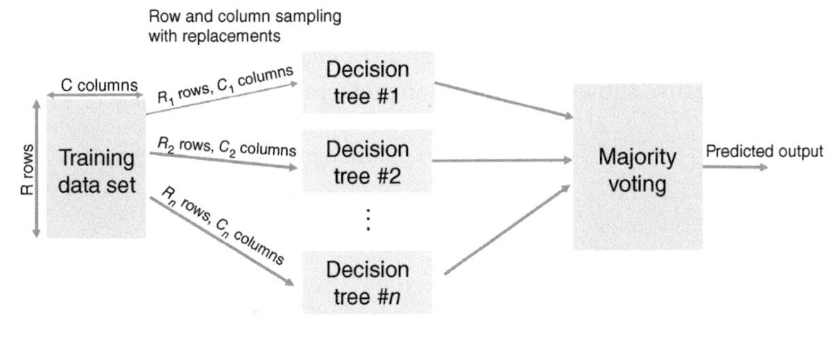

Figure 11.9 Random forest architecture.

Random forest architecture consists of several decision trees as shown in Figure 11.9. Each decision tree uses a randomly selected subset of the training data set and features. Assume a training data set with R rows and C columns. Random forest randomly selects some rows and some columns from the training data set and uses them for the training of the decision trees. For example, R_1 rows and C_1 columns have been randomly chosen and are used to train the first decision tree in Figure 11.9. The number of rows and columns for the second decision tree is R_2 and C_2, respectively. It is possible that some of the rows and columns that are applied to different decision trees will be the same. In other words, in a random forest algorithm, for each decision tree, a part of a training data set is randomly chosen with replacement, and therefore, each decision tree becomes trained for part of the training data set. The outputs of all decision trees then get aggregated. For discrete output values, a majority vote can be applied to determine the predicted output of a random forest classifier. The output is calculated by averaging the individual decision tree predictions for continuous outputs. Random forests are fast and additionally very robust to overfitting.

11.5.1.2.1 Performance Metrics in Classification There are several performance metrics that can be used to determine the efficiency of a classifier. These metrics are usually described using a confusion matrix. A confusion matrix is a table in which its columns represent the possible actual output values and its rows show the possible predicted output values. For a binary output, the confusion matrix is shown in Figure 11.10.

Where TP, TN, FP, and FN stand for True Positive, True Negative, False Positive, and False Negative, respectively. TP shows situations where the actual and predicted values are both positive. TN shows situations where both the actual and predicted values are negative. FP shows the situations where the model has predicted a positive output (1), while the actual value is negative (0). This is also called a Type I error. FN shows situations where the model predicted a negative output (0), while the actual output value is positive (1). This is called a Type II error. Let us represent the number of TP, TN, FP, and FN cases in a data set as T_P, T_N, F_P, and F_N, respectively. We will now explain some important performance metrics:

Accuracy: It represents the probability that a classifier has predicted the output value correctly. This is a good metric for balanced data sets. A balanced data set is a data set in which the number of positive outputs is, to a certain extent, similar to the number of negative outputs. Otherwise, the data set is called unbalanced. For example, if the percentage of negative outputs in the data set is 10%, the accuracy of the classifier that predicts all outputs as positive would be 90%. Therefore, using accuracy as a metric for an unbalanced data set is misleading.

$$\text{Accuracy} = \frac{T_P + T_N}{T_P + T_N + F_P + F_N} \tag{11.14}$$

Recall: It indicates the ratio of correctly predicted positive outputs to the total actual positive outputs. This metric is usually used for unbalanced data sets where the false negative outputs are more important than false positive outputs. For example, if an IoT device is used in a structural health monitoring system, a false negative shows that there is no problem with the structure, while

Predicted output value \ Actual output value	1	0
1	TP	FP
0	FN	TN

Figure 11.10 Confusion matrix for a binary output.

there is a problem with the structure. It is clear that the results of a structural failure can be very devastating. Therefore, recall is a suitable metric for performance evaluation.

$$\text{Recall} = \frac{T_P}{T_P + F_N} \tag{11.15}$$

Precision: It indicates the ratio of correctly predicted positive outputs to the total predicted positive outputs. This metric is usually used for unbalanced data sets where false positive outputs are more important than false negative outputs. For example, in an IoT-based smart parking system, the system can predict whether parking is full or if it has spaces available for parking. Predicting the system as full, while it has empty spaces, is a false positive. It is clear that a large number of false positive outputs can decrease the efficiency of a smart parking system. Therefore, precision is a suitable metric for performance evaluation.

$$\text{Precision} = \frac{T_P}{T_P + F_P} \tag{11.16}$$

F-Beta: It is a combination of both precession and recall. This metric is usually used for unbalanced data sets where the effect of both false positive outputs and false negative outputs is important.

$$F_{\text{Betta}} = (1 + \beta^2)\frac{\text{Precession} \times \text{Recall}}{\beta^2 \times \text{Precession} + \text{Recall}} \tag{11.17}$$

When beta is one ($\beta = 1$), F_1 is the harmonic mean of the precession and recall of a model. Therefore, F_1 is defined as:

$$F_1 = \frac{2 \times \text{Precession} \times \text{Recall}}{\text{Precession} + \text{Recall}} \tag{11.18}$$

F_1 is a good metric for unbalanced data sets when both false positive and false negative are equally important. Parameter β indicates to what degree the recall metric is more important than the precision metric. For example, if recall is twice as important as precision, we should set β to 2. Similarly, if precession is twice as important as recall, we should set β to 0.5.

11.5.1.3 K Nearest Neighbor (KNN)

KNN is one of the simplest supervised machine learning algorithms, and therefore, the training data set has labels that show two or more categories. K represents the number of neighbors, and this value should be selected according to the application. The first step for implementing the KNN algorithm is to find K nearest neighbors to a new data point. This is done by calculating the distance from a new data point to all other data points and choosing K neighbors with the shortest distance. Then, among these K neighboring data points, a KNN algorithm that is designed to act as a classifier counts the number of the data points in each category. Finally, it assigns the new data point to the category with the maximum number of data points. In other words, by using KNN as a classifier, a majority vote must be performed to determine the output. For regression, the mean value of the K nearest neighbors should be used to calculate the output.

Let us bring a simple example to explain the KNN algorithm. Assume a training data set has 1000 data points. Each data point belongs to one of the possible classes of A or B. Now assume that there is a new data point and you need to determine whether this new data point belongs to class A or class B. Assume that $K = 6$. Therefore, we should find six data points from the training data set that have the smallest distance to the new data point. Let us assume that among these six data points, two data points belong to class A and four data points belong to class B. Since most of these six data points belong to class B, the new data should be labeled as class B.

11.5.1.4 Support Vector Machine (SVM)

We explain the concept of SVM as a classifier in this section. However, it should be noted that SVM can also be used for classification as well as regression.

The main aim of an SVM classifier is to separate the data into different classes where each class represents a specific output value. Let us explain SVM with a simple example that has two-dimensional inputs (x_1, x_2) and a binary output as shown in Figure 11.11a. Using SVM, we try to separate the data into two classes using a hyperplane. Since the input data in this figure is two dimensional, the hyperplane is a line. If the number of inputs is more than two, we need one or more planes to separate the data sets into different classes. The line in Figure 11.11a divides the data into two classes as class A and B.

Apart from hyperplanes, SVM defines marginal hyperplanes. A marginal hyperplane is parallel to the hyperplane and passes through one or more data points of the class that are closest to the hyperplane. For example, two marginal lines are shown in Figure 11.11b (dashed lines). In this figure, one of the marginal hyperplanes passes through one data point of class A, while the other marginal line passes through two data points of class B. The distance between the hyperplane and these two marginal hyperplanes is shown as d_+, d_-. The distance between these two marginal hyperplanes is called the marginal distance.

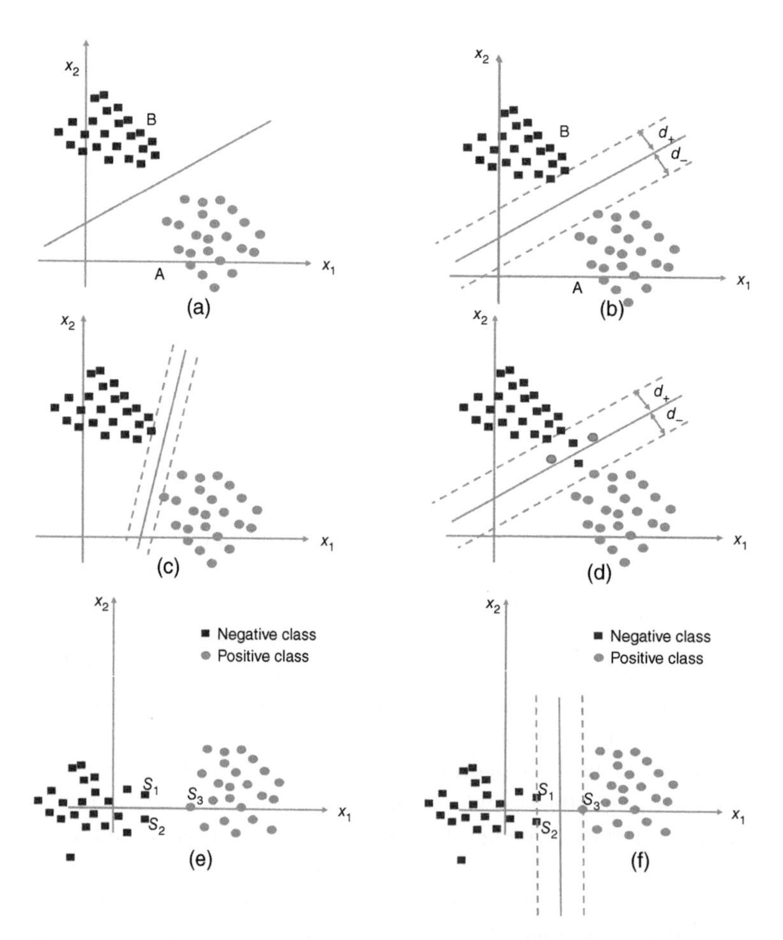

Figure 11.11 2D linear SVM: (a) hyperplane; (b) an example of a hyperplane and marginal planes; (c) another example of a hyperplane and marginal planes with a smaller margin; (d) an example of errors in SVM; (e) an SVM example; (f) hyperplanes of the example (e).

We could also design the hyperplane and marginal hyperplanes for this example as shown in Figure 11.11c. However, this would result in a smaller marginal distance as compared to the marginal distance of Figure 11.11b. The SVM should be designed in such a way that we can achieve the maximum possible marginal distance.

The data set in our example was completely separated using marginal hyperplanes. To make a generalized model for SVM, we should consider that we cannot always find hyperplanes that separate the data perfectly for all training data points as shown in Figure 11.11d. Even if that is possible, it may result in an overfitting problem, and the trained model may not be a good model for the new data. Therefore, when considering hyperplanes, we should calculate the error for those data points that do not classify the output correctly. The design of hyperplanes is an optimization problem that tries to maximize the marginal distance and minimize the error.

Let us show how to find the hyperplane for a simple example that is shown in Figure 11.11e. This example involves two features (x_1 and x_2). Let us assume that support vectors are s_1, s_2, and s_3, where $s_1 = (2, 1)$, $s_2 = (2, -1)$, and $s_3 = (4, 0)$. In this example, s_1 and s_2 belong to the negative class, and s_3 belongs to the positive class. The hyperplane for a training data set that has one, two, or three features is a point, line, and plane, respectively. Therefore, in this example that involves two features, the hyperplane is a line. A simple method to find the equation of the hyperplane is as follows:

1. Add a bias of 1 to the support vectors. Therefore, in this case, $s_1 = (2, 1, 1)$, $s_2 = (2, -1, 1)$, and $s_3 = (4, 0, 1)$.
2. Associate a support vector that belongs to the positive class to 1 and the one that belongs to the negative class to −1. If the operator C shows the association of a class to a support vector, then $\mathrm{C}(s_1) = -1$, $\mathrm{C}(s_2) = -1$, while $\mathrm{C}(s_3) = 1$.
3. Since we have three support vectors, determine three coefficients a_1, a_2, and a_3 that satisfy the following equations:

$$\begin{bmatrix} a_1 \; a_2 \; a_3 \end{bmatrix} \begin{bmatrix} s_1.s_1 & s_1.s_2 & s_1.s_3 \\ s_2.s_1 & s_2.s_2 & s_2.s_3 \\ s_3.s_1 & s_3.s_2 & s_3.s_3 \end{bmatrix} = \begin{bmatrix} \mathrm{C}(s_1) & \mathrm{C}(s_2) & \mathrm{C}(s_3) \end{bmatrix} \tag{11.19}$$

where $s_i.\ s_j$ is the inner product of vector s_i and vector s_j. Therefore, for our example:

$$\begin{bmatrix} a_1 \; a_2 \; a_3 \end{bmatrix} \begin{bmatrix} 6 & 4 & 9 \\ 4 & 6 & 9 \\ 9 & 9 & 17 \end{bmatrix} = [-1 - 1 + 1)] \tag{11.20}$$

By solving Eq. (11.20), the values of unknown coefficients can be determined. The values of coefficients a_1, a_2, and a_3 are −3.25, −3.25, and 3.5, respectively.

4. Determine vector $\omega = \sum_{i=1}^{n+1} a_i s_i$, where n is the number of features. This example represents two features (x_1, x_2), and therefore,

$$\omega = -3.25 \begin{bmatrix} 2 \\ 1 \\ 1 \end{bmatrix} - 3.25 \begin{bmatrix} 2 \\ -1 \\ 1 \end{bmatrix} + 3.5 \begin{bmatrix} 4 \\ 0 \\ 1 \end{bmatrix} = \begin{bmatrix} \omega_1 = 1 \\ \omega_2 = 0 \\ \omega_3 = -3 \end{bmatrix} \tag{11.21}$$

5. Vector ω represents the hyperplane. The hyperplane is a line in this example, and a line can be represented by its slope and intercept as $mx + b$. Use the last element of ω as the intercept of the hyperplane ($b = \omega_3$).The first two elements show the slope of the line $m = \frac{\omega_2}{\omega_1}$. This line is shown in Figure 11.11f.

Figure 11.12a shows a situation where the data are not linearly separable. As you can see for this two-dimensional input data, we cannot draw a line to separate the data points into two classes correctly. In this situation, we may be able to transfer the data to a higher dimension in such a way that the data can be separated into different classes. Figure 11.12b shows a situation where

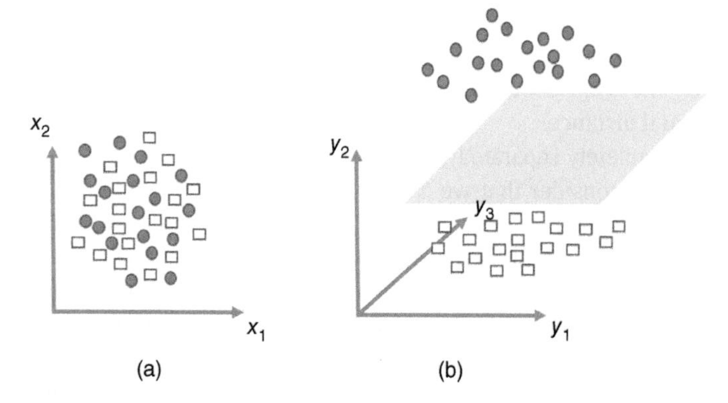

Figure 11.12 (a) Not linearly separable SVM in 2D. (b) Linearly separable SVM after transformation to a higher-order dimension.

by transferring the training data sets to a three-dimensional space, the data set has been separated into two classes using a linear hyperplane.

11.5.2 Regression

Classification is concerned with the prediction of a discrete value, whereas regression techniques are used to predict a continuous value. For example, regression could be used to predict the maximum distance that a smart car can travel based on the amount of gas in the tank. It may also be used to predict the maximum distance that a smart car can travel based on several parameters such as the amount of gas in the tank, the roads that the car travels on, and the time of travel. In other words, a regression model predicts the value of a dependent variable based on the value of one or several independent variables.

In general, there are two types of regression depending on the number of independent variables used in the model. If the number of independent variables is one, the regression is called simple regression. If the dependent variable is modeled based on more than one independent variable, the regression is called multiple regression. In this section, we discuss the simplest form of regression, called linear regression, in which the relationship between the dependent variable and independent variables is a linear equation. In linear regression, we use the training data set to estimate the coefficients of a linear equation that best predicts the value of an output variable.

If the training data set, which contains n data points (x_i, y_i), is used to model a system in the form of a linear equation ($y = mx + b$), the error in each prediction would be e_i:

$$e_i = y_i - mx_i - b \tag{11.22}$$

The Sum of Squared Error (SSE) for all data points can be calculated as:

$$SSE = \sum_{i=1}^{n} e_i^{\,2} = \sum_{i=1}^{n} (y_i - mx_i - b)^2 \tag{11.23}$$

To find the best line to represent the data, we should minimize the SSE function. This can be achieved by considering $\frac{dSSE}{dm} = 0$ and $\frac{dSSE}{db} = 0$.

Therefore,

$$\frac{dSSE}{dm} = 0 \rightarrow \sum_{i=1}^{n} x_i\,(y_i - mx_i - b) = 0 \rightarrow \sum_{i=1}^{n} x_i y_i - m\sum_{i=1}^{n} x_i^{\,2} - \sum_{i=1}^{n} bx_i = 0 \tag{11.24}$$

$$\frac{dSSE}{db} = 0 \rightarrow \sum_{i=1}^{n}(y_i - mx_i - b) = 0 \rightarrow \sum_{i=1}^{n} y_i - m\sum_{i=1}^{n} x_i - \sum_{i=1}^{n} b = 0 \quad (11.25)$$

From Eqs. (11.24) and (11.25), we can find the slope, m, and the intercept, b, of the best line, which are:

$$m = \frac{n\sum_{i=1}^{n}(x_i y_i) - \left(\sum_{i=1}^{n} x_i\right)\left(\sum_{i=1}^{n} y_i\right)}{n\sum_{i=1}^{n} x_i^2 - \left(\sum_{i=1}^{n} x_i\right)^2}, \quad b = \frac{\sum_{i=1}^{n} y_i - m\sum_{i=1}^{n} x_i}{n} \quad (11.26)$$

R-squared (R^2) is a measure that indicates how close the data points are to the regression line. In other words, it shows a value that represents the amount of the output variable variations from the linear regression model. The R-squared usually generates a value between zero and one. While not always true, usually the higher the R-squared, the better the model fits the data. We can use Excel data analysis tools to find the value of R-squared, as well as the equation of the regression model.

In multiple regression, there are multiple factors that predict an output or an event. A multiple linear regression model fits a surface that minimizes the error between predicted and actual output values and can be modeled as:

$$y_i = m_1 x_{1i} + m_2 x_{2i} + \ldots + m_p x_{pi} + b \quad (11.27)$$

where p is the number of independent variables. The training data points are expressed as $(x_{1i}, x_{2i}, \ldots, x_{pi}, y_i)$. The error in each prediction would be e_i:

$$e_i = y_i - m_1 x_{1i} - m_2 x_{2i} - \ldots - m_p x_{pi} - b \quad (11.28)$$

If n is the number of data points, the SSE of all data points can be calculated as:

$$SSE = \sum_{i=1}^{n} e_i^2 = \sum_{i=1}^{n}(y_i - m_1 x_{1i} - m_2 x_{2i} - \ldots - m_p x_{pi} - b)^2 \quad (11.29)$$

To find the coefficients of a multiple linear regression model, we should minimize SSE function. This can be achieved by calculating:

$$\frac{\partial SSE}{\partial m_1} = 0, \frac{\partial SSE}{\partial m_2} = 0, \ \ldots \ \frac{\partial SSE}{\partial m_p} = 0, \text{and} \ \frac{\partial SSE}{\partial b} = 0 \quad (11.30)$$

An independent variable should not be highly correlated with other independent variables used in a multiple regression model. If one or more independent variables are redundant, then it can result in a model that is very sensitive to minor variations.

Similar to a simple regression model, we can use Excel data analysis tools to find the value of R-squared, as well as the equations, of the multiple regression model.

Example 11.1 Simple regression

The torque, speed, current, and power of a DC motor are all measured by an IoT device. Table 11.2 depicts the training data set. We would like to create three simple linear regression models. The first model will be used to predict speed based on torque, the second model will be used to predict current based on torque, and the last model will be used to predict power based on torque. The unit for torque is ounce-inch that is equal to 0.007 062 Newton meter. The units for speed, current, and power are Revolution Per Minute (RPM), milliamperes, and Watts, respectively.

(Continued)

Example 11.1 (Continued)

Table 11.2 Training data set for Example 11.1.

Torque (oz-in)	Speed (RPM)	Current (mA)	Power (Watts)
0.025	11200	0.02	0.21
0.125	9400	0.12	0.87
0.225	7500	0.22	1.26
0.325	5700	0.31	1.37
0.425	3800	0.41	1.21
0.525	2020	0.5	0.78
0.625	175	0.6	0.081

a. Calculate these three simple regression models.
b. Using these models, predict torque, when speed is 5000 rpm. Also, predict the current when the torque is 0.275 Oz-in. What is the torque when power is 0.8 Watts?
c. Do you think these models are accurate models for predicting torque? Use Excel data analysis tool to find the R-squared value.

Solution

a. By using Eq. (11.26), we can find the slope and the intercept of these three models as:
   - To model speed and torque relationship: Speed = m (Torque) + b, where $m = -18405$ and $b = 11667$.
   - To model current and torque relationship: Current = m (Torque) + b, where $m = 0.9589$ and $b = -0.008$.
   - To model power and torque relationship: Power = m (Torque) + b, where $m = -0.22$ and $b = 0.897$.
b. 5000 = −18405(Torque) + 11667, Torque = 0.362
   - Current = 0.9589(0.275) − 0.008 = 0.256
   - 0.8 = −0.22(Torque) + 0.897, Torque = 0.441
c. We could use Excel data analysis tools to find the R-squared and regression equations. The results are shown in Figure 11.13. As can be seen, simple linear regression is not a good model for predicting power from torque (R-squared value = 0.0087).

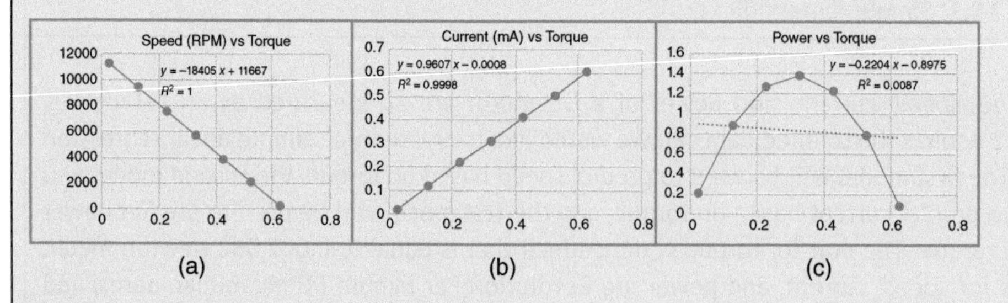

Figure 11.13 Linear regression models: (a) speed vs. torque; (b) current vs. torque; (c) power vs. torque.

Example 11.2 Multiple regression

To monitor the location of patients inside a building, people are asked to wear an IoT-based wearable device that uses Bluetooth Low Energy (BLE) technology. Each wearable device has a unique Identification number. The BLE chip inside each IoT device is configured to operate in advertising mode, and sends advertising messages every 10 seconds with constant transmit power. There are three IoT gateways inside the building that listen to the advertising messages received from the wearable IoT devices and measure Received Signal Strength Indicator (RSSI) of the advertising messages. The location of the IoT gateways is shown in Figure 11.14. Each IoT gateway sends the RSSI value to the cloud with the wearable identification number and a time stamp. Therefore, for each wearable device and at a specific time, the cloud backend program receives three RSSI values. Let us call these three values as d_1, d_2, and d_3. To predict the location of a patient inside the building, the backend program uses two multiple regression models with three values of d_1, d_2, and d_3 as its independent variables. These models are:

$$x = m_{1x}d_1 + m_{2x}d_2 + m_{3x}d_3 + b_x \tag{11.31a}$$

$$y = m_{1y}d_1 + m_{2y}d_2 + m_{3y}d_3 + b_y \tag{11.31b}$$

where x and y represent the coordinates of the location of a patient inside the building, and d_1, d_2, and d_3 are the RSSI values received by each IoT gateway for a specific point (x_i, y_i).

This prediction model is trained using the data points in Table 11.3, which are based on RSSI values. These RSSI values are measured in dBm. It should be noted that a smaller negative value indicates a stronger signal.

a. Find the equations of the multiple regression lines using the data analysis tool in Excel.
b. Do you think the three independent variables (d_1, d_2, and d_3) that are used in this example are correlated?
c. Do you think that the regression model used in this scenario is suitable for this application?

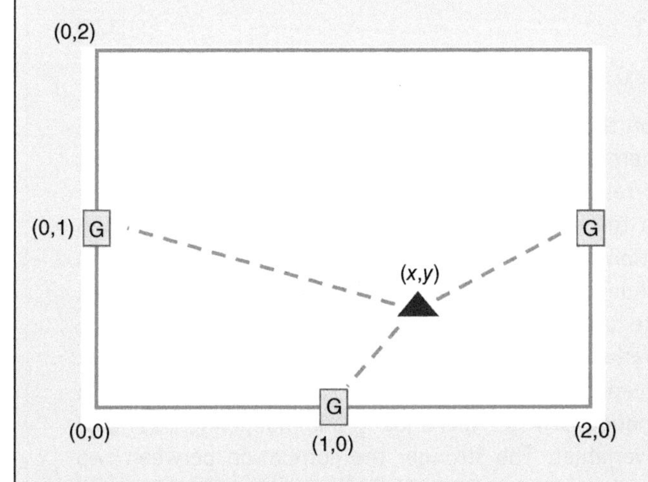

Figure 11.14 The location of IoT gateways inside the building.

(Continued)

Example 11.2 (Continued)

Table 11.3 Training data set in Example 11.2.

d_1	d_2	d_3	x_i	y_i
−63	−63	−70	0	0
−63	−70	−63	2	0
−66	−65	−65	1	2
−66	−66	−63	2	2
−66	−63	−66	0	2
−63	−63	−63	1	1

Solution

a. Using the data analysis tools in Excel, we can find the coefficients of the multiple regression model. The prediction of y is shown in Figure 11.15, which is actually the result of fitting the regression model between y and d_1, d_2, and d_3. The first table shows the regression statistics, which indicate how well the model fits the data. Adjusted R-squared is a good performance metric. The closer the adjusted R-squared is to one, the better the model fits the training data. The second table shows the results of the ANalysis Of VAriance (ANOVA) test. Simply put, the last value (significant F) shows how confident we are that the output y is related to at least one of the inputs (d_1, d_2, or d_3). Smaller values show a higher probability of such a relation. The third table shows the estimated values for coefficients of d_1, d_2, and d_3 (the first column), along with the confidence of the model on these values (p-values). A small p-value means a higher chance that the output of the model is actually dependent on that input parameter. The last table (residual output) just shows the residual error between the estimated and the actual y values for different data points. We can perform the same analysis for modeling x. Therefore, the location of coordinates (x, y) can be written as:

$$x = -0.17\ d_2 + 0.17 d_3 + 1 \tag{11.32a}$$

$$y = -0.493\, d_1 + 0.14\ d_2 + 0.14 d_3 - 12.45 \tag{11.32b}$$

b. An RSSI value highly depends on the structure of the environment between the BLE transmitter and the receiver. Therefore, as all three RSSI values are captured in the same environment, they are correlated with each other. This can be calculated by finding the correlation between the data sets. The objective of multiple regression analysis is to separate the relationship between each independent variable and the dependent variable. When an independent variable is changed by one unit and all other independent variables are constant, then the output changes based on the coefficient of that independent variable. When you change the value of an independent variable, the other independent variables should not change. However, if there exists a correlation between the independent variables, a change in one independent variable affects the other independent variables. The stronger the correlation between two independent variables, the more dependency between these variables. This is called multicollinearity, which causes the model to not be able to estimate the relationship between each independent variable and the dependent variable correctly.

c. The model assumes a linear relation between RSSI values, which is not accurate. It should be noted that the values of transmit and receive power are not related to each other linearly. Thus, it is better to consider a more complex model that can capture such relations.

Regression Statistics	
Multiple R	0.998619737
R Square	0.997241379
Adjusted R	0.993103448
Standard E	0.081649658
Observatio	6

ANOVA

	df	SS	MS	F	Significance F
Regression	3	4.82	1.606667	241	0.004135076
Residual	2	0.013333333	0.006667		
Total	5	4.833333333			

	Coefficients	Standard Error	t Stat	P-value	Lower 95%	Upper 95%	Lower 95.0%	Upper 95.0%
Intercept	−12.45333333	2.662813883	−4.67676	0.042806	−23.91049676	−0.99616991	−23.91049676	−0.99616991
d_1	−0.493333333	0.023094011	−21.362	0.002184	−0.592698842	−0.39396782	−0.592698842	−0.39396782
d_2	0.14	0.016045911	8.724964	0.012883	0.070960017	0.209039983	0.070960017	0.209039983
d_3	0.14	0.016045911	8.724964	0.012883	0.070960017	0.209039983	0.070960017	0.209039983

RESIDUAL OUTPUT

Observation	Predicted y	Residuals
1	0.006666667	−0.006666667
2	0.006666667	−0.006666667
3	1.906666667	0.093333333
4	2.046666667	−0.046666667
5	2.046666667	−0.046666667
6	0.986666667	0.013333333

Figure 11.15 Regression statistics from Microsoft Excel data analysis tool.

11.6 Unsupervised Machine Learning Techniques

11.6.1 Clustering

Clustering techniques are part of unsupervised machine learning algorithms. This means that the training data set does not have any labels. The objective of clustering techniques is to group the data into different clusters based on their similarities or patterns. In other words, data points that are similar should go to the same cluster. There are many techniques for clustering. In this section, we explain the K-Means method.

11.6.1.1 K-Means

K-Means clustering is an unsupervised technique that is very popular. K indicates the number of clusters that a clustering algorithm needs to generate. For example, if $K = 3$, the algorithm groups the data points into three clusters. For a given data set, there are methods to determine an optimum value for K. The objective of K-Means algorithm is to group the data in K clusters in such a way that the distance between the data points and their cluster's mean value is minimized.

Let us consider a simple scenario in which there are two inputs and each data point belongs to one of the two possible clusters ($K = 2$). For simplicity, we only consider two clusters in this scenario, but the idea can easily be extended to incorporate additional clusters.

In K-Means clustering with $K = 2$, the algorithm assigns two centroids randomly. For each data point, the distance between the data point and each centroid is calculated. Then, each data point will be assigned to the centroid that has a smaller distance to the data point. After completing the process for all data points, the data is separated into two clusters. At this point, these two clusters might not be the best clusters for separating the data. Therefore, the algorithm calculates the centroids of each of these two clusters and performs the process again with the new centroid values. K-Means clustering is an iterating process, and in each iteration, the centroids are replaced with the actual centroids of the grouped clusters. The process continues until the centroids in an iteration are almost the same as the centroids of the previous iteration.

Different distance metrics can be used in machine learning algorithms. Examples of these metrics are Euclidean distance, squared Euclidean distance, or Manhattan distance. For two points represented by x_i, y_i of a data set with P data inputs (P dimensional scenario), these distance metrics can be calculated as:

$$\text{Euclidean distance} = \sqrt{\sum_{i=1}^{p} (x_i - y_i)^2} \tag{11.33}$$

$$\text{Squared Euclidean distance} = \sum_{i=1}^{p} (x_i - y_i)^2 \tag{11.34}$$

$$\text{Manhattan distance} = \sum_{i=1}^{p} |x_i - y_i| \tag{11.35}$$

In K-Means clustering, the Euclidean distance is used as the distance metric. It is guaranteed that the K-Means method converges when Euclidean distance is used as the distance metric. The method also guarantees that each data point belongs to one of the clusters. In other words, there is no possibility for overlapping clusters or having a data point that is not assigned to one of the clusters. There is another type of clustering, called Fuzzy K-Means, in which a data point might belong to two or more clusters. A variation of K-Means is called K-Medians, in which the median is used instead of the mean value to determine the centroid of each cluster. K-Median clustering uses Manhattan distance as its distance metric. Generally speaking, there are many clustering techniques. In this section we briefly explained K-Means, which is one of the most popular clustering techniques.

Example 11.3 Simple K-Means example

Consider the three training data points $a(0, 1)$, $b(1,0)$, and $c(0, 0)$ as shown in Figure 11.6. The initial centroids are chosen randomly and are $(-1, 0)$ and $(2, 0)$. Using the K-Means algorithm, group these three points into two clusters.

Solution: we first calculate the distance from each data point to centroid 1 and centroid 2. We classify a data point to the first cluster (cluster 1) if the distance between that data point and centroid 1 produces a smaller number as compared to the distance between the data point and centroid 2. We then find new centroids and continue the process until the results converge. Figure 11.16 shows the clustering process for this example.

			Distance from		Round 1		Distance from		Round 2
	x	y	Centroid 1	Centroid 2	Class		Centroid 1	Centroid 2	Class
a	0	1	$\sqrt{2}$	$\sqrt{5}$	1		0.5	$\sqrt{2}$	1
b	1	0	2	1	2		$\sqrt{1.25}$	0	2
c	0	0	1	2	1		0.5	1	1
Centroid 1	−1	0	New centroid 1		$0.5(x_a + x_c) = 0$	$0.5(y_a + y_c) = 0.5$			
Centroid 2	2	0	New centroid 2		$x_b = 1$	$y_a = 1$			

Figure 11.16 K-Means process in Example 11.3.

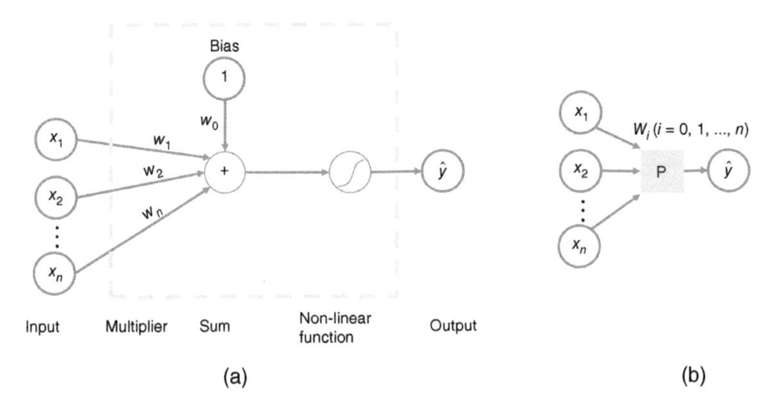

Figure 11.17 (a) Perceptron architecture; (b) using a block to represent the internal architecture of a perceptron.

11.7 Deep Learning Techniques

Deep Learning is currently the most successful AI technique. In order to give readers a high-level understanding of deep learning, we strive to discuss deep learning algorithms in this section at a very high level, without getting into details, and in simple language. The history of deep learning and a general description of its operation can be found in [4]. The fundamental building block of a deep learning algorithm is a single neuron called a perceptron as shown in Figure 11.17a. A perceptron contains input elements, weight, sum, nonlinearity element, bias, and output elements. The propagation of data in a perceptron is as follows: the input data is multiplied by corresponding weights and added together. The result is passed through a nonlinear activation function. The real world data is usually not linear and the nonlinear activation function introduces nonlinearity for the perceptron. The output from the nonlinear function is the output of the perceptron. The bias function introduces a bias to the summed weighted input data. To simplify the operation, the weights, bias, sum, and the nonlinear function are shown as a block represented by P in Figure 11.17b. The corresponding weights of the perceptron are shown on top of the block.

The mathematical representation of a perceptron is:

$$\hat{y} = g\left[w_0 + \sum_{i=1}^{n} x_i\, w_i\right] \tag{11.36}$$

We can consider different types of nonlinear functions for a perceptron. Examples of these nonlinear functions are Rectified Linear Unit (ReLU), sigmoid function, and hyperbolic tangent function as shown in Figures 11.18a–c, respectively. The mathematical formula for these nonlinear

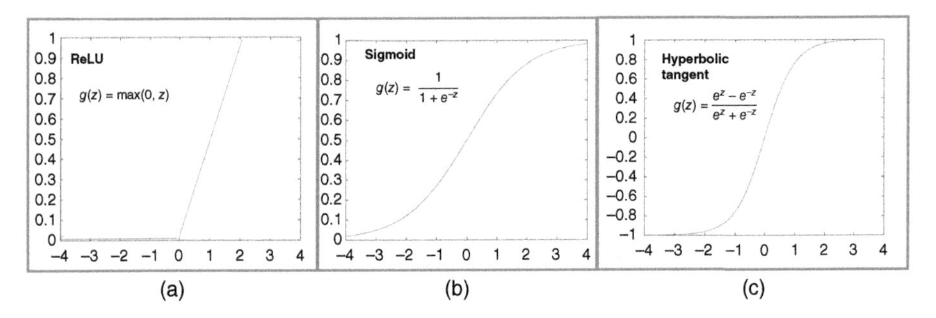

Figure 11.18 Examples of nonlinear activation functions: (a) rectified linear unit, (b) sigmoid, (c) hyperbolic tangent.

activation functions g(z) is shown in each of these figures, where z is the input to the nonlinear activation function.

Example 11.4 Input-output relationship in a perceptron with two inputs

During the training process, a perceptron determines the initial weights of its multipliers, and as more data arrives, it updates the multiplier weights. Assume a perceptron has two inputs and uses a sigmoid nonlinear activation function. Let us also consider that the perceptron is trained, has a bias with a weight of 1, and the weight values corresponding to its inputs (x_1, x_2) are −2 and 3, respectively. In this case the output of the perceptron can be expressed as:

$$\hat{y} = g[1 - 2x_1 + 3x_2] \tag{11.37}$$

By looking at Eq. (11.37), one can see that the input to the nonlinearity activation function is a linear function. The line $1 - 2x_1 + 3x_2$, as shown in Figure 11.19, represents the decision boundary of this trained single neuron. Assume that there is a new data point such as (2,−1). The output of the system based on this new data point would be:

$$\hat{y} = g(z) = g[1 - 4 - 3] = g(-6) = \frac{1}{1 + e^6} = 0.0025 \tag{11.38}$$

The line $1 - 2x_1 + 3x_2$ divides the space into two parts. Remember that the sigmoid function can be divided into two regions: the region $z < 0$, in which the output of the function is smaller than 0.5, and the region $z > 0$, in which the output is greater than 0.5.

It should be noted that in Example 11.4, there are just two inputs, and therefore, drawing the space in a two-dimensional format is possible. When there are thousands of inputs and weight values, the space is not possible to be shown as a picture, but readers can imagine a line in n-dimensional space.

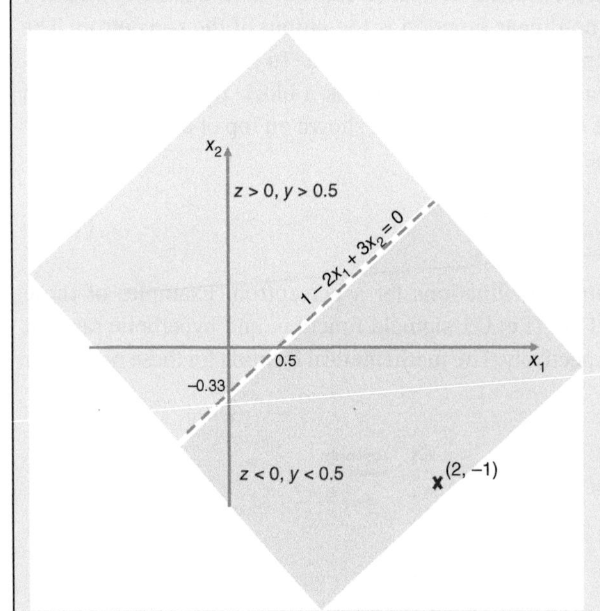

Figure 11.19 Decision boundaries of the trained single neuron in Example 11.4.

Example 11.5 Simple explanation of the loss function

An IoT device measures temperature and humidity and transmits these values to the cloud. A researcher would like to build a simple neural network to determine whether there is any possibility to identify a fire based on the collected temperature and humidity values. The researcher has collected data that can be used to train its network. Each training data indicates a temperature value, a humidity value, and a binary value indicating whether those temperature and humidity values can indicate a fire. Let us see how the researcher can use the training data to train the neural network. Due to the simplicity of this example, the researcher can use a perceptron as a suitable neural network that has two inputs (temperature and humidity) and one output (which predicts whether the possibility of fire exists or not). To design this perceptron, the researcher needs to find the two weights of this perceptron. Assume the researcher estimates two values as the initial weight values for the perceptron and considers a specific nonlinear function. The researcher then applies the training data points and finds whether the predicted values by the perceptron are the same as the actual label values or not. The researcher defines a loss function that shows the cost of the difference between the predicted value and the actual value. The loss function measures the error that occurred from each incorrect prediction and is expressed in terms of the weights of the perceptron. The researcher calculates the average loss over all training data points and tries to minimize the average loss function that results in finding the best weight values. In other words, finding the weights of a neural network is an optimization problem and can be achieved by minimizing the average loss function. The loss function is defined as $\frac{1}{2}(y-\hat{y})^2$, where y is the actual output and $\hat{y}$ is predicted by the perceptron's output layer. An optimal algorithm to minimize the loss function is gradient descent. This algorithm determines the local or global minima of a function and gives the direction that the model should take in order to reduce the error.

We can use several simple perceptrons to build a perceptron that has several outputs. A perceptron that has two outputs is shown in Figure 11.20a. The simplified version using block P is shown in Figure 11.20b. Since the inputs are densely connected to the outputs, the connections between the inputs and block P are called a dense layer.

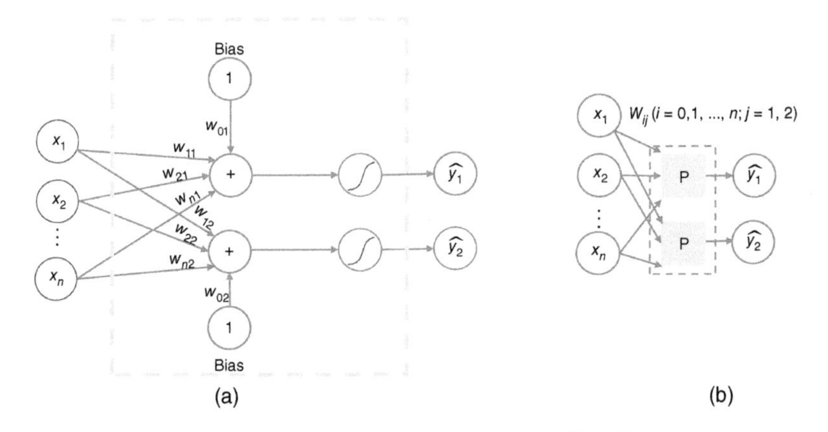

Figure 11.20 (a) A perceptron that has two outputs. (b) Simplified block representation of part a.

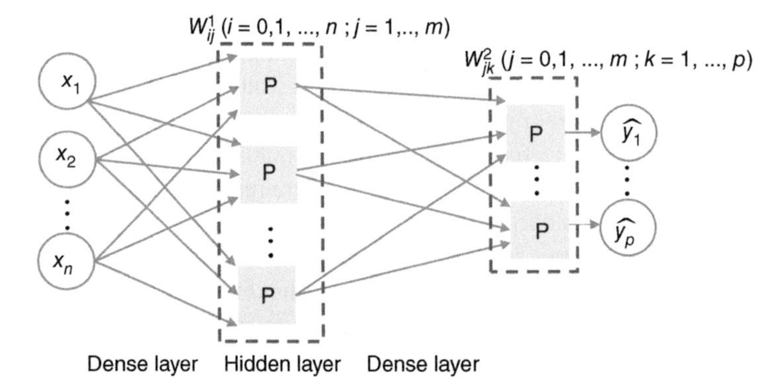

Figure 11.21 A single layer neural network.

Now that we have an idea on how to build a perceptron, let us extend our idea and build a neural network. A single layer neural network is shown in Figure 11.21. This single layer neural network consists of a hidden layer and two dense layers. The hidden layer is a layer that is between two dense layers and unlike input or output, its states are unobservable. There are two sets of weights defined in a single layer neural network that are W^1 and W^2 as shown in Figure 11.21. The hidden layer has m output signals. The Lth output signal of the hidden layer can be expressed as $g\left(W^1_{0L} + \sum_{i=1}^{n} x_i W^1_{iL}\right)$. The output signals of this single layer neural network can be expressed as:

$$\widehat{y_k} = g\left[W^2_{0k} + \sum_{j=1}^{m} g\left(W^1_{0j} + \sum_{i=1}^{n} x_i W^1_{ij}\right) W^2_{jk}\right] \quad k = 1, \ldots, p \tag{11.39}$$

If we want to extend this model and make a deep neural network, we need to add more and more hidden layers. Figure 11.22a shows the simplified version of Figure 11.21, and 11.22b shows a deep layer neural network with several hidden layers. As the number of hidden layers increases, the neural network becomes deeper.

Training a neural network with several hidden layers and many inputs and outputs is a complex and computationally expensive task. As explained earlier, training a complex neural network and finding the weights are achieved by minimizing the loss function calculated based on all training data points. This process can lead to a local minimum instead of a global minimum of the loss function. The optimum weights can be calculated and the best trained system can be designed when the process can find the global minimum of the loss function.

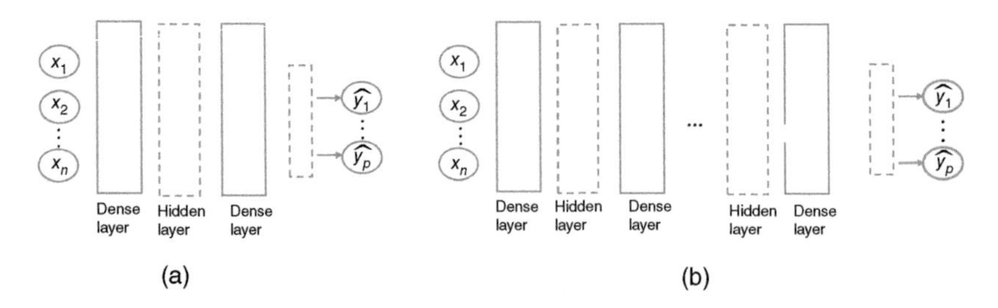

Figure 11.22 (a) Simplified version of Figure 11.21. (b) Deep layer neural network with several hidden layers.

11.7.1 Recurrent Neural Networks (RNN)

Many IoT applications have time-based data points. For example, a smart microphone may measure the feeling associated with a voice signal. In this IoT application, the input is a time-based voice signal, while the output is not sequential. Now assume the smart microphone can translate the voice from one language to another language, and can transmit the translated voice signal to a speaker. In this application, both input and output data are time-based.

In this section, we discuss the idea behind the design of neural networks that are suited when we are working with time-based inputs and outputs. In other words, we would like to extend the neural network that we discussed in the previous section and consider temporal information as part of the design of neural networks. For the neural networks that we discussed earlier including perceptron, single layer neural network, or multi-layer deep neural network, we present the input signal as discrete values. These discrete values can be considered as the sampled values of a time-based signal. One may think that we can apply the same neural networks to time-based signals by applying the sample values of the time-based input signal one by one. In other words, we treat the time-based input signal as a sequence of isolated data points. Let us consider a simple scenario in which we get n samples from $x(t)$ and apply these samples as isolated data points to a neural network shown as a box in Figure 11.23a.

But since $x_1, x_2, \ldots, x_{n-1}$ are prior samples of $x(t)$, it would make sense if we could say that predicted output $\hat{y}_n$ is dependent on $x_1, x_2, \ldots, x_{n-1.}$ There exists an inherent relationship in time-based data that we are not considering if we treat all data points of a time-based input as isolated data points.

To address this issue, we can link the computations at each point in time to each other as shown in Figure 11.23b. Since the predicted output at a particular time is dependent not only on inputs at that particular time but also on prior inputs, we can add the concept of memory to the neural network. Therefore, the main idea behind RNN is the addition of memory, which represents the prior states to the neural network architecture. To build an RNN system, the memory element (h_t) is updated at each time step. In RNNs, h_t can be expressed as:

$$h_t = f_w(x_t, h_{t-1}) \tag{11.40}$$

where f_w represents a nonlinear function. At each individual time step, we predict the output, and we can find the error in the prediction at each step in time. If we calculate the sum of error of all

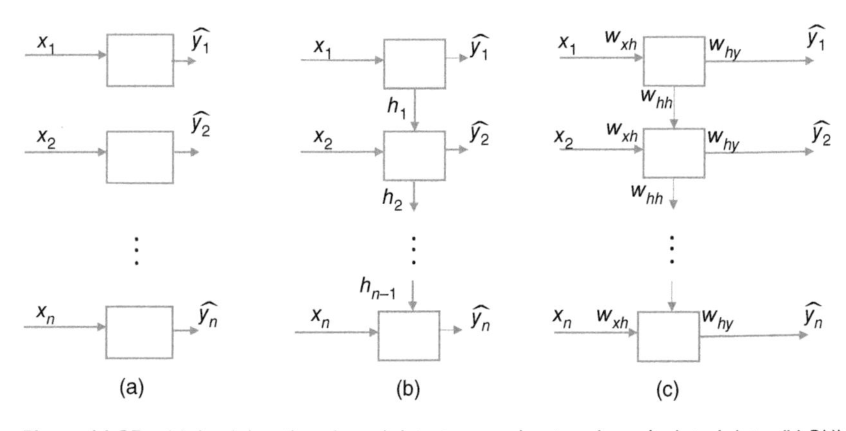

Figure 11.23 (a) Applying time-based data to neural networks as isolated data; (b) RNN network architecture by considering memory; (c) representing the weight coefficients of an RNN network.

predicted outputs at each time step and minimize the total error, we can effectively train the RNN. The implementation of RNN is shown in Figure 11.23c.

A more complex RNN is called Long Short-Term Memory (LSTM) that can track longer dependencies in time-based data. In other words, LSTM better understands what information needs to be forgotten and what information needs to be used in memory. Therefore, it can track the information within several time steps. LSTM is more effective at handling time-based data.

11.7.2 Convolutional Neural Network (CNN)

CNN is another class of deep learning algorithms that use the convolution operation to detect features from the data inputs. Convolution is a mathematical operation and convolving two functions produces another function that can be expressed in continuous and discrete formats as:

$$\begin{cases} x(t) * g(t) = \int_0^t x(t)g(t-\tau)d\tau \\ x[n] * g[n] = \sum_{i=1}^{n} x[i]g[n-i] \end{cases} \tag{11.41}$$

To apply convolution to input data, matrix multiplication is performed element by element on two signals. The first signal is a convolution filter, and the second signal is a small part of the input data. The sum of the results is then calculated, which generates a feature map. The process continues by sliding the convolution filter over the input data and performing the same operations over and over again. The process ends when we reach the end of the input data.

The convolution of a convolution filter $w(x, y)$ of size $m \times n$ and an image $f(x, y)$, denoted by $w(x, y) * f(x, y)$, can be expressed as:

$$\begin{aligned} w(x,y) * f(x,y) &= \sum_{s=\frac{-(m-1)}{2}}^{\frac{(m-1)}{2}} \sum_{t=\frac{-(n-1)}{2}}^{\frac{(n-1)}{2}} f(s,t)w(s-x,y-t) \\ &= \sum_{s=\frac{-(m-1)}{2}}^{\frac{(m-1)}{2}} \sum_{t=\frac{-(n-1)}{2}}^{\frac{(n-1)}{2}} f(s-x,t-y)w(s,y) \end{aligned} \tag{11.42}$$

To perform convolution operation using Eq. (11.42), every element of the convolution filter should visit every pixel in the image for all the displacement values of x and y. To be able to calculate the convolution values for the boundary data points in a data set such as pixels of an image, the data (image) needs to be padded with a minimum of $m-1$ rows at the top and bottom, and $n-1$ columns on the left and right. For example, for a 3×3 convolution filter, and a small image of 3×3, the convolution operation is shown in Figure 11.24. An example of an image and a convolution filter is shown in Figure 11.24a. In this figure, the image is shown as a small matrix just to show the operation. However, an actual image has more pixels. Figure 11.24b shows the padded image and the convolutional filter that has been rotated by 180°. According to Eq. (11.42), the convolutional filter needs to be flipped and shifted in the convolutional operation. That is why the term $w(x-s, y-t)$ is part of the definition of convolution. The image is zero-padded, which means two rows at the top and bottom, and two columns on the left and right, are padded with zeros. Figure 11.24c shows the calculation of the two elements of convolution. It should be clear that the operation should continue for all x and y displacements in order to determine all of the convolution matrix. As you can see, at any point (x, y) in the image, the result of convolving an image and a 3×3 convolution filter can be expressed as:

$$\begin{aligned} w(x,y) * f(x,y) = {} & w(-1,-1)f(x-1,y-1) + w(-1,0)f(x-1,y) + \dots \\ & + w(0,0)f(x,y) + \dots w(1,1)f(x+1,y+1) \end{aligned} \tag{11.43}$$

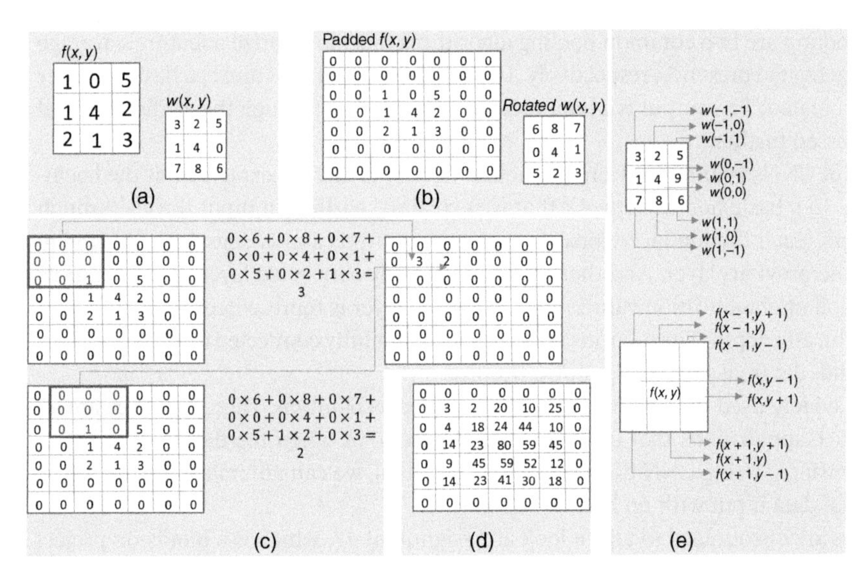

Figure 11.24 (a) An image and a convolution filter; (b) padded image and the rotated convolution filter; (c) the calculation of two elements of the convolution process; (d) the notations of elements for 3 × 3 image and a 3 × 3 convolution filter; (e) final convolution result.

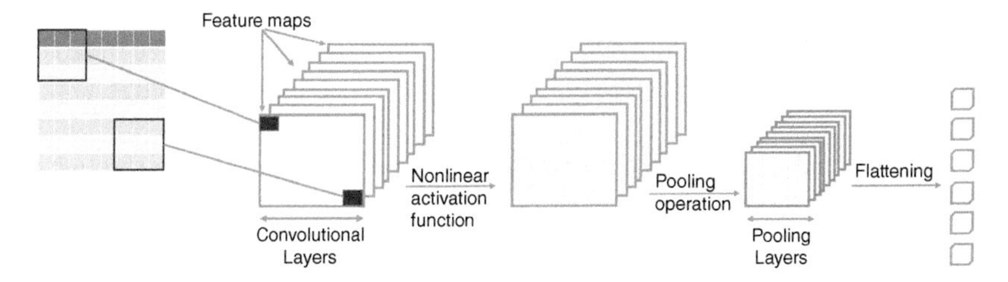

Figure 11.25 CNN architecture.

The result of the convolution is shown in Figure 11.24d. Figure 11.24e shows the notations of elements for a 3 × 3 image and a 3 × 3 convolution filter, where $w(0, 0)$ is aligned with $f(x, y)$.

CNN uses three core operations to build a machine learning classification or prediction algorithm as shown in Figure 11.25. The first operation is to perform various convolution operations to data inputs in order to detect the features of the input data. Convolution is a fundamental building block of CNNs, as the name implies. CNN creates a series of feature detectors and uses them to create feature maps known as convolutional layers. The network identifies which features are essential through training.

The convolutional layers are then subjected to a nonlinear function, as these features are frequently nonlinear. The goal of introducing the activation function is to give neural networks nonlinear expression capabilities, allowing them to better suit the results and enhance accuracy. In different neural networks, different activation functions behave differently. Several nonlinear activation functions, such as ReLU and sigmoid, have been discussed earlier and can be utilized in CNN.

The final operation is to perform a pooling operation, which reduces the size of the feature maps. The pooling layers are the result of the pooling operation, and they are used to down-sample feature maps by summarizing the existence of features in different regions of the feature map. Average pooling and max pooling are two common pooling algorithms that summarize a feature's average presence and most activated presence, respectively. The pooled feature maps must be flattened after the pooling layer is created. The output is a long vector of input to run through the artificial neural network to be processed further.

The architecture of CNNs is different from the neural networks that we explained at the beginning of this section. In a basic neural network that we explained earlier, an input travels through several hidden layers. Each hidden layer consists of a set of neurons and is densely connected to all the neurons in the previous layer. And there is a final densely connected layer that its outputs present the prediction or classification values. In CNNs, each layer is represented in three dimensions of width, height, and depth. The neurons in one layer are not fully connected to all the neurons in the next layer. And, the final stage scales down the size.

CNNs have been widely used in computer vision and image recognition. Therefore, they are an important part of IoT applications that utilize imaging sensors, or when the data can be represented in two dimensions. Using CNN deep learning algorithms, we can automatically detect the important features of data input with no human supervision.

Interested readers are encouraged to take a look at Assignment #7, which is a hands-on project on this topic.

11.8 Summary

Smart decision making can be achieved by performing analytics on the data that is collected by IoT devices. Generally speaking, there are three different types of analytics; descriptive, predictive, and perspective.

Since data may be lost, corrupted, or arrive out of order to the cloud, the data must go through several data processing operations that prepare the data to be fed to analytical tools. The transportation of data from one or more sources to a destination, where it can be stored and further analyzed, is called data ingestion. Three types of data ingestion include batch processing, stream processing, and micro-batch processing. Batch processing is a scheduled processing and relies on stored data, while stream processing is based on real-time processing and usually relies on continuous data.

The field of AI is the science that focuses on building algorithms that process information and make future decisions. An AI algorithm might be very simple, consisting of a series of if-then statements. On the other hand, an AI algorithm may use a very complex statistical model to process the data.

Machine learning is a subset of AI that focuses on teaching an algorithm to learn from experiences without being explicitly programmed. Machine learning algorithms can be classified into five categories. These categories are supervised learning, unsupervised learning, semi-supervised learning, reinforcement learning, and deep learning.

In supervised learning, a machine learning algorithm is trained using some training data sets, which include both the inputs and their corresponding outputs (also called labels). We discussed many classification techniques such as decision trees, random forests, K nearest neighbor, and SVM as well as regression to explain supervised learning methods.

Unsupervised learning refers to a group of machine learning algorithms that are trained with data sets that do not have any labels. In unsupervised learning, a machine learning algorithm attempts to learn from input in order to train itself. We discussed K-Means as an example of clustering techniques which belong to the unsupervised machine learning methods.

Semi-supervised learning is a hybrid scheme, in which the training data contains data sets that have labels, similar to supervised learning, and data sets that, are similar to unsupervised learning, do not have any labels.

Reinforcement learning refers to a group of machine learning algorithms in which a physical system (called an agent) interacts with its surroundings (called an environment) to train itself. The agent is not provided with any training data, and the agent trains itself based on a trial and reward scheme.

Deep learning focuses on using neural networks to automatically extract useful patterns or features from data and learn from these features to train the model. Therefore, deep learning means teaching algorithms how to learn a task directly from data using the concept of neural networks. The fundamental building block of a deep learning algorithm is a single neuron, called a perceptron. We showed that the extension of this building block can result in a single layer neural network with a hidden layer and two dense layers. The model of single layer neural networks can be extended to deep neural networks with many hidden layers. RNN and CNN are two important deep learning categories. RNN is suited to time-based signals and considers temporal information as part of the neural network design. CNN is another class of deep learning algorithms that uses the convolution operation in order to detect the features of data inputs.

References

1 Ortiz, G., Castillo, I., Garcia-de-Prado, A. et al. (2022). Evaluating a flow-based programming approach as an alternative for developing CEP applications in IoT. *IEEE Internet of Things Journal*, 9 (13), pp. 11489–11499, doi: https://doi.org/10.1109/JIOT.2021.3130498.

2 Larose, D.T. (2014). *Discovering Knowledge in Data: An Introduction to Data Mining*, 174–179. Hoboken, New Jersey: Wiley. ISBN 9780470908747.

3 Dian, F. J., Vahidnia, R., Rahmati, A. (2020). Wearables and the Internet of Things (IoT), applications, opportunities, and challenges: a survey. *IEEE Access.* 8: 69200–69211, https://doi.org/10.1109/ACCESS.2020.2986329.

4 Strawn, G. O. (2022). Masterminds of Deep Learning. *IT Professional.* 24 (3), pp. 13–15, doi: https://doi.org/10.1109/MITP.2022.3172838.

Exercises

11.1 In what types of IoT applications the role of AI algorithms is more prominent?

11.2 Cloud computing is a suitable choice for performing analytics in many IoT applications. The main reasons are the processing power as well as the scalability of cloud infrastructure. What problems performing cloud computing may have that prevent it from being suitable for some IoT applications?

11.3 How can IoT help analytics to advance?

11.4 What type of analytics is used in the following scenarios?

a. A driver wearing smart glasses that determines the drowsiness of the driver and sends the information to the cloud. The analytics is used to determine whether the drowsiness of the driver was the reason for an accident.

b. A smart printer can measure ink levels and transmit this information to the cloud over the Internet. The analytics is used to predict the date that the printer requires a new ink cartridge, and schedule a time for the ink cartridge to be shipped to the customer.
c. A tennis player wears a wearable device that measures the angle of his wrist at the time that he hits the ball, as well as the speed of the ball, and sends the data to the cloud. The analytics is used to train the tennis player by sending the player the angle that the player should have made the best hit.
d. A pump is sending several parameters indicating its performance to the cloud. The analytics is used to find the time to failure for this pump.

11.5 What is an IoT data pipeline? Why should an IoT data pipeline be scalable?

11.6 Explain different types of data ingestion. Which type has more applications in IoT?

11.7 Why unsupervised machine learning algorithms are more cost-effective as compared to the supervised algorithms?

11.8 Reinforcement learning refers to a group of machine learning algorithms in which a physical system interacts with its surroundings to train itself.
a. Give an example of an IoT application that uses reinforcement machine learning.
b. Draw a block diagram that shows how reinforcement learning operates.

11.9 The process of identifying unexpected events (outliers) in a data set is called anomaly detection. For example, anomaly detection can be used on the received data from an IoT-based medical device to detect critical and life-threatening situations.
a. Explain how the concept of anomaly detection can be used to reduce energy consumption in a smart building.
b. What do you think the output of an anomaly detection algorithm would be?

11.10 An IoT device is used to monitor the performance of a system. The IoT device sends the information from its sensors to the cloud. You need to write a machine learning classifier to detect whether IoT data indicates any fault in the system or not.
a. Do you think the data in these types of scenarios are balanced or unbalanced?
b. Assume a data set is unbalanced, what technique you think can be used to design a machine learning algorithm which uses an unbalanced data set?
c. Do you think *accuracy* is a good performance metric when dealing with unbalanced data sets?
d. Give examples of performance metrics that are good for unbalanced data sets.

11.11 Consider four 3D points $a(0,1,0)$, $b(1,0,0)$, $c(0,0,1)$, and $d(1,1,1)$. Run K-Means with two clusters. Assume the initial cluster centers are (0,1,1) and (2,1,0). Determine which cluster each data point belongs to?

11.12 Is it possible that a centroid in K-Means algorithm does not become associated with any data points?

11.13 What are overfitting and underfitting in the context of a machine learning algorithm?

11.14 What is the difference between a feedforward neural network and RNN?

11.15 What are the different layers of CNN? Briefly explain the functionality of each layer?

11.16 K Nearest Neighbor (KNN) is a simple method for supervised machine learning. It can be used for classification and regression. In this method, to find the label of a new data, the algorithm looks at the K neighbors with the shortest distance to a new data. A KNN classifier can use a majority vote to determine a label for the data. For regression, the mean value of the K nearest neighbors should be calculated. Assume the training data set has four data points. These data points, in the form of (x_1, x_2, label) are A(8,13,1), B(5,9,1), C(14,7,2), and D(1,7,2). Using the Manhattan distance metric, what is the label for a new point at (8, 7, label) if $K = 3$?

Advanced Exercises

11.17 Different distance metrics are used in machine learning algorithms.

a. Minkowski distance metric is a generalized distance metric. This distance metric is expressed in terms of n. You can manipulate this metric in different ways by changing the value of n. The Minkowski distance for two data points in a p-dimensional scenario is:

$$\left(\sum_{i=1}^{p} |x_i - y_i|^n\right)^{\frac{1}{n}}$$

b. For what value of n this metric represents Euclidean distance? What about Manhattan distance?

c. What is the Manhattan distance between two points (1, 2, 3, 4) and (5, 6, 7, 8) in a four-dimensional space?

11.18 A data set contains 500 training data points expressed as (x_i, y_i), which its label is expressed as $y_i = (-1)^i$. A classier simply predicts $y = 1$ all the time. What is the accuracy of this classifier?

11.19 Assume a decision tree with the depth of D, where all its internal nodes have M children. The tree is also complete, meaning all leaf nodes are at depth D. If we require each leaf node to contain at least P training examples, what is the minimum size of the training set?

11.20 Pruning is a process that reduces the size of decision trees by removing parts of the tree that are non-critical. What do you think is the main reason for pruning a decision tree?

11.21 The sigmoid nonlinear activation function is defined as:

$$g(x) = \frac{e^x}{1 + e^x}$$

Another widely used nonlinear activation function is the hyperbolic tangent function, which is defined as:

$$tanh(x) = \frac{e^x - e^{-x}}{e^x + e^{-x}}$$

Explain the mathematical relationship between these two functions.

11.22 A nonlinear activation function that is used in many deep learning algorithms (often in the output layer) is called the softmax function. If there are n inputs to softmax function, the value of the i-th output can be expressed as:

$$s(x_i) = \frac{e^{x_i}}{\sum_{j=1}^{n} e^{x_j}}$$

The value of each output is between zero and one, and the sum of all outputs is equal to one. If the input to a softmax function is (4, 3, 1, 2), what is the output?

11.23 In a simple neural network, the input layer has a single input (x). The network has 6 hidden layers and each hidden layer has 10 P blocks. The output layer has a single P block, which outputs the signal expressed as y. Between any two layers, the network is fully connected. All P blocks have a bias of 1 and use an identity function (there is no nonlinear activation function). What kind of functions can this network compute?

11.24 Consider a linear perceptron $a = w_0 + w_1x_1 + \ldots + w_dx_d$. Given a single training point ($x_i = 1, i = 1\ldots d, y = 100$).

a. Calculate the loss function.
b. Compute the gradient at weight vector ($w_i = 2, \quad i = 0\ldots d$).

11.25 The main aim of an SVM classifier is to separate data to different classes where each class represents a specific output value.

a. A hyperplane can be expressed as $\vec{w}.\vec{x} + b = 0$, where $\vec{w}$ is a vector representing the hyperplane parameters and b is the intercept. Calculate the margin for a binary SVM classifier between $\vec{w}.\vec{x} + b = +1$ and $\vec{w}.\vec{x} + b = -1$.
b. Suppose an SVM has $\vec{w} = (1\ 2\ 3)$, and the value of the intercept is -1. What is the distance between the two marginal hyperplanes?
c. A linear SVM in $w = (1, 0, 1, 0, 1)$ and $b = -10$. What class does the data point $x = (5, 4, 0, 1, 0)$ belong to?

11.26 A neural network is shown in Figure 11.E26. The network uses a step function as its nonlinear activation function. A step function is a threshold-based activation function, in which if the input value to the activation function is smaller than zero, the output is zero,

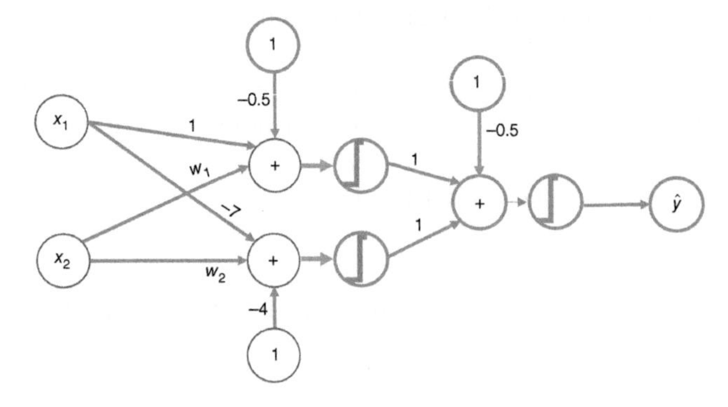

Figure 11.E26 The neural network used in Exercise 11.26.

otherwise, the output is one. Assume this neural network has two inputs (x_1 and x_2). Determine the range of possible values for w_1 and w_2 for the following outputs:

a. x_1 OR x_2
b. x_1 XOR x_2

11.27 Calculate the convolution of an image $f(x, y)$ by itself, expressed as $f(x, y) * f(x, y)$. Use zero padding to calculate boundary values.

$$f(x, y) = [0 \ -2 \quad 1; 0 \quad 0 \quad 0; 0 \quad 2 \quad 0]$$

Chapter 12 IoT Security and Privacy

12.1 Introduction

Today, there are massive numbers of Internet of Things (IoT) devices readily available and the number of these devices is growing at a rapid pace. Therefore, hackers and threat actors attempt to exploit the vulnerabilities of these devices. The hackers may try to compromise an IoT device or a group of IoT devices to access data. They may also target a large number of these devices as an army of nodes in order to make them attack other systems. For these reasons, securing IoT devices is absolutely crucial. As more and more IoT devices are used in different applications, ensuring the security of these devices becomes an absolute necessity.

To be able to compromise a system, threat actors need to identify a vulnerability in that system or the systems that are connected to that system. When a vulnerability is identified, bad actors can plan to launch an attack against the system. If the operation is successful, the attacker can compromise the system or even try to pivot to other parts of the system. In other words, the hackers first try to open a door and enter the system. Once they are in, they may try to find a way to execute an attack against other parts of that system or connected systems. A threat actor might be an application, a process, a system, or a person. A threat actor can be a person outside of an organization that plans to launch an attack against IoT devices of that organization in order to gain access to confidential data, or it can be an insider with good knowledge of the systems of that organization.

Threat modeling is the process of identifying threats and calculating the risk of each threat. An IoT designer should perform comprehensive threat modeling in order to determine potential threats and estimate risks associated with these threats. In this chapter, we discuss IoT threats and vulnerabilities and explain the process of threat modeling and risk calculation for each threat. We will also discuss existing IoT security regulations.

IoT devices may gather huge amounts of personal data and may cause privacy issues if it is not clear how the data is collected, analyzed, used, or stored. We also discuss privacy concerns and regulations as it relates to IoT devices in this chapter.

IoT security and privacy protection methods can protect a system from security or privacy threats. In this chapter, readers become familiar with some of these protection schemes.

12.2 IoT Threats

Threat actors may try to compromise an IoT system or a group of IoT systems to access the data. They are interested in finding the content of data, modifying the data, sending new data, or disrupting the operation of the IoT system partially or entirely. To discuss threats and their

Fundamentals of Internet of Things: For Students and Professionals, First Edition. F. John Dian.

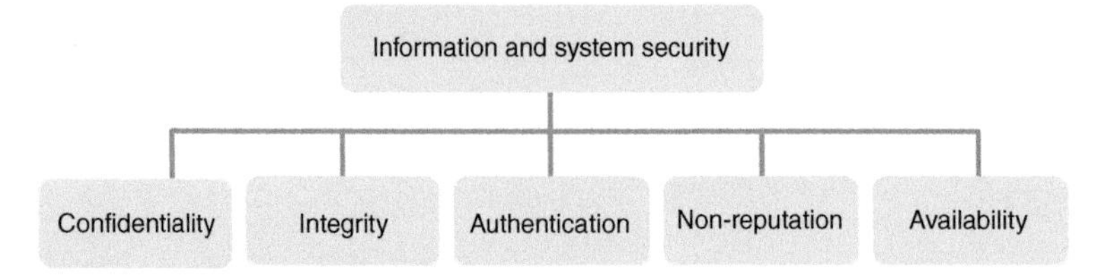

Figure 12.1 Different types of security threats.

potential security issues, we explain information and system security in terms of confidentiality, integrity, authentication, non-repudiation, and availability, as shown in Figure 12.1.

12.2.1 Confidentiality

Threats against confidentiality occur when the threat actor that has gained access to data can understand its content. Therefore, it is crucial to put in place adequate control mechanisms to make sure that the content of a message is confidential. The more sensitive the data, the more sophisticated protection methods need to be implemented to keep the content of the data confidential. The confidentiality of data content should be maintained no matter if the data is stored in a database (data at rest), being processed (data in use), or it is being transmitted (data in motion). The primary tool for protecting the confidentiality of data is encryption. We will discuss data encryption in Section 12.8.1. For now, consider that threat actors are interested in finding the content of data after gaining access to it, and the most important protection mechanism against this threat is encryption.

12.2.2 Integrity

Threats against integrity occur when the threat actor is able to modify data. With integrity protections in place, we can be assured that the data within a system has not been modified or tampered with. The more sensitive the data, the more sophisticated protection methods need to be implemented to ensure the integrity of the data. For example, a threat actor may be very motivated to change the financial information of a transaction. It is also important to put mechanisms in place to detect when the integrity of data has been violated. The primary tools for protecting the integrity of data are based on the hash algorithm. We will discuss protection for data integrity in Section 12.8.2. For now, consider that threat actors are interested in changing the content of data after gaining access to it, and the most important protection mechanism against this threat is the hash algorithm.

12.2.3 Authentication

Threats against authentication occur when the threat actor is able to declare itself as a legitimate user of a system and gains access to its data. Generally, authentication is the process of finding out if someone or something is, in actual fact, who or what it declares itself to be. The more sensitive the data, the more sophisticated authentication methods should be used to ensure that access to data is only available to authorized users. Using a username and password is the simplest mechanism for authentication. There exists many more advanced mechanisms for authorization. We will discuss protection against unauthorized access to a system in Section 12.8.3. For now, consider that threat actors are interested in gaining access to data by identifying themselves as authorized users of the system, and there are many mechanisms that ensure an unauthorized user cannot access the system.

12.2.4 Non-Repudiation

Threats against non-repudiation occur when the threat actor is able to deny the validity of data. Non-repudiation makes it completely difficult to deny who has sent the message. Digital signatures combined with other measures are the primary methods for protection against non-repudiation. For example, in online transactions, it is completely important to ensure that a party to a contract is not able to deny the authenticity of its signature on that transaction. We will discuss protection against non-repudiation of data in Section 12.8.4. For now, consider that threat actors might be interested in denying the authenticity of transaction data. The primary tool for protection against non-repudiation is a digital signature that provides evidence for the existence of a message or transaction and ensures that its contents cannot be disputed once sent.

12.2.5 Availability

Threats against availability occur when the threat actor is able to disrupt the operation of a system. IoT availability ensures that an IoT system will be available to provide service to its customers. The attackers may want to shut down the operation of an IoT system in such a way that the system becomes unavailable, or the system performs its operations at a lower speed or capacity than it has been designed to do. To launch an attack against availability, attackers may try to saturate the network or system in order to make the system unavailable to service requests. This type of threat is called Denial of Service (DoS) attack, which has been around for a long time, and has become more and more sophisticated over the years. One type of DoS is called Distributed Denial of Service (DDoS) in which a large number of distributed devices participate in DoS attacks. Due to the existence of a massive number of IoT devices, IoT has the potential to be utilized in DDoS attacks, assuming that the attacker can initially compromise enough IoT devices and then instruct them to saturate a target system.

Denial of sleep attack (DoSA) is another threat against availability of IoT systems. As mentioned earlier, there is a need to reduce the power consumption especially in battery-powered IoT devices. In IoT applications that use battery-powered devices, an IoT device usually sends a small amount of data and goes to sleep to reduce power consumption. This increases the lifetime of the battery and allows the device to operate for a long service time. In a DoSA, the attacker tries to prevent the IoT device from going to sleep mode. It is actually an attack on the power supply of the device with an objective to increase the power consumption in order to reduce the lifetime of the device. A simple way to launch a DoSA attack is to use a fake IoT device that constantly transmits data to legitimate IoT devices. The objective of using a fake device is to make IoT devices within the range of the fake node consume more energy by communicating with the fake IoT device. This reduces the lifetime of their batteries and eventually makes them unavailable.

12.3 IoT Vulnerabilities

IoT systems may have many vulnerabilities in terms of security and privacy. In this section, we discuss some of the important vulnerabilities of IoT systems.

12.3.1 Insufficient Authentication

Lack of the existence of sophisticated authentication and authorization procedures in some IoT systems makes them vulnerable to attacks. Some IoT devices need to be low cost, and therefore, manufacturers as well as system designers mostly concentrate on performance and cost rather than security. For example, if IoT devices are connected to the Internet through an IoT gateway, an IoT system designer may decide to connect devices to the IoT gateway without any authentication or may use a very simple method of authentication. At first, this might not seem like an immense issue due to the fact that these IoT devices are usually local and the number of these devices is limited in the system. However, unauthenticated messages between IoT devices and the IoT gateway, or even between the IoT gateway and the cloud, make the system vulnerable and attackers can use them to compromise the IoT system.

12.3.2 Insecure Ports and Interfaces

IoT devices usually have interfaces such as USB ports or other I/O ports. In general, an IoT device might have two types of I/O ports. The first type consists of I/O ports that are needed for operation of the IoT device. The second type consists of the ones that are used for testing or initial provisioning, and therefore, they are not part of the operations of the IoT device. Access to the second type should be restricted, and these interfaces must be disabled in order to avoid attackers from gaining access to the IoT device. Insecure ports and interfaces can make an IoT device vulnerable and allow attackers to enter the system. The threat actors use port scanning techniques to identify open ports of a system and use these ports to enter the system for malicious purposes.

Let us look at a practical example. At the design phase, an embedded system may have a Joint Test Access Group (JTAG) interface. This interface is a common interface that provides the embedded system with a direct way of communication among the chips that exist on a circuit board. It is designed by JTAG for testing Printed Circuit Boards (PCBs). JTAG has been widely used in microprocessors, microcontrollers, and embedded systems. It is mainly used for debugging, programming, and testing. Since the JTAG interface is used for testing and has an inherent capability to disrupt system operations, it must be secured. Usually, semiconductor manufacturers disable JTAG access entirely during the production phase. This is often performed by fusing off JTAG inputs. If JTAG interfaces are accessible, it may allow threat actors to gain access to various parts of the system.

12.3.3 Lack of a Secure Update Mechanism

Securing firmware and software update processes are important tasks, especially when it comes to IoT devices. Over-the-air updates are also necessary for some IoT applications, which makes the need for a secure updating process more essential. The threat actors are very interested in placing malware to infect the system during firmware updates, which can ultimately compromise the system.

12.3.4 Insufficient Encryption

Insufficient encryption of data is an important problem that many IoT devices are facing today. In general, to have good security hygiene, data transfer among IoT devices, between an IoT device and an IoT gateway, between an IoT gateway and the cloud, or directly between an IoT device and the cloud, need to be encrypted. IoT protocols should also support encryption and have built-in encryption mechanisms. Some protocols used in the IoT ecosystem do not provide strong encryption techniques or provide an option to configure the protocol without encryption. The designers of IoT systems may decide to use this option, which makes IoT systems vulnerable to attacks.

12.3.5 Insecure Network Connectivity

The threat actors are interested in launching routing attacks to spoof, redirect, or drop packets at the network layer. For example, in a Black hole attack, the threat actor tries to use a fake

IoT device that pretends to be able to provide the shortest path to other nodes of the network. Once the traffic is redirected to the fake IoT device, it can redirect the data packets or drop them. Many IoT system implementations do not use network security properly, which thus makes them vulnerable to attacks. The use of secure Layer 3 and Layer 4 protocols, as well as implementing firewall rules across the cloud, databases, and servers, should always be considered when it comes to safeguarding IoT systems. One of the most important Layer 3 security protocols is Internet Security (IPSec). A brief discussion on IPSec is provided in Appendix A. One of the most important Layer 4 security protocols is Transport Layer Protocol (TLS). A brief discussion of this protocol is provided in Appendix B.

12.3.6 Insecure Mobile Connection

Many IoT devices support short range connectivity schemes such as WiFi, as well as long range connectivity schemes such as cellular networks. To configure these IoT devices, they usually need to be paired with an app that needs to be downloaded on a smartphone. During this process, the app requests for the WiFi's password and information. If the app does not securely store the information, it can create a huge problem in terms of security. If attackers can access the login information that is stored on a database by this app, they can authenticate themselves as legitimate users and easily enter the IoT system.

12.3.7 Not Utilizing Whitelist

A blacklist is a list of users, systems, or applications that are dangerous and need to be blocked from accessing data. To protect a system, blacklists need to be constantly updated. A whitelist is the inversion of a blacklist. A whitelist blocks everything except the users, systems, or applications that are on the list. Neglecting to use whitelist is another issue in securing IoT systems. For example, an IoT gateway or cloud backend system usually accepts whitelisting. It is important to enable whitelisting in such a way that it ensures only valid IoT clients can gain access to IoT gateways or the cloud backend system.

12.3.8 Insecure IoT Device Chip Manufacturing

Every IoT system is designed to operate within a specific range of voltage, current, or temperature. A threat actor may want to increase or decrease these parameters in order to force the device to operate in a situation that it is not originally designed to operate. This may cause the IoT system to malfunction and go to an insecure state, which allows an attacker to take advantage of the situation and gain access to the system. The threat actors may also create a substantial amount of noise to make the IoT system behave strangely. For instance, the threat actor may try to zap the power supply and apply electrostatics to make the IoT system start the booting process in an insecure state. Adding noise and disturbance to a system in order to make it insecure is called glitch injection. The chip manufacturers should have countermeasures for protection against environmental threats or glitch injection attacks at the chip level.

12.3.9 Configuration Issues

Even if the components of an IoT system are designed with security in mind, vulnerabilities can be created during configuration of an IoT device. Neglecting to use a proper password, utilizing default settings and configuration, using a weak encryption scheme despite stronger ones being available, or disabling security event logging are a few examples of security issues that can arise as a result of poor configurations.

12.3.10 Privacy Issues

Privacy concerns can arise when an IoT device collects, analyzes, or stores data without consent. The IoT device may also cause a privacy breach if data is used to identify a person or is used maliciously against that person. Sometimes, the collected data from an IoT device is not sensitive and does not cause any privacy concerns. However, if the data is integrated with other information from other systems, it may cause serious privacy concerns.

12.4 IoT Threat Modeling and Risk

Threat modeling is the process of identifying the important threats to a given system and determining the impact that these threats have on the system. Threat modeling can be performed by rating each threat according to its impact as well as its probability of occurrence. It is clear that the probability of an attack is higher in situations where the data is more valuable to an individual or a group of people. For this reason, and to better perform threat modeling, it is good practice to identify any possible attackers and their potential motivations. A hacker may want to draw attention to a social or political cause, or may have grievances against an individual or an organization. On many occasions, the attackers are motivated by financial factors and there are situations where they are interested in gaining access to technical knowledge. Through the threat modeling process, we can identify security risks and establish a deeper understanding of threats that require more attention or the components of an IoT system that need more protection. We use the following four steps to perform threat modeling:

1. Identifying the assets: each IoT system contains many assets. An asset might be a physical asset, such as a sensor, or might be a data asset. The first step in threat modeling is to identify assets and their security requirements in terms of confidentiality, integrity, authentication, non-repudiation, and availability. We should determine the high value assets and understand who owns these assets, who maintains them, and whether external parties have access to them.
2. Identifying the message flow: in this step, the flow of data and messages among assets and between assets and external parties will be identified.
3. Identifying threat types: in this step, we identify threats and assign them to different threat types. Identity spoofing, data tampering, repudiation, information disclosure, DoS, escalation of privilege, and bypassing of physical security are some examples of different threat types. We will discuss each of these categories in this chapter.
4. Rating threats and risk calculations: in this step, we rate threats based on their probabilities and their impacts. To be able to do that, we also need to calculate the risk involved in each threat.

12.4.1 Threat Modeling for Smart Gas Station

To demonstrate how we can perform threat modeling, we use a practical scenario and perform threat modeling for this scenario based on the four steps explained earlier. Let us consider a smart electric car charging station system with several charging stations that provide service to the owners of electric cars. Customers can charge their cars, and pay the cost of charging using different methods of payment such as credit cards or an online mobile app. The collected data from this system can be processed in order to determine an efficient pricing strategy. Pricing might increase during periods when the station is busy, and can decrease when the charging station is more vacant. The system provides customers with the status of charging stations as well as pricing information. It can also provide the license plate numbers of the cars entering the station to police authorities. The police can compare these license plate numbers with the list of stolen cars and can then send a notification to a police car in the neighborhood for further investigations.

The system uses a sensor in each charging station to determine whether it is vacant or if it is currently serving a consumer. This sensor is placed on the ground and uses Bluetooth Low Energy (BLE) technology to communicate data to an IoT gateway. The IoT gateway is connected to a router using WiFi technology. The router is located inside a building, and it is connected to the cloud. The charging station system is equipped with an IP camera that sends information to the router using WiFi technology. The IP camera is used to capture license plate information, which is sent to the router. All data from the router is transferred to a cloud-based application, which processes the data and also sends it to a cloud-based database. Using cellular technology, each charging station communicates electricity usage and payment data directly to a cloud-based payment processing center. The payment data from the mobile app is also sent to the cloud using a cellular technology. The cloud app can provide customers with information about the pricing and status of all charging stations. It is also worth noting that the system app may interact with other cloud-based apps. For example, a police notification app may take license plate information and, if necessary, send a notification to a police officer who is close to the charging station system's location. Let us consider that the security objectives in this scenario are to ensure the integrity of data, the confidentiality of sensitive data, and the availability of the system. Using the four phases we have outlined previously, we will now begin the threat modeling process for this case.

12.4.1.1 Identifying the Assets

The assets in this example can be physical assets such as sensors mounted on the ground, the IP camera, charging stations with payment capabilities, the IoT gateway, and the router. The data assets include sensor data, video data, and license plate information transmitted by the IP camera, payment data, charging station data, mobile app data, and the analyzed data related to the system status and pricing, which are provided to the customers.

12.4.1.2 Identifying the Message Flow

Figure 12.2 shows the message flow for this example. From this figure, we would like to determine the data types, as well as the points of entry for authorized or unauthorized people. In this example, the information of the highest value within the system consists of the payment data, license plate data, and potentially the raw video footage from the IP camera.

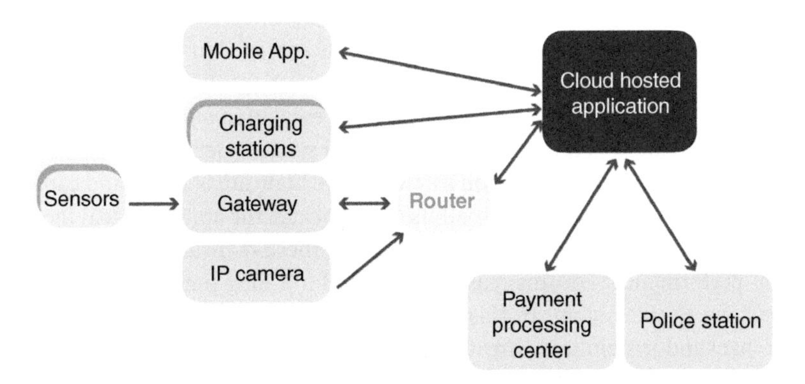

Figure 12.2 Message flow for the electric car charging station example.

12.4.1.3 Identifying the Threat Types

We can classify threats in different categories such as:

- Identity spoofing: this category deals with the threats in which an individual or a system uses a fake identity to gain unauthorized access to an IoT system. Some of the possible ways to perform identity spoofing are social engineering, database compromise, and Man In The Middle (MITM) attacks. Social engineering deals with situations where attackers attempt to manipulate people to give up confidential information. Sometimes, it is easier for a threat actor to find information using social engineering rather than using other techniques. For instance, it might be much easier to deceive people into giving them their password rather than trying to hack a system to find the password. A special type of social engineering attack is called phishing, in which the threat actor masquerading as a trusted entity asks the victim to open an email or a text message. An MITM attack is a general term for situations where the threat actor attempts to eavesdrop or impersonate one of the parties in a communication.
- Data tampering: this category includes situations where attackers attempt to tamper with data. For example, if the integrity of data for controlling a ship, a plane, or a car is not maintained, catastrophic events can occur, and an attacker can sink a ship, make a plane crash, or control a car to veer off the road.
- Repudiation: this category is related to threats in which an attacker repudiates a transaction.
- Information disclosure: this category relates to situations where the confidentiality of the data is threatened.
- Denial of service: this category of threats considers situations where a threat actor compromises the availability of a system.
- Escalation of privilege: most systems are designed to be used with multiple user accounts. These accounts provide different abilities to users known as privileges. While some users can just see the system information, other users with higher privileges may be able to make modifications to a part of the system or the entire system. The category of escalation of privilege deals with threats where an attacker exploits a bug or designs a flaw in a software application to gain elevated access to a part of the system that is available to users or systems with higher privileges.
- Bypassing physical security: this category deals with the physical security of a system or network. For example, an attacker may break into a server room and get access to confidential data, access to the Internet drop line from outside of the building to intercept data, cut a communication link, or gain access to devices that might be exposed in the field in order to utilize them as part of an attack.

In our electric car charging station example, some of possible threats are listed in Table 12.1. As can be seen from this table, there are many threats that could compromise the electric car charging station system. For example, the last row in Table 12.1 is related to the escalation of privilege. This threat can be launched by the installation of a rootkit on the backend system. Generally, rootkits are a type of malware that are designed to stay hidden on a system. The malware is active and can help hackers remotely access an IoT device. Installing rootkits gives hackers the ability to steal the user's credentials or even disable security features. It is hard to detect these malwares, since they usually do not lower system performance. For this reason, they live for a long time and may be able to cause significant damage to the system. In this example, we assume that a threat actor can install a rootkit on the backend system and threatens the operation of the system by finding elevated access to parts of the system that are only available to users or systems with higher privileges.

Table 12.1 Possible threats for the electric car charging station system.

Threat example	Threat type	Security control	Threat target	Possible attack techniques	Security measures
The threat actors identifies themself as legitimate customers of the system by accessing the customer accounts	Identify spoofing	Authentication	Customer account	Social engineering, phishing, database compromise, MITM attack	Use of better authentication
The attacker receives free service by unauthorized access to the backend application	Data tampering	Authentication Integrity	Cloud system application	Social engineering, web server compromise	Use of better integrity schemes
The malicious actor attempts to change the license plate numbers to take the attention of a police officer to something else	Data tampering	Integrity	IoT gateway, router, or backend system	Web server or backend compromise	Use of more advanced integrity schemes
The attacker receives free service by asserting the system malfunction	Denial of transaction	Non-repudiation	Cloud system application	Mobile app backend compromise	Use of digital signature by each charging station
The attacker covers the sensors to show that the charging stations are not vacant	Bypassing physical security	Availability	Sensors	Access to physical assets	Using the IP camera to also monitor the occupacy
The attacker accesses customers' financial information through backend application	Information disclosure	Authentication confidentiality	Cloud system application	Financial app compromise	Use of better authentication or encryption
The attacker shuts down service through DoS attack	Denial of service	Availability	Multiple factors	DoS on gateway, charging stations, router, or servers	Use of whitelisting, port securing, or firewall
The attacker disrupts the operation of the system by implementing a rootkit on backend system	Escalation of privilege	Authentication	Backend system	Backend compromise	Use of better authentication method

12.4.1.4 Rating Threats and Risk Calculations

To have an absolutely secure system, system designers need to close all the doors on hackers. This is almost impossible or very costly. On the other hand, hackers only need to find one vulnerability in the system to open one door! For this reason, business leaders with the help of security experts need to make a decision on the level of risks related to each threat. Let us discuss how to perform risk analysis for one of the threats in Table 12.1.

To find the risk level for each threat, a number from 1 to 10 should be assigned to each threat, where 1 represents a low-risk threat and 10 shows a high-risk one. This number represents the risk level, and it is a good quantitative measure for business leaders and decision-makers.

To assign a number to a threat in order to represent its risk level, we should consider the damage that can be caused by a threat, the skill sets that are needed to launch the attack, the possibility of repeating the attack, and the complexity of detecting the attack as examples of important factors in determining the risk level associated with a threat.

DREAD is a risk assessment model designed by Microsoft that can be used for this purpose. DREAD is an abbreviation of the five letters: Damage, Reproducibility, Exploitability, Affected users, and Discoverability. Damage indicates how catastrophic an attack can be. Reproducibility explains how easily an attack can be reproduced. Exploitability indicates the complexity, skill sets, and time that are needed to launch the attack. "Affected users" is another factor that is an indication of the number of people or systems that will be impacted. Discoverability shows how easy it will be to discover the threat.

Let us find a risk number for our first threat in Table 12.1. In this case, the bad actor has found the information of a legitimate customer, and has used the information to compromise the system. Therefore, our assumption is that the attack is limited, and therefore the damage will also be limited. We may consider a low-risk number such as 2 for the amount of damage that the threat can cause. The possibility of repeating the attack depends on the way that the attacker has gained access to customer information in the first place. Let us assume that access to the customer information by compromising the customer account database is not easily possible and that the database is secure. The attacker may use social engineering to get account information. Even though it is possible to repeat this process, it is not something that can be repeated very easily. We consider number 3 for reproducibility. The skill sets that are needed to launch this attack are fairly simple and therefore

Table 12.2 Average risk score for the first item in Table 12.1.

Damage	Reproducibility	Exploitability	Affected users	Discoverability	Average score
2	3	8	1	1	3

we assign a score of 8. The attack compromises only one account and one customer, and therefore we consider 1 for affected users. Due to the fact that this attack can easily be detected, we give a number as low as 1 for discoverability to this case. Table 12.2 shows that an average number can be calculated to represent the risk of the first item in Table 12.1. The average score number for other threats in Table 12.1 should be calculated with the same method that we explained earlier.

The average score number corresponding to the threat in the first row of Table 12.1 is 3, which represents a low risk threat. To complete the process, the risk score for each row of Table 12.1 needs to be calculated in a similar fashion. Readers should understand that the risk numbers must be determined by security experts that have a good understanding of threats. The simplified threat modeling example explained in this chapter only tries to give readers a basic understanding about the threats, the threat modeling process, and the risk calculation. Business leaders compare the risk scores associated with different threats and decide on what sequence of actions need to be taken to lower security risks. Some threats may require mitigation, and some may be deferred. In general, an organization needs to analyze the risk associated with each potential threat, and decide on threats that need immediate attention. The threat modeling gives leadership the opportunity to prioritize their risk mitigation scheme.

It should be noted that Microsoft used DREAD as part of its internal software development lifecycle, and stopped using it due to the fact that the DREAD ratings were very subjective [1]. However, DREAD is still in use for threat modeling in practice. Besides DREAD, there are many other structured threat modeling methods. For its simplicity, we briefly explained DREAD to give readers a basic understanding of the procedures involved in the threat modeling process.

12.5 IoT Security Regulations

The true power of IoT can be seen when IoT is integrated everywhere and when IoT is regulated. At this time, IoT regulations lag the IoT technology. For example, the use of smart helmets by hockey players can bring many advantages for players as well as their coaches. However, due to the lack of regulations, the use of smart helmets is not allowed in hockey leagues. IoT security regulations are also behind technological advances in security. Actually, IoT security regulations are at their starting point. A lot of work needs to be done in order to see the effectiveness of security regulations for IoT.

An IoT system consists of various sections such as sensors, the IoT module, the IoT device, connectivity schemes, cloud, and analytics. Each of these sections is vulnerable to threads and each one needs its own regulations in terms of security. There also must be security regulations that discuss the connection between the various sections. In the United States, the Department of Homeland Security (DHS) has published a set of IoT security guidelines [2]. These guidelines ask the IoT system developers to consider and incorporate security during the design phase, develop the system based on best practices, perform risk assessment based on potential security issues and their impact, design processes regarding security updates, and establish transparency across IoT platforms.

Due to the use of IoT technology in various industries, other sets of rules and regulations might be needed for the use of IoT in a specific industry. For example, the US Food and Drug Administration (FDA) needs to make regulations for smart medical devices.

There are many organizations around the world working to make IoT more secure. For example, European Commission (EC) is one of the entities in Europe that works toward regulation guidelines for a more secure IoT system. The EC has created a list of minimal security requirements, which is needed for most IoT devices. The EC has also established a set of best security practices for developing IoT systems, and has worked to design a labeling system for IoT devices.

In 2017, the IoT Cyber Security Improvement Act (CSIA) was established to provide guidelines for the use of IoT devices in federal buildings, which enacted the IoT cyber security improvement act of 2020 (the IoT Act) [3]. Even though CSIA is related to federal buildings, the same principle can be used for other environments as well. The legislation mainly focuses on vendors verifying that their IoT devices do not contain any known security vulnerabilities. They also want vendors to ensure that their devices use advanced protocols, do not include hard-coded credentials, and utilize notifications in case of a security incident. The legislation recognizes the existing challenges for providing robust security features for some IoT systems that might have limited functionalities and processing power capabilities. However, the legislation requires federal agencies to create a publicly accessible database of devices and their manufacturers, for which such devices have limitations of liability under the Act.

12.6 IoT Privacy Concerns and Regulations

IoT privacy can be defined as the ability of IoT users to have control on dissemination of their data; to have a say on how their data is collected, stored, analyzed, passed along, or even sold; and to control the degree in which a person or a system can be identified.

There are many IoT systems installed in outdoor locations that intend to monitor people's activities, behaviors, and emotions. People may get captured on cameras installed on buildings, street corners, and intersections; the ones mounted inside cars and motorcycles; or even the cameras installed on drones that fly overhead. The use of voice-activated personal assistance devices in homes or inside cars that may use always-on cameras and microphones are growing. Examples of these devices are Amazon Echo, Google Home, or even Smart TVs. People are using more and more IoT-based health tracking devices that record and transmit the medical information of users. With connected vehicle technology, cars talk with other cars on the road as well as infrastructures. Many IoT devices that are used inside a car have a GPS tracker that can track the movement of the users and show their location data. To find out whether there is any privacy concern with a system, we may need to find the answer to the following questions: Who collects the data, for what purpose the data is being used for, at what locations the data is being stored, who is authorized to access the data, is the data shared with any third parties, and for how long the data is kept. The answers to these questions may indicate possible privacy issues. Examples of these issues can happen when there is no consent for data collection, in situations where people do not have any understanding on why their personal data is being analyzed, or when there is no process in place in case of a privacy breach. Concerns in terms of privacy can be identified if data can be used to identify a person, track one's movement, or be used against an individual by the ones who are collecting the data or by unauthorized people that can access the data.

De-identified information is a new category of personal information, and it is part of privacy law. This information can be used for research purposes, and be disclosed to government or healthcare institutions. Information is considered to be personal when there is a high possibility that the information might be used to identify an individual. Personal information becomes de-identified when there is no possibility that the data can be used to identify an individual. Therefore, it raises privacy concerns if there is a possibility that the information that is tagged as de-identified can be identified.

To address privacy concerns at the time that an IoT system is being designed and deployed, IoT system developers must perform a process called Privacy Impact Assessment (PIA). The process aims to address concerns before privacy issues arise. PIA is a review process, and its objective is to protect personal information that a system is collecting. The process should be performed by experts in the area of privacy. These experts should identify, evaluate, and control privacy risks.

Some of the important parts of the PIA process are related to addressing the following subjects:

1. Finding who can access the data and what laws and regulations exist for accessing the data in the location where the data resides. If the data is stored in various data centers, some locations may have softer or stricter policies and regulations in regard to accessing the data, which may cause privacy issues.
2. Finding out what information originates, terminates, or transits through an IoT system. IoT systems should always limit the collection of information to what is absolutely required, and minimize the collection of personally identifiable information. The data must be destroyed when it is no longer needed.
3. Finding out how data is shared. The PIA considers the data collection within the system as well as any third parties with whom the data is shared. It should be clear how the data has been shared, what types of agreements exist or need to be established with the third parties, and the privacy issues that may arise as a result of sharing data.
4. Finding out what policies and procedures exist in situations where private information is compromised, what compensation policies and redress exist, and what processes have been established for notifying people that their private information is compromised.
5. Finding out what data collection and privacy policies exist in order to ensure that the developers have consent for data collection, have given options to the consumers to opt out, and whether the consumers are allowed to see their own data. It is clear that data should be used for the purposes that users have consented to. PIA ensures that IoT system developers understand a duty of care for protecting private information.

It is clear that PIA is a dynamic process and needs to be updated when there is any change in the system. Let us consider a smart toy manufacturer as an example. In recent years, connected toys are becoming more and more available in stores. These IoT systems are examples of consumer devices on the market that require an evaluation of privacy assessment impacts. When a smart toy manufacturer wants to design a new product, it should address privacy issues from the beginning by conducting a PIA. A child usually talks to their smart toys. These conversations are usually recorded and transmitted over the Internet. As a result, they may pose privacy concerns if privacy issues are not addressed properly. Also, smart toy manufacturers should respect the Children Online Protection Act among others. The manufacturers may like to keep the data for as long as possible to be able to constantly evaluate their products and make more efficient and robust algorithms. However, this may cause privacy concerns if unauthorized people can access the stored data.

12.7 IoT Security and Privacy Examples

In this section, we discuss some of the IoT-related security and privacy examples.

12.7.1 Threat Against Availability – Mirai Bonnet

In October 2016, a DoS cyberattack was performed against Dyn, a DNS service provider, and disrupted its operation. This attack first was performed against the operation of the service provider on the US east coast, but later became a global disruption and affected more than 13 000 Internet domains. The affected domains are then stopped using DNS services. In this example, the attackers first infected many IoT devices with Mirai malware. These IoT devices were mostly cameras, baby monitors, and residential gateways, which usually had hard-coded or factory-default usernames and passwords. Once a large number of these IoT devices became infected, the attackers had an army of nodes in place to execute their cyberattack. The attackers then coordinated these nodes, which are also called zombies, to initiate an attack against Dyn. The Mirai malware instructed zombies on how to compromise the availability of the DNS servers by sending traffic toward the targeted system. In this case, the zombies were instructed to connect to a command and control server to get instructions on how to initiate the attack.

The Mirai botnet attack showed that it is possible to take advantage of a large number of IoT devices with low security configurations in order to establish a distributed DoS attack. In this attack, the threat actors did not use their own computers to launch the attack and instead tried to launch the attack using IoT devices with low security hygiene that are easy targets. As the number of IoT devices increases, attackers may look at these devices as excellent potentials to be used in DDoS attacks.

12.7.2 Threat Against Integrity – LockState

In 2017, a smart lock manufacturer, LockState, made a firmware update on one of its products. The new firmware had a bug in its design that caused the smart lock not only to stop working but also to not accept any further software updates. Even though no security attack was involved in this case, readers can imagine similar results could happen if attackers could tamper with and corrupt the software update code. This scenario would not be a huge issue in normal situations, since the users of the smart lock could also use their physical keys to enter their property. However, in this case, the smart lock had been used in some Airbnb rental properties. The renters did not have a physical key and only had received a code that allowed them access to their rental property. In this case, many renters became stranded and were locked out. Since no factory reset capability was considered for the smart lock by the manufacturer, the property owners needed to ship the locks back to the manufacturer and wait to receive a new lock. To protect a system from a threat actor who intends to compromise a system at the time of updating its firmware code, integrity control protections must be in place to secure the system from these types of malicious attacks.

12.7.3 Threat Against Software Update – Jeep

In 2015, a group of researchers hacked into the Controller Area Network (CAN) bus system inside a Jeep car. This group tried to find a vulnerability in the firmware update process that is used by Jeep.

They not only succeeded in gaining access to the CAN bus but also found a way to have complete control over the vehicle. The researchers could control the cars' speed and veer the car off the road. Even though there was no malicious attacker involved in this case, the team of researchers showed that it is possible to compromise the system if IoT security issues are not taken seriously.

12.7.4 Threat Against Confidentiality – TRENDnet Webcam

In 2010 and 2011, TRENDnet, the manufacturer of IP cameras for applications such as home security, sent login information without any encryption over the Internet. The company also stored all login information in clear text (with no encryption) in its database. TRENDnet also had a bug in its software that allowed attackers to access the device, watch the video content, or listen to its audio data with knowing only the IP address of the device. Even though large malicious attacks did not happen in this case, it showed how important it is for the designers of an IoT system to take security issues seriously. The IoT module manufacturers must introduce security measures in their module, the consumers should configure their security parameters properly, and the IoT system designers should know that they have a large stake in the blame, in case of a security breach.

12.7.5 Threat Against Availability and Integrity – St. Jude Medical's Implantable Cardiac Devices

In 2017, it became clear that implantable cardiac devices such as pacemakers and defibrillators designed by St. Jude Medical are vulnerable to attack. These devices are designed to monitor the heart and prevent heart attacks. The attackers could deplete the battery or administer wrong pacing or shocks by controlling the device's transceivers. The manufacturing company ignored the problems at first, but released a set of cybersecurity updates later.

12.7.6 Threat Against Availability – Cyberattack on the Ukrainian Smart Grid

In 2016, an attack on a smart grid in Ukraine disrupted power to more than 200 000 customers. Generally speaking, electric grids are controlled and monitored by the Supervisory Control And Data Acquisition (SCADA) system. The SCADA architecture has evolved during the past four decades along with advancements in computer technology. Simply put, the SCADA system is a computer system that communicates with many remote devices that are called Remote Terminal Units (RTU). An RTU is a microcontroller-based device that interfaces with field equipment on one side and the SCADA control system on the other side. An SCADA system is similar to a corporate computer network, but it is more resilient. An attack on the SCADA system can have not only serious but also widespread effects on the power system. An attacker can get access to the SCADA system, control the switches, and disrupt power flows. For example, it can change the state of a switch breaker from closed (allowing power to pass) to open. Also, it can continuously switch between open and closed states. An example of these attacks happened on Ukrainian SCADA system that disrupted power and delayed recovery operations in one of the coldest months of the year. This attack was caused by Crashoverride malware, which is specifically designed to attack electrical grids.

12.7.7 Privacy Concern – DJI

In 2016, Da-Jiang Innovations (DJI), the world's largest civilian drone and aerial imaging company, shared its customer data from around the world with the government of China. The US banned the use of DJI drones in response to this privacy violation. The company later decided to add a new feature called local mode or easy-to-use mode in order to address privacy issues. This mode prevented transmission of drone data over the Internet and solved some of the privacy concerns at the cost of reducing the ability of drone operations.

12.8 Threat Protection Methods

In this section, we briefly discuss some of the protection methods against confidentiality, integrity, authentication, and non-repudiation threats.

12.8.1 Confidentiality Protection

The primary tool for the confidentiality of data is cryptography, in which a plaintext is encrypted to create what is known as a ciphertext. In other words, plaintext is the original message and ciphertext is the message after being encrypted using an encryption algorithm. The algorithm used for performing encryption or decryption is called a cipher. The cipher uses a number or a set of numbers called key(*s*) to be able to encrypt or decrypt messages. Generally speaking, ciphers can be classified into two main groups. The first group contains methods that have symmetric keys. This means that the same key is used for both encryption and decryption. The second group contains those methods that use asymmetric keys. In this case, there are two types of keys: a private key that is kept for decryption and a public key that is open to the public and can be used by everyone. We discuss some of the encryption methods that utilize symmetric and asymmetric keys in the next section.

12.8.1.1 Methods Based on Symmetric Key

Let us first consider simple methods such as the ones that are based on character-based or bitwise ciphers. We will then extend these methods and discuss more advanced symmetric ciphers that consist of multiple rounds of operations.

The idea behind character-based ciphers is to change each character in the plaintext to another character in the ciphertext. Some character-based ciphers are shown in Figure 12.3. In Figure 12.3a, a constant value (i.e. 3) is added to the ASCII value of each character in the plaintext to construct the ciphertext. The constant value needs to be subtracted from the ciphertext for decryption. In this method, there is a unique mapping between a character in plaintext and a character in ciphertext, and the constant value represents the symmetric key. In Figure 12.3b, each character in plaintext is mapped to another character in ciphertext using a look-up table. In this case, the mapping table can be coded as the symmetric key. The problem with the character-based ciphers in Figure 12.3a, b is that the relationship between each character in the plaintext and its encrypted character in the ciphertext is a one-to-one relationship. For example, in Figure 12.3a, a character like "e" always changes to "h," and in Figure 12.3b this character always changes to "j." This type of encryption can be easily decrypted by looking at the probability of the occurrence of any single character in a language. To create a one-to-many relationship between a character in the plaintext and its encrypted character in the ciphertext, we can use a set of keys to perform the encryption. The first key in this set is a value that both parties must agree upon. The other keys in this set must be calculated based on a rule. For example, Figure 12.3c shows a situation where the first key is 3, and the other keys are calculated by adding one to the value of the previous key in the set. If the value of the key reaches 9, then we start from the first value again. As can be seen from Figure 12.3c, the character "e" in the plaintext has changed to "o" and "h" based on the position of "e" in the plaintext.

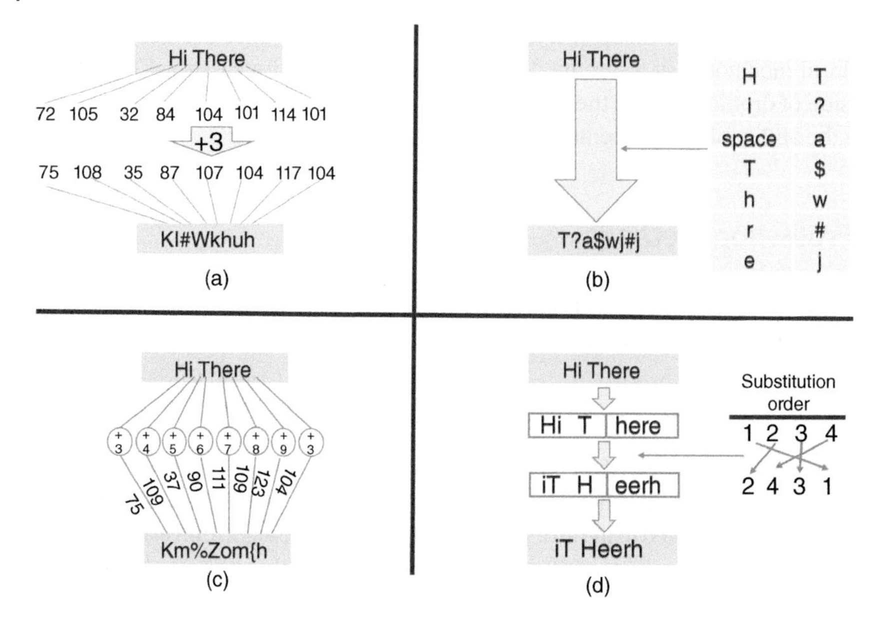

Figure 12.3 Examples of character-based ciphers.

We used a very simple rule to show the dependency of a key to the position of a character in the plaintext, but more complex rules can be made to create a better one-to-many relationship between characters in the plaintext and the ciphertext. Figure 12.3d shows a situation where the position of the characters in the plaintext has been changed based on a substitution order in order to calculate the ciphertext. In this case, the plaintext is first divided into blocks of n characters. Then, the location of each character within the block is changed based on a substitution order. For example, in Figure 12.3d, the plaintext is first divided into blocks of four characters. Then, the location of each character is changed based on the substitution order that is shown in this figure. The value of n and the substitution order makes the symmetric key in this case.

Some examples of bitwise-based ciphers are shown in Figure 12.4. In Figure 12.4a, the plaintext has been xored with a constant binary value to make the ciphertext. Figure 12.4b shows a situation where the bits of each character are rotated based on a rotation scheme that is shown in this figure. In this scenario, the key shows the rotation scheme. It should be noted that there are different variations of ASCII tables for values above 127. The characters shown in Figure 12.4b are based on ISO 8859-1.

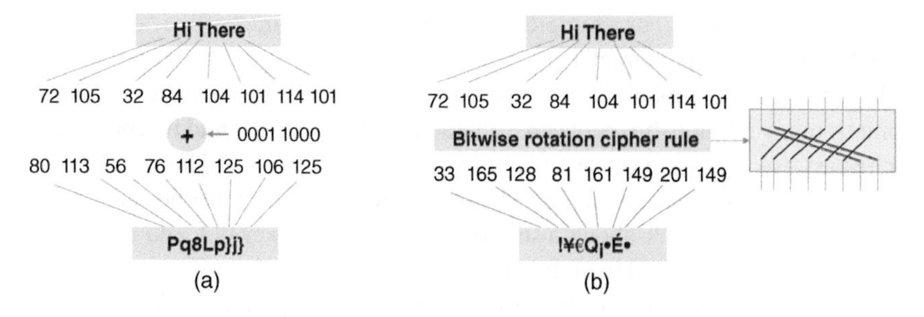

Figure 12.4 Examples of bitwise-based ciphers.

S-Box table		Inner bits															
		0000	0001	0010	0011	0100	0101	0110	0111	1000	1001	1010	1011	1100	1101	1110	1111
Outer bits	00	0010	1100	0100	0001	0111	1010	1011	0110	1000	0101	0011	1111	1101	0000	1110	1001
	01	1110	1011	0010	1100	0100	0111	1101	0001	0101	0000	1111	1010	0011	1001	1000	0110
	10	0100	0010	0001	1011	1010	1101	0111	1000	1111	1001	1100	0101	0110	0011	0000	1110
	11	1011	1000	1100	0111	0001	1110	0010	1101	0110	1111	0000	1001	1010	0100	0101	0011

011011 → S-Box table → 1001

m = 6 n = 4

Figure 12.5 An example of an S-Box with $m = 6$ and $n = 4$ [4].

The Substitution-Box (S-Box) and Permutation-Box (P-Box) are two important components of advanced symmetric key algorithms that use bitwise operations and can be used in order to make the relation between the ciphertext and plaintext difficult to detect. In cryptography, an S-Box performs bitwise substitution based on a function, and is often used as part of advanced ciphers to obscure the relationship between the ciphertext and the key. An S-Box can be represented by a block that transforms m input bits to n output bits based on a function. If $m = n$, then only the locations of bits are changed. In this case, the relationship between input and output can be shown using a look-up table. Both fixed tables and key-dependent look-up tables are used in the architecture of different S-Boxes. If $m > n$, then some information is lost. However, the designers of advanced ciphers use S-Boxes in addition to other operations, in such a way that lost information can somehow be found from other parts of the cipher, and therefore, the whole operation is invertible. For example, an S-Box that uses a fixed look-up table and has been used in some advanced ciphers is shown in Figure 12.5. This S-Box maps a 6-bit input to a 4-bit output. Let us explain this S-Box in more detail. The S-Box selects the row of the look-up table based on the outer two bits (the first bit and last bit of the inputs) and the column of the look-up table based on the inner four bits. For example, an input "011011" has the outer bits that are "01" and the inner bits that are "1101"; therefore, the output of the look-up table would be "1001."

P-Boxes basically shuffle the input bits around, and are used as part and in the middle stage of advanced ciphers. Similar to S-Boxes, P-Boxes are keyless. If the number of input bits of a P-Box is the same as its output, it is called a straight P-Box. A P-Box is called compression or expansion depending on whether the number of its output bits is less than or greater than the number of its input bits, respectively.

Advanced encryption methods have multiple rounds in which each round is complex, and it is made of other ciphers. Data Encryption Standard (DES), Triple DES, and Advanced Encryption Standard (AES) are examples of more advanced ciphers [5]. Figure 12.6a shows the DES encryption method. In this method, each plaintext is divided into blocks of 64 bits. Each block is then passed through a P-Box. The output of the P-Box goes through 16 rounds of operation (represented as Round i ($i = 1, \ldots, 16$)). The operation performed in each round is shown in Figure 12.6b. The input to each round is 64 bits, which are divided into two 32-bit numbers. The most significant 32 bits are called L_i. The least significant 32 bits are called R_i, which on one hand go directly to the output of the round block without going through any further operation (L_{i+1}), and on the other hand, go through a P-Box. The output from this P-Box gets xored with a 48-bit key, and then passes through an S-Box and a P-Box before xored with L_i to make the least significant 32 bits of the output (R_{i+1}). After passing through all 16 rounds, the results pass through a P-Box. The output of this P-Box is the ciphertext corresponding to the 64-bit plaintext. Each round stage Round i ($i = 1, \ldots, 16$) needs a 48-bit key. For DES, the key is 64 bits (actually 56 bits plus 8 bits of parity). A key generator is used to generate sixteen 48-bit keys from the 64-bit key. These 48-bit keys are used in each round.

The main problem with DES is related to the size of its symmetric key, which is too short. Two solutions to solve this problem are Triple DES and AES. Triple DES uses three single DES

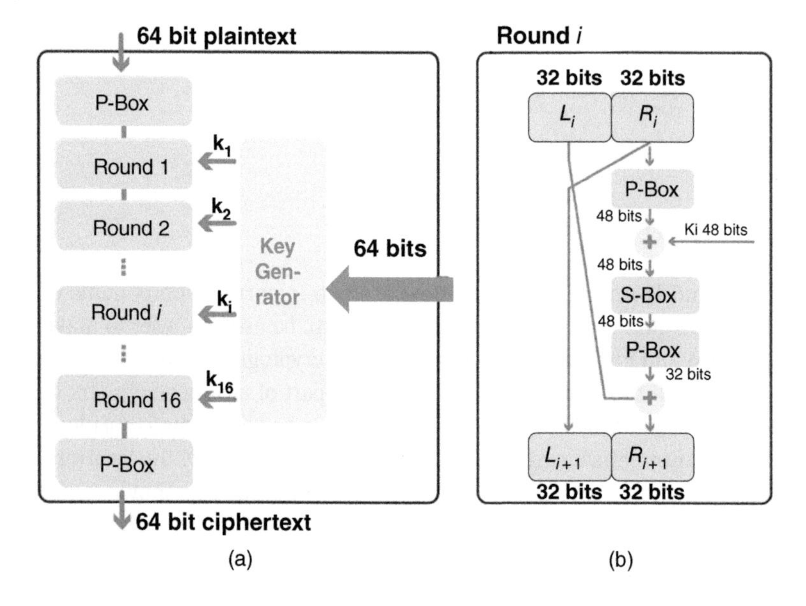

Figure 12.6 DES cipher: (a) encryption of a 64-bit plaintext and (b) operation in each round.

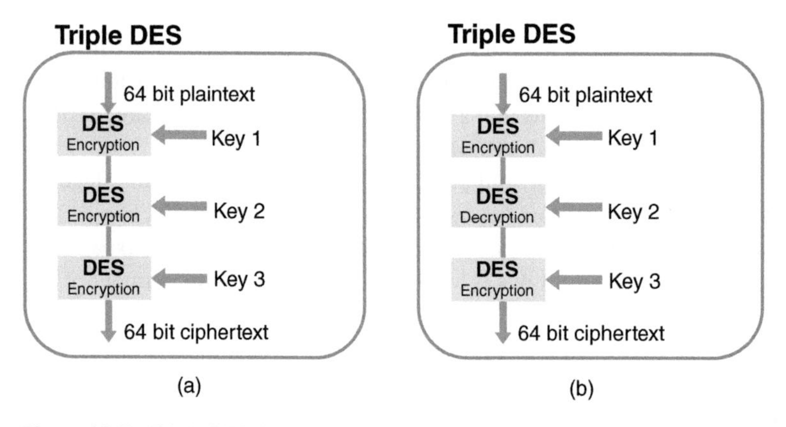

Figure 12.7 Triple DES: (a) EEE3 format and (b) EDE 3 format.

sequentially. Generally, we can define n-DES because a single DES is not secure enough. As mentioned earlier, the size of the symmetric key used in DES is too short, and can be brute-forced by motivated threat actors. In order to improve security, the value of n in n-DES should be greater than or equal to 3, and therefore 2-DES is not secure either. Of course, every additional DES requires more computational power. Triple DES can be implemented in different formats. Figure 12.7a shows a format that is called EEE3 (three times DES encryption using three different keys). Figure 12.7b shows a format that is called EDE3 (a combination of DES encryption, DES decryption, and DES encryption using three different keys). This format uses three keys of key 1, key 2, and key 3. In the first step, it encrypts with key 1, then it decrypts with key 2, and finally encrypts with key 3. Therefore, the ciphertext can be expressed as:

$$\text{Ciphertext} = E_{\text{key }3}(D_{\text{key }2}(E_{\text{key }1}(\text{Plaintext}))) \tag{12.1}$$

Obviously, to decrypt the ciphertext, the operation should be performed in reverse order. Equation (12.2) shows the decryption equation.

$$\text{Plaintext} = D_{\text{key }1}(E_{\text{key }2}(D_{\text{key }3}(\text{Ciphertext}))) \quad (12.2)$$

The National Institute of Standards and Technology (NIST) has deprecated the use of both DES and triple DES due to their vulnerabilities from 2023 and suggested the use of AES instead. There are many similarities between triple DES and AES methods. However, AES is faster and provides substantially better protection as compared to triple DES.

AES divides the plaintext into blocks of 128 bits, instead of the 64 bits used in triple DES. It also arranges these 128 bits in a two-dimensional array of 4×4. Therefore, there are 4 bytes (32 bits) in each row, and there are four rows. Similar to triple DES, the architecture of AES is based on three blocks. However, each block uses less number of rounds and larger keys. AES can be implemented using blocks that have 10 rounds and use a 128-bit key, 12 rounds and use a 192-bit key, or 14 rounds and use a 256-bit key. In other words, the number of rounds depends on the size of the key. Simply put, the AES algorithm for a 128-bit plaintext has the following steps:

1. Key addition: a key is added to the plaintext, and the data is sent to the next step.
2. Byte substitution: an S-Box is used to shuffle the data.
3. Row shifting: the second row of data is shifted by one, the third row by two, and the last row by three.
4. Column mixing: a predefined matrix is used to multiply by the columns and make a new matrix.
5. Key addition: a key is added to the results.

Another version of AES is AES-256 that divides the plaintext into blocks of 256 bits. A detailed explanation of AES is outside the scope of this book. However, readers should understand that the AES algorithm is very secure, and using current technology, it takes many years to decrypt a ciphertext without having the corresponding keys. It should be mentioned that no encryption method is entirely secure, but if a method takes a very long time to be broken by a threat actor, it would be considered as a secure method.

12.8.1.2 Methods Based on Asymmetric Key

In asymmetric key ciphers, there are two keys: a private key, which is kept for decryption, and a public key that is open to the public. An example of ciphers with asymmetric keys is Rivest–Shamir–Adleman (RSA). The algorithm in RSA is shown in Figure 12.8a.

The following steps need to be performed in order to calculate public and private keys using the RSA algorithm:

1. Choosing two large prime numbers: p and q
2. Calculating $n = p \times q$ and $\varphi = (p-1)(q-1)$
3. Choosing a random number e, where $1 < e < \varphi$
4. Calculating d so that $(e \times d)\ mod\ \varphi = 1$
5. Announcing e and n as public keys, and keeping d as a private key for decryption

To encrypt a plaintext, each character of the plaintext needs to be represented by a decimal number. Then these decimal numbers should be grouped as blocks of numbers. For example, two decimal numbers may be grouped to make a block of 16 bits. If we represent one of these blocks as m_i, the encrypted block in ciphertext, c_i, can be calculated as:

$$c_i = m_i\hat{}e\ \ mod\, n\ \ ; m < n \quad (12.3)$$

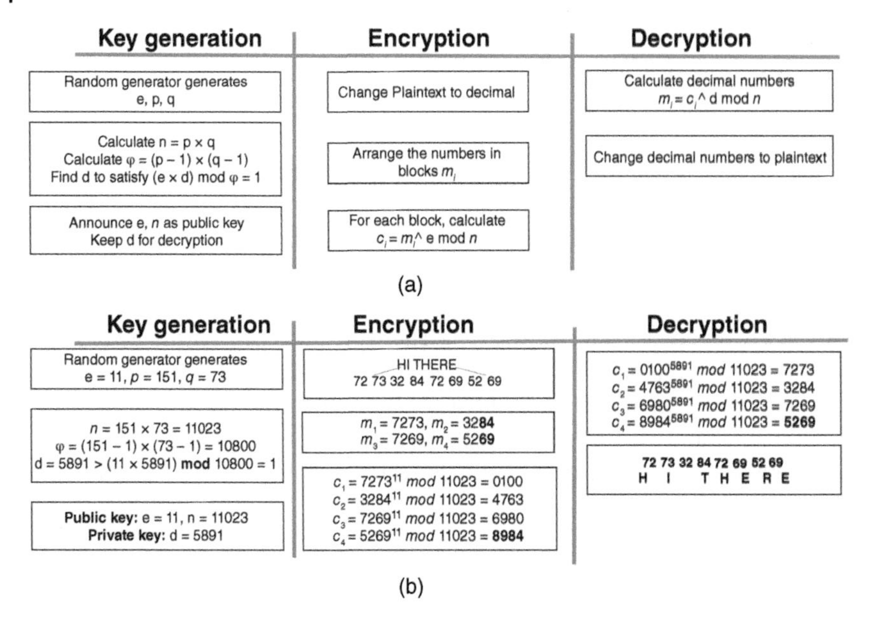

Figure 12.8 RSA algorithm: (a) RSA processes for key generation, encryption, and decryption and (b) a simple example indicating RSA calculations.

To decrypt a block of ciphertext such as c_i, the decimal value of this block in plaintext, m_i, can be calculated as:

$$m_i = c_i\hat{}d \bmod n \tag{12.4}$$

Figure 12.8b shows an example of performing the RSA algorithm.

12.8.2 Integrity Protection

Cryptography methods are used to protect the confidentiality of a plaintext. Integrity ensures that the data has not been changed by unauthorized actors. To provide integrity, plaintext is passed through an algorithm called the hash function. The hash function generates what is called a digest, which is similar to a fingerprint. Secure Hash Algorithm (SHA) is a standard method for performing the hash function. Hash functions always generate fixed-size digests, out of a variable size plaintext. Once the digest is generated, it will be added to the ciphertext as shown in Figure 12.9. To check for the integrity of data during decryption, we need to calculate the digest again, and compare it with the original digest. If the two digests are not the same, it shows that the plaintext is modified.

One of the most widely used hashing algorithms is secure hash algorithm-1 (SHA-1). SHA-1 creates a digest of N bits, where $N = 160$ bits (20 bytes). In SHA-1, data is divided into multiple

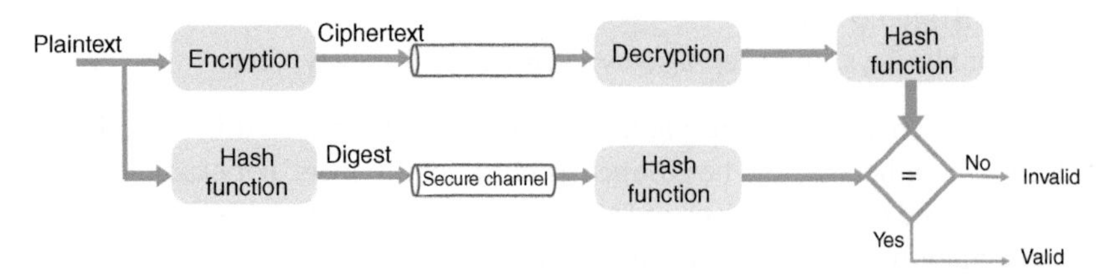

Figure 12.9 Integrity protection using hash function.

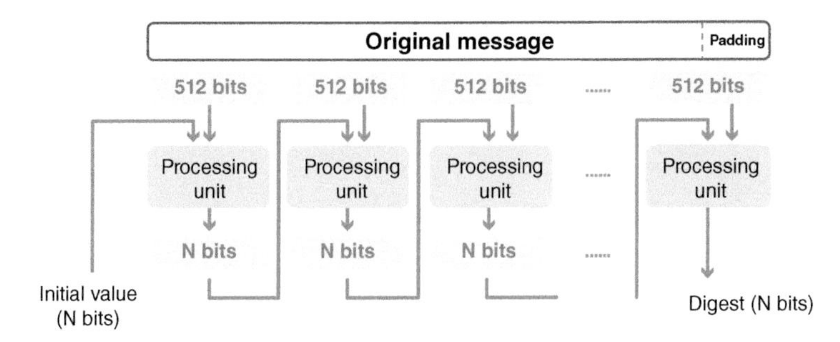

Figure 12.10 SHA-1 hashing algorithm.

blocks of 512 bits. If the last block is not 512 bits, padding is needed to make the last block 512 bits. The block diagram of SHA-1 is shown in Figure 12.10. As can be seen from this figure, the algorithm needs an initial value for initialization, and a predetermined value is used for this purpose. The processing unit is at the heart of the SHA-1 algorithm. The explanation of this unit is outside the scope of this book. Readers can use an online SHA-1 calculator to find the digest generated by using the SHA-1 algorithm. Let us see the digest generated for a simple message such as "Internet of things." The 20-byte digest for this simple message in hexadecimal format is:

Internet of things > 842b08c5b0ea5323782e05d6f4bd5511adb01f93

Now let us change the message slightly, and calculate the digest again. For example, let us change the uppercase I to lowercase in the same message. Therefore, the new message is "internet of things." The 20-byte digest for this message in hexadecimal format is:

Internet of things > 39f6f53ec86e80d2489bc705e10a925aca608ad0

As can be seen, an entirely different digest is generated as a result of a small change in the message.

Another hashing algorithm, SHA-2, is a collection of hashing algorithms including SHA-224, SHA-256, SHA-384, and SHA-512. The three digits shown as part of the names of these algorithms are the length of the hash output. Since the SHA-2 algorithm uses a larger hash size as compared to SHA-1, it has better performance, and it is used for the protection of more sensitive data.

12.8.3 Authentication Protection

As mentioned earlier, the use of username and password is the main method for protection against authentication. The threat actors may try to find out the username or password using different methods. One of these methods is called Brute Force attack, in which a threat actor tries to find login information using trial and error. To do that, the threat actor tries to apply all possible combinations or the most probable ones. In most cases, these attacks are automated in which a software program automatically tries possible credentials in order to login to the system. Increasing password complexity, or setting a limit for the number of login failures, are some of the possible methods to prevent this attack.

There are a few alternative schemes that can be used separately or in addition to username and password to further protect a system in terms of authentication. One of these methods is to ask users to answer security questions in addition to their username and password. An authentication server

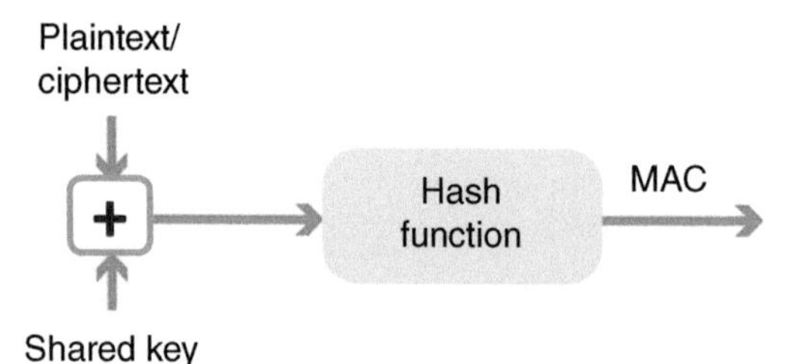

Figure 12.11 Authentication protection.

has several security questions, as well as the answers to the questions. The questions and answers are designed by the user. The system randomly selects a question from a list of questions and does not allow a user to access the system, or data, unless that question has been answered correctly. The use of a security token is another method that can guarantee better authentication. Software and hardware tokens that are not connected to the Internet can be used to generate one-time passwords. In this case, the user usually does not need to provide any username and password, and only uses the password generated by the token. The use of mobile tokens instead of hardware tokens is also possible. In this method, the smartphone is used to generate a one-time password. This eliminates the use of a physical token. However, it may create a potential vulnerability, since the smartphone is connected to the Internet at the time of generating the password. Biometrics can also provide another set of methods for authentication and are more convenient as compared to the use of a physical token. Some of the disadvantages of biometric-oriented methods are their accuracy and cost. Also, the attackers might be interested in finding biometric data from high-resolution photos. For example, if an attacker finds a high-resolution photo of a user showing V-sign by their fingers, the attacker may be able to find biometric fingerprint information of that user by using image processing techniques. Another method for performing authentication is to use notification messages. In this method, the user needs to respond to a notification sent to an email address or mobile device before they are allowed to access the system. Text messaging and SMS are other methods in which a one-time code is sent to the user's mobile device. It should be noted that sending these codes through SMS or text can also create potential risks. However, using these methods in addition to the username and password can make it more difficult for the threat actors to be able to identify themselves as legitimate users of the system. Some of the aforementioned methods are only applicable if a user wants to access an IoT system.

If two IoT systems are exchanging information or performing a transaction, authentication can be provided by using Message Authentication Code (MAC). To provide integrity and authentication, the plaintext and the shared key can be passed through a hash function as shown in Figure 12.11. The hash function produces what is called MAC, which protects against message forgery by an authorized actor who does not know the shared key. MACs can be created using unkeyed hashes and MAC algorithms. The message (plaintext or ciphertext) and MAC can be sent through the same channel. This is not similar to sending a digest that needs to be sent through a secure channel separately. When the combination of the message and MAC arrives at the destination, the message is used to calculate a new MAC using the hash algorithm and the secret shared key. The new MAC is then compared to the received MAC. If these two MACs are not similar, then the message will be discarded.

12.8.4 Non-Repudiation Protection

Cryptography methods using symmetric or asymmetric keys assure that no one in between can see the clear plaintext. Integrity of a plaintext wants to ensure that the plaintext has not been

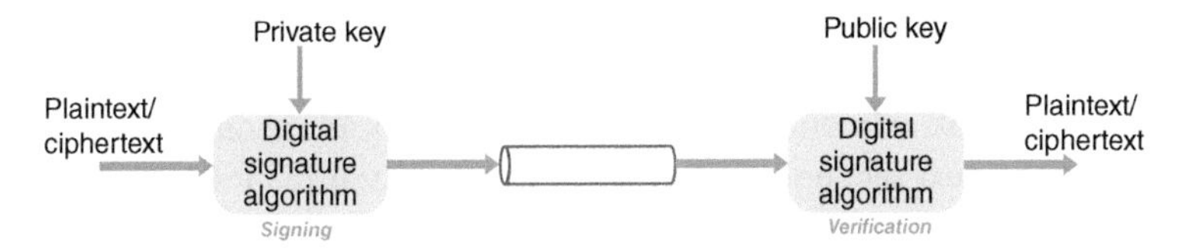

Figure 12.12 Digital signature process.

changed for any reason. Authentication ensures that the sender is the correct one. To protect against non-repudiation, a digital signature can be used. To create a digital signature, a private key is required. However, a digital signature can be verified with the corresponding public key of an asymmetric key-pair. The holder of the private key is the only entity that can create its own signature, and normally anyone knowing the public key can verify it. A digital signature is different from the MAC method explained in the previous section, since the MAC method uses a shared key. With digital signatures, the recipient of a message cannot forge a message or deny the integrity of a transaction. Figure 12.12 shows the processes used for signing and verification, where the signer uses a private key to sign, and the verifier uses a public key to verify the message.

12.9 IoT and Blockchain

Current IoT systems rely on the client-server model, in which an IoT device that plays the role of a client is authenticated by a server located in the cloud. Also, IoT devices are connected through centralized cloud servers for processing and storage. As the number of IoT devices grows, the client-server model will not be efficient for the IoT ecosystem anymore. As massive numbers of IoT devices send their traffic to centralized servers in the cloud, or receive information from them, each of these servers becomes a bottleneck of the network. This limits the maximum achievable data rate in a client-server model, especially when more and more IoT devices start to communicate with these servers, which are located in centralized data centers. It should be mentioned that the installation and maintenance of large centralized data centers are costly, and it would be interesting to distribute computing and storage among massive numbers of IoT devices. Also, centralized servers and data centers are points of failure in a network. Since the threat actors know that data is stored in a centralized location, they try to find a vulnerability in the system that allows them to access data. While the client-server model has been around for decades and can be used to support small-scale IoT networks, it will not be capable of providing the requirements of the IoT ecosystem in the future.

A decentralized approach can be considered as a solution to the problems created by the client-server architecture for IoT systems. Distributing computation and storage requirements across massive numbers of devices that form an IoT network can solve the problems of the centralized model. It definitely increases the data rate, since IoT devices do not need to go to centralized locations for storage and computing. It can distribute traffic to other IoT devices that can provide the same services in terms of computing and storage. The decentralized model also prevents situations in which a failure in a single node of the network can shut down the entire network.

Let us have an example that shows the difference between a centralized and decentralized model. For years, banks were the main financial institutions that played the role of servers. People go to the bank for all their financial needs. The architecture of the banking system is based on a client-server model. If the number of customers increases, the number of banks should also increase, which is very costly. On the other hand, as more and more customers need to use the services provided by the banking system, the line-up in the banks grows and the speed of receiving the services decreases. Thieves know that there are huge amounts of money in a bank (centralized location) and they are very motivated to rob the bank. In the next section, you can see how a decentralized model using Blockchain technology is different than the centralized model for this example.

It should be mentioned that establishing a decentralized model has its own set of challenges. One of the most important issues is security. The IoT ecosystem is extremely diverse and comprises devices with very different computing and storage capabilities. For example, an IoT system with high computing power can execute advanced encryption methods, whereas another IoT system with low computing power cannot perform complex encryption algorithms.

Blockchain technology provides an excellent solution to security and privacy concerns in the IoT ecosystem. It uses a decentralized solution and enables secure processing of transactions and coordination among IoT devices. We will discuss blockchain technology in the next section.

12.9.1 Blockchain Technology

Blockchain is based on the concept of a distributed database. This database maintains a constantly growing set of records. The data records are identified as blocks. The blocks cannot be deleted easily and therefore, the database is growing constantly. The blocks are chained to each other. The chained blocks of data are not processed or stored inside centralized servers and instead are distributed among many nodes in the network. These nodes have a copy of the chain. As more and more data is added to the database, the size of the chain will be increased. A blockchain has two main elements. One is called transactions, which are actions made by users or systems. The other one is called a block, which stores a set of transactions in an efficient way. Blocks should be recorded in a correct sequence in such a way that they cannot be modified or tampered with in the future. Since the blockchain is based on a decentralized model, there is no single entity that can approve a transaction. Instead, the participating nodes have to reach an agreement to accept and validate a transaction. Therefore, there must be a sense of trust among the nodes of the network. Once a transaction is accepted, it is chained to the previous records. As we mentioned earlier, previous transactions cannot be easily changed. A set of approved transactions makes a block. When a block is formed, the block will be sent to all nodes in the network to be validated. Each successive block contains a fingerprint of the previous block calculated using a hash algorithm.

Figure 12.13 shows a simple example that consists of four blocks. The first block is called the genesis block, which does not have any predecessor. Each block has the hash value of the previous block, in addition to its own hash value. The fact that each block contains the hash of its previous block makes the blockchain very secure.

If a threat actor somehow is able to change the data in Block 2 in this example, the value of Hash#2 changes. Since Block 3 has the old hash value of Block 2, then Block 3 and the succeeding blocks become invalid. Changing all hash values is a time-consuming task, especially if the chain is long. Now, assume that the threat actor tampers with the data in one block such as Block 2, and is willing to calculate the hashes of all other blocks to make blockchain valid again. This can create

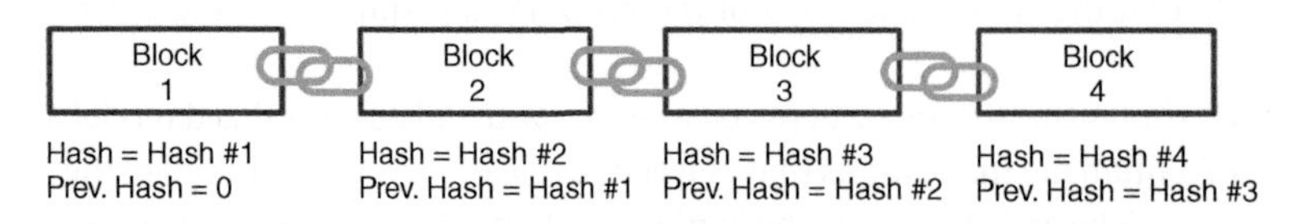

Figure 12.13 Four-block blockchain link list structure.

a security problem and needs to be fixed. To solve this problem, blockchain introduces a concept called proof-of-work.

A proof-of-work is a computational problem that requires a certain amount of effort to be solved. Therefore, it takes some time to solve a computational problem. The amount of time and effort that it takes to verify the results of a computational problem is substantially lower as compared to the amount of time and effort that is required to solve the same computational problem. Performing a proof-of-work may take several minutes, and creating a new block requires the proof-of-work. If the threat actor in our example tampers with the data in Block 2, it needs to perform proof-of-work, which is a very time-consuming process by itself, and then it needs to make changes to all other succeeding blocks. Therefore, hashing and proof-of-work make blockchain very secure.

The readers should not forget that blockchain does not have a centralized architecture to manage the chain. When a new block is created, the block is sent to all other devices or users on the network. Each device or user requires to verify the block and ensure that it is correct. There must be a consensus among these devices or users about the validity of the block. Therefore, to make changes to a block, three tasks need to be performed:

1. Changing the hash value of the block and all other succeeding blocks in the chain
2. Performing proof-of-work
3. Getting consensus of a majority of the nodes in the distributed network

Therefore, blockchain technology is robust and the data stored using blockchain technology cannot be modified and tampered with by threat actors. That is why the technology has been used in financial services, and it is the building block of Bitcoin cryptocurrency, which provides a model for peer-to-peer payment services without the need for a centralized center or a third party.

In the IoT ecosystem, the blockchain can establish secure data transfer between IoT nodes. Blockchain technology can also add transparency to the system. Every participant node in the network can see the blocks and the list of transactions stored in each block. It goes without saying that if a node wants to read the content of a transaction, it must have the private key associated with it.

The need for high processing power in order to perform encryption, and to calculate hash values, as well as the need for large storage capacity to keep constantly growing blocks are the main problems with using blockchain for some IoT applications. Many IoT systems have limited capabilities in terms of processing power and storage, or require fast data transfer, which makes the use of blockchain very challenging.

It should be noted that blockchain technology can be categorized into two main types: public and private blockchain. The difference between a public and private blockchain is related to who is permitted to participate in the network, maintains the ledgers, and executes the consensus algorithm. In a public blockchain, everyone is allowed to join the network, and therefore, this type of blockchain requires a large amount of computational power to maintain the distributed ledger. Also, in order to achieve consensus, each node should perform proof-of-work. As mentioned earlier, proof-of-work is a time-consuming, resource-intensive, and complex problem. A private blockchain is based on invitations that can be validated by the network starter. It has restrictions on who is permitted to join the network. This means that a node requires to receive an invitation to participate in the activities. Different private blockchain networks may use different invitation strategies. A private blockchain network, also called a managed or permissioned blockchain, needs less processing time per transaction, since there are fewer nodes in the network, and it uses simpler consensus algorithms. There are many challenges in using public blockchain for resource-constraint IoT devices. However, private blockchain is a popular technology, and it is used in industrial IoT [6].

12.9.2 A Practical Example of IoT and Blockchain for Smart Grid

The traditional method for generation and distribution of energy, used by utility companies, was to generate energy in centralized locations such as power generation facilities, and distribute the energy to consumers using transmission and distribution lines. Traditional grid systems were based on unidirectional distribution of energy where a utility company provided the energy, and the residential and industrial customers consumed the energy. In other words, utilities would sell the energy, and consumers would buy the energy.

Smart grid architecture is far more complex than the traditional grid, and the energy can move in a bidirectional way. Small companies or even houses can generate energy for their consumption, and can sell their excess energy to the grid or their neighbors. The term prosumer (producer and consumer) can be used to name an entity that both consumes and produces. For example, a house may have a solar roof that generates energy, and the owner of the house may like to sell its excess energy. One may have a charged electric car and decide to sell the energy during peak times to the grid, and re-charge the car's battery when the price of energy is cheaper. We would like to discuss the advantages that bidirectional movement of energy can provide to both the utility companies and the consumers:

1. The traditional grid was based on a centralized method where energy was generated in centralized locations, while the smart grid uses a distributed energy generation scheme. In this case, a house is capable of producing and consuming energy, or may want to sell its excess energy.
2. Centralized power generators are usually far from the consumers, and during transmission of energy between a central location and the location of a consumer, a huge amount of energy would be lost. However, if the energy is generated at the consumer's location, there will not be any energy loss.
3. In order to provide the demand during peak hours, a utility company needs to deploy additional infrastructure or use some of its old, less efficient plants during its peak times. If a consumer can generate energy during this period, and sell the energy to the grid, the utility companies do not need to deploy additional infrastructure or use their older plants. This creates a win-win scenario for both the utility companies and the energy consumers.
4. Some small companies may start to generate energy and become independent power producers. These companies can compete with the large utility companies, provide energy to consumers, or sell their energy to the larger utilities.
5. In current smart grid systems, a house can sell its excess energy to the grid, but it cannot sell its energy to other consumers. Generally speaking, the grid would be smarter if there is the possibility that a house that has excess energy can trade its energy with other houses or consumers as well.
6. To understand the supply and demand of energy, and ensure a reliable distribution of electricity on the power grid, utilities have used software platforms such as Energy Management Systems (EMS) and Advanced Distribution Management Systems (ADMS). Simply put, EMS tries to control, monitor, and optimize the performance of the generation or transmission of a smart grid system. ADMS software is an integrated platform that supports distribution management and optimization, including outage restoration and management and the distribution grid performance optimization. However, these platforms cannot be scaled to consider the transaction between houses or the many IoT devices that are installed in these houses.

Trading energy between a house that has excess energy and another house in a neighborhood that needs energy in a peer-to-peer manner should be established in a secure manner. There is no utility company involved in this trade to distribute the energy, measure the usage, and charge the consumer. Therefore, there must be a sense of trust among these smart meters in order to trade energy. The distributed trading of energy is not possible unless there are methods in place to verify and ensure that each device is doing exactly what it is supposed to do. Since the utility companies are not involved in this trade, everything requires to be done without a need for a central control. This is where blockchain comes in. For example, two houses in a neighborhood can connect via a blockchain in such a way that one can sell its excess energy, generated by its solar roof, to another house that needs energy. To accomplish this task, we can establish two blockchain transactions: one for the amount of generated and transmitted energy, and one for a payment.

The question here is how two neighbors can trade energy without involving a central entity. If a central entity is involved, a neighbor (let us say, Neighbor A) could send the money to the central entity. When the energy is provided by the other neighbor (let us say, Neighbor B), the central entity would pay the money to Neighbor B. If Neighbor A does not receive the energy for any reason, the central entity sends the money back to Neighbor A. In this case, both neighbors trust the central entity. However, it is not practical to have a central secure architecture for trading power among a massive number of neighbors.

We can use smart contracts and blockchain to accomplish this task. Generally speaking, smart contracts are like real contracts, but in a digital format. There are tiny programs that are stored inside a blockchain. A smart contract can be configured to perform the trading, when certain criteria are met. For example, Neighbor A has a term to buy energy at a certain price. Neighbor B also may have a term for the selling price of its energy. When all the terms of a smart contract are met, then the contract will be executed. Since smart contracts are inside the blockchain, they have a distributed architecture, and they do not need a central entity for their operations. When a smart contract is created, the context of the contract cannot be changed. As we explained earlier, using blockchain technology, it is almost impossible to tamper with the contract. The output of the contract is validated by everyone in the network, and no one can force anyone to change the contract.

By using IoT technology, there is a good understanding of real-time energy generation and demand, and by using blockchain, a market can be securely set up to trade energy in a peer-to-peer manner between two neighbors.

12.10 Summary

It is important to make IoT systems secure. Threat actors are interested in launching their malicious attacks against IoT systems to access the collected data by these devices. They may also look at the possibility to take advantage of insecure IoT systems to launch their attacks against connected systems to an IoT system. Since there are massive numbers of IoT devices in the world, the threat actors are also interested in getting unauthorized access to a large group of IoT systems and ordering this group to initiate an attack against other systems. On the other hand, many IoT devices have limitations in terms of power consumption, processing power, or storage. These limitations do not allow systems to run sophisticated protection schemes for security. Also, many designers, manufacturers, or installers of IoT systems do not take the security of IoT systems very seriously. For these reasons, IoT systems have the potential to become victims of attacks.

The threat actors may want to get access to an IoT system (threat against authentication), find content of its data (threat against confidentiality), modify its data (threat against integrity), deny the validity of data (threat against non-repudiation), or disrupt the operation of an IoT system (threat against availability).

In this chapter, we discussed the vulnerabilities of an IoT system in terms of insufficient authentication, insecure ports and interfaces, lack of secure update mechanism, insufficient encryption, insecure network connectivity, lack of utilizing whitelists, insecure IoT chips, and lack of using proper configurations.

We explained the concept of threat modeling and risk analysis for prioritizing threats and calculating the risks associated with each threat in an IoT system. To perform threat modeling, the physical and data assets of an IoT system need to be identified. The potential threats need to be determined and the flow of messages among data assets within an IoT system needs to be identified. The threats must then be rated and the risks associated with each threat calculated. Threat modeling helps decision makers in determining how to handle threats against IoT systems.

To protect IoT systems, we need to use protection schemes to secure the systems regarding threats against confidentiality, integrity, authentication, non-repudiation, and availability. We briefly discussed some of the existing protection methods to give readers some basic understanding of these protection schemes.

The concept of blockchain was briefly presented in the last section of this chapter. Blockchain is based on the idea of distributed databases. The integration of IoT and blockchain creates lots of new opportunities such as peer-peer energy trading in a smart grade system as discussed in this chapter.

References

1 Bodeau, D.J., McCollum, C.D. and Fox, D.B. (2018). Cyber Threat Modeling: Survey, Assessment, and Representative Framework.

2 Department of Homeland Security (2016). Strategic Principles for Securing the Internet of Things (IoT). https://www.dhs.gov/sites/default/files/publications/ Strategic_Principles_for_Securing_the_Internet_of_Things-2016-1115-FINAL....pdf (accessed 30 June 2022).

3 116th Congress Public Law 207 (2022). https://www.congress.gov/bill/116th-congress/house-bill/1668/text (accessed 30 June 2022).

4 Buchmann, J.A. (2001). *DES- Introduction to cryptography*, vol. 5, 119–120. New York, NY: Springer. ISBN 978-0-387-95034-1.

5 ISO/IEC 18033-3 standard (2016, 2010). Information technology—Security techniques—Encryption algorithms—Part 3: Block ciphers.

6 Assaqty, M. I. S. et al. (2020). Private-blockchain-based industrial iot for material and product tracking in smart manufacturing. *IEEE Network.* 34 (5), doi: https://doi.org/10.1109/MNET.011.1900537.

Exercises

12.1 Smart toy manufacturing companies produce toys that may record the audio and video data and send the collected data to the Internet.

a. Give examples of data that a smart toy manufacturer may collect from its consumers.

b. Give an example of a regulation that a toy company should consider when designing an IoT-based toy for children.

c. In the case of a Barbie doll, there is a feature built into the doll that requires a child to push a button on the belt of the doll to activate the session before the start of a conversation with the doll. Why do you think that the manufacturer has developed this feature?

d. Provide a couple of backend security protection schemes associated with the services provided by smart toys that transmit data to the cloud?

12.2 What is thread modeling?

12.3 Explain the Brute Force attack for threat against authentication. How can the system be protected against this attack?

12.4 What is the port scanning threat?

12.5 What is MITM attack?

12.6 Explain what a botnet is.

12.7 Explain what does the term "social engineering" mean in the context of cybersecurity.

12.8 Give examples of some protection methods that can make software updates more secure.

12.9 The National Institute of Standards does not suggest the use of biometrics as the only authentication method for getting access to a system. Do you think their recommendation is correct? Why?

12.10 An automobile insurance company has a product that collects data regarding the driving habits of its customers. By collecting and analyzing the data, the insurance company can understand how the driver drives the car, the distance, speed, or even the amount of force the driver applies to the brake pedal to stop the car. The insurance company intends to use the data to customize insurance premiums. The product can be installed into the on-board diagnostic system on a car that can give the insurance company potential access to more information. Do you think that the customer should have privacy concerns for this product?

12.11 A random generator generates the following numbers ($p = 97$, $q = 83$, and $e = 13$) to initialize the RSA algorithm. If you want to encrypt the word CIoT:

a. What are the values of public keys and private keys?

b. Using the public keys encrypt the word CIoT.

c. Using the private key calculated in part (a), decrypt the results in part (b). (Hint: $C = 67$, $I = 73$, $O = 79$, $T = 84$)

12.12 What is Denial of Sleep Attack (DoSA)? Give a simple example of launching this attack using a fake IoT device.

12.13 Figure 12.6a shows the building block of DES cipher.

a. What type of P-Box (straight, expansion, or compression) is used in the DES cipher as shown in Figure 12.6a? Explain your reasoning.

b. How does a P-Box work? Does P-Box need a key to operate?

c. Each of the sixteen rounds of the DES cipher (Figure 12.6b) consists of two P-Boxes and one S-Box. What type are these P-Boxes and the S-Box (straight, expansion, or compression)?
d. Due to the loss of bits in a compression S-Box or P-Box, is it possible to use them in a DES cipher?

12.14 Can two different inputs that are applied to a hash function generate the same hash output?

12.15 Can hash function be used for passwords? Do you think there is any advantage in hashing passwords?

12.16 Assume that there are two ways for an authorized individual to get information about a marriage that happened several years ago. In the first method, the individual goes to the register office to get the information. In the second method, the individual asks people who attended the wedding of the married couple.
a. Which method uses a centralized approach and which one uses a decentralized approach in acquiring the information?
b. In this example, what security problems can happen in situations that a centralized method is used?
c. How does a decentralized method work for this example? Compare the centralized and decentralized methods in terms of security.
d. An IoT device needs to obtain some data. Similar to the example in this question, explain how the IoT device can use centralized or decentralized methods to obtain data.

12.17 What is de-identified information?

12.18 Give an example of using IoT and blockchain in supply-chain applications.

12.19 In symmetric key encryption techniques, the data is encrypted using a key. To decrypt the data, the same key needs to be used. Therefore, the encryption process and the decryption process need to share the key. A method for sharing a symmetric key is called Diffie–Hellman. In this method, assume that g and p are two public numbers, and therefore, they are available to both encryption and decryption processes. The encryption process (process A) chooses a large random number, x, and calculates $R_1 = g^x \ mod \ p$. The decryption process (process B) also chooses a large random number, y, and calculates $R_2 = g^y \ mod \ p$. Then the two processes exchange R_1 and R_2 publicly. On the reception of R_2, process A computes the symmetric key, k, by calculating $k = R_2{}^x \ mod \ p$. Process B performs a similar operation and finds the symmetric key by calculating $k = R_1{}^y \ mod \ p$. In the method, both processes have found the shared symmetric key, k. However, process A does not know y and process B does not know x. By knowing the values of R_1 and R_2, it takes years for a hacker to find the shared key. Assume $g = 7$ and $p = 37$. Find R_1 and R_2, and calculate the shared symmetric key. (Assume $x = 5$ and $y = 8$. Generally, $\underline{x}$ and y are large random numbers. For simplicity, small numbers are considered for this question.)

12.20 Physically Unclonable Functions (PUFs) are a class of hardware security solution for authentication of IoT devices. PUFs are the fingerprint of electronic devices, and they can generate unique identifications for these devices. For example, two identical IoT devices usually have small electrical differences at the chip level due to variations in silicon as well as the manufacturing process. A simple, low-cost hardware can use these small differences and generate a unique output that represents an IoT device. During the authentication phase, a server sends challenge questions to the IoT device and the device responds with an answer to the questions. Explain how PUFs can be beneficial in authentication for IoT devices.

12.21 A replay attack is a form of attack in which the threat actor can capture data during its transmission and maliciously repeat the message at a later time or delay the transmission of the message. For launching a replay attack, the hacker does not require to have state-of-the-art knowledge and sophisticated skills to decrypt a message after the message is captured from the network.

a. Do you think using a sophisticated encryption scheme can be a good protection strategy against replay attacks?
b. Can a digital signature be considered a protection strategy against this attack?
c. A threat actor may capture a message issued to a protection relay indicating a fault in part of the smart grid system. When the message is received by the protection relay, the relay trips a circuit breaker after milliseconds to avoid damaging the smart grid equipment. Explain how a replay attack can be damaging for smart grid applications.
d. What method do you think could be used to protect systems against this attack?

12.22 Blacklists and whitelists might be used to make systems more secure.

a. What is the difference between a blacklist and a whitelist?
b. An IoT gateway (central node) uses BLE technology and always communicates with three BLE devices (peripheral nodes), which have the following three Media Access Control (MAC) addresses.

MAC1 = "453115800700"

MAC2 = "8bc6c2800700," MAC3 = "92c6c2800700."

Assume that the central BLE (IoT gateway) has an MAC address of "25136c230900." Explain how you can make this system more secure.

12.23 Typical power ratings of some electric devices are shown in Table 12.E23. Assume that these devices are smart, and that they can be controlled remotely. With an example explain how a DDoS attack can be executed to disrupt the entire power grid in an area by utilizing these smart devices as part of the attack.

Table 12.E23 Examples of smart devices and their typical power ratings.

Smart device	Watts
Hair dryer	900
Heater	1500
Air conditioner	3000
Electric oven	4000
Water heater	4500

12.24 An IoT device does not have the capability to generate large random numbers. But it can generate small random numbers. Do you think this creates security issues for this IoT device?

12.25 A constraint IoT device (MQTT client A), which is a thermostat, sends its messages to an MQTT broker. Another IoT device (MQTT client B) is subscribed to the MQTT broker to receive the messages published by MQTT client A. Assume MQTT client B is an air conditioning unit.

a. How can you make the communication in this system secure?
b. Why do we need to make this system secure?
c. Do you think a very secure MQTT is possible for constraint IoT devices such as MQTT client A?
d. Conduct research and find out how a sense of trust can be implemented for this application. You could use Appendix B to help you answer this question. (Explain building trust by creating a private certificate authority.)

Chapter 13 IoT Solution Developments

13.1 Introduction

Many companies which are exploring the new opportunities for their organizations see the potential benefit of connected products and identify building Internet of Things (IoT) solutions as part of their strategic plans.

When a company intends to perform an IoT implementation, it requires a systematic approach for developing an IoT solution for its organization. Even though there are similarities between the high-level development architecture of non-IoT solutions and IoT solutions, some differences exist due to distinct characteristics of IoT systems. For example, scalability is an important factor in the design of most IoT solutions, while it might not be an important factor in many non-IoT solutions. There is a high level of complexity and risk when considering scalability in an IoT solution, and the degree of scaling an IoT system might require. An IoT solution involves sensors, IoT devices, connectivity schemes, cloud architecture, computation, analytics, and security. Therefore, developing an end-to-end solution involves a high level of complexity and specific expertise. There is not much information available regarding the implementation of enterprise IoT solutions, their successes and failures, and their lessons learnt, in order to assist with the development of new IoT solutions. Also, there is a level of confusion in the implementation of IoT-based systems for companies that are forced to use IoT to stay competitive, but are not clear on how to utilize and take advantage of IoT for their organizations.

One of the most valuable outcomes of performing an IoT solution development is the collected data. An organization may collect data to improve its performance, increase its revenues, reduce its costs, or provide better service to its customers. However, there might be many other organizations that may be willing to pay for access to the collected data. In some cases, the value of the collected data by an organization might be worth even more than the collective value of all other assets of the organization! Therefore, it is always important to know who owns the collected data. Some organizations might give permission to vendors to own the data or even share their data publicly.

In this chapter, we discuss a nine-step process as a framework for high level IoT solution development. These nine steps are IoT business case development, strategy implementation, detailed design, building configuration and testing, pilot implementation, regulation acceptance, deployment, sustainment, and continuous improvements.

Fundamentals of Internet of Things: For Students and Professionals, First Edition. F. John Dian.

13.2 IoT Solution Development Methodology

There are many organizations which have realized the benefits that IoT can provide them, and intend to start their IoT implementation journey. These companies need to have a clear approach on how an IoT solution should be developed, and understand the risks involved in this process. A five-step model for IoT solution development is discussed in [1]. However, this model is not detailed enough to be used as a practical guideline for IoT solution development. Here, we discuss a nine-step approach that is suitable for most IoT solutions as shown in Figure 13.1.

Besides these steps, there are three important considerations that need to be factored in when performing these nine steps. These considerations include security and privacy, Change Management (CM), as well as business case updates and benefit realizations. In this section, we briefly discuss these nine steps and three considerations in order to give readers a general idea of our methodology. We will then discuss each of these steps in more detail in the next section. We explained the security and privacy issues related to IoT applications in Chapter 12. Those concepts should be considered in all steps of our methodology during IoT solution development. We discuss CM in Section 13.4 of this chapter.

An IoT business case indicates the potential benefits that an IoT solution can bring for an organization in terms of factors such as financial benefits, operation efficiency, economic growth, customer convenience, or operational safety. An IoT business case must be prepared by a team of experts and requires sufficient collaboration among different divisions across an organization. The business case must then be approved by the executives of the organization and sometimes the board of directors. Preparing a business case document is the first step toward building an IoT solution.

After the business case is approved by business leaders, the organization needs to find an appropriate strategy for implementation. The organization should determine whether it has the required expertise to develop the IoT solution in-house or not. Most of the time, organizations do not have the required skill sets to implement an end-to-end IoT solution, due to the fact that IoT requires a broad range of skill sets. Therefore, organizations should usually choose between finding an external partner to fully implement the solution, or a combination of an external partner and their internal expert team. It should be noted that vendor evaluation and selection is an important part of this step.

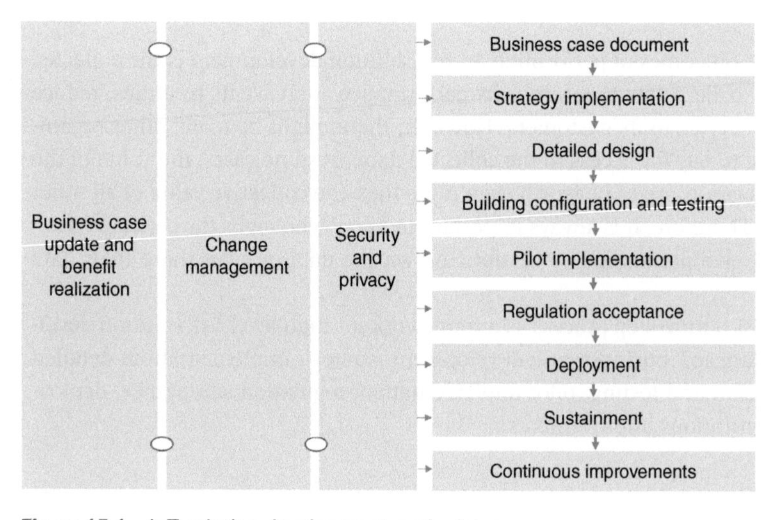

Figure 13.1 IoT solution development methodology.

The organization, with the help of the external partner, should then work on a detailed design, which is the third step in our methodology. The detailed design shows the processes, technologies, and integration points among various systems that need to be built, as well as a detailed implementation plan with options which show how the IoT solution would be deployed.

After the detailed design is complete, the IoT solution needs to be built, and the organization should perform the configuration and system integration testing, as well as the user acceptance testing.

The next step is pilot implementation. This step is needed to test the project design on a limited scale in order to lower deployment risk. It also validates the benefits of the IoT solution before deployment. The outcomes of this step can be used to measure customer adoption and satisfaction, and to enhance the IoT solution before deployment.

Since different industries have their own set of regulations, there is a need for IoT applications to apply for the required certifications or perform additional testing to satisfy the requirements of the regulatory bodies. Different geographical areas or countries may have different regulations, and therefore, it is important to have regulation acceptance before going to the market and starting the deployment phase.

After a successful pilot implementation and the acquisition of all required regulatory certifications, the IoT solution will be completely deployed to all customers. It is important to encourage, educate, and train customers to understand and take advantage of the deployed IoT solution's benefits. Scalability is an important part of this step since we need to scale the solution to all existing customers. If scalability has been addressed properly, we should be able to scale the solution even if the number of customers increases drastically.

Deployment of an IoT solution is not the end of the road. After the solution is deployed on a large-scale basis, there is a need to support the solution in order to remove minor bugs and provide help desk support for the customers of the IoT solution.

The last step is continuous improvements. To improve the efficiency of a deployed IoT solution, there is a need for an ongoing effort to enhance the systems, processes, and services of an IoT solution. This can be accomplished by identifying the improvement opportunities, and working toward a plan to implement the necessary actions for solution enhancements. This is especially important for IoT solutions in which vast amounts of data are collected. As time passes and more data is available, the organization may be able to use the collected data for solution enhancements.

As an organization is going through this journey, there are three important concepts that must be considered at all times. We have shown each of these considerations as a vertical block in Figure 13.1. In each step of the way, as the organization goes through different steps of the solution, each of these considerations must always be taken into account.

The first consideration is related to security and privacy, which are important topics in IoT solution development and must be considered from the start to the end of a solution. The second consideration is related to CM. Each IoT solution is unique and should be customized for the organization which is implementing it. Therefore, managing the changes that the organization needs to perform in order to have a successful solution deployment becomes an important task. The third consideration is related to business case document updates. It should be clear that a business case is a living document and as an organization goes through different steps of an IoT solution, it needs to be updated accordingly. As an organization finishes the deployment of an IoT solution, the organization should create the benefit realization, that is also part of and related to the business case document.

Let us explain our IoT solution development strategy with an example. Assume Company A, which is making regular tennis racquets, has decided to manufacture smart racquets, if they are more profitable. Company A partners with Company B, an IoT chip designer, to design a specific IoT chip for its product. Company A has enough expertise in-house to design an electronic board using the required sensors and an IoT chip. It then installs the electronic board on the racquets. To make smart racquets, Company A should go through the nine-step solution development process explained earlier. However, in this example, Company A is not making an end-to-end IoT solution, and its operations are only related to the manufacturing of IoT-based racquets. Now assume Company A intends to make an end-to-end IoT solution for this scenario. Company A partners with Company C, an application provider, as well as a cloud provider and a cellular connectivity provider, to create a solution that can be used by the users of smart racquets. The collected data from this solution can be used by the manufacturer to improve the quality of its product, by an athlete for training purposes, by a coach to identify performance patterns, and by a team of physicians to predict and prevent injuries. To be able to build this IoT solution, Company A should go through the nine-step sequential methodology and pay attention to the three considerations at each step along the way.

13.3 Further Details on IoT Solution Development

We discuss the IoT solution development methodology in more detail in this section.

13.3.1 Business Case Document

An IoT business case provides justification for the deployment of an IoT solution. It involves cost–benefit analysis, Return of Investment (RoI) calculation, assumptions regarding the impact of the solution, and the estimated timeline for the deployment of the solution. In other words, it evaluates the benefit, cost, and risk associated with the alternative options for performing the solution and provides the reasoning for the best possible solution.

The organization should not underestimate the time that it takes to go through each step in their journey. There are many technical and non-technical issues that may prolong the solution development process, and those issues should be taken into account when preparing a business case. Since a business case is based on estimated costs and revenues, the assumptions for the estimation must be clearly stated in the business case, and the organization should plan contingencies. A business case is a living document and must be updated along the way. The organization should prepare the assumptions and estimates for the business case by using good assessments. Let us consider the scenario that we explained earlier, in which a company intends to prepare a business case to see whether manufacturing smart tennis racquets is better than making traditional ones. In order to determine how many potential customers would purchase smart racquets, a market assessment should be conducted. By incorporating key market sizing factors such as geographical outreach, customer age, level of tennis expertise, and budget, a target demographic can be established from which a select estimated percentage of customers will purchase the product. However, an inaccurate target demographic can be calculated as a result of wrong estimates, flawed sampling, and inaccurate market sizing measurements. Determining a target demographic is not as simple of a process as it might seem, and without careful calculations, flawed market sizing can lead to incorrectly estimating current and future benefits.

13.3.2 Implementation Strategy

In this step, the organization should decide which components of a solution can be done in-house and which components require external partners. As explained earlier, most IoT solutions need external partners, due to the diverse range of expertise required for developing an IoT solution. Also, a solution might be built completely in-house, or if available, the organization might be able to integrate some existing solution products in order to build its own IoT solution. Utilizing the available Commercial Off-The-Shelf (COTS) products and infrastructure has many advantages. First of all, it shortens the IoT solution development timeline, since the product already exists. Secondly, most COTS products are designed by experts who have good knowledge of the solution. Thirdly, they are usually built with security and scalability in mind. And finally, they are mostly designed based on industry standards. However, an organization might find out that there is no COTS product available or that the existing ones do not support the required features and standards.

It is clear many companies that were pioneers in building IoT solutions for their organizations started building the solution from scratch. The main reason was the fact that there were not many COTS products available at the time. Since more COTS products exist today, most companies and their partners rely on utilizing these products as part of their design.

It is important to understand that if an IoT solution uses a specific COTS product from a specific vendor, the organization has limited itself to the features that the product is offering. It is also important to clearly understand the cost of the licenses, maintenance, and upgrades as these costs play vital roles in the overall cost of an IoT solution.

There are many reasons for an organization to select one vendor over another vendor. For example, IoT cloud providers are popular vendors, since they can offer many components of a solution such as processing power, storage, databases, computing, development environment, operating systems, and even applications. A vendor who has a well-established partner network can definitely help to implement an IoT solution more effectively. If there exists a good partnership among a cloud provider, a communications provider, and a hardware vendor, then you can expect a smooth development for the IoT solution.

13.3.3 Detailed Design

Detailed design is the third step in our methodology. In this step the design is clarified, and specifications for the solution are created. The output of this phase is the processes, the technologies, and a detailed implementation plan. It also shows the design of interfaces and integration points among various systems. After the detailed design is complete, the organization can have a very good estimate of the cost and the architecture of the solution. In other words, the functionality of various components of an IoT solution becomes defined and detailed solutions to carry out those functions are designed in this phase.

Let us bring an example to demonstrate the outputs of a detailed design. A transportation company has a large fleet of buses and intends to put a variety of sensors inside each bus to measure many quantities including the location of the bus, engine data, and the number of passengers who gets on and off the bus in each station. The information from the sensors is transmitted to a gateway inside each bus and then sent to the cloud using a cellular connection. In this case, the detailed design includes the following:

- The specifications of sensors and gateways.
- The protocols for transmitting data to gateways, and from gateways to the cloud.
- The data structure of data between the sensors and the gateway, as well the data structure between the gateway and the cloud.
- The processing of data inside each gateway.
- The batch or stream processing inside the cloud.
- The specifications of the database and database management system that need to be used.

- A detailed flow chart of all the functionalities of the services and processes, or systems, that will use the data.
- The format of data that needs to be visualized or sent to user interface applications.
- The details of data flow at the interfaces and integration points where different systems, or different processes, connect to each other.

The outcome of the detailed design is used in the next step to practically build and test the solution. A more thoughtful detailed design makes the build, configuration, and testing step to be smoother.

13.3.4 Building, Configuration, and Testing (BCT)

The conventional approach to build a solution was to start with crafting a perfect detailed design for the solution. Upon the completion of the design, the organization would move to build the solution based on the provided detailed design. Even though the conventional method is straightforward, it might not be the best approach for solutions that may need more flexibility in terms of their ability to easily add new features or update existing ones.

Generally speaking, there are two main methodologies for the solution development. The first is called Waterfall and the second is called Agile methodology.

Waterfall methodology is based on a sequential approach in which the organization divides the work into several distinct phases. Each phase should be built in sequence, and therefore, the organization should finish building the first phase before starting the second phase. There is often a gate between these phases. The work does not continue until it satisfies the conditions defined as gating criteria. For example, the organization should sign off a phase, before the next phase can start. The methodology in Figure 13.1 is an example of a Waterfall approach. With the Waterfall approach, it would be troublesome if a problem reveals itself at its final phases, which requires modifications to previous phases of this sequential approach. Changing and updating previous phases is time consuming and due to the lack of flexibility in the architecture, the organization may fall behind in building the solution. The Waterfall approach is easy to understand and suits many solutions. However, due to the lack of flexibility, it is not suited for software developments or applications with high levels of adjustments.

As previously stated, the methodology in Figure 13.1 is an end-to-end solution development based on the Waterfall approach. We can even think of implementing each step of this solution using the Waterfall approach. Let us look at a simple example of building an IoT system based on the Waterfall methodology. A team of experienced engineers with solid knowledge in sensors, electronics, and hardware builds an IoT device. Once the hardware is installed on the IoT device and satisfies the design requirements, another team builds the communication protocol among IoT devices and the cloud. When the communication with the cloud is completed, reviewed, and verified, we can move to the next phase and start building a part of the processing in the cloud. As you can see, Waterfall creates silos in which a team of developers focuses on building specific functions without a need to communicate with other developers. The approach does not either focus on IoT users or testing results. It mostly focuses on how building a solution goes from one phase to the next phase in an efficient manner. In the event of a major change, the Waterfall approach may require the project to start from the beginning.

To solve the flexibility issue of a Waterfall methodology, the Agile approach was created. In the Agile methodology, the development of different phases happens simultaneously. In other words, instead of planning for the entire IoT solution development, the Agile methodology divides the

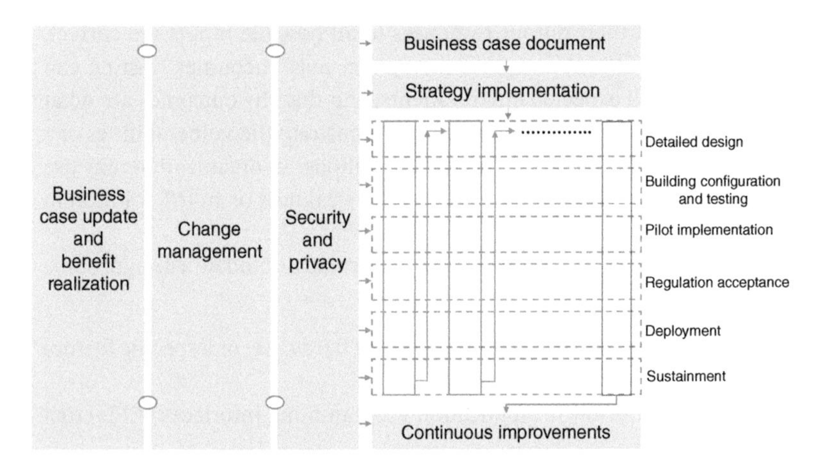

Figure 13.2 Agile IoT solution development methodology.

development work into small increments which are completed in iterations. Therefore, the solution gets completed as it goes through multiple iterations and updates. In the Agile approach, there is a need for various teams of developers to collaborate, and therefore, communication among different teams is an important necessity for a successful Agile implementation. The Agile representation of our nine-step methodology is shown in Figure 13.2.

The choice between the Waterfall or Agile methodologies for an IoT solution development is dependent on the IoT application. For projects where objectives are clear, the level of complexity is low, and developers are familiar with the implementation, Waterfall is a better choice. If the IoT project is more complex, the objectives are not clear at the start of the solution development, and developers need to incorporate the results of their testing to the project, then using the Agile methodology makes more sense.

In the Waterfall approach, there is not much learning occurring during the solution developments and the project has usually gathered all possible requirements at the beginning, while the Agile approach allows multiple cycles of the design, build, and test phases of the solution continuously. The testing outcome can be used as feedback in order to improve the quality of a solution. Table 13.1 compares the features of the Waterfall and Agile approaches.

When we build a solution, it needs to be tested in order to ensure that it validates the design requirements. To perform testing, developers need to prepare a test plan. A test plan is a document that outlines the test strategy, objectives, and schedule, as well as the resources that are needed to perform the tests. The objective of a test plan is to ensure that developed hardware and software

Table 13.1 Comparing the Waterfall and Agile approaches.

Waterfall	Agile
Easier to understand and implement	More difficult to implement
Requires low level of communication among different teams	Requires a high level of collaboration and communication among different teams
Major changes cannot be done easily	Adapts to change, adjustments can easily be done
End result is completely clear at the beginning of the solution development	End result is less certain at the beginning of the solution development
Slower solution development, but final solution will be deployed without any iterations	Faster solution development, but the final solution will be deployed after multiple iterations
Does not apply feedback from testing	Uses feedback information obtained from the testing or deployment data

meet the design specifications and that their output responses to all possible inputs are correct. It also tries to detect defects, errors, and problems that the system may encounter. Testing can ensure that the system will operate in all expected environments, and that the outcomes are what the organization and its users should expect. Testing can also demonstrate the vulnerabilities of a solution in terms of security. For example, to test security vulnerabilities, an organization may use ethical hackers to apply hacking schemes. This can help detect the weakness of an IoT solution in terms of security.

There are different testing operations that usually need to be performed to validate a design. Some examples of these testing operations are:

- Unit testing: testing one or a group of software functions, a part of hardware, or a specific feature of an IoT solution.
- Application interface testing: testing the entire Application Programming Interfaces (APIs) that are designed for an IoT solution.
- Interface testing: testing the integration points of an IoT solution.
- System testing: testing the complete system to validate its performance against design requirements.
- Configuration testing: testing various configurations used in an IoT solution.
- Installation testing: testing the install, uninstall, and upgrade processes.
- Interoperability testing: testing the solution using the devices from different manufacturers.
- Compatibility testing: testing the solution using different hardware devices, different operating systems, and in various environments.
- Stress testing: testing the solution beyond its normal conditions, such as when the level of noise in the operating environment is high, the temperature increases sharply, or a sudden increase in the number of devices, traffic, or processing workload.

Similar to the discussion that we had regarding the building of a solution, testing using the Waterfall or Agile approaches would be different. In the Waterfall approach, testing is a separate activity and would be performed after the hardware is ready, and the coding phase is complete. But in Agile methodology, testing is a continuous task, and testers continuously communicate with the organization and the solution developers. This creates a feedback loop which assists developers to improve their system or their code.

13.3.5 Pilot Implementation

A pilot validates how an IoT solution scheme performs on a small scale. A pilot should have a pilot test plan and a formal feedback channel for obtaining the pilot outcome from data, users, or systems.

A pilot can be applied to targeted users, systems, or sensors in order to detect or mitigate issues, validate technical performance, ensure organization readiness, and find the customer's reactions.

To have a successful pilot implementation, the organization should know how to make a pilot plan which includes pilot timelines, as well as the metrics and performance indicators which represent the success of a pilot. Also, the organization should target a group of users, systems, or sensors to participate in a pilot. A formal test plan for the pilot allows users and systems to perform real-world tasks that are aligned with IoT applications. There should be a feedback loop which is efficient and convenient for sharing the results of the pilot outcomes.

When an organization is performing pilot testing, the pilots, outcome must be monitored very closely to remedy any issues that may arise in order to keep the pilot on track. Also, at the end of a pilot, the results should be carefully assessed.

13.3.6 Regulation Acceptance

It is important to apply and receive the necessary certifications for an IoT solution before deployment. For example, IoT-based wearable medical devices may need to receive certifications from medical authorities before they can be used in a specific country. Different industries have their own set of regulations, and therefore, there is a need for IoT applications to apply for the required certifications or perform additional testing to satisfy the requirements of regulatory bodies. These regulations may be different in various geographical areas. Therefore, for an IoT device which might be used in multiple countries or regions, the organization needs to apply and receive the required certifications from all those countries or regions.

13.3.7 Deployment

Depending on the IoT solution, an organization may go through full deployment, or decide to implement the solution in phases. For example, an organization may go through a three-phase deployment in which through each phase, a set of the solution's features will be added to the list of existing features. Also, it is possible to deploy the solution with full features, but to serve additional geographical areas in each phase. If the organization encounters many problems in a deployment phase, it may delay the deployment of the next phase, until it can resolve its existing problems.

When companies move to full deployment and as the number of IoT devices increases, the most important question is how the overall IoT solution deals with the scalability and how it can handle the increase in the devices that intend to use the solution.

13.3.8 Sustainment

After the deployment of the project comes the sustainment step. The purpose of this step is to maximize readiness and deliver the best possible products, systems, services, or features by solving minor problems that users or customers may face in a fully deployed scenario. Also, the organization should be able to answer user questions, educate and train the users, and help them to use proper configurations in order to maximize the performance or usage of the deployed IoT solution.

13.3.9 Continuous Improvements

There are many reasons to continue improving an IoT solution even after the complete deployment of the solution. The first reason is the rapid advancements in the tools, techniques, and algorithms of an IoT solution. It would be advantageous, if there is a possibility to take advantage of these advancements in order to improve the quality of an existing IoT solution. The second reason is that both the organization and its customers may feel a need for additional new features, processes, or services when they realize the benefits of a deployed IoT solution. The third reason is related to the possibility of improving the IoT system through software updates. Therefore, we can continue updating the IoT system software in order to improve the solution with ease. The fourth reason is related to the IoT solution design itself, which enables the organization to use the data to improve the solution. For example, in the Agile implementation of an IoT solution, we need to continuously improve the initial implementation.

13.4 Change Management

The CM process is a series of tasks which need to be performed for a smooth transition from a current state to a desired future state, also called the target state, without obstructing the actual operations of an organization or creating adverse impacts. CM is a gradual process created by an organization in order to ensure that modifications made in affected areas are completely managed without jeopardizing the organization's actual operations.

CM is about managing the people of the organization during the implementation of a new solution. It is important to identify any affected process and the impact it may have on people who are involved in that process. The adoption rate and acceptance of the impacted staff will have a significant implication on the success of any type of solution development, including an IoT one. It is common for people to be set in their ways. So, staff members who have been doing the same job for a long time are inclined to continue doing it in the same fashion. Why? This mostly boils down to concerns regarding a lack of understanding on the reasoning behind the change. When an organization introduces a change, and forgets or chooses not to include the impacts of that change on the staff, the entire process backfires. Staff need to be consulted, involved, and trained continually from the time a change is being conceived until past sustainment. This enhances the chances of a successful solution development to a greater extent. Also, it helps increase staff morale, productivity, and overall culture.

Many companies understand the necessity for change, but they either underestimate or poorly design the CM plan. As a result, the desired outcomes are therefore not reached. It is possible that an organization leaves the communication about change to the very end and then forces new processes onto its staff. In this case, the staff may unwillingly apply the new processes, which causes poor adoption rates and consequently results in poor benefit realization.

Successful CM needs to include anyone in the organization who will be impacted in some capacity. In IoT projects, change is inevitable, and therefore having a sound change plan at the onset and the execution of that plan are keys to make that IoT project successful.

As the business case for the IoT project gets approved, the organization needs to communicate the "case for change" to the impacted staff members. In other words, simple and straightforward communication needs to be made to staff members so that they understand the reason related to the change. Examples of such reasons might be economic growth, business survival, or customer demand. Once that initial awareness is made, it is important to involve a subset of the staff in the detailed design stage as business stakeholders. Not only do they have a rich understanding of the organization at hand, but they can also be designated as change ambassadors among their peer group.

During the subsequent phases of a solution development, it is imperative to keep on leveraging this group. In parallel, regular and meaningful updates should be provided to staff at large so they are kept in the loop on the evolution of the IoT project. A proper training plan should be put in place, one that is tailored to the needs of the various staff groups. Remember, not all staff groups will be affected in the same way. So, the training curriculum needs to be modular and a combination of in-person and e-Learning modules geared toward the various affected groups. Finally, it is important to probe, poll, and ask for feedback after any training. This data is precious in ensuring how training should get improved for future releases of the IoT solution.

13.5 Summary

In this chapter, we discussed how companies can develop IoT solutions for their organizations. We explained a nine-step solution development methodology that can be used as a guideline for this purpose. Generally speaking, there are two main approaches for the solution development: the Waterfall and Agile methodologies.

The Waterfall is based on a sequential approach in which the organization divides the work into several distinct steps. Each step should be built in sequence, and therefore, the organization should finish building the first step before starting the second one.

In the Agile methodology, the development of different pieces occurs simultaneously. In other words, instead of planning for the entire solution, the Agile methodology divides the development work into small increments which are completed in iterations. Therefore, the solution gets completed as it goes through multiple iterations and updates.

The choice between the Waterfall or Agile methodologies for an IoT solution development is dependent on the IoT application. But generally speaking, Agile fits better with most IoT solution developments.

IoT solution development usually introduces a large amount of change to an organization. The CM process is a collection of tasks that need to be performed for a smooth transition from a current state to a desired future state without obstructing the organization's main operations and workflow. CM is about managing people within the organization during the implementation of a new solution and should be taken very seriously by the organization as it goes through the solution development journey.

Reference

1 Scully, P. and Lueth, K. L. (2016). *Guide to IoT solution*. https://iot-analytics.com/wp/wp-content/uploads/2016/09/White-paper-Guide-to-IoT-Solution-Development-September-2016-vf.pdf (accessed 30 June 2022).

Exercises

13.1 Why is there a need in most IoT solution developments to partner with a vendor?

13.2 What is the reason for considering a sustainment step as part of an IoT solution development methodology?

13.3 Give an example in which the collected data by a company is very valuable to other companies. Who do you think owns the data that a company collects?

13.4 A company sells filtered water in many different locations in a city. This company uses an IoT-based system in each of these locations. The system transmits data regarding its performance and the quality of the water to the cloud. The data can be used to determine the time that the filter needs to be replaced. Give examples of some organizations that might be interested in having this data.

13.5 What are the integration points of a detailed design?

13.6 Business case document provides justification for the deployment of a solution.

a. What does it mean to say that a business case is a living document?

b. Who approves a business case?

c. How does a business case have information pertaining to the cost, even though the details of the design are not available when a business case is being prepared?

13.7 An IoT solution for a company producing smart medical devices took one year to go from business case development to successfully finishing its pilot trial. The company was very eager to go to the market as soon as possible. However, it took over two years to actually deploy the solution. What do you think the reason(s) might be?

13.8 What is the role of system integrators? How is this role different from the role of a product vendor?

13.9 In an IoT solution developed for a smart factory, the solution considers the integration of IoT systems and non-IoT systems in its plan. What do you think must be the reason for considering non-IoT systems?

13.10 What is the idea behind Agile implementation? How is it different from Waterfall implementation?

13.11 What are the reasons to continue improving an IoT solution even after complete deployment?

13.12 Who is an ethical hacker?

Advanced Exercises

13.13 What are some of the criteria for vendor evaluation?

13.14 A company which manufactures floor cleaning machines goes through an IoT solution development. The company now manufactures machines that are connected to the Internet and sends information about the square footage of the cleaned floor to the cloud. The company changes its business model, and instead of selling the machines, it provides its customers with these machines free of charge, and charges its customers based on the square footage of the cleaned floor. Explain some of the CM challenges that this company may face?

13.15 Do you think most of the companies today are ready for Agile implementation?

13.16 In industrial IoT, many IoT solutions should be provided to machines which are using non-IoT protocols. Therefore, protocol translation is necessary for these situations. In which step during the proposed nine-step IoT solution development process, should we work on protocol translation?

13.17 Four vendors offer the following components of an IoT solution as shown in Table 13.E17. In the following cases, which vendor(s) do you think the organization must partner with?

a. A transportation company wants to build a cloud-based asset management tracking system for all its assets (stationary or mobile), by connecting the assets to the Internet.

b. A company intends to provide IoT-based Augmented Reality (AR) and Virtual Reality (VR) training to its customers.

Table 13.E17 Exercise 13.E17.

Solution's component	Company A	Company B	Company C	Company D
System integration	✓	✓	✓	
Cloud database	✓			
Advanced analytics	✓			
Operating system	✓		✓	
Edge gateway	✓			
Proprietary enterprise application		✓		✓
VR/ AR CTOS solutions		✓		
Dedicated IoT development team	✓	✓	✓	

13.18 Give two examples of IoT solution developments in which adding new features might be problematic.

13.19 Why should a company choose to deploy an IoT solution in phases instead of a full deployment?

13.20 What is the difference between the design testing in step three of our methodology and the pilot testing in step four?

13.21 A company has many departments such as manufacturing, hardware engineering, software engineering, and analytics. There is not much interaction among these departments. The company wants to perform an IoT solution. Explain a problem that the company may face if it chooses to use an Agile approach?

Practical Assignments

This book provides readers with theoretical knowledge about the fundamentals of Internet of Things (IoT). We encourage readers to perform some experiments in order to obtain hands-on experience in the design of IoT systems and to better grasp the theoretical concepts addressed in this book. In this section, we explain several assignments that students and instructors can use for this purpose.

These assignments are intended to offer a series of hands-on experiments that students can perform in the IoT field. The assignments try to offer some hints and shed some light on how they should be completed, rather than providing detailed step-by-step instructions. In order to complete the assignments, students need to perform their own research. Additionally, students (with the help of their instructor) can define a more challenging version of these assignments to be completed as capstone projects, or simplify the assignments to be used as their course projects.

Assignment #1: Connecting an IoT Device to the Cloud

In this assignment, you will use a microcontroller to read a temperature sensor and communicate the temperature data to an IoT gateway using WiFi technology. The data must then be transferred to the cloud (ThingSpeak server) for analysis and visualization. In addition, when the temperature exceeds a certain level, the cloud application must send an email notification.

ThingSpeak is a cloud-based IoT analytics platform that allows users to aggregate, compute, and visualize IoT data. As a result, IoT devices can send their data to ThingSpeak, which can analyze it, build live data visualizations, and send notifications as needed. Because ThingSpeak is based on MATLAB®, students who are familiar with MATLAB® can use this platform to perform analytics without having extensive knowledge of cloud infrastructure and services.

ThingSpeak can be accessed through https://thingspeak.com. On the main page of the ThingSpeak site, you will find three menus. These menus are **Channels**, **Apps**, and **Support**. Using the **Channels** menu, you can add a new channel to ThingSpeak. An IoT device can send its data to a defined channel. Each channel has a name, a description, and several fields. Channels can be private where no one can see the data, public where data can be accessed by everyone, or shared in which the data can be shared with specific users. An IoT device can import its data to a channel or export data from public or shared channels. Each channel is identified using a unique API key and a channel ID. To be able to send temperature data to the ThingSpeak server on the cloud, you need to create an account in ThingSpeak and make a channel for the purpose of transmitting temperature data.

Fundamentals of Internet of Things: For Students and Professionals, First Edition. F. John Dian.

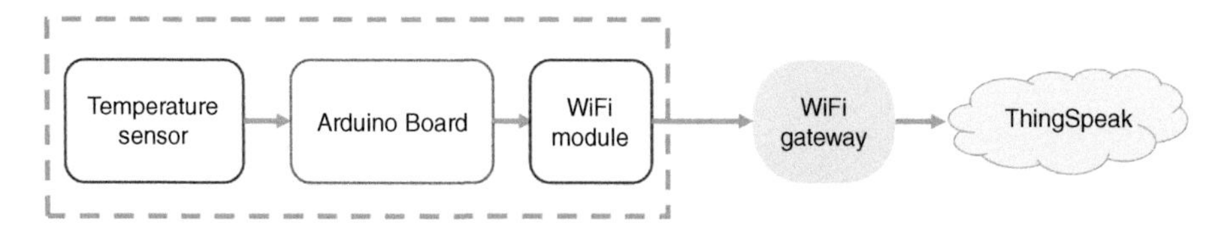

Figure A1.1 Circuit diagram of the IoT system in Assignment #1.

To practically build a circuit for transmitting data to ThingSpeak server, you should design a system using a low-cost microcontroller such as Arduino (uno R3 or nano), a WiFi module such as ESP8266-01, and a temperature sensor such as LM355 as shown in Figure A1.1.

Build the circuit shown in Figure A1.1. You need to program the Arduino board to read the temperature sensor, and send the temperature data wirelessly through a WiFi gateway to a channel in ThingSpeak. To do this, you can use an open source Arduino software Integrated Development Environment (IDE) in order to write your code and upload it to the Arduino board. Since you need to use ThingSpeak for this assignment, the ThingSpeak library needs to be included in the Arduino IDE. Write a code for Arduino to read the temperature sensor every 20 seconds, and send its data to the ThingSpeak channel that you previously made.

You need to visualize the data on the ThingSpeak channel in order to observe the collected temperature data. For this purpose, MATLAB® analysis apps in ThingSpeak can be used. To do this, you need to use the **Apps** menu in ThingSpeak. You should also write a MATLAB® code which sends an email when the temperature values are higher than a certain threshold value. Write a MATLAB® code which analyzes the data on the ThingSpeak channel that you made, and send an email when the temperature is above 25° for more than 5 minutes.

Assignment #2: Building a Battery-Powered Vision-Based System

In this assignment, you need to investigate whether it is possible to develop a battery-powered vision-based IoT system which does not have any energy harvesting capabilities. Because the system is not directly connected to power, you should concentrate on lowering the system's energy use. Assume you are designing an IoT-based automatic rodent recognition system that can capture videos of rodents entering the trap. By analyzing the captured data from this camera, you should be able to determine the type, size, and color of rodents. Also, you should be able to determine the frequency and the average number of rodents that enter the trap every day. In addition, you need to find out which method of computing (edge processing or cloud processing) is more suitable for this application.

For this assignment, you need an imaging sensor. The OV2640 camera module, which is compact and has a high sensitivity to low-light operations, is one option. This camera uses Serial Camera Control Bus (SCCB) interface and supports many image formats including Joint Photographic Experts Group (JPEG). To be able to read the data, analyze it, and send the data to the cloud, there is a need for an embedded system. One choice can be the ESP32-S module, which supports built-in WiFi and Bluetooth Low Energy (BLE) connectivity schemes and provides enough memory and processing power in order to perform the required analysis of this application.

You can activate the camera only when a motion is detected inside the trap and turn off the camera at other times. Actually, this can be a good idea in order to reduce power consumption

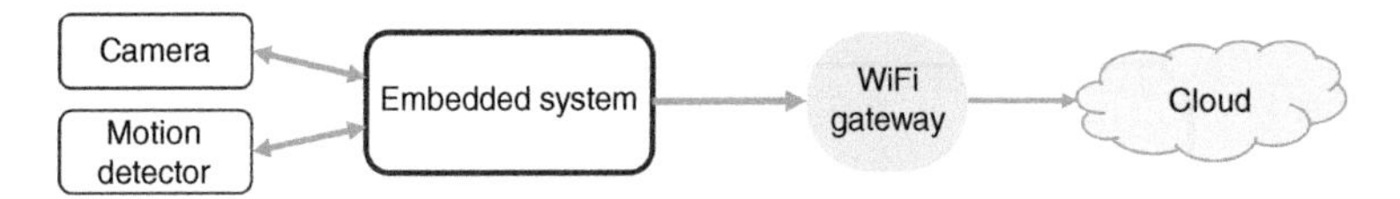

Figure A2.1 Block diagram of the system in Assignment #2.

substantially for this application. Use a low-power infrared sensor to detect motion, its direction, and its velocity. When a motion is detected, you can take a video, or several pictures. Taking a few good-quality pictures might be enough for this application and can result in reducing power consumption as compared to capturing video signals. If your embedded CPU has the processing power, you can perform edge processing and send out a time-stamped message to the cloud via an IoT gateway which uses technologies such as WiFi or BLE. This message should contain information such as the type, size, and color of the rodent, or the amount of bait left inside the trap.

For this assignment, you need to build an IoT-based system using a camera, a low-power motion detector, and an embedded CPU as shown in Figure A2.1, and install this system inside a box (rodent trap). Write the following codes for your embedded CPU: a code to initialize the camera, one to detect the motion, a code to take a few images when a motion is detected, a function to save the images in a memory, a code to analyze the images, and a code to send the analyzed data to the IoT gateway. The embedded CPU should go to sleep mode when it does not need to perform any of these functions.

Choose a battery for this application. For example, you can choose a lithium polymer battery with an acceptable form factor and a capacity of around 20 Ah. Measure the current draw and the power consumption of this application. Calculate (estimate) the duration of time this system can work without changing its battery (assume rodents enter the trap five times per day on average).

Assignment #3: Configuring an LTE-M module using AT Commands

For this assignment, choose a cellular IoT (CIoT) module from the list of modules that we provided in Chapter 7 and configure the selected CIoT module using ATtention (AT) commands. For example, let us consider that you choose the Quectel BG96 LTE-M/NB-IoT module.

There are many companies that provide electronic boards that contain a CIoT module. These electronic boards usually contain several other parts such as eSIM, SIM card holder, antenna, GPS, and interface ports such as USB, in addition to the CIoT module. Examples of these electronic boards are the Avnet's Qucetel BG96 board and Raspberry Pi cellular IoT kit.

You must receive a SIM card from your service provider and activate the SIM card in order to start working with your CIoT module. Each service provider has its own method of activation of the SIM card, and therefore, you must follow the instructions provided by your service provider to activate your SIM card. Service providers usually provide you with a website where you can manage your CIoT module, find out the status of your SIM card, and see the information and charges on your plan.

You can connect your electronic board to a computer using the available interface ports. For example, if your electronic board has a USB port, you can connect a computer to your board using a USB cable. Then you need to install the USB driver for the CIoT module on the computer.

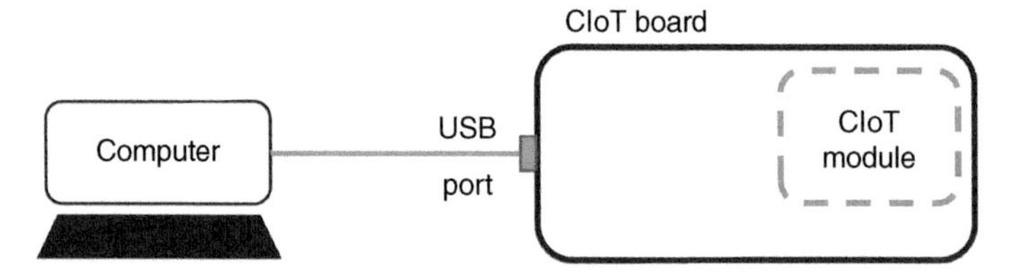

Figure A3.1 Block diagram of the system in Assignment #3.

For example, Quectel-LTE-Windows-USB-Driver-V1.0 or higher versions should be installed on a computer which uses the Windows operating system in order to interface the computer to an electronic board that uses Quectel CIoT module. By doing this, the electronic board appears as a USB AT port on the computer. You can look at the device manager on your computer in order to see this USB port. When the connection between the computer and the CIoT module is established, you can use AT commands to configure the CIoT module. The block diagram of the system for this assignment is shown in Figure A3.1.

Find the list of supported AT commands for the CIoT module that you selected and try to configure your CIoT module using those AT commands. For example, for the Quectel BG96 module, you can configure the Radio Access Technology (RAT) as LTE-M or NB-IoT. You can also configure the network category depending on the available technologies of your service provider. You may also be able to choose between LTE-M Cat M1 and Cat M2. You need to configure the frequency bands that the CIoT module can use for its operation. Using the AT commands, you can get information about your CIoT module or the SIM card. Try to configure other parts of the electronic board such as the GPS, GNSS system, or the SMS text messaging system. Find a list of AT commands from the complete list of AT commands which are supported by your CIoT module. Enter those AT commands and see the corresponding reply messages from the CIoT module on your computer.

Assignment #4: Connecting an IoT Device to an MQTT Broker

For this assignment, you need to program a computer to play the role of a Message Queue Telemetry Transport (MQTT) broker. For this purpose, use NODE-RED programming tool to build an MQTT broker on a computer. You can also program a computer to play the role of an MQTT client by installing an MQTT client software on the computer.

You can also set up the CIoT board in Assignment #3 to act as an MQTT client. The CIoT board publishes messages with a specific topic to an MQTT broker. A computer can be used to play the role of an MQTT broker. An MQTT client which is installed on a computer can subscribe to receive the messages published by the CIoT board. The computer which plays the role of an MQTT client should also publish messages to the MQTT broker. These messages can be received by the CIoT boards, if the CIoT board has subscribed to the topic of those messages.

The architecture that needs to be designed for this assignment is shown in Figure A4.1. You can install the MQTT broker and MQTT client on two computers as shown in Figure A4.1a or on one computer as shown in Figure A4.1b.

To build an MQTT broker on a computer using NODE-RED, you need to download and install javascript runtime Node.js and install NODE-RED on your computer as a global module with all its dependencies. The NODE-RED editor can be accessed at http://localhost:1880 on your computer where you can design an MQTT broker graphically. Install an MQTT broker such as Mosca MQTT broker on NODE-RED in order to make your computer to act as an MQTT broker. By dragging

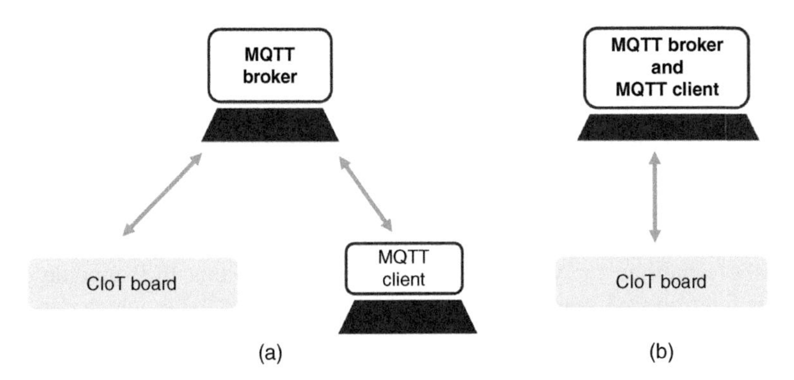

Figure A4.1 Block diagram of the system in Assignment #4 (a) MQTT broker and MQTT client are installed on two different computers (b) MQTT broker and MQTT client are installed on one computer.

and dropping the following nodes in the NODE-RED editor, you can build an MQTT broker and client.

1. Mosca out; to connect an MQTT broker and publish messages
2. MQTT in; to connect MQTT broker and subscriber to a specific topic
3. Debug; to see messages in debug side bar
4. Inject; to manually inject messages to the flow or automatically inject them at constant interval

Also, you need to install MQTT.fx to make a computer to act as an MQTT client. Let us assume that the MQTT client which is installed on the computer publishes messages with the topic "Computer-generated," and the CIoT board publishes messages with the topic "CIoT-generated." Build the interaction shown in Figure A4.2 in the NODE-RED editor.

To program your CIoT module to subscribe to an MQTT topic, you can use AT commands. Therefore, you should find out what AT commands need to be used for this purpose. Build the system in Figure A4.1a or Figure A4.1b for this assignment, and experiment with the message exchange using the MQTT protocol. The block represented as MQTT Client-Computer in Figure A4.2 generates messages with the topic Computer-generated. The broker sends these messages to the CIoT borad. Connect your CIoT board to a computer and run a terminal program such as HyperTerminal or Putty to see the computer-generated messages.

Interested readers are encouraged to look at [1] for more practical experiments using CIOT devices.

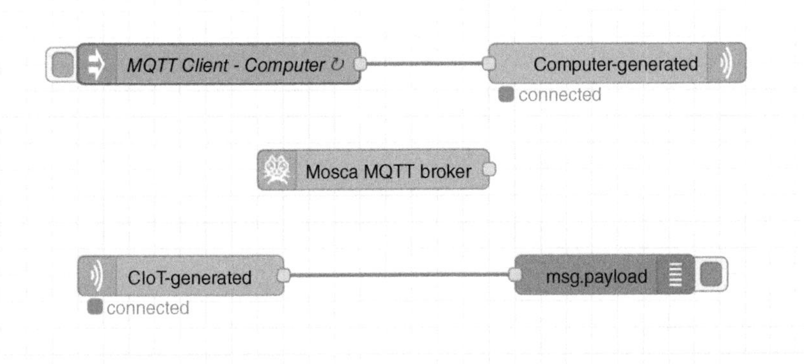

Figure A4.2 Interactions among MQTT broker, MQTT client, and CIOT board in NODE-RED editor.

Assignment #5: Connecting an IoT Device to an IoT Gateway Using BLE

In this assignment, you need to connect a BLE module which resides in an IoT system with another BLE module which resides in an IoT gateway. You can use any BLE module for this experiment. TI CC2540 is a popular BLE module that has been widely used in different IoT applications. However, programming of this module may be somehow complex for this assignment, and it may take sometime for you to learn how to program this BLE module. Bluegiga (now Silicon Labs) introduced some BLE modules such as BLE121LR which are based on TI CC2540. It also introduced a high-level programming language in order to make it easier to program these BLE modules. In this assignment, let us connect two BLE121LR modules to each other and transfer data between them. To program these BLE modules, you need to have the following files: hardware.xml, config.xml, GATT.xml, script.bgs, and project.bgproj.

The file hardware.xml is the hardware configuration file. As the name suggests, the configuration of the hardware settings can be written in this file. For example, you can set up the transmit power, hardware timers, or interfaces of the BLE module such as UART or Serial Peripheral Interface (SPI). It is obvious that configuring transmit power of a wireless device is crucial for its operation, if you want to configure the maximum allowable transmit power of a BLE module, then the hardware.xml file should contain a line as <txpower power = “7” bias = “5”/>. If there is nothing related to the power in this file, then a default value will be used for the maximum power. For BLE121LR, the highest transmit power setting is 9 (around + 8 dBm), while 0 is the lowest value (around −10 dBm). The bias indicates the amplifier bias setting.

The file config.xml is a configuration file which is used to program some of the BLE features such as the number of connections, or the throughput performance. For example, if you have one master node and four slave nodes (four connections to a master node), then the config.xml file should contain <connections value = “4”/>. The maximum number of connections for BLE121LR is eight. To change the number of data packets that are sent over the air during each connection interval, you should add a line indicating <throughput optimize = “performance”/> to the config.xml file. You can use “power,” or “balanced” instead of “performance” in this line. Optimizing the throughput as power allows the BLE module to send only one packet during a connection interval, while optimizing the throughput as balanced allows the BLE module to send three or four packets during a connection interval. Optimization based on “performance” allows BLE to send as many packets as possible in a connection interval.

The GATT.xml file specifies the structure of data which will be exchanged between a Generic ATTribute (GATT) server and a GATT client. Generally speaking, to transfer data between two BLE devices, the data should be transferred from a GATT server to a GATT client. The GATT server and clients are relational databases, and the data which is placed in a GATT server by a BLE module can be sent to the database that resides in the other BLE module. It should be noted that the role of the GATT server and GATT client is independent from the master and slave roles. The data is stored in a GATT server based on a clear hierarchy in order to enable efficient transfer of data between the GATT server and client. The data is typically arranged in a profile-service-characteristic-attribute hierarchy. You need to understand this hierarchy and define the data structure required for your application program in the GATT.xml file.

Write your application code in script.bgs file using a text editor. The syntax of the language, BGscript, can be found in [2].

The file project.bgsproj is a project file. To run a project, you should write the names of all the files that need to be compiled inside the project file. Examples of these files are script.bgs, hardware.xml, GATT.xml, and config.xml files.

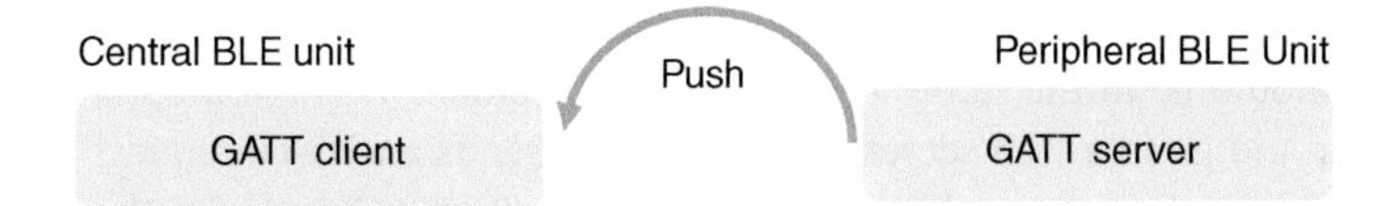

Figure A5.1 Block diagram of the system in Assignment #5.

For this assignment, you need to program both the master BLE and slave BLE nodes by writing the code in BGscript. Build your system as shown in Figure A5.1 and program the peripheral BLE node to write one byte of arbitrary data to an attribute every 200 ms. Also, program the master node to enable notifications. The GATT server resides on the slave node and the GATT client resides on the master node. Activate the notification on the central BLE node so that the data from the GATT server is pushed to the GATT client.

Use a packet sniffer to see the exchange of messages between these two nodes. You can use the packet sniffer from Texas Instrument (TI) for this purpose. You need to identify the address of the master (initiator), connect the packet sniffer BLE dongle to your computer, and run the packet sniffer software program. Configure your program to have the MAC address of the master node as the initiator address, and choose an advertising channel. Observe the exchange of packets between the master and slave nodes. Change the connection parameters and see the changes in the transmission of packets between the master and slave nodes. Change the throughput optimization method in config.xml file, and observe the changes in the transmission of packets.

Assignment #6: Building an IoT-Based Home Automation System

In this assignment, you need to build a home automation IoT-based system. Using this system, when a light switch inside a house is turned on or off, you can see a change in the status of the light switch on your computer or smartphone by going to a web page that you design for this home automation application. Also, you can remotely turn on or off a light bulb inside your home by sending a command to the light bulb using this application program.

To perform this assignment, you can use a push button and a Light-Emitting Diode (LED) to represent the light switch and light bulb, respectively. Figure A6.1a shows a typical method to build this project. However, we are interested to build this project based on the block diagram shown in Figure A6.1b, in which the processing board not only is connected to the LED and the switch but also acts as a web server. Assume the connection between the web server and the computer is an Ethernet interface. You can use any CPU board for this project. If your CPU board does not have an Ethernet interface, you can use an Ethernet module such as the W5500 Ethernet network module.

Build your system based on Figure A6.1b. If you decide to build this system using an Arduino microcontroller, you can use the Ethernet library and Ethernet shield. You also need to design a

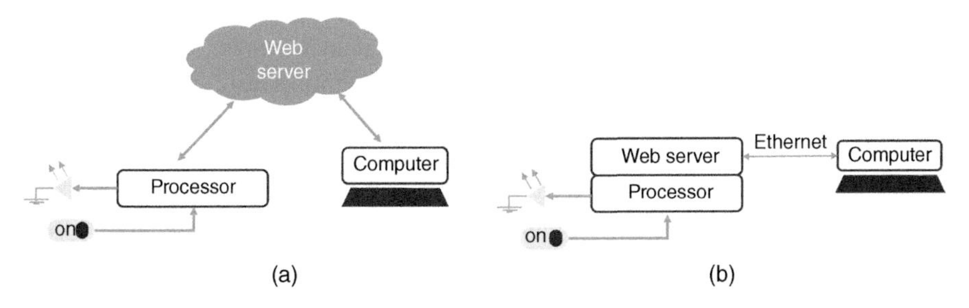

Figure A6.1 Home automation assignment: (a) a typical method (b) using a local network and Ethernet connection.

web server for this assignment. You need to assign the Ethernet shield with an IP address and a MAC address. The IP address should be in the same subnet as your computer. When you open a web browser on your computer and go to a web server, your computer sends an HTTP request to the web server. Your web server (Arduino) should respond with a small HTML code for the web browser to display the status value of the light switch. On the web page, you should design a push button. When the position of this push button is changed, the LED, which is connected to the processor should be turned on or off accordingly.

Design a web page for this application. The connection between the computer and the web server is HTTP. Open a packet sniffer such as Wireshark and observe the packet transfer between your web page and the web server.

Assignment #7: Designing a Smart Toy System

In this assignment, you need to add a notice to a smart toy. Smart toy manufacturing companies design toys that may record audio and video data and send the collected data to the Internet. Manufacturers are normally obligated to provide consumers with notice to guarantee that the toy is not passively listening or recording any conversations. A toy manufacturer, for example, may include a push button on its device to provide notification. The toy begins its main functions after the button is pressed. In this assignment, we want a toy to detect a predefined spoken word, which will be used as notice, before the toy starts its main operations.

This is an edge processing operation, and you should use a machine learning algorithm to perform this task. As we explained in Chapter 11, you can use TinyML to run machine learning algorithms on embedded systems. Find a microcontroller for deploying your machine learning model. One choice could be the Arduino Nano 33 BLE sense which has a 32-bit Advanced RISC Machines (ARM) processor, 256 kb of random access memory (RAM), and 1 MB of program memory. It also has several sensors such as proximity, gesture, motion, vibration, orientation, temperature, humidity, and a digital microphone.

One of the most popular machine learning frameworks that works with TinyML is TensorFlow Lite (TFL), which supports microcontrollers such as the Ardunio Nano 33 BLE sense. Generally speaking, TensorFlow (TF) is an open source project which provides a set of tools for developing machine learning algorithms. It was originally developed by Google, and it is now maintained by many contributors. Google started the project with the intent of building a library to enable mobile devices to run neural network models. TFL is a simpler version of TF and does not support all functions of TF, but it requires substantially less memory and occupies only a few hundred kilobytes.

For this assignment, you need to design an embedded system which is capable of listening to an audio signal and recognizing a predefined word that you choose for this assignment. Your design is part of the smart toy system and acts as a binary classifier in which the predefined word belongs to one class, and any other words, background noise, and silence belong to the other class.

To perform this assignment, you should capture the voice signal continuously, extract features that are suitable to feed into your model, and train your system by collecting many samples of different words including the one that you chose for this assignment. You can also use the available voice data sets for this purpose. For example, Speech command data set, which is an audio data set of spoken words, can be used for this purpose. Transform the time-domain voice signal into the frequency domain, find the frequency information of different words, and attempt to build your classifier system. Use the 2D spectrogram of the voice samples to model and train the system. Since you use a 2D spectrogram, you should also use a 2D tensor in TF.

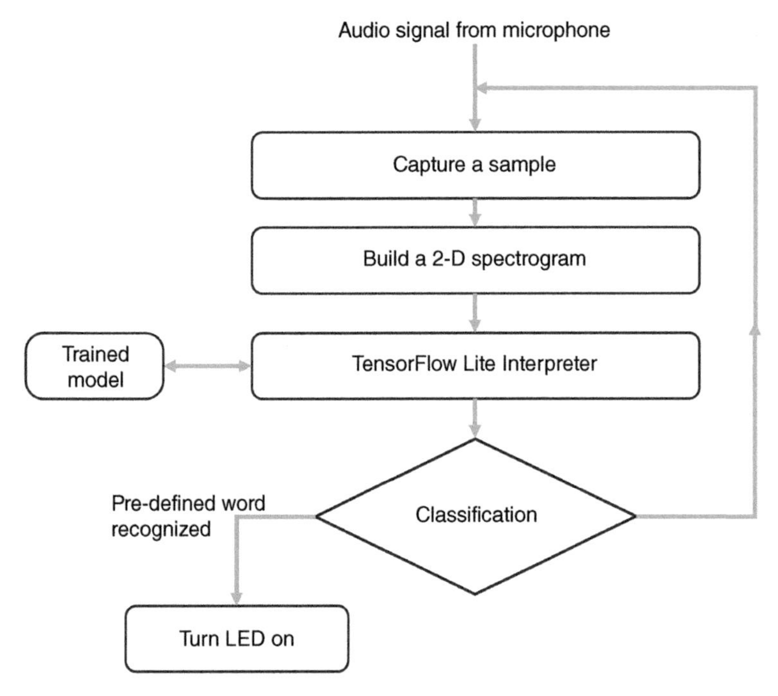

Figure A7.1 Flow chart of the system in Assignment #7.

Build a Convolutional Neural Network (CNN) model for this assignment. A 2D spectrogram is like a 2D image and a CNN model can work well in this situation. The building block of your system is shown in Figure A7.1. You can turn on an LED to show that you have detected the predefined word.

Assignment #8: Controlling a Smart Tank System Using LoRaWAN Technology

In this assignment, you need to build and control a smart tank system using LoRaWAN connectivity scheme. Your design should show how you can remotely monitor and control the smart tank system. Assume that a tank is equipped with a sensor which measures the level of water in the tank. If the tank becomes empty, a remotely-controlled valve needs to be turned on in order to increase the amount of water in the tank. When the amount of water in the tank reaches a threshold value, the valve should be turned off.

You can either physically construct a tank and a valve for this project, or you can simulate the tank and valve and focus solely on the communication method. If you wish to build a physical tank and valve, you need to determine the tank's size and select a suitable sensor for measuring the water level inside the tank. You also need to look up the specifications for the remote-controlled valve you will be using.

Figure A8.1 depicts a general block diagram of the system in Assignment #8. The tank and valve are connected to a LoRaWAN node that is far from a controller. The LoRaWAN node is connected to a LoRaWAN gateway in the area which connects the node to the controller through the LoRaWAN network.

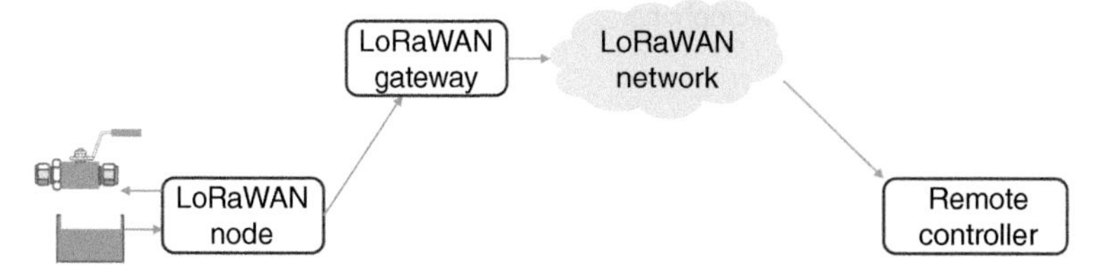

Figure A8.1 A typical LoRaWAN network for Assignment #8.

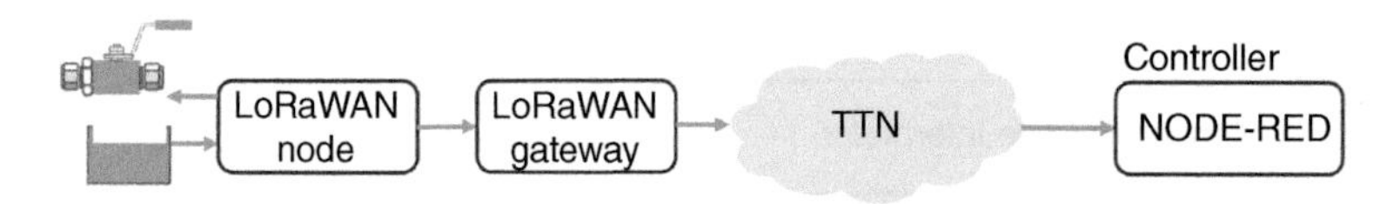

Figure A8.2 The block diagram of the system in Assignment #8.

Build the system based on Figure A8.2, assuming there is no LoRaWAN gateway in the area and that you should build your own LoRaWAN gateway to complete this assignment. Also, connect to the controller using The Things Network (TTN).

Let us take a closer look at Figure A8.2. To complete this assignment, you need a LoRaWAN node and a LoRaWAN gateway. Choose a suitable node and gateway. A good option for this purpose could be LoPy4 from Pycom, which is a compact development board and a good IoT platform for connected things. It supports LoRa, Sigfox, WiFi, and Bluetooth technologies. We also use TTN, which is a community-based LoRaWAN network. You can go to TTN, register your own gateway, and allow other people to use it, or use the freely available gateways that exist in TTN. Check the LoRaWAN coverage map to see whether there is LoRaWAN coverage in your area or not. If you see good coverage, you may be able to use one of the available gateways. Otherwise, the only solution is to use your own gateway. The gateway should have a wireless connection to the LoRaWAN node and a WiFi connection to TTN. To simulate the controller, you can install NODE-RED on a computer.

To simulate the tank and valve, you can program your LoRaWAN node to generate some values as the water levels. The values should be reduced when the valve is off and increased when the valve is on. To show the status of the valve (open or closed), you can turn on/off an LED on the LoRaWAN node to represent the position of the valve. This is shown in Figure A8.3. The valve should be opened or closed based on the commands that it receives from the controller.

To be able to use TTN, you need to create an application for this assignment. Let us call this application "SmartTank_app." Register the LoRaWAN gateway and the LoRaWAN node with TTN.

As previously noted, the controller can be built using NODE-RED. Design the NODE-RED to read the water level values (sent by the LoRaWAN node), and send the appropriate commands to open or close the valve. Observe the exchange of messages on TTN to ensure that the system is working properly. You can format the payload of the messages in TTN.

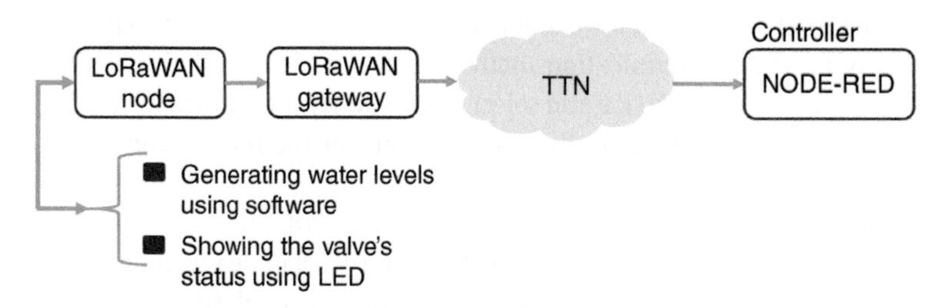

Figure A8.3 The simplified block diagram of the system in Assignment #8.

Assignment #9: Building IoT Systems Using Cisco Packet Tracer

In this assignment, you need to design an IoT system using Cisco Packet Tracer tool. Generally speaking, this tool can be used to create simulated networks by dragging and dropping various types of network devices such as switches and routers. Newer versions of Cisco Packet Tracer introduce two new entities: smart things and components. Smart things are IoT devices which are used in smart homes, smart cities, smart factories, and power grid systems. These IoT devices can be connected to IoT gateways or register servers. Components are physical objects that do not have any networking capabilities, and they should be connected to a Micro-Controller Unit (MCU-PT) or a Single Boarded Computers (SBC-PT) in order to be connected to an IoT network. The SBC-PT is essentially a programmable board. It has several I/O pins that can be connected to the sensors and actuators. Examples of MCU-PT and SCB-PT are the Arduino microcontroller and Raspberry Pi, respectively.

The block diagram of a simple IoT network for home automation is shown in Figure A9.1, where a home gateway is directly connected to some IoT devices. Basically, a home gateway is an IoT gateway which supports different types of wired and wireless connectivity schemes, and can be connected to wide area network interfaces. A computer or smartphone can be used to manage an IoT network (IoT devices and components) through a web interface hosted by the home gateway. As can be seen in Figure A9.1, the home gateway is also connected to several sensors and actuators through MCU-PT. A computer is also connected to the home gateway in order to manage all the smart things and components.

The home gateway supports a programming editor which allows an MCU-PT or SCB-PT to be programmed using Java script or Python and can publish the code to the MCU-PT or SCB-PT boards. For example, Packet Tracer can emulate Ardunio IDE for the programming of an Arduino board.

The new version of Packet Tracer tools also supports dynamic environment management, which enables more realistic results from IoT device simulation. For example, when an air conditioning unit (smart thing) is turned on, the temperature of the environment decreases. Therefore, you can program the Packet Tracer to perform various tasks according to the environmental conditions. For example, if there is a car in an environment, it can increase the amount of gases which consequently can trigger a smoke detector in that environment. To be able to perform these types of tasks in Packet Tracer, a network can be separated at the physical layer. You can define environments such as a city, building, room, container, or cabinet at the physical layer. This is important since the dynamic changes of an environment can influence the behavior of IoT devices.

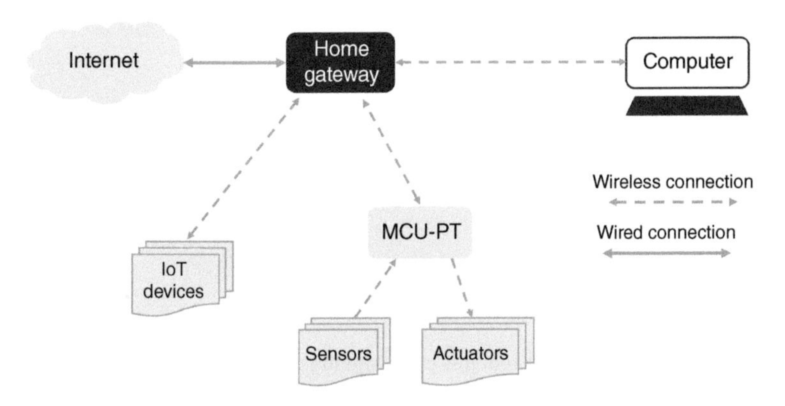

Figure A9.1 A simple IoT network for home automation.

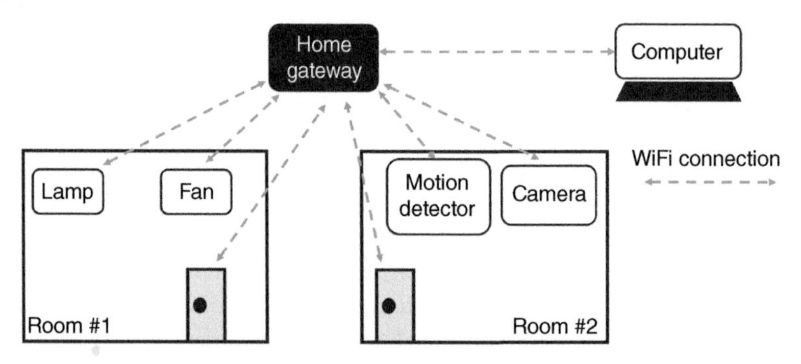

Figure A9.2 A smart home automation network which requires configuration of smart things.

For this assignment, you need to become familiar with the Packet Tracer tool, and build several IoT-based systems. Figure A9.2 shows a home gateway which is not connected to the Internet. The environment shows two rooms: the first room contains a smart door and a smart lamp, while the second room contains a motion detector and a camera. The home gateway is also connected to a computer. You must run a web browser on this computer to monitor and control all the IoT devices on this network. For example, from the computer, you can open or close the doors and turn on or off the devices. You also must be able to configure these devices to perform different tasks when certain conditions are met. Perform the following tasks for this assignment:

1. Turn on the camera when motion is detected in the second room, and turn it off when there is no motion. In Packet Tracer, you can simulate a motion by pressing ALT and moving your mouse over the motion detector. You must be able to see that the camera turns on and off from the computer.
2. Assume that both the lamp and the fan are turned off. When the smart door in the first room opens, turn on the lamp and the fan.

Figure A9.3 shows an IoT network that uses an MCU-PT device. You need to program the MCU-PT to read the temperature sensor and show the temperature value on the LCD display. It should also turn on the fan when the temperature is above 27°, and turn off the fan when it is below 21°.

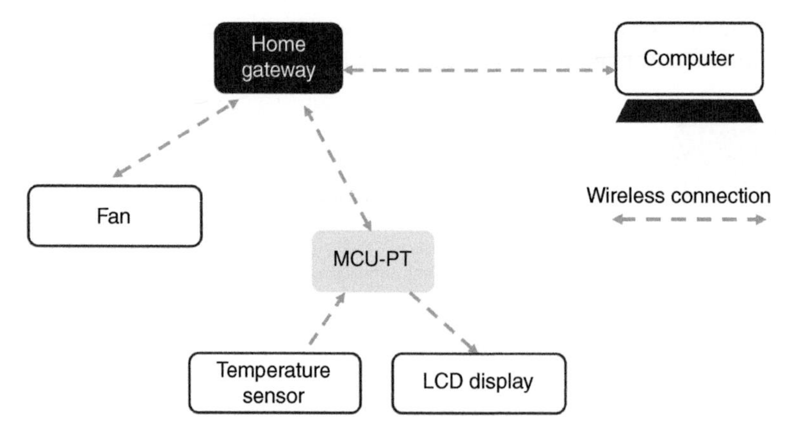

Figure A9.3 A smart home automation network which requires programming of MCU-TP board.

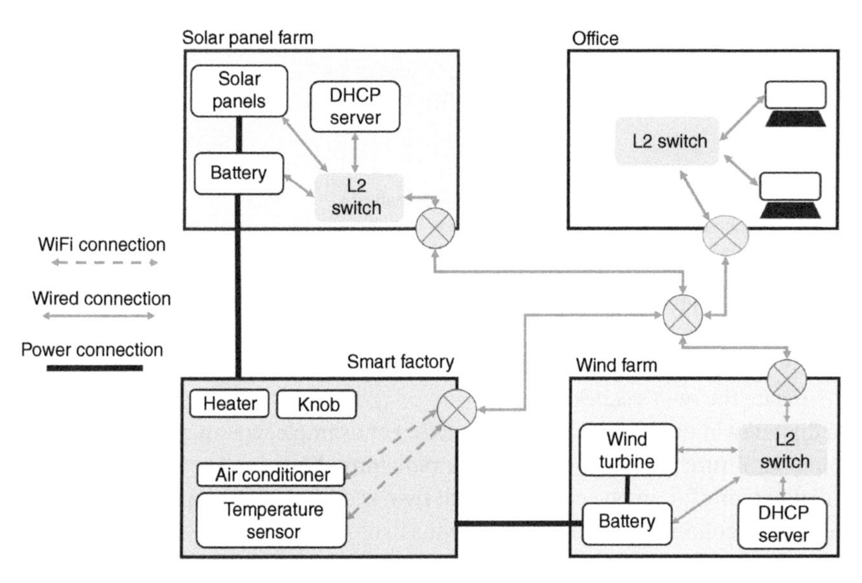

Figure A9.4 An example of an IoT system to be analyzed using Packet Tracer.

Figure A9.4 shows a more complex situation where there are four different environments: a solar panel farm, a wind farm, a smart factory, and an office environment. Electricity is granted by the solar panel farm and wind farm. The generated power is stored in batteries and is used in the smart factory for a heater and an air conditioning unit. Besides the heater and the air conditioning, there is a switch knob and an IoT-based temperature sensor in this smart factory. The heater is not IoT-based, and it is controlled by the knob which does not have any connectivity scheme. Assume by using a computer in the office, you are able to monitor all of the environments. Make some assumptions for the specifications of this network, and design the network based on your assumptions. Find the peak time of energy consumption in the smart factory for this system. To change the knob, press ALT and move the knob with your mouse. You must configure the IP addresses and the routing protocol between the routers, and configure the wireless connection between the IoT-based devices and the router in the smart factory environment. Also, you need to add wireless connectivity to the air conditioning unit and the temperature sensor by adding a wireless card in the I/O config tab.

Assignment #10: Building a Digital Twin in the Cloud

A digital twin is a virtual copy of a physical object. An object can be a system, an integrated system, or a subsystem. For example, a pump can be considered an object, and you can build its digital twin. Also, you can create the digital twin of an object that contains several pumps. You can also build the digital twin of a component of a pump. In addition, a digital twin can include processes such as the process control used in the operation of a smart factory.

As mentioned earlier, a digital twin is a digital representation of an object. To make a digital twin, you need to make a model of the object and provide the model with the data from the physical object. By doing this, the digital twin accurately represents the current state of the object. The data which is obtained from the object can be stored, and it can create a history of the object states over

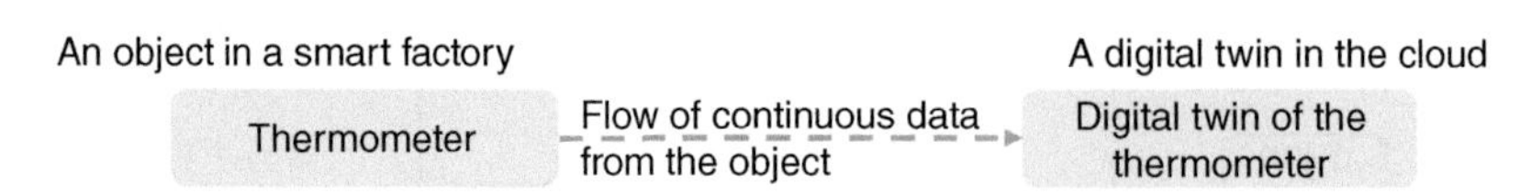

- Write the model of a thermometer in DTDL
- Upload the DTDL code to the cloud and create a digital twin instance
- Interact with the instance using Azure digital twins explorer
- Stream the data to an IoT hub which is associated with the digital twin instance

Figure A10.1 Steps to build the digital twin of a thermometer.

time. The digital twin is an IoT-based system since the data from the physical object should be continuously used to update the twin's states.

The modeling of a digital twin depends on your objectives. For example, you may want to monitor an object, simulate its future states, or calculate its remaining lifetime. Therefore, you can even define several digital twins for an object, where each twin is designed for a specific purpose. A digital twin can be built by connecting several digital twins in order to represent a more complex object.

In this assignment, you will need to build a cloud-based digital twin for a simple thermometer as shown in Figure A10.1. To do this assignment, you need a platform in the cloud that enables you to create and interact with this object (thermometer) in the digital space, and store the digital twin. You can use any platform for this assignment. One choice could be Microsoft Azure. Microsoft has created a language, which is called Digital Twins Definition Language (DTDL). Using this language, you can define a model for your digital twin. This model has four elements which are:

- Properties: this element specifies the properties of an object such as name, size, shape, color, serial number, and power rating.
- Telemetry: this element shows the measurement data.
- Components: this element shows the components of an object.
- Relationship: this element shows how an object is related to other objects. Based on the relationship, you can build a digital twin graph, which can show the relationship of all the digital twins in an environment.

In this assignment, the physical object is simple and does not have any component or relationship with other digital twins. Therefore, you only need to define the properties of the object and its data. You can write the model of this object in any text editor and save it in JavaScript Object Notation (JSON) extension. When the model is ready, you need to upload it to the Azure digital twins in the cloud in order to create the model. Create a resource for this digital twin instance in the Azure digital twins. Then you need to interact with the digital twin instance using Azure APIs. Microsoft has created Azure digital twins explorer for this purpose.

You also need to send the temperature values to this digital twin instance. For this purpose, you can use a microcontroller such as Arduino or Raspberry Pi to publish temperature messages to the Azure IoT hub. These messages then need to be received by an Azure function app and sent to Azure digital twins instance. Therefore, you need to create an IoT-hub and create a resource as well as a device in which its device ID will be your digital twin instance.

References

1 Vahidnia, R. and Dian, F.J. (2021). *Cellular Internet of Things for practitioners.* Vancouver: BCcampus https://pressbooks.bccampus.ca/cellulariot/.

2 Bluegiga (2015). BGscript scripting language-developer guide-version 4.1. https://www.silabs.com/documents/public/user-guides/ug209-bgscript-scripting-language.pdf (accessed 30 June 2022).

Appendix A Internet Protocol Security (IPSec)

As discussed in Chapter 12, an Internet Protocol (IP) packet must be protected against various threats as it travels through the network. All Open Systems Interconnection (OSI) layers should be secured in order to create a secure network for data communications. The security at the network layer (L3) is briefly discussed in this appendix.

A threat actor can create and send an IP packet with the source IP address of an IoT device or an IoT gateway in order to masquerade itself as these devices. This is called IP spoofing. A threat attacker can also intercept an IP packet in order to understand its contents and make a copy of it. The attacker can then send the same IP packet toward the packet's original recipient. This is called IP packet sniffing. Following the capture of a packet, the threat actor can alter the content of the IP packet before transmitting it to its intended recipient, or divert the message to another IP address. To protect IP packets from these attacks, the network layer needs to be secured. In other words, network layer security is required anytime an IP packet is transmitted to a router by an IoT device or an IoT gateway, or when a router delivers a packet to an IoT device directly or via an IoT gateway. In addition, there is a need for network layer security when an IP packet travels between two routers in a network.

One of the most essential protocols for providing security at the network layer is Internet Protocol Security (IPSec). IPSec provides protection against confidentiality, integrity, and authentication. In other words, IPSec creates a secure link with agreed security keys between two network entities (an IoT device and a router, an IoT gateway and a router, or two routers). IPSec is designed by the Internet Engineering Task Force (IETF) and operates in two modes: transport mode and tunnel mode. We discuss these two operation modes in this section.

The concept of packet encapsulation is introduced in Chapter 1. When L3 gets data from its higher layer (L4), it appends a header to the L4 data to create an L3 packet. The packet encapsulation for IPSec in transport mode is shown in Figure A.1a. As can be seen, the IPSec layer in the transport mode is located between the L3 and L4 layers. The IPSec layer receives the L4 data, adds a header and a trailer to the data, and passes the new packet to the L3 layer. Then the network layer (L3) adds an IP header to the packet received from the IPSec layer. IPSec does not provide any protection for the IP header in transport mode, but it does protect the L4 data.

The operation of IPSec in tunnel mode is shown in Figure A.1b. In this mode, the IPsec layer is located inside the L3 layer. The network layer receives the L4 data and adds an IP header (let us call it the first IP header) to the L4 data in order to make an IP packet. It then passes the entire L3 packet to the IPSec layer. At this point, the IPSec layer adds a header and a trailer to the packet and gives the packet back to the L3 layer to add the second IP header to the packet. The format of the second IP header is different than the first IP header.

Fundamentals of Internet of Things: For Students and Professionals, First Edition. F. John Dian.

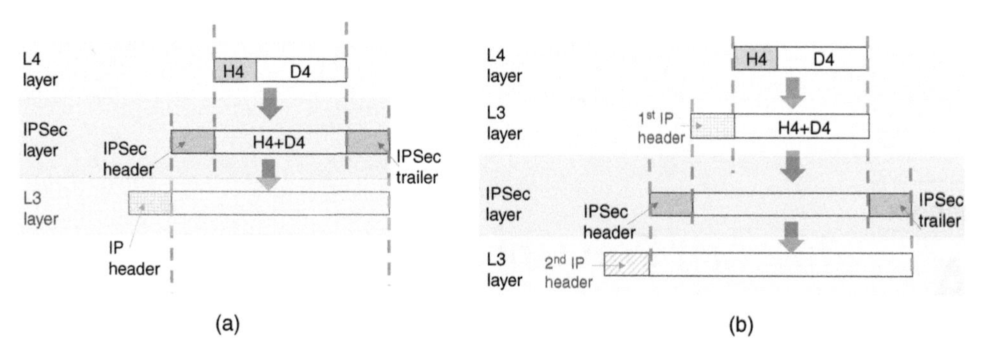

Figure A.1 IPSec modes of operation: (a) Transport mode and (b) tunnel mode.

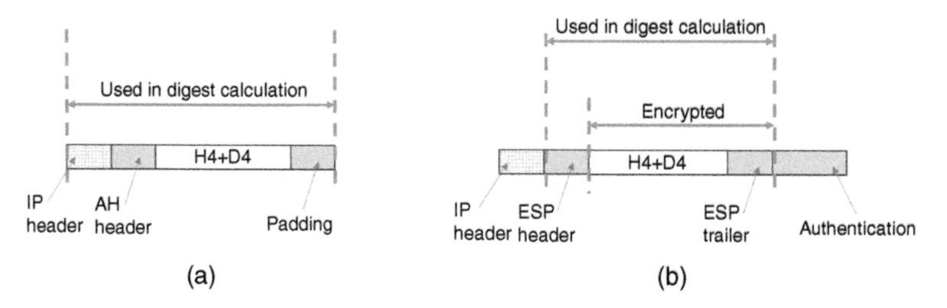

Figure A.2 IPSec in the transport mode using (a) AH protocol and (b) ESP protocol.

To provide authentication and encryption, IPSec uses two protocols to create the IPSec header and trailer. These two protocols are the Authentication Header (AH), and the Encryption Security Payload (ESP) protocols.

The use of the AH as well as the ESP protocol as part of IPSec in the transport mode is shown in Figure A.2. AH is designed to protect the integrity of data and provide authentication. It uses a hash function and a symmetric key to create a digest which is a field in the AH header. However, AH does not provide protection against threats to confidentiality. ESP is designed to provide protection against threats to confidentiality, integrity, and authentication.

An AH packet as shown in Figure A.2a is created using the following steps:

1. The AH header is added to the padded data. The AH header has several fields. One of the fields is a digest. The digest field is set to zero at this step.
2. The padding section might be needed in order to make the total length appropriate for a specific hashing algorithm. The padding section is added to AH + H4 + D4.
3. The IP header is added to the AH + H4 + D4 + padding.
4. A digest is calculated for the entire packet made in step 3, except for the fields in the IP header that change during transmission, such as Time To Live (TTL). The digest is inserted in the AH header.
5. The IP header is modified. For example, the value of the protocol field is changed to 51 from its original value. This indicates that the IP packet uses the AH protocol. A field inside the AH header shows the original value of the protocol field.

An ESP packet as shown in Figure A.2b is created using the following steps:

1. The ESP trailer is added to H4 + D4.
2. The packet composed in step 1 is encrypted. That is the reason that the ESP protocol provides protection against threats to confidentiality.

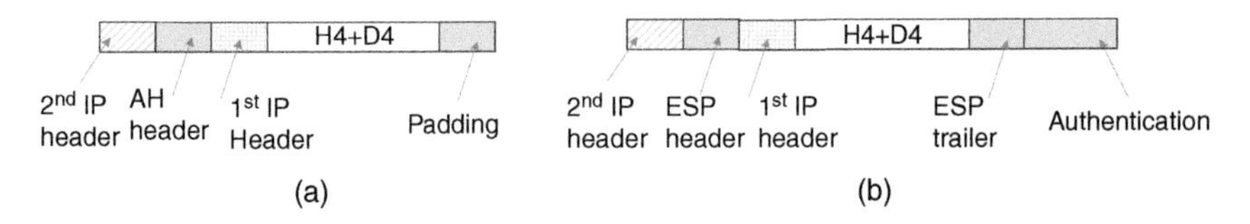

Figure A.3 IPSec in the tunnel mode using (a) AH protocol and (b) ESP protocol.

3. The ESP header is added to the encrypted section.
4. The entire packet build in step 3 is used to create a digest. The digest is added to the end of the packet as authentication field.
5. The IP header is added to the packet created in step 4.

The use of the AH protocol as well as the ESP protocol as part of IPSec in tunnel mode is shown in Figure A.3.

The detailed information about the AH and ESP headers and comprehensive explanations of how confidentiality, integrity, and authentication by using these headers as part of IPSec can be achieved are outside the scope of this book. However, readers can understand that by using the hash function and creating a digest as part of the AH or ESP protocol, and embedding this digest as part of the IPSec header, network security has been achieved.

Appendix B Transport Layer Security (TLS)

A threat actor might be interested in modifying the content of the application data exchanged between a sender and a receiver. It is important to provide protection against threats to authentication, data confidentiality, and integrity of the application data. The two most important protocols for providing security for the data entering to the transport layer are Secure Socket Layer (SSL), and Transport Layer Security (TLS). Although SSL and TLS are quite similar, TLS is the one that is designed and published by the Internet Engineering Task Force (IETF). Despite the fact that the name TLS implies that the transport layer is secure, TLS does not protect the transport-layer header. Actually, TLS protects data in the application layer, and therefore, the TLS layer can be thought of as a presentation and session layer of the OSI model. The TLS layer, as well as the layers in the OSI model that represent it, are depicted in Figure B.1.

The principal operations of SSL/TLS are shown in Figure B.2. The TLS layer, as shown in this figure, performs the following tasks:

1. Data fragmentation: application data is fragmented to blocks of 16 KB (Kilobyte) or less.
2. Compression: the fragmented data is compressed. This is an optional task.
3. Integrity protection: using a hash function, an MAC is calculated and is added to the compressed fragmented application data.
4. Data encryption: the compressed fragmented data and the MAC data are encrypted using a symmetric key.
5. Header insertion: a header is added to the encrypted payload.

The client and the server should agree on the hashing algorithm, encryption technique, and key-exchanged scheme. These are important parts of SSL/TLS layers, and they are the reasons these protocols can provide protection for threats against authentication, data confidentiality, and integrity for the application data. A client needs a key for authentication and one for encryption. The server also requires its own keys for authentication and encryption. To securely exchange data and generate these keys, SSL/TLS employs a variety of additional protocols.

The combination of the encryption algorithm, hash function, and key exchange algorithm is called a cipher suite. SSL/TLS uses a protocol called the Handshake protocol to negotiate the cipher suite and perform the required client–server authentication. This protocol is slightly different in various TLS versions. The following is a high-level overview of the Handshake protocol.

1. The client sends a message, called *Client Hello* message to the server. This message contains the version of TLS, and a list of the cryptographic algorithms and compression methods that are supported by the client. It is self-evident that IoT devices with limited processing power and memory can only employ fairly basic encryption algorithms and compression techniques.

Fundamentals of Internet of Things: For Students and Professionals, First Edition. F. John Dian.

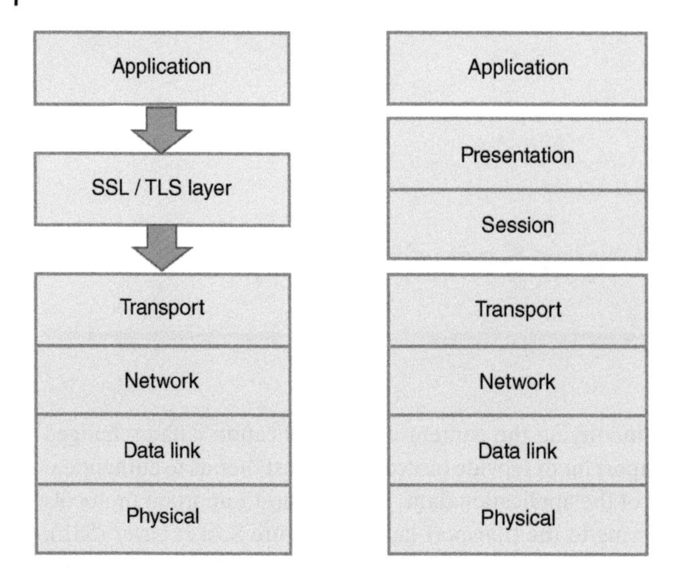

Figure B.1 The location of SSL/TLS layer as compared to the OSI model; (a) SSL/TLS layer and (b) OSI model.

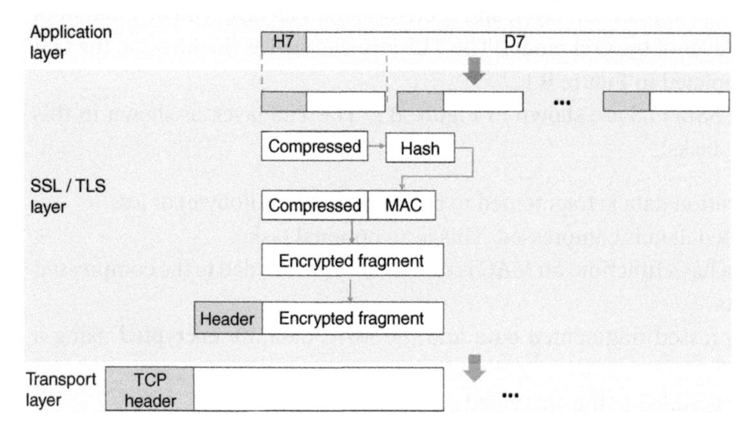

Figure B.2 Principal operations of SSL/TLS.

2. The server sends a message, called *Server Hello* message to the client. The encryption algorithm and compression mechanism that the server has consented to use are contained in this message. In other words, the server selects the best appropriate encryption technique and compression method from the client's list of supported methods. It goes without saying that the server must be able to support the chosen encryption algorithm and compression method. A session ID, the server's certificate, and the server's public key are also included in the server's Hello message. It is worth noting that the client and server agree on the cipher suite using asymmetric encryption methods during the Handshake protocol. However, the client and server use a symmetric key to exchange data when the Handshake protocol is completed.
3. The client contacts the assigned certificate authority and verifies the server's certificate in order to confirm the authority of the server. In one-way TLS, the client verifies the server, but the server does not verify the client. In two-way TLS, the server also needs to contact the certificate authority in order to verify the client's certificate. In this section, we explain one-way TLS.

4. The client sends a shared secret key to the server. This secret key is encrypted with the server's public key.
5. The client sends a message called *Client Finished* message to the server. This message is encrypted with the shared secret key.
6. The server sends a message called *Server Finished* message to the client. This message is encrypted with the shared secret key.

It is obvious that executing the Handshake protocol by constrained IoT devices offers numerous challenges. Besides the Handshake protocol, TLS uses three other protocols. These protocols are called the ChangeCipherSpec protocol (for determining the time that the client and server can use the secret key), Alert protocol (for reporting unusual conditions), and Record protocol (for merging messages that come from the Handshake protocol, ChangeCipherSpec protocol, and Alert protocol as well as the application data). The detailed information about these three protocols is outside the scope of this book. However, readers should know that the most important protocol in SSL/TLS is the Handshake protocol.

TLS or SSL can be used to secure application-layer protocols that run over TCP. Several application-layer protocols that are employed in IoT system design and that run over TCP were addressed in Chapter 9. HTTP and MQTT are two examples of these protocols. These protocols can be used with TLS to create a more secure communication protocol. Message Queue Telemetry Transport Secure (MQTTS) is the secure version of MQTT which is also referred to as MQTT over TLS. Similarly, HyperText Transfer Protocol Secure (HTTPS) is an extension of HTTP that provides secure communication over a computer network. HTTP over TLS or HTTP over SSL are other names for HTTPS.

Appendix C Satellite IoT

Traditional satellite systems are not able to provide connectivity for Internet of Things (IoT) applications that require low latency, low power consumption, and consistent availability. They are also quite expensive, which makes the use of satellite technology in most IoT applications impractical. Traditional satellites impose two main challenges on satellite operators as well. First, the cost of launching these large satellites is high, and it is not financially feasible to replace them as satellite technology advances. Therefore, these satellites often operate between 10 and 15 years in space. The traditional satellites have usually been launched into Geostationary Earth Orbit (GEO) fixed positions that keep up with the Earth's rotation, and are positioned at a great distance from the Earth, roughly 22 000 miles. A a result, each GEO satellite is able to cover larger areas of land or sea. Some of the main limitations of GEO satellites are: there is a need for mobile antennas to point directly at a GEO satellite and there is no GEO coverage near the polar caps. In addition, their complete frequency spectrum is shared across the entire coverage area. GEO has a latency of roughly 250 ms, which is too large for many IoT applications. Satellites can also be launched into a Medium Earth Orbit (MEO), which is closer to the Earth and has a lower latency of around 120 ms. The demand for low-latency, high-capacity, secure, and always-on Internet connectivity has fueled rapid expansion in the satellite communications sector in Low-Earth Orbit (LEO) which is around 200–400 miles above the Earth's surface. The coverage area of LEO satellites is significantly smaller than that of MEO and GEO satellites. In other words, a huge number of smaller satellites need to be launched into low orbit in order for them to work together and function as a single unit. To attain global coverage using satellites launched at LEO, large constellations are required, especially given their proximity to Earth. The latency at LEO is around 25 ms, which is significantly smaller as compared to the MEO and GEO satellites. This makes LEO satellites a better option for IoT applications. A simple comparison between LEO, MEO, and GEO satellites in terms of distance from Earth and latency is provided in Figure C.1a.

A significant shift from the design of large and complex satellites is the innovative design of small, low-cost satellites. While large satellites weigh more than 1000 kg, small satellites are quite lighter. The weight of these small satellites is commonly used to classify them and they come in numerous types such as femto satellites (less than 100 g), pico satellites (0.1–1 kg), nano satellites (1–10 kg), micro satellites (10–100 kg), and mini satellites (100–1000 kg). For example, Starlink is a satellite Internet constellation operated by SpaceX, which wants to provide satellite Internet access globally everywhere on Earth by launching mini satellites at LEO. This certainly is useful for some IoT applications.

Nano satellites, also known as CubeSats, have recently become very popular for IoT applications. The CubeSat project started at Stanford University in 1999 with the intention of creating a low-cost, light-weight satellite. Cubesats have a cubic shape, as the names suggest, and are available in

Fundamentals of Internet of Things: For Students and Professionals, First Edition. F. John Dian.

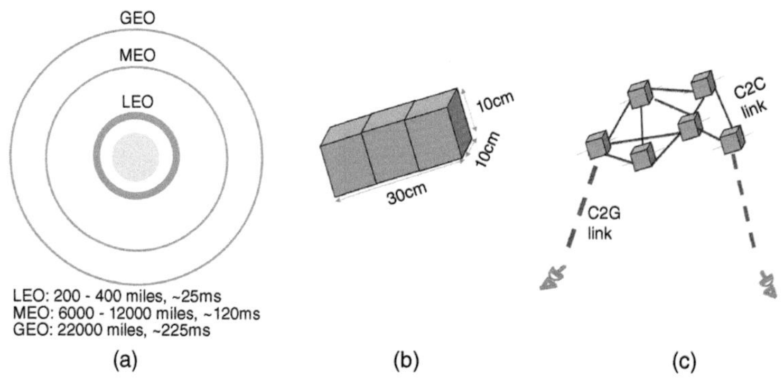

Figure C.1 Satellite for IoT (a) a comparison of GEO, MEO, and LEO orbits (b) shape and size of a 3U-CubeSats (c) CubeSat communication links.

a variety of weights and sizes expressed in units or U's, often up to 12U. Their mass per unit is around 1.3 kg, and their unit size is 10 cm × 10 cm × 10 cm. For example, 3U CubeSats would weigh 3.9 kg and have dimensions of 10 cm × 10 cm × 30 cm as shown in Figure C.1b. Cubesats are low-cost, low-power satellite systems made from commercially available components. These satellites are designed in a cubic shape to maximize the amount of surface area available for solar energy generation. Cubesats are tough enough to stay in orbit for two to five years and are launched at LEO. CubeSats are the IoT's future, and they are now in the research, development, and deployment stages by many companies. It is self-evident that LEO satellites are less expensive to launch than traditional GEO satellites. CubeSats are routinely sent into orbit utilizing spare space on rockets, providing a diverse range of launch options and low launch costs.

Generally speaking, CubeSat communication links can be separated into two categories: C2G (CubeSat-to-Ground) and C2C (CubeSat-to-CubeSat) as shown in Figure C.1c. The C2G link can use different communication technologies such as, RF, laser, or visible light communication. Currently, RF is widely used to establish C2G links. Using laser signals for transmitting signals would be very interesting. However, at this point, the laser introduces many losses such as atmospheric-turbulence or absorption and scattering losses, which can cause a severe degradation in performance if they are not well compensated for.

Solutions

Chapter 1

E1.1 With 8 nodes, the number of links is $\frac{8\times7}{2} = 28$, and with 10 nodes, the number of links becomes $\frac{10\times9}{2} = 45$.

E1.2 If one link fails:

a. In a mesh network, we lose the connection between only two nodes in the network.
b. In a star network, we lose the connection between one of the clients and the central unit.
c. In a single unidirectional ring, there might be a good link between some of the nodes in the network. However, since the data must go through the ring, if a source node sends its data, the acknowledgment does not have a path to go back to the source node.
d. In a tree network, the location of the faulty link is important. While something toward the leaf nodes at the bottom of a tree may cause two or a few nodes to lose connection, a faulty link close to the root may cause many nodes to lose connection with each other.
e. In a bus topology, the network will be divided into two separate sections, and therefore, the nodes in each section can communicate with each other. However, if there is a need for a terminator in a bus network for matching the impedance, the entire network goes down as a result of a link failure.

E1.3 The topology is shown in Figure 1.SE3.

E1.4 Duplexing: Layer 1; routing: Layer 3; access control: Layer 2; end-to-end error control: Layer 4; compression: Layer 6

E1.5 Framing, addressing, flow control, error control, and access control

E1.6 The method to determine which node in a network is allowed to access the network, at a specific time or certain frequency, is called access control.

E1.7 The occurrence of several consecutive corrupted bits is called a burst error, while a single error happens when only a bit among many bits becomes corrupted. The possibility of a burst error is usually higher as compared to a single bit error.

Fundamentals of Internet of Things: For Students and Professionals, First Edition. F. John Dian.

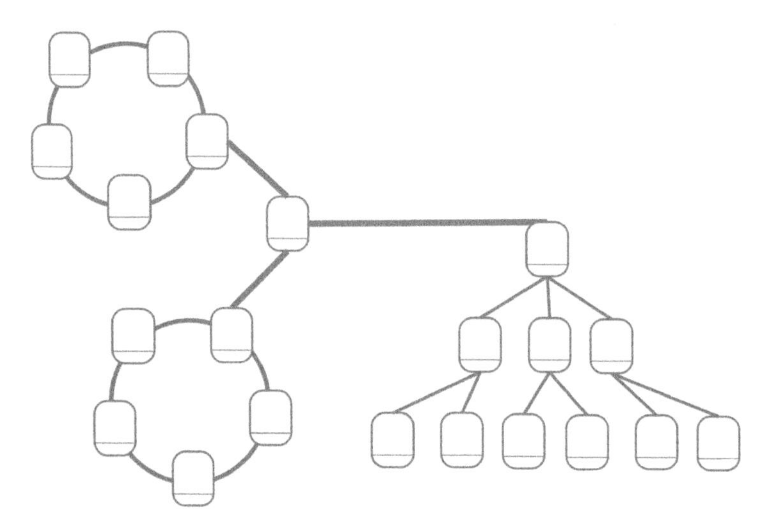

Figure 1.SE3 A hybrid network.

E1.8 First of all, two or more nodes may find that the link is free and send their data at the same time. This can cause collisions. Also, due to propagation delay, if two nodes are far from each other, a node may detect that the link is free, even though another node has already started its transmission. This can happen due to propagation delay.

E1.9 The router drops both packets.

a. The total length of the header is 0011 (which is $3 \times 4 = 12$ bytes). This is not possible. The minimum length of header should be 20 bytes. Therefore, the router drops this packet.
b. The version of the header is 0110 (which is 6). An IPV4 router should receive the packets with version 4; otherwise, it drops the packet.

E1.10 $\frac{1100 - 1072}{4} = 7 = 0b0111$

E1.11 The address is 140.120.84.24/20. The number of addresses in the block is $2^{32-20} = 2^{12} = 4096$. In this network, the first 20 bits (host ID) of the IP addresses do not change, and only the last 12 bits change. The IP address is 10001100.11110000.01010100.00011000. To find the first address of the block, consider that the least significant 12 bits are 0. This changes the address to:
10001100.11110000.01010000.00000000 = 140.120.80.00
So, the first address in this block is 140.120.80.00, and the number of addresses is 4096. The last address in the block is 140.120.95.255.

E1.12 The organization cannot use a private address as the address of its web server, since the web server must be accessible to people from outside. The organization should allocate one of its non-private addresses, such as 176.10.10.1 to the web server.

E1.13 The main responsibilities of Layer 4 are: process-to-process communication, end-to-end error control, end-to-end flow control, congestion control, and connection control.

E1.14 Physical addressing (L2) such as Ethernet address is used in local area settings, logical addressing (L3) such as IP address is used to route the packets in a wide area network, and port addressing (L4) is used to find the correct process within a node.

E1.15 TCP is based on a client-server model. A client node chooses a port number randomly and uses this number as its source port number. For this communication, there is also a need for a destination port number at the server side. The destination port number is based on the type of server. For example, port number 80 is for the HTTP protocol and is used to access most web servers.

E1.16 A socket is a combination of an IP address and a port number.

E1.17 If a segment is not acknowledged during a specific time, TCP retransmits the segment. For this purpose, when a TCP sending process sends a segment, it starts a timer at the same time. If the timer expires while the receiving process has not acknowledged the receipt of the segment, TCP retransmits the unacknowledged segment again. Also, when the TCP sending process sends a segment and receives three packets from the receiving process, in which none of them acknowledges the sent segment, TCP retransmits the segment.

E1.18 It should advertise 6000 (7000–1000).

E1.19 The organization should use VLAN. It can make the required changes without touching the physical wiring and only with changing the switch configuration that can be done remotely.

E1.20 There are many advantages to using VLANs. Generally speaking, physical reconfiguration and wiring are time-consuming tasks that are quite costly. Applying changes through software reduces the time and costs. VLANs can also be used to separate different types of traffic such as a data VLAN and a voice VLAN, where voice VLAN packets have higher priority than the data VLAN traffic.

E1.21 When VLAN traffic goes from one switch (A) to another switch (B) using a tag link, a tag is added to the frame. When the frame arrives at switch B, the tag is removed, but the switch knows what VLAN the frame belongs to.

E1.22 S-IP and D-IP will not change as the packet goes through the network from one router to another. The router changes S-MAC and D-MAC, based on the link information.

E1.23 A hub cannot change the Ethernet address, since it is a Layer 1 device. The MAC header belongs to Layer 2. A router can read or change an Ethernet header, since a router is a Layer 3 device, which can read up to Layer 3.

E1.24
a. Gateway
b. A packet in L2 is a frame, a packet in L3 is a datagram, and a packet in L4 is a segment.
c. Best-effort router
d. L2 switch

E1.25 $2^{32-8} + 2^{32-12} + 2^{32-16} = 2^{24} + 2^{20} + 2^{16} = 16{,}777{,}216 + 1{,}048{,}576 + 65{,}536 = 17{,}891{,}328$

E1.26 UDP does not perform any error, flow, or congestion control. It is a connectionless protocol that makes it simpler and faster than TCP. Therefore, it is better for real-time applications.

E1.27 The OSI model has become a theoretical model that is mostly used for understanding and the study of a network, while TCP/IP is a practical model and most data communications are designed based on this model. Research and technical communities usually explain data communications and networks based on the OSI model, and therefore, it is required to learn about this model.

E1.28 Ethernet addresses are 48 bits or 6 bytes long. Therefore, the number of addresses can be up to 2^{48}, which is a very large number. There is no need to have such a large addressing space in a local area network, since the number of nodes in a local area setting is limited. This is not efficient, since each packet has 6 bytes of overhead for each MAC address. Each Ethernet address is related to a specific NIC card, and therefore, an Ethernet address is a hardware-based address. This makes Ethernet addresses unique globally, even though there was no need for that. However, Ethernet now can be used in wider area networks (metro Ethernet) and having this large addressing space is an advantage.

E1.29
a. Higher layers must retransmit the lost frame. For example, if Layer 2 does not retransmit a lost or corrupted frame, Layer 4 should retransmit that frame. If Layer 2 of a source node ensures that each frame is reached by Layer 2 of the destination node, the work of Layer 4 would be simpler.
b. Layer 2 may not perform error control if the frames are rarely corrupted and arrive at the destination node safe and sound most of the time. For example, Layer 2 may be designed without error control, when the network has a very robust Layer 1, and uses a good quality channel.

E1.30 Yes. Besides error control, Layer 2 is also responsible for flow control, framing, addressing, and access control. So, if the channel is error-free, we still need the data link layer.

E1.31 Ethernet packets must be limited in length for different reasons. First, we do not want a large frame to occupy the link and not allow other frames to pass through the network. Second, large frames can cause jitter in the network. Third, in case of error, the transmission of a large frame makes the network inefficient. Also, the maximum length of a frame should be determined based on the access control method that is used in that network. Considering the possibility of frequent collisions in a network, the data link layer may put limitations on the maximum size of a frame. In other words, the access control might not be able to handle frames larger than a specific size.

E1.32 $0 \times 180 = 384$ bytes or 3072 bits. Since the network is 10Mbps, the duration of 3072 bits is 307.2 μs.

E1.33 The provided block of address (132.19.219.64/24) has 256 addresses. We need to make 5 subnets of 32, 16, 16, 4, and 4 nodes. These subnets are: 132.19.219.64/27, 132.19.219.96/28, 132.19.219.112/28, 132.19.219.128/30, and 132.19.219.132/30.

E1.34 $x = 5501, y = 7301, z = 5501, t = 5926$

E1.35 In many real-time applications, there is no need for the retransmission of packets that arrive late. For example, in a live audio or video transmission, the quality of the audio or video will be reduced due to packet loss. But since the application is time-sensitive, retransmitting of packets is not required. In other words, there is no benefit in retransmitting a real-time packet at a later time.

E1.36
a. Connectionless-oriented
b. Connectionless-oriented
c. Connection-oriented
d. Connectionless-oriented

E1.37 Packet delivery in local area networks is based on MAC addresses, while in a wide area network, it is based on IP addresses.

E1.38
a. International Organization for Standardization (ISO), International Telecommunication Union-Telecommunication Standards (ITU-T), American National Standards Institute (ANSI), Institute of Electrical and Electronics Engineers (IEEE)
b. The OSI model is published by ISO, and Ethernet is published by IEEE.

Chapter 2

E2.1 Table 2.SE1 shows the use cases and applications for the two verticals of energy and agriculture.

E2.2 The notation of vertical, use case, and application are not used properly in different literatures. In general, an IoT vertical often relates the use of IoT to a specific industry segment or a specific class of users. A specific vertical has unique regulatory bodies and supports specific standards, specialized policies, procedures, and protocols. Use cases subdivide each vertical into sections that can often be served by the same platform, and they need similar processing, storing, and analyzing of data. Table 2.SE2 shows the verticals and use cases associated with Example 2.1–2.12.

Table 2.SE1 Some use cases and applications for the two verticals of energy and agriculture.

Vertical	Use case	Application
Energy	Oil/Gas	Predictive maintenanace of oil and gas equipment
		Remote monitoring
	Renewable	Optimizing renewable and fossil fuel power generation
		Residential generation of renewable power
	Smart grid	Smart meter
		Fault detection
Agriculture	Green house	Enhancement of irrigation and fertilization processes
		Control infection and avoid disease outbreak
	Precision farming	Yield improvement
		Improving water use efficiency
	Agriculture drones	Crop health imaging
		Surveying of agricultural land

Table 2.SE2 Verticals and use cases associated with Example 2.1–2.12.

	Vertical	Use case
Example 2.1	Smart cities	Health monitoring
Example 2.2	Energy	Smart grid - Metering
Example 2.3	Smart cities	Waste management
Example 2.4	Environment	Disaster detection
Example 2.5	Automotive	Vehicle software update
Example 2.6	Environment	Metrological station
Example 2.7	Agriculture	Pest management
Example 2.8	Sport	Activity detection
Example 2.9	Health	Health monitoring
Example 2.10	Gaming	Haptic
Example 2.11	Retail	Food supply chain
Example 2.12	Energy	Smart grid – Fault detection

E2.3 The number of houses in each cell (with the area of 0.866 km^2) is $1517 \times 0.866 = 1313.67$. Since there are 40 IoT devices in each household, on average the total number of devices are $1313.67 \times 40 = 52{,}547$.

E2.4 $\frac{50 \times 800 \times 10 \times 2 \times 10^3}{8} = 10^6 \text{ bits} = \frac{10^6}{1024 \times 1024} = 0.954\,\text{MB}$. It should be noted that usually each kbps means 1000 bps, while each kbyte means 1024 bytes. In other words, the data rate (kbps) is not expressed as 1024 bps.

E2.5 The large amount of collected data from sensors would be analyzed in order to estimate the demands of the engine and adjust thrust levels accordingly. This results in a reduction in fuel consumption and performance improvements in engine noise and emissions. The amount of collected data per day is $10\ \text{GB} \times 3600 \times 24 = 864000\ \text{GB} = 864\ \text{TB} = 0.864\ \text{PB}$. After one month (assuming 30 days per month), the amount of data is $864\ \text{TB} \times 30 = 25290\ \text{TB} = 25.29\ \text{PB}$.

E2.6 Smart factory, predictive maintenance, and smart robotics.

E2.7 To improve efficiency, have a better compliance with government regulations, and reduce costs, we should coordinate and manage vehicles in a fleet to work in an efficient way. Fleet management performs these tasks by tracking vehicles, monitoring mechanical diagnostics, and drivers' behavior. Fleet management allows managers to gain insight into the performance of the fleet as well as drivers. Since managers know the location of the vehicles and possible problems, they can mitigate or eliminate possible risks.

E2.8 Firefighters need to know the location of the fire and the area that the fire encompasses. Since these IoT devices need to be installed in the woods, they need to be battery operated. Therefore, low power consumption is an important requirement for this application.

E2.9 Workers would find their tools faster, which would increase productivity. The resolution of the tracking is an important factor in the effectiveness of this application. More accurate information about the location of the tools can further increase productivity, since less amount of time is needed to find the tools. As resolution becomes more precise, it can be used to guide assembly operations.

E2.10 To solve the problem with early/late frost, farms use giant fans. In a non-IoT-based system, these fans need to be turned on manually. By monitoring the weather conditions, an IoT-based system can sense the temperature and accordingly turn on or off the fans.

E2.11 For beekeepers with thousands of beehives in different locations, it is a challenging task to find out which hives or colonies have problems. It is certainly difficult to drive to a site potentially hundreds of kilometers away. Therefore, the more information you can get about the health of the hive can be helpful. Opening up a beehive reduces the pollination capacity of that hive for several days. Collecting more meaningful data provides more accurate information about the hives. This information can be used by analytical tools to provide smarter decisions.

E2.12 The new model not only provides a better forecast for HP ink production and distribution but also provides better customer satisfaction. First of all, customers do not go without ink, since the ink is automatically shipped based on customer usage. Also, customers do not need to go to the store or order the ink when they need it. This application does not need to be battery operated, since usually printers are connected to electricity.

E2.13 Traditional methods that use traps are not very effective since rodents are smart enough to avoid going to traps to eat the bait. IoT-based systems can be used to monitor rodent activity or find their hotspots, type, size, and behavior. Some companies are using IoT technology to keep track of pest populations. Some of these companies use sensors and vision technology to track pest populations, collect data about their behavior, and give information about the amount of bait left in the trap.

E2.14 Figure E2.14 shows all the steps for the Amazon key. The person who delivers the parcel connects to Amazon, when they are next to the door. Then Amazon checks the parcel and home address and authorizes the delivery. It then turns on a cloud-based camera, records the delivery, and also sends a message for the homeowner to watch the delivery online. It then unlocks the door. After the parcel is put inside the house, Amazon locks the door. There are many security concerns with this application, and Amazon has tried to solve them by implementing security measures.

E2.15

a. There are several differences between digital twin and simulation. The most important ones are: 1-simulations are not based on real-time data, while digital twins are based on real-time data. 2-Simulation usually studies a particular component of an object, or certain process in an operation, while digital twin is about the copy of the complete object and entire process.
b. For predictive maintenance, improving the performance of the systems, fault detection, and estimating the remaining lifecycle of the products.
c. When there is a problem with a conventional car, the driver is the first entity that understands the problem. It may take some time for the driver to take the car for the service, and at the time, the manufacturer and the repair shops do not have much visibility of when exactly the problem started or in which environmental conditions the car has been driven. Often manufacturers require the customer to bring their car for the service at predetermined times or after certain distances, and that is the time that they identify a problem with the car. A digital twin is related to capturing the real-time data of each car, which can provide visibility to the reason behind a car's problem in a fast

and efficient way. Smart cars constantly send data about the environment and their performance to the digital twins that usually live in the cloud. By analyzing this data, it is possible to identify whether the car works as expected, or not. It should be noted that some of the problems might be fixed by sending over-the-air software updates.

d. Yes, this is true.

E2.16

a. The utility company can add another backup transformer.
b. Since real estate is expensive in the area, it requires a huge capital investment in order to add another transformer.
c. Many distributed energy sources are connected to the grid. These sources of energy can be used to generate energy and manage the supply and demand of energy in the area. When the first transformer fails to operate, the backup transformer comes up. Then the smart grid system communicates with the smart appliances and energy sources in order to put them on high alert. If the second transformer fails to operate, the grid can manage the situation since:
   i. The energy sources start to add distributed energy to the grid. Examples of these energy sources include batteries of electric cars, solar roofs, and home generators. These energy sources can store energy during off-peak times, and use them in these situations.
   ii. Adjust the scheduling of the smart devices. These devices might be turned off during this situation and turned on at a later time.
d. Yes. Smart grid systems can adjust the capacity with demand. Therefore, the same result as adding a physical transformer can be achieved by utilizing distributed energy sources and controlling the smart appliances and machineries. So, it can be considered a virtual transformer.

Chapter 3

E3.1 Business integration is not considered in this IoT architecture model. We can see the true power of IoT when IoT applications become integrated. So, the business integration and collaboration layer should be part of any IoT architecture model.

E3.2 Not exactly. A blueprint is made for a specific building and shows the architecture of that building. The architectural model is not for a specific IoT system. We can design the architecture of a specific IoT system based on a specific architectural model. The architectural diagram of a specific IoT system is similar to the blueprint of a specific building.

E3.3 IoT applications need edge computing for the benefits it provides in terms of latency, bandwidth, and security.

E3.4 Using edge computing, each hydrant is capable of analyzing the data, and therefore, if the water pressure is lower than a threshold value, it can quickly generate an alarm or send information to a water management system.

Using fog computing, a fog server is installed somewhere close to several hydrants. Therefore, by processing the data coming from several hydrants, it is possible to find the problem in a wider area and among several hydrants.

E3.5 Edge computing is suitable for very time-sensitive applications. Fog computing is suitable for applications that can tolerate larger delays as compared to edge computing. Cloud computing is suitable for the ones that can accept higher delay values.

E3.6 Any device with computing power, storage, and network connectivity can become a fog computing node. IoT gateways and industrial controllers are the most suitable fog computing nodes. Even networking devices such as switches and routers can play the role of fog computing nodes. However, the main responsibility of these networking devices is to route the packets as fast as they can, and usually we do not want these devices to open the packets and perform computation. In some applications, a smartphone or a tablet can play the role of a fog computing node.

E3.7 Some part of the transmitted data to the cloud might not be very useful, and storing and analyzing all data that is sent to the cloud might be impractical, or not be necessary.

E3.8 When dealing with massive numbers of IoT devices in IoT ecosystem, the bandwidth of the cloud is shared among all these devices. Therefore, there would be a huge problem if all devices send data to the cloud simultaneously. Fortunately, the traffic of many IoT applications is extremely low.

E3.9 Data accumulation is related to storage and preparation of data for processing in the cloud.

E3.10 User layer is similar to Layer 1 of the IoTWF architecture model. Proximity network and public network layers are equivalent to Layer 2 of the IoTWF architecture model. Provider cloud layer is similar to Layers 3–6 of the IoTWF model. Enterprise network is the same as Layer 7 of the IoTWF architecture model.

E3.11 Data from IoT devices can be unstructured or structured. For example, surveillance data or audio data are unstructured. A system may send temperature data with time information inside a packet that it can be easily saved as structured data in a database.

E3.12 Consider a met mast that collects environmental data (wind speed, wind direction, temperature, humidity, etc.) every three seconds, and sends the collected data to the cloud. Assume an IoT application that needs minute by minute environment data. The application can average out the data every one minute, or use one of the samples during each minute, and filter out the rest of data.

E3.13 Tier I data centers have the lowest performance. The annual down time for Tier I data centers is around 30 hours. They have a single path for power distribution and a cooling system without any redundant components. To increase the performance, the higher tiers use redundant components. Tier II data centers have better performance in terms of availability. This is achieved by using redundant components and raised floors. Tier II data centers reduce the downtime to around two hours per year. This is achieved by using multiple paths for the cabling distribution, as well as the power distribution. Also, Tier III data centers are connected to more than two telecommunication service providers. Tier IV data centers are the best in terms of performance and can achieve less than 30 minutes of downtime each year. By utilizing two active power and cooling distribution paths, Tier IV data centers can tolerate any single equipment failure. These data centers also have redundant components in each power and cooling distribution path.

E3.14 In the past, the PUE of data centers was very poor and many data centers had a PUE of greater than two. This means that electrical and mechanical systems consumed more than twice as much power as servers and storage elements. There have been many efforts to achieve better PUE, and the PUE of newer data centers is extremely better.

E3.15 Fog has limited scalability. If there is a need to increase the number of fog computing nodes, these devices need to be purchased, installed, and tested. These tasks are very difficult and time consuming.

E3.16 Fog computing sounds like a good option for this case. Edge computing might not be a viable solution based on the size of data and the complexity of the processing. Cloud access when an IoT device is located deep underground can be problematic. A pump failure would be very costly, and therefore, monitoring of the pump's condition is an important task. A small latency introduced by fog computing might be fine, since a pump is an electromechanical device and can tolerate small delays.

E3.17 For this scenario, a fog computing node can be installed to accumulate data from several smart meters. The data from several fog computing nodes (first-stage fog computing nodes) can be processed by second-stage fog computing nodes for data reduction.

E3.18 $\frac{1}{24} = \frac{z}{240 \times 10}$; therefore, $z = 10$ GB

E3.19 The simple answer is no. One may think that when we talk about the fog, large fog system with many fog computing nodes can be considered as a small cloud. However, the vision of building a fog system is different than the one for building a small cloud. A cloud is the same as a multi-tenant data center, which is usually designed to fulfill the computing requirements of many users. It has many resources, and users can easily scale resources based on their needs. On the other hand, fog systems are usually designed to fulfill the requirements of a single or a few IoT applications.

E3.20 Better cloud performance especially in terms of latency and security can certainly solve the technical feasibility of cloud computing for many IoT applications. However, the high cost of cloud computing can be a huge issue.

Chapter 4

E4.1 Speakers, DC motors, and LEDs are examples of actuators. Microphones, cameras, and gyroscopes are examples of sensors.

E4.2 Signal conditioning is the method used to make the sensor's output signal suitable for processing in the next stages by a microprocessor or a data acquisition system. For example, a sensor's output might be too low or high to be used by the next system. Signal conditioning ensures that the sensor's output is suitable for the system connected to the sensor.

E4.3 The scale factor of the sensor is 5 volts per inch (10/2). Nonlinearity can cause an error of $\pm 25\,\text{mV}\left(\frac{0.25}{100}\times 10\text{V}\right)$, which can be translated to an error of ± 0.005 inch $\left(\frac{25\text{mv}}{5\text{v}}\times 1\text{ inch}\right)$. Therefore, the sensor is not suitable for this application.

E4.4 The production cost of MEMS sensors is low for mass production. However, there is an upfront cost associated with the design and fabrication of MEMS sensors. Therefore, there is less interest in developing low volume MEMS sensors.

E4.5 (a) Good accuracy and good repeatability, (b) poor accuracy and poor repeatability, (c) poor accuracy and good repeatability.

E4.6 A smart sensor is a sensing device that performs many other operations besides sensing. A smart sensor consists of a base sensor and several other functional blocks to perform filtering, signal conditioning, A/D conversion, calibration, data processing, power management, diagnostic, and communication. A smart sensor has computing power, storage element, and connectivity capability, in addition to one or more base sensing units.

E4.7 Device C is a low power, device B is a low noise, and device A is an ultralow noise device. The new device has the lowest noise among all sensors of the other manufacturers. It has almost 28 times lower noise than device C. However, it has much higher power consumption as compared to other devices. The new design is good for IoT applications where power consumption is not very important, but noise is an important factor.

E4.8
a. Even though the selected sensor might not have the best accuracy, it might satisfy the design requirements. Also, NTCs are commonly used to detect small temperature changes, since they have high sensitivity.
b. Thermocouples and RTDs can measure temperature above 200 °C.

E4.9 Hysteresis means that the electrical measurement value for the same amount of pressure differs depending on whether pressure is being released or applied. When a sensor has high hysteresis, the measurement under the same pressure becomes unreliable. Having a low-hysteresis sensor indicates better reliability in measurements.

E4.10 We can use a smart ingestible sensor that can sense and measure the medication, and send a signal outside of the body. For example, it can send the data to a wearable patch on the skin. The wearable patch transmits the data to an IoT gateway or directly to the cloud for processing.

E4.11 Self-powered sensors can use energy harvesting to get their energy. This can be obtained from energy sources such as solar, thermal, wind, chemical, mechanical, biomass, or electromagnetic waves. Hybrid energy harvesting that can use several sources of energy might be interesting. However, a hybrid model increases the cost, and there are many situations where multiple sources of energy might not be available. The generation of power using a solar cell is dependent on the weather conditions, and for this reason might not be a good choice for all IoT applications. If mechanical energy can be found where the sensor is located, it can be a good choice for energy harvesting. Finding energy from environmental vibrations would be possible for many sensors.

E4.12 An absolute pressure sensor is used to measure pressure against a pressure of zero bar (vacuum). A relative pressure sensor is used to compare a pressure with the ambient pressure.

E4.13 Sensor fusion is the process of combining the sensory data from several sensors in order to get a better understanding of the system that uses these sensors, make a better model of the system, or make a more accurate, more reliable, and more functional sensors output as compared to the output of a single sensor. Sensor fusion can be used to improve the accuracy, reliability, and capability of a sensor. Also, it can enable us to measure a physical quantity that none of the single sensors used in the sensor fusing process is able to measure.

E4.14

$$a_0\frac{dy}{dt} + a_1y(t) = b_1x(t) \tag{4.s1}$$

Let us divide both sides of the equation by a_0,

$$\frac{dy}{dt} + \frac{a_1}{a_0}y(t) = \frac{b_1}{a_0}x(t) \tag{4.s2}$$

The input to the sensor is a step function with the amplitude of A. So, $x(t) = Au(t)$, and the Laplace transform of (4.s2) gives us:

$$sY(s) - y(0) + \frac{a_1}{a_0}Y(s) = \frac{b_1A}{a_0S} \tag{4.s3}$$

We can simplify the equation as:

$$Y(s) = \frac{\frac{b_1}{a_0}A + y(0)s}{s\left(s + \frac{a_1}{a_0}\right)} = \frac{\frac{b_1}{a_1}A}{s} + \frac{\frac{a_1y(0)-b_1A}{a_1}}{s + \frac{a_1}{a_0}} \tag{4.s4}$$

By calculating the inverse Laplace transform, we can find the output, $y(t)$ as:

$$y(t) = \frac{b_1}{a_1}Au(t) + \frac{a_1y(0) - b_1A}{a_1}\ e^{-\frac{a_1}{a_0}t} \tag{4.s5}$$

In this question,

$$a_0 = 1, a_1 = 2, b_1 = 2, A = 30, \text{and } y(0) = 20 \tag{4.s6}$$

Therefore,

$$y(t) = 30\,u(t) - 10\ \ e^{-2t} \tag{4.s7}$$

Next we need to find the time when $y(t)$ is 25,

$$25 = 30 - 10\ \ e^{-2t} \rightarrow ln\left(\frac{25 - 30}{-10}\right) = -2t \tag{4.s8}$$

Therefore, $t = \frac{-ln\,(0.5)}{2} = 0.346$ ms

E4.15 a. A larger object far from the camera has the same number of pixels as a small object that is closer to the camera.
b. The fusion algorithm can make a 3D model of the scene using the data from two cameras, and therefore it can calculate the distance.

E4.16 A magnetometer is a sensor for measuring the strength and direction of the magnetic field. For example, smartphones usually have a magnetometer to sense the orientation of the device. Many apps, such as the ones used to work similar to a compass, can use the data to show the orientation with respect to magnetic North. Fusing the data from gyroscope, magnetometer, and accelerometer can measure motion data.

E4.17 To make the output signal of a base sensor band limited, low pass filtering can be applied. This can be beneficial, when we want to sample the sensor's output signal and encode it to a digital format. By making the signal band limited, we can avoid aliasing. Generally speaking, the A/D conversion is accomplished by sampling, quantization, and encoding processes. The sampling rate must be at least twice of the highest frequency of the signal in order to avoid aliasing. Making the signal band limited reduces the possibility of aliasing.

E4.18 A chemical sensor measures the chemical information such as absorption, concentration, pressure, or activities of particles of a chemical substance. Optical chemical sensing is performed based on interaction of light with the chemical substance, and subsequently conversion of the optical signal to an electrical signal that represents a chemical property of the substance.

E4.19 a. Piezoelectric pressure sensors are passive sensors.
b. A LIDAR is an active sensor.

E4.20 The average output signal from a gyroscope has a small offset even when there are no movements or vibrations. This is called sensor bias.

E4.21 Figure 4.SE21 shows this situation.

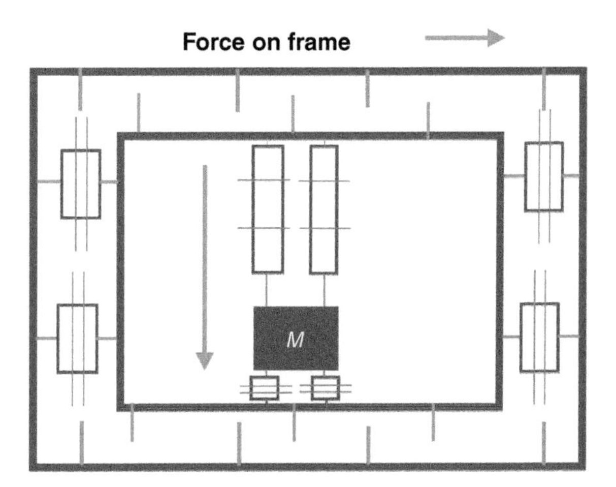

Figure 4.SE21 MEMS gyroscope movement of mass based on the direction of force on inner frame.

E4.22 Repeatability is defined as the variance of the data values that are collected by repeating the same measurements multiple times. The variance of 2 mv, 2.05 mv, 2.3 mv, 2.2 mv, and 2.1 mv is 0.0145. Therefore, the repeatability is 0.0145.

E4.23 There is a dielectric material between the two plates of a capacitor. When the voltage across the two plates of the capacitor is constant, the dielectric, which is a chemical substance, can increase the charge of the capacitor as compared to the charge of a capacitor without a dielectric. Dielectric constant or relative permittivity (ε_r) is a measure that indicates how easily a dielectric material can become polarized by an electric field. Capacitance is directly proportional to the dielectric constant. Therefore, by measuring capacitance, we can distinguish between different chemical substances.

E4.24 Aging, drift, or excessive noise and vibrations that are applied to the sensor.

E4.25 a. $x = \frac{2 \times 21 + 4 \times 24}{(2 + 4)} = 23$

b. $x = \frac{21 + 24}{2} = 23.5$

Chapter 5

E5.1 The Ethernet port of the IoT device must be enabled, and it should use the same technology as the one used for the switch port (the one that the IoT device is connected to). The IoT device should also have an IP address. The default gateway address of the IoT device should be assigned to the IP address of the Ethernet port of the router that is connected to the Ethernet switch.

E5.2 Proxy ARP is a device that can respond to an ARP request, and provide the MAC address associated with an IP address. A server or a router can play the role of a proxy ARP.

E5.3 When configuring a VLAN, we can assign the VLAN to a priority class value from 0 to 7. The lower the traffic class value, the lower is the priority. For example, VLAN data assigned to the priority level 0 indicates the lowest priority. A VLAN data that is assigned to level 7 indicates the packets that require high priority. The packets with VLAN level 7 usually belong to time-sensitive applications.

E5.4 In deterministic communications, the duration of time that it takes for data to move between two nodes of a network is predictable and can be determined. To be able to have a deterministic communication, all the nodes of the network including IoT devices, non-IoT devices, and networking devices should have a common sense of time.

E5.5 Yes, because it can provide security.

E5.6 a. It cannot be connected to a local area network.

b. It can be used inside a wide area network to route the packets.

E5.7 PROFINET defines an application layer (L7) on top of the Ethernet layer (without TCP/IP) to transmit packets in real time. However, it uses TCP/IP for configuration and diagnostics. Ethernet defines Layer 2 and 1 of the OSI model.

E5.8 Since standard Ethernet did not provide the requirements of IIoT applications, many Ethernet-based solutions are designed by different vendors to satisfy the various requirements of industrial applications. Industrial Ethernet is the common name used for these solutions. These solutions are provided by different vendors in order to control or monitor applications in many industries such as automation, automotive, and manufacturing. These solutions extended standard Ethernet to provide deterministic latency, reliable communication, and real-time operations.

E5.9 Cisco IE-4000 switch is a TSN switch, while Cisco catalyst 2970 is not a TSN switch. Cisco IE-4000 series have an FPGA that enables the support for TSN using cisco IOS release 15.2(5) E2.

E5.10
a. Jitter is defined as the variations in delays of the consecutive received packets.
b. Network congestion and queuing policies can create jitter.
c. By averaging the delays of reception in consecutive packets

E5.11 The source node should advertise a message with its required bandwidth and other QoS requirements to a connected switch. The switch then considers the reservation of the bandwidth and QoS parameters, and sends a message to the next switch until it reaches the destination node. The destination node sends back a message toward the source node. As the message comes back toward the source node, the bandwidth will be reserved on the switches along the way.

E5.12 (a) Switch 1 (b) Switch 1 (c) Switch 2 (d) Switch 2.

E5.13 Applications that are very sensitive to latency.

E5.14 PLC has been very popular with many utility companies, since the connectivity is provided by the infrastructure that they control, and can be used for data transmission among substations.

E5.15 Narrowband PLC technologies operate at lower frequencies, and they support lower data rates and shorter distances as compared to the broadband PLC technologies.

E5.16 HomePlug

E5.17 An Ethernet jumbo frame has a length that is longer than the length defined by the legacy Ethernet. A jumbo frame can be up to 9000 bytes. TAS defines a guard band to solve the problem of sending low priority frames during the time allocated for the high priority traffic. During the guard band period, TAS prohibits any traffic to be sent by any node. The duration of the guard band should be larger than the maximum length of an Ethernet frame. So, jumbo frames increase the length of the guard band.

E5.18 a. HomePlugAV2
b. HomePlug Green PHY
c. HomePlug 1.0

E5.19 CBS can be used to reduce jitter. CBS removes bursts in traffic, generates an almost constant bitrate traffic, and therefore, it is very effective in jitter reduction. TAS and pre-emption ensure that the latency is within the required range, while TAS and CBS can guarantee optimized latency and jitter.

E5.20 Due to the extensive cost of implementation, it takes a very long time before all substations will be upgraded. When this happens, PLC can still play the role of a backup solution in case of a fiber break or a fault in an optical network.

E5.21 a. If a high frequency signal enters the substation's equipment that is designed for 50–60 Hz, it may damage the equipment. Therefore, there is a wave trap, also called line trap, at the entrance of substations to filter the high frequency signal.
b. The filtered high frequency signals by a wave trap should then be sent to (injected onto) other parts of communication systems.
c. No, mostly the substations that use PLC technology.

E5.22 Since PLC signals can travel through very long transmission lines, they might be attenuated or become noisy. In these situations, the PLC signals might need to be refreshed. This can be done by filtering out the signal from the transmission line, demodulating the signal, modulating it, and reinjecting onto the transmission line again. These operations are performed in PLC carrier repeating stations.

E5.23 Those time critical applications should receive their packets within their required latency values. The best-effort traffic might see longer delays, or can face packet loss and substantial performance degradation.

E5.24 Yes, it can. PoE does not need to use free twisted pair wires. It can send power and data on one twisted pair at the same time.

E5.25 Ethernet is a Layer 2 technology. It reads Layer 2 frames and makes a decision on how to switch a frame to an output port based on the MAC address, not IP address. Therefore, it can work with any Layer 3 protocol (IP or non-IP).

E5.26 The devices that only support wireless communication can now become part of PLC network. Each G3-PLC node can connect to the devices that only support wireless connectivity and route their data toward destination using power line cables.

E5.27 a. IEEE 802.3cg, also known as 10BASE-T1S
b. Overall, the SPE cables are lighter, thinner, and smaller in size. This can reduce the cost of cabling in some IoT implementations, such as industrial IoT applications.
c. SPE supports Power over DataLine (PoDL), a technology similar to PoE in order to provide power to devices via the same single twisted pair cable.

d. SPE is an interesting technology for industrial IoT where many field devices can be connected to an IoT gateway that is located far from these devices. In the past, sensors or actuators in the field were connected to gateways that needed to be close to the device.

Chapter 6

E6.1 a. No, ISM band is part of unlicensed bands, and therefore, there is no need for a license.
b. No it cannot. The maximum power output is 1 W in the United States. Also, different technologies have their own restrictions.
c. Yes, 902–928 MHz is a sub-GHz unlicensed band that can be used in the United States.

E6.2 a. The bandwidth is 100 MHz based on Table 6.1. For the IoT device, the bandwidth is $2483 - 2400 = 83$ MHz.
b. Since there would be two guard bands of 5 MHz, then there are three non-overlapping channels. $83 - (2 \times 5) = 73$ MHz is the available bandwidth that can be allocated to three 20 MHz channels.
c. 41 channels can be defined in 83 MHz spectrum with the guard band of 1 MHz.

E6.3 a. Mesh
b. Router R_1 keeps the message, and E_2 can request the message when it wakes up at a later time.
c. The coordinator and the routers are FDDs. The endpoints are RFDs.
d. R_1 sends it to R_2 and the messages get relayed to the gateway.

E6.4 Home controller can play the role of a coordinator. Light bulbs, thermostats, and fans can play the role of routers. These devices are usually connected to the mains. The portable light switches, door/window sensing devices, smoke detectors, and motion security devices are endpoints.

E6.5 a. $83/7 = 11.85$. Therefore, there are 11 channels.
b. Zigbee uses the same channel for the entire network. Therefore, the same channel will be used for communication between R_1 and E_2.
c. Zigbee uses the same channel for the entire network. Therefore, the same channel will be used.

E6.6 The first method is called contention-free method or beacon mode. In this method, the coordinator sends beacons periodically for time synchronization, and gives points of time that each endpoint can start its transmission. This time is called guaranteed time slot (GTS). The second method is called contention-based method or non-beacon mode, in which each endpoint competes to access the channel. In this method, the coordinator and routers should not sleep, since any endpoints can wake up at any time and start transmission. In the contention-free method, the coordinator and routers can sleep if there is no traffic in the network.

E6.7 No. Zigbee is a short-range technology. If the range becomes large, the routing protocol in Zigbee is not able to provide the routing.

E6.8 No. They both use the same topology. However, they have different architecture in terms of factors such as the physical layer and the routing strategy. Bluetooth mesh uses flooding in its routing, which is simple but not very efficient.

E6.9 $15/1.25 = 12$

E6.10 $150\ \text{m} \times 4 = 600\ \text{m}$

E6.11 No, it does not. The central unit connects every 30 ms, and the peripheral BLE does not have data to transmit most of the times. Changing *CI* to 100 ms, or if $CI = 30$ ms, changing the value of latency to at least 4, can be considered as suitable configurations for this situation.

E6.12 First, they support huge data rates that are not needed for many IoT applications. For example, 802.11ac supports data rates in Gbps range. Second, these standards use technologies such as MIMO that make them expensive. Third, the overhead of the traditional WiFi is high. Fourth, they are not designed to work with a large number of devices.

E6.13 WiFi HaLow operates in the unlicensed sub-1-GHz spectrum, where the traditional WiFi standards mostly operate in 2.4 and 5 GHz spectrum. The lower the frequency, the longer the signal can travel in the air. Also, it has a better penetration through walls and barriers.

E6.14 When a WiFi-enabled device wants to transmit, it listens to the channel, and if the channel is not free, the device waits and does not transmit. The device checks the channel again at a later time, and it starts to send its data if it finds that the channel is free. There are many situations where a device thinks that the channel is not free, due to the fact that the devices are sending their data with high power. To solve this problem, WiFi 6 and WiFi HaLow introduce BSS coloring in which the access point asks its associate devices to use a lower power in their small cell.

E6.15 LoRaWAN is a low-power wide area technology, and it is over a physical layer called LoRa. This physical layer is based on the spread spectrum modulation technology, and operates in the unlicensed ISM band.

E6.16 Class A allows for bidirectional data transfer, in which an end nodes uplink data transmission is followed by two short periods of data reception in the downlink direction. This class is an excellent choice for IoT end nodes that only need the downlink transmission from the network server shortly after the IoT device has transmitted an uplink message. This class is suitable for the most energy-efficient end nodes. Class B opens additional reception windows at scheduled times in addition to the ones existing in class A. To be able for class B end nodes to open their reception windows at the scheduled times, they should receive a time synchronization message from the gateway. Class C end nodes have their reception window open at all times except the time that they are transmitting.

E6.17 Bluetooth 5 improves the discovery process significantly compared to BLE. First, in addition to the three channels (37, 38, and 39), it uses all possible data channels for advertising. Bluetooth 5 uses the term primary channels for these three channels, while it uses the term secondary channels for all other channels. This is called advertising extension. Second, Bluetooth 5 advertising messages can be longer as compared to BLE. A BLE advertising message can be up to 32 bytes, while for Bluetooth 5, the size of advertising messages can be up to 255 bytes. Finally, the advertisement can be chained to increase the length of advertisements even further.

E6.18
a. The wearable device can be an advertiser to send an ID showing a specific visitor. The BLE devices next to each painting can be scanners. The system can work in advertising mode. The scanners can send their data to the gateway.
b. The system can make connections with the devices in the range. During each connection interval, it can check whether the wearable IoT device is there or not. If not, the connection times out. By doing this, the designer can get a sense of the amount of time that each visitor has spent to see a specific painting.
c. BLE does not specify a number, but most BLE module manufacturers allow up to 8 concurrent connections.
d. As the number of visitors close to a painting increases, the possibility of collision of advertisement messages increases as well. Also, it is possible that the advertiser is configured to send advertisement messages less often. In these situations, it is possible that a scanner misses the advertisement sent from a visitor.

E6.19
a. 5 MHz
b. Channel 39 uses 2.480–2.482 and does not overlap with any of the WiFi channels.

E6.20 The channels are 14, 20, 32, and 3.
$(4+10)\ mod\ 37 = 14, (14+6)\ mod\ 37 = 20, (20+12)\ mod\ 37 = 32, (32+8)\ mod\ 37 = 3$

E6.21 The maximum power of a cellular technology is an important factor in determining the maximum range of communication that the technology can provide. The higher the power, the further the signal can travel. Energy shows the amount of power consumed. In other words, it shows how much current the system draws. It is possible to have high power and low energy systems. In this case, for a short amount of time, we should transmit with high power. This can give us the range, without draining lots of current. The definition of power and energy is not correctly used in some books and research papers.

Chapter 7

E7.1 No. 3GPP is a technical body, and it is in charge of developing technical specifications for mobile communication networks. 3GPP is responsible for writing technical specifications that will be transferred into standards by 3GPP organizational partners.

E7.2 We cannot say that a specific release number belongs to a specific generation of cellular networks. For instance, the third generation of cellular networks continued to evolve in 3GPP Release 8+ in parallel with the 4G network. Similarly, 3GPP introduced the 5G network from Release 15+, but will continue to develop 4G in parallel.

E7.3 GSM is globally available. It uses frequency bands of either 850 or 900 MHz, and provides a good coverage. GSM is simpler as compared to LTE technology, and therefore, the cost of manufacturing devices with GSM technology would be cheaper than LTE or 5G technologies.

E7.4 The 14 dBm power class would be useful especially for wearable IoT technology. The current draw for both the 20 and 23 dBm power classes is more than 100 mA. The current draw in the 14 dBm power class is substantially less as compared to the 20 and 23 dBm power classes, and therefore, an IoT device can use cell-size batteries.

E7.5 20 MHz × 1.5 bps/Hz = 30 Mbps

E7.6 In situations where the selected CIoT module does not work properly on a network, the network operator and their customers may need to spend a lot of time to address and resolve possible technical problems and issues. That is the reason that the use of certified IoT modules by operators is always recommended.

E7.7 The frequency bands B39-41 are defined for TDD duplexing. Therefore, uplink and downlink use all the available frequency bands for data transmission but at different times.

E7.8 NB-IoT does not support the handover mechanism. But if the IoT device is connected to a cell and is moving slowly within the cell, there is no issue with using NB-IoT technology.

E7.9 NB-IoT has better density than LTE-M. LTE-M is far superior to NB-IoT when it comes to bandwidth and data rate.

E7.10 No, that is not possible. The cell batteries have internal resistance and voltage drops substantially as the current draw increases. Therefore, it is not possible to draw 100 mA current. The power amplifier current draw for the 20 dBm (100 mw) IoT device is expected to easily exceed 100 mA. So, it is not possible to use coin-cell batteries for NB-IoT devices in a 20 dBm power class.

E7.11 LTE-M in different 3GPP releases can use up to 10 HARQ processes, while this range for NB-IoT is 2 processes. Low complexity of UE is an important consideration in NB-IoT design.

E7.12 NB-IoT defines half-duplex FDD in its design. The reason is to make UE implementation easier. It added TDD in its 3GPP Release 15. However, NB-IoT does not support full duplex FDD.

E7.13 Since a maximum of 300 Kbytes needs to be downloaded within 1 minute (60 seconds), then the DL data rate is $\frac{3\times8\times10^5}{60}$ = 40 kbps. The maximum data rate for NB-IoT in 3GPP Release 13 in the downlink direction is 25 kbps. Using NB-IoT in 3GPP Release 13 is not possible to perform the software update. NB-IoT in 3GPP Release 14 has the maximum DL data rate of 127 kbps, and it is good for this application.

E7.14 The maximum TBS in 3GPP Release 13 for NB-IoT is 680 bits. Sending 680 bits during 3 ms gives a peak data rate of $\frac{680}{3\text{ms}}$ = 226.7 kbps. The peak throughputs of downlink is lower than

226.7 kbps, when the time DCI, ACK, T_D, T_{DUS}, and T_{UDS} are also considered. Assuming $T_{Ack} = 1$ms, the peak data rate is 29.1 kbps $\left(\frac{680}{4+3+12+1+3}\right)$.

E7.15 Not only the spectrum of unlicensed LPWAN technologies are different in different regions of the world, but they also work under different regulatory bodies. For example, in the United States, the Federal Communications Commission (FCC) defines the operation regulations for unlicensed LPWAN technologies, while in Europe it is the European Telecom Standards Institute (ETSI).

E7.16 The LTE-M CAT M1 in 3GPP Release 14 uses higher bandwidth and consequently can support higher data rates as compared to LTE-M CAT M1 in 3GPP Release 13.

E7.17 The graph is shown in Figure 7.SE17.

E7.18 Since with 32 repetitions, MCL of 155 dB can be achieved, it can expect to achieve MCL of 158, 161, 164, 167, 170, and 173 dB for repetitions of 64, 128, 256, 512, 1024, and 2048, respectively. However, we know that coverage enhancement mode B achieves an MCL of 164 dB. This shows that accurate channel estimation and frequency tracking is not possible.

E7.19 Four frequency bands are added for NB-IoT in 3GPP Release 14, which are B11, B25, B31, and B40.

E7.20 B27 frequency band is not supported by NB-IoT. The module must be an LTE-M module. It is usually deployed by operators in North America but not in Europe.

E7.21 Having a telecommunications-specific industry certification shows that the module has been tested, and it is designed based on the required standard. This can solve interoperability issues among various equipment. However, network operators may need to perform more testing specific to their network configuration and network parameter settings to ensure that there is no issue when using this module in their network.

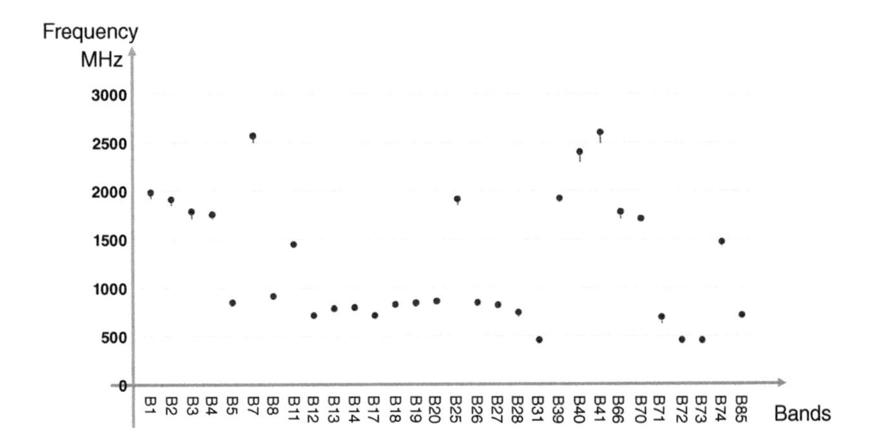

Figure 7.SE17 Frequency band versus frequency range of CIoT modules in Table 7.5.

E7.22 The transmit power of a 14 dBm CIoT device is 9 dBm less than a device with the transmit power of 23 dBm. (23–14 = 9 dBm). For every dB the TX power is decreased, and there is 1 dB decrease in the MCL value. So, the MCL for a 14 dBm device is expected to be 164–9 = 155 dB.

E7.23 The area of a miniSIM is $15 \times 25 = 375\ \text{mm}^2$. The area of an eSIM is equal to $5 \times 6 = 30\ \text{mm}^2$. Therefore, the percentage of area reduction is $\frac{375-30}{375} \times 100 = 92\%$.

E7.24 Since it supports some of the frequency bands supported in 3GPP Release 14 and 15, then the device should support LTE-M or NB-IoT in 3GPP Release 14 and 15.

E7.25

a. If the IoT module just supports a specific frequency band, the IoT module developer has limited its applications to specific regions and operators. Obviously, this is not desirable for manufacturers.
b. Supporting LTE-M and NB-IoT with the same module gives the flexibility to deploy your application without having to create multiple different products for different regions.
c. Adding GSM in addition to LTE-IoT can add the support of GSM fallback that can be used to offer reliability and coverage. A module that supports GSM fallback ensures connectivity in situations where the LTE-IoT coverage is poor or not available. Also, if for some reason there is a problem with the LTE-IoT link, the application can switch to GSM.

E7.26 ARIB (The Association of Radio Industries and Businesses, Japan), ATIS (The Alliance for Telecommunications Industry Solutions, United States), CCSA (China Communications Standards Association), ETSI (The European Telecommunications Standards Institute), TSDSI (Telecommunications Standards Development Society, India), TTA (Telecommunications Technology Association, Korea), and TTC (Telecommunication Technology Committee, Japan).

E7.27 In poor radio condition, a less-efficient modulation technique, as well as a less efficient error coding scheme (which uses more error correction bits), is needed. In excellent radio condition, the radio technology uses high spectral efficiency modulation and low error coding bits. If the radio condition is not poor or excellent, the system may use the same modulation technique as used in the poor radio condition, but with less error correction bits, it may use a more efficient modulation scheme with higher number of bits for error correction (less efficient coding scheme). The maximum spectral efficiency can be achieved in excellent radio conditions, since the number of bits used for encoding and modulation would be less.

E7.28 It shows HD-FDD. The HD-FDD technique is more cost effective as compared to FDD and TDD.

E7.29

a. CE Mode B works better in a basement. However, mode B is optional, and not all devices support coverage enhancement mode B.
b. CE Mode A is a better choice in this case, since for voice communication, CE mode B may result in excessive delay.

E7.30 No. All the LTE-M modules should support CE mode A, since it is a mandatory mode. However, CE mode B is optional, and not all LTE-M modules support this mode.

E7.31
a. It is 680 bits for NB-IoT in 3GPP Release 13. It is 1352 bits for NB-IoT in 3GPP Release 14 with two HARQ processes. It is 2536 bits for NB-IoT in 3GPP Release 14 with one HARQ process.
b. The packet needs to be padded in such a way that the packet size becomes the same as the TBS size. In this situation, the header indicates the actual size that enables the receiver to discard the padded bits.
c. 6 options (based on MCS = 4, 5, 7, 8, 10, and 12).
d. $4+10+12+1+3=30$ ms. Therefore, the system sends 680 bits in 30 ms. The data rate is $680/30=26$ kbps
e. Yes. Table 7.4 considers the maximum data rate as 0.025 Mbps, which is approximately 26 kbps.
f. $4+10+12+1+3=30$ ms. Therefore, the system sends 2536 bits in 30 ms. The data rate is $2536/30=84$ kbps.
g. Yes. Table 7.4 considers the maximum data rate as 0.085 Mbps, which is approximately 85 kbps.
h. $4+6+6+12+1+3=32$ ms. Therefore, the system sends 2704 bits in 32 ms. Therefore, the data rate is $2704/32=84.5$ kbps.

Chapter 8

E8.1.1 The amount of energy required for the transmission of the additional signaling messages when the CIoT device turns on its radio again might be more than the amount of energy that is saved as a result of turning off the radio.

E8.1.2
a. It is more beneficial to put the IoT device in an idle state instead of turning it completely off. This way the reconnection to the network can be performed with less amount of signaling.
b. This influences latency. The less sensitive the application is to latency, the longer the deactivation period can be.

E8.1.3 When a CIoT device is initially powered on, it goes to a mode that is called an RRC idle mode. The device does not have an established physical connection to the base station at this point. Therefore, it needs to set up a connection with the base station. After the network connection is established, the CIoT devices transition into a mode, which is called an RRC connected mode. For saving power, the base station can release the RRC connection that causes the UE to move to the RRC idle mode again. However, this time it stores the current context. By doing that, UE may later resume to the RRC connected mode and avoid the connection setup.

E8.1.4 When the inactivity timer expires, the IoT device goes to RRC idle mode. The inactivity timer is controlled by the base station.

E8.1.5 No, PSM is part of the RRC idle mode. There is no need to negotiate security again when the CIoT device transitions from the RRC idle mode to the RRC connected mode.

E8.1.6 No, because its radio is off.

E8.1.7 Yes, it can. The value of both timers must be the same.

E8.1.8 Network entry procedures are not needed. Examples of these procedures include scanning, key exchange, authentication, and getting an IP address.

E8.1.9 In RAI, an IoT device can indicate to the base station that neither it has more uplink data to transmit nor it expects to receive any more downlink data. This causes an early transition from the RRC connected mode to the RRC idle mode. Therefore, RAI is efficient when after sending a packet, the application does not have anything else to send, and does not expect to receive any packet.

E8.1.10
a. 3GPP Release 15
b. Using a matched filter, which is a type of low power receiver.
c. The benefit of WUS is that it reduces the unnecessary power consumption related to monitoring of the control channels. Without WUS, the UE needs to monitor the signaling messages at each paging event. With the WUS approach, the UE only needs to decode this channel when WUS is detected.

E8.1.11 After a node transmits data or performs a transaction, there is a possibility to receive downlink information shortly after its transmission. This is very typical in most applications. Therefore, after the inactivity timer expires, we start with short cycle DRX, and after a while, we switch to long cycle DRX.

E8.1.12
a. The IoT device is transmitting 375 mw for 4 seconds and consuming 300 mw for the next 4 seconds. So, the average power consumption would be $(375 \times 4 + 300 \times 4)/8 = 337.5$ mw.
b. In this case, the average power consumption is 375 mw during a 4 second period.
c. No, since NB-IoT in 3GPP Release 13 does not support RAI feature.
d. Yes, this is possible for 3GPP Release 14 since the RAI functionality can be activated.

E8.2.1 Some of the possible situations are:
- base station does not receive the preamble due to low SNR,
- base station does not have sufficient resources for the IoT device,
- base station detects a collision during transmission of Msg1 since multiple IoT devices have used the same preamble,
- base station receives the same preamble from multiple IoT devices (collision will not be detected) and all IoT devices receive the same Msg2 that causes collision in the Msg3 transmission,
- base station does not receive Msg3 due to low SNR, and
- IoT device does not receive Msg4 within the contention resolution window.

E8.2.2 Device #1 can initiate the RA process. The other two devices are barred. The most significant index of the bitmap pattern corresponds to class 0. There are classes 0–9. The least significant of the bitmap patterns correspond to class 9.

Table 8.2.SE4 Completed table for Exercise 8.2.E4.

Coverage level	First attempt	Second attempt	Third attempt
Extreme	Device#3	Device#3	Device#2 & #3
Extended	Device#2	Device#1 & #2	Device#1
Normal	Device#1		

E8.2.3 No. LTE-M supports both ACB and EAB, while NB-IoT only supports EAB.

E8.2.4 Device #1 moves to extended coverage level after one failure. After two failed attempts, Device #2 moves to extreme coverage class. In each coverage class, the amount of allowed repetitions is different. This is shown in Table 8.2.SE4.

E8.2.5 In its first attempt, the power level for transmitting the preamble is determined by the IoT device according to the strength of the received reference signals from the base station. If the transmission of Msg1 fails, the IoT device performs power ramping that increases the transmit power after each failure of the RA process until it reaches the maximum allowable power transmission.

E8.2.6 The duration of a preamble is the submission of T_{CP}, T_{Seq}, and T_{GB}. Since each radio subframe is 1 ms, then Format 0,1,2,3 needs 1,2,2,3 subframes, respectively.

Preamble Format	T_{cp} (μs)	T_{Seq} (μs)	T_{GB} (μs)	Duration (ms)
Format 0	103	800	97	1
Format 1	684	800	516	2
Format 2	203	1600	197	2
Format 3	684	1600	716	3

E8.2.7 3GPP has introduced two main features in 3GPP Release 15 and 16. Early Data Transmission (EDT) is introduced in 3GPP Release 15, and Preconfigured Uplink Resources (PUR) is introduced as part of 3GPP Release 16.

E8.2.8 Both the IoT device and base station can enable PUR. An IoT device can request the network by sending a request message, while the network can enable PUR based on the traffic pattern of the IoT device.

E8.2.9 The following are some of the conditions that need to be satisfied:

1. The IoT device needs to validate the TA to ensure that the serving cell has not been changed.
2. The IoT device should ensure that the RSRP signal has not been changed substantially.
3. The IoT device must ensure that the TAT has not expired.

Normal random access process (Release 13) should be used as the RA strategy, if the aforementioned conditions are not satisfied. Simply put, PUR cannot be implemented if these conditions and the validity of the TA are not satisfied.

E8.2.10 An IoT device is allowed to skip transmitting a certain number of consecutive allocated occasions. The maximum number of these consecutive occasions is 8.

E8.3.1 No. AGNSS can reduce start-up acquisition time, the amount of signaling, and the consumption of power during the start-up process. However, the device still needs to be able to receive satellite signals. So, AGNSS does not work in indoor locations.

E8.3.2 In CID, the location information only shows the cell that the device resides in. Location accuracy is dependent on cell size. So, the position accuracy is ±5 km in this case.

E8.3.3 CID is more accurate in urban areas since the cell size is smaller. OTDOA is more accurate in rural areas since the *RSTD* measurement error is less. The position accuracy of GNSS is the same for urban and rural areas. It is in the 15—100 m range.

E8.3.4 ECID is introduced in 3GPP Release 9 for LTE. 3GPP introduced LTE-IoT technologies (LTE-M and NB-IoT) in 3GPP Release 13. But ECID is used in 3GPP Release 14 for IoT applications.

E8.3.5 If a reference cell and a neighbor cell use the same frequency band (the center of the frequency bands are the same) what we measure is intra-frequency measurement. If the center of the frequency bands of the reference cell and the neighbor cell are not the same, then what we measure is inter-frequency measurements.

E8.3.6 The PRS bandwidth has a direct relationship with the accuracy of OTDOA positioning. The higher the bandwidth, the better accuracy is expected.

E8.3.7 The most important reason is multipath. Other reasons could be time synchronization errors among base stations, noise and interference, or mobility.

E8.3.8
a. Diffraction
b. Reflection
c. Scattering

E8.3.9 The formula for *RSTD* is

$$RSTD_{i,R} = \frac{\sqrt{(x_t - x_i)^2 + (y_t - y_i)^2}}{c} - \frac{\sqrt{(x_t - x_R)^2 + (y_t - y_R)^2}}{c}$$

Given $x_R = 0, y_R = 0, x_i = 2, y_i = 1, RSTD = 2.91\mu s$, and $c = 3 \times 10^5$ km/s, the formula for calculating *RSTD* is:

$$2.91\mu s = \frac{\sqrt{(x_t - 2)^2 + (y_t - 1)^2}}{3 \times 10^5} - \frac{\sqrt{(x_t)^2 + (y_t)^2}}{3 \times 10^5}$$

We can put the location ($x_t = 0.5, y_t = 0.5$) in the preceding formula, since both sides are equal, then the location ($x_t = 0.5, y_t = 0.5$) would be part of hyperbola. Therefore, it is possible for the device to be located at this location.

E8.3.10

$$RSTD_{1,R} = \frac{\sqrt{(x_t - 1)^2 + (y_t - 4)^2}}{c} - \frac{\sqrt{(x_t)^2 + (y_t)^2}}{c}$$

$$RSTD_{2,R} = \frac{\sqrt{(x_t - 4)^2 + (y_t - 1)^2}}{c} - \frac{\sqrt{(x_t)^2 + (y_t)^2}}{c}$$

Since $RSTD_{1,R} = RSTD_{2,R}$, we can write:

$$\frac{\sqrt{(x_t-1)^2+(y_t-4)^2}}{c} - \frac{\sqrt{(x_t)^2+(y_t)^2}}{c} = \frac{\sqrt{(x_t-4)^2+(y_t-1)^2}}{c} - \frac{\sqrt{(x_t)^2+(y_t)^2}}{c}$$

simplifying this equation results in:

$$\frac{\sqrt{(x_t-1)^2+(y_t-4)^2}}{c} = \frac{\sqrt{(x_t-4)^2+(y_t-1)^2}}{c} \rightarrow 6x_t - 6y_t = 0 \rightarrow x_t = y_t$$

$RSTD_{1,R} = 2.91\ \mu s$, and we know $x_t = y_t$. Therefore, based on one of the $RSTD$ equations, the values of x_t or y_t can be determined:

$$2.91 = RSTD_{1,R} = \frac{\sqrt{(x_t-1)^2+(x_t-4)^2}}{3\times 10^5} - \frac{\sqrt{(x_t)^2+(x_t)^2}}{3\times 10^5} \rightarrow$$

$$\sqrt{2x_t^{\,2}-10x_t+17} - \sqrt{2x_t^{\,2}} = 0.873$$

$$2x_t^{\,2} - 10x_t + 17 = 2x_t^{\,2} + 0.873^2 + 2\times 0.873 \times \sqrt{2}\,x_t$$

$$x_t = \frac{16.24}{12.47} = 1.3$$

Therefore, the coordinates of the device are (1.3 km, 1.3 km).

E8.3.11 If T_i and T_R represent the start time of transmitting the positioning reference signals by the base station i and the reference base station, respectively, then $T_i - T_R$ is the time offset between the two base stations, referred to as real time differences (RTDs). Since the PRS signal of the reference cell has arrived 1 μs earlier, $T_i - T_R = -1$, and therefore the hyperbola would be:

$$\text{RSTD}_{i,R} = \frac{\sqrt{(x_t-3)^2+(y_t-1)^2}}{c} - \frac{\sqrt{(x_t-0)^2+(y_t-1)^2}}{c} - 1$$

E8.3.12 It is shown as Figure 8.3.SE12c.

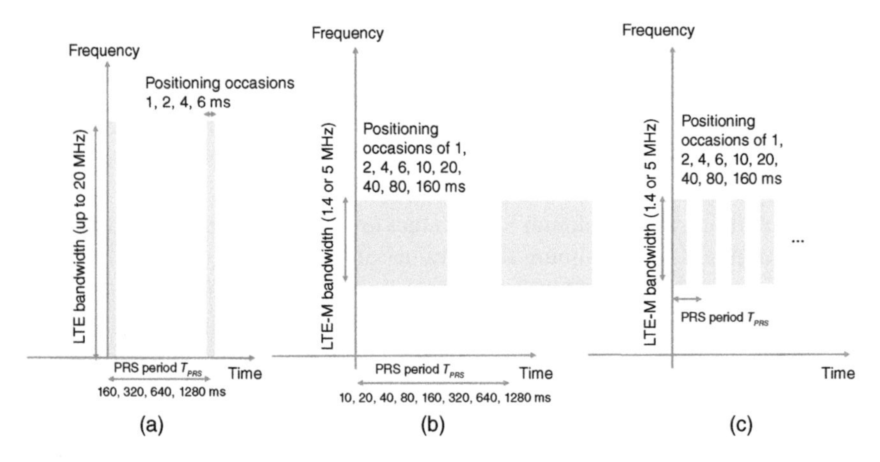

Figure 8.3.SE12 (a) An example of LTE PRS with wide bandwidth and short positioning occasion. (b) An example of LTE-M PRS with narrow bandwidth and wide positioning occasion. (c) An example of LTE-M PRS with narrow bandwidth and narrow positioning occasion and small PRS period.

E8.4.1 In both cell selection and reselection processes, the IoT device needs to find the best suitable cell to camp on (based on measuring the power levels).

E8.4.2 The two downlink synchronization signals that are used in the cell selection process are the Primary Synchronization Signal (PSS) and the Secondary Synchronization Signal (SSS).

E8.4.3 The period for a low-speed device is 5 second. For a high-speed device, the period is 1 s ($1 = 5 \times 0.2$). Therefore, this period is dependent on the speed of the CIoT device.

E8.4.4 Once a mobile device registers with the network and selects a cell, MME provides the mobile device with a specific TAL. TAL consists of a list of tracking areas that are close to the current location of the mobile device. When the mobile device leaves the current TAL, it receives a new TAL by a process called Tracking Area Update (TAU).

E8.4.5 Inter-frequency measurements. In general, a mobile device performs inter-frequency measurements when the center of the carrier frequency of the neighboring cell is different as compared to the center of the carrier frequency of the serving cell. If two cells do not have any overlapping bandwidth, therefore, the center frequency of these cells is not the same.

E8.4.6
a. When a train passes through a TAL border, there are a large number of IoT devices inside the train that need to initiate tracking area updates at the same time. This creates an excessive TAU from these devices in a short period of time. This is not a desirable situation since it may create signaling congestion.
b. If mobile devices inside the train are given different TALs, not all of them perform TAU at the same time.

E8.4.7
a. Mobile Device #1 does not need to perform any TAU while moving from TA9 to TA7.
b. Mobile Device #2 performs tracking area update when going from TA2 to TA7, since it does not have TA7 in its TAL.
c. The paging message will be sent to all the cells inside TAL that is assigned to the mobile device.

E8.4.8 TAs are grouped to form TAL in order to reduce mobility-related signaling overhead. When a mobile device moves inside TAs of a TAL, it does not need to update its location.

E8.4.9 The minimum and maximum RSRP values are −125 dBm and −66.1 dBm, respectively. The minimum and maximum RSRQ values are around −15 dB and −2 dB, respectively. The average RSRP and RSRQ values are around −90 dBm and −8dB, respectively. This shows a good quality signal most of the time during this period.

E8.4.10 Handover happens around measurement slot #150, as shown in Figure 8.4.SE10.

E8.4.11 Since B2 frequency band operates at 1.9 GHz, then

$$\text{Doppler shift} = v \times \frac{f}{c} = \frac{100\,\text{km} \times 1.9\,\text{GHz}}{3600\,\text{s} \times 300{,}000\,\text{km/s}} = 176\,\text{Hz}$$

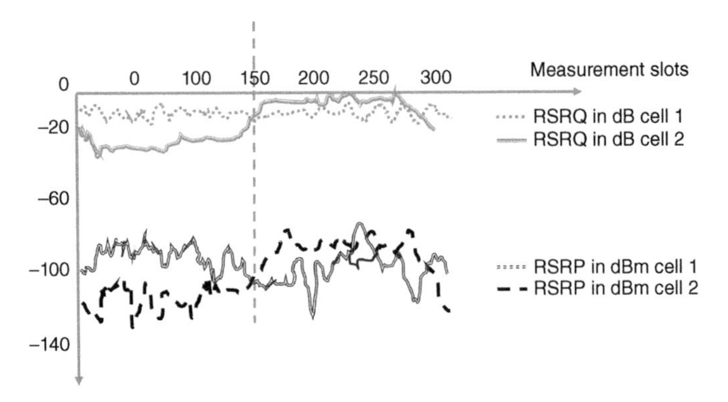

Figure 8.4.SE10 Handover in Exercise 8.4.E10.

E8.4.12 Both the size of a TA and the number of TAs in a TAL have an impact on the tracking update procedure and the paging process. Having smaller TAs and TALs increases tracking area updates, but decreases the number of paging. On the other hand, with a larger TA and TAL, paging-related signaling increases, while TAU-related signaling decreases.

E8.4.13
a. Metering devices are usually stationary and are frequently installed in basements. In these situations, utilizing coverage enhancement techniques is necessary. A good paging strategy would be to page the metering device with the same number of repetitions as where the mobile device previously accessed the network.
b. A mobile device in RRC idle mode does not usually inform the network of changes related to coverage or the quality of its reception. If in this mode, the mobile device moves constantly, MME may not have enough information about the last location and the coverage conditions, such as the number of required repetitions. Therefore, finding a good paging strategy might be very difficult. Paging a large number of cells with excessive numbers of repetitions would not be very practical. In this case, designing the system in such a way that the device updates its location information more often might be a better solution

Chapter 9

E9.1
a. MQTT is very reliable but has minimal authentication features.
b. CoAP is preferred because it relies on UDP that is a connectionless protocol. UDP is not as reliable as TCP, but it has better performance in terms of latency (no connection needs to be established or torn down, and no error, flow, and congestion control is needed).

E9.2 CoAP is more suitable for NB-IoT because it is designed for resource-constrained networks. MQTT can be used for LTE-M, because it has a higher bandwidth compared to NB-IoT.

E9.3 No, HTTP requires a high bandwidth communication channel and is over TCP. Therefore, it is not suitable for IoT devices with limited resources. Since it is over TCP, and requires lots of overhead and signaling information, it is not a low-power protocol.

E9.4 Observers in CoAP reduce the need for constant polling of data by allowing a client to receive updates based on the state of an object.

E9.5
a. The standard port number is 1883.
b. Yes.
c. No. This is only possible if the MQTT client includes its identity in the topic or payload.
d. They are discarded by the broker.
e. Once an MQTT broker sends the message to all subscribers, it discards the message.

E9.6 MQTT uses TCP, while CoAP uses UDP as its transport layer protocol.

E9.7 If a hacker can subscribe to the wildcard topics, the hacker can receive every published message. The hacker can find out about the status of window or door sensors, or the time that a light switch has been pressed. If the hacker can subscribe and publish on behalf of the devices, then attacks become more dangerous. For example, the hacked device can publish a message to the server in which the server thinks it has come from the front door smart lock. The receiver cannot find the source of an MQTT message, since MQTT messages do not have a sender field.

E9.8 CoAP provides higher data rates and lower latencies as compared to MQTT. The main reason is the fact that MQTT is over TCP, while CoAP is over UDP. TCP has more overhead and has more signaling since it is designed to be a reliable transport protocol. For this reason, CoAP can have better performance metrics in terms of data rates, latency, or even power consumption.

E9.9
a. The sensor detects sunlight or sunset, and publishes a message to the broker. The smart curtain receives the published data sensor and opens/closes accordingly.
b. Multicast

E9.10 The IoT device should be configured to operate at MQTT QoS level 2. In QoS level 0, the message might be lost and if a message is lost then the readings are not accurate. In QoS level 1, a message might be sent several times and the readings become inaccurate in this example.

E9.11 The broker acts like a post office. Instead of a device, sending the messages to or getting them from another device directly (peer-to-peer), the client (publisher) sends the messages to the post office (broker) and then it is forwarded to everyone who needs the message (subscribers). The difference is that MQTT uses a subject line called "topic" instead of mail addresses. Everyone who needs a copy of that message should subscribe to that topic.

E9.12 Yes, it can. Even though the MQTT client certificate is very secure, by combining two methods of authentication such as both client certificate as well as username and password, a more secure authentication is achievable.

E9.13 CoAP, unlike MQTT, does not provide an explicit QoS level model. However, the confirmable/non-confirmable message types can be mapped to the MQTT QoS semantics. The QoS mapping of CoAP and MQTT is shown in Table 9.SE13.

Table 9.SE13 Table 9.SE13 CoAP and MQTT QoS mapping.

Delivery method	CoAP	MQTT
Unreliable	CON (Confirmable)	QoS Level 0
Reliable	NON (Non-confirmable)	QoS Level 1

E9.14 It can be in the cloud or locally in the gas station.

E9.15 If many MQTT clients publish exactly the same topic, and there are multiple MQTT clients that have subscribed to that topic, a many-to-many communication model has been established.

E9.16
a. Device#1/indoorTemp, and Device#2/outdoorTemp
b. Temp

E9.17 To provide multicasting capability, CoAP supports group communications. In this case, a single request can be transmitted to several recipients at once. This enables CoAP to create a one-to-many model.

E9.18 MQTT can manage messages asynchronously. If the connection between an MQTT broker and an MQTT subscriber gets broken, the broker can hold the message and forward it when the connection is re-established. This is not similar to the HTTP that is a synchronous protocol in which a client must wait for the server to send its response.

E9.19 Originally, CoAP was designed based on a request-response model. Later, a capability called observer was added to CoAP in order to reduce the need for constant polling of a CoAP server by a CoAP client. With this new capability, we can say that CoAP supports both request-response as well as publish-subscribe architectures. So, the CoAP client registers as an observer and the server pushes the message to the CoAP observer.

E9.20
a. False
b. False
c. False
d. False
e. True

E9.21
a. MQTT QoS level 1 can be used for this IoT application. In this case, if a duplicate message is sent, it does not cause any problems.
b. MQTT QoS level 1 cannot be used for this IoT application. In this case, if a duplicate message is sent, it makes the robot to go forward with each message, and therefore, it can cause problems.

E9.22 This means that the MQTT subscriber has subscribed to receive all the broker's topics.

E9.23 Turn off the equipment, trigger an alarm, or send an email to a supervisor.

E9.24 Eclipse mqtt.eclipse.org, Hive MQ, broker.hivemq.com, Mosquito test.mosquitto.org, and EMQX broker.emqx.io

E9.25 IoT hub allows IoT devices to use these protocols: MQTT, AMQP, and HTTP. IoT hub does not support CoAP.

Chapter 10

E10.1 24 GHz is equivalent to 12.49 mm. 100 GHz is equivalent to 3 mm. The range of the wavelength is from 3 to 12.49 mm. Therefore, it is correct to call the FR2 band as a millimeter band.

E10.2 Due to its importance, 5G wants to increase network efficiency 100-fold as compared to 4G. We know that 5G traffic would be substantially higher than the traffic on the 4G network. If 5G wants to work with the same network energy efficiency defined by the 4G network, then it consumes significantly higher energy. Since this cannot be sustainable, there have been enormous efforts to increase the number of bits that can be transmitted in 1 J of energy.

E10.3 The radio access technology for 5G is called 5G NR, which stands for new radio. 5G NR is designed as a standard for the air interface of the 5G network. 5G is the complete radio technology for the fifth generation of cellular networks.

E10.4 It is not possible for the network to allow a user to transmit continuously using the maximum possible data rate at all times. The network should share its resources among all the active users in a cell. Therefore, resources (time and frequency slots for data transfer) are shared among all the mobile devices in a cell. The data rate that a user can experience is dependent on many factors such as the location of the mobile device within a cell, the number of active mobile devices, and their traffic patterns.

E10.5 Latency has been defined differently. Therefore, you should ask the person how he defines latency. For example, it is important to know whether network access has been considered part of latency or not.

E10.6 Three main areas are: enhanced Mobile BroadBand (eMBB), massive Machine Type Communications (mMTC), and Ultra-Reliable and Low-Latency Communications (URLLC).

E10.7 No, it is isolated from the 5G public network. Standalone private networks are the ones that can be purchased from an operator, and they are managed by a customer.

E10.8
a. SDN stands for software-defined networking.
b. SDN separates the data plane from the control plane. The hardware is responsible for fast data transmission of data, and its functionality is determined by software.
c. A traditional router performs signaling as well as forwarding the packets. For example, it sends routing packets to update its routing table. An SDN router, actually, is a fast hardware that forwards the packets. A software-based controller provides the routing table to this hardware. The router is not involved in signaling operations any more.
d. Yes, 5G uses SDN architecture.

E10.9 Simply put, 5G RAN contains the radio access technologies and the base stations, while 5G core is responsible for operations such as authentication, signaling, paging, mobility, and security.

E10.10 If 5G deployments are based on the implementation of 5G RAN for data communications, while the signaling and control plane uses 4G core, we call the implementation Non-StandAlone (NSA) deployment.

E10.11
a. $30\frac{\text{bps}}{\text{Hz}} \times 300\text{ MHz} = 9\text{Gbps}$, achievable in a larger spectrum of 300 MHz in the downlink direction.
b. $4.4\frac{\text{bps}}{\text{Hz}} \times 200\text{ MHz} = 880\text{ Mbps}$
c. $0.06\frac{\text{bps}}{\text{Hz}} \times 200\text{ MHz} = 12\text{ Mbps}$

E10.12 Downlink spectral efficiency $= 2.3 \times \frac{10^9\text{bps}}{100\times10^6\text{Hz}} = 23\text{ bps/Hz}$

E10.13 A bundle slice can be defined for a taxi company that has customized authentication with the other requirements needed for this use case.

E10.14
a. Broadcasting needs a one-to-many connectivity scheme that is different from eMBB.
b. This use case requires special network functions to be able to perform multimedia broadcast services at core and a multiple cell coordinating function for multicast transmissions. The assigned slice should also support high throughput and low latency below 40 ms.

E10.15 Yes, operators can deploy different slices for different users.

E10.16 5G architecture establishes a distributed data center scheme in which the edge data centers can be created in customer's locations as private 5G networks, on top of the base stations towers, or in telecommunication offices close to the base stations, as well as in the centralized cloud. This architecture can provide the latency, reliability, throughput, and privacy requirements of various customers and businesses.

Chapter 11

E11.1 AI algorithms are more needed when there are a larger number of sensors (especially, when there are sensors that are of different types) and when finding the patterns among the generated data from these sensors are extremely complex. For a small data set in which the relationship among attributes is not complex, a simple algorithm might be able to perform the necessary computation.

E11.2 Cloud data centers often suffer from high and variable latency, as well as limited upload bandwidth. Moreover, extensive cloud processing incurs considerable cost.

E11.3 For the analytics community to advance their algorithms today, there is an opportunity to advance them using IoT data. IoT can provide the analytics community with an abundance of real-time data that can be used to advance analytical algorithms designed for various applications.

E11.4
a. Descriptive analytics
b. Prescriptive analytics
c. Descriptive analytics
d. Predictive analytics

E11.5 An IoT data pipeline represents the flow of data between a group of sensors and the cloud. This data pipeline should be scalable to be able to handle the flow of data as the number of IoT systems grows. The cloud should also have the capacity to receive and process data, and therefore, needs to know the demand that is placed on a cloud backend.

E11.6 Three types of data ingestion include batch processing, stream processing, and micro-batch processing. Batch processing is scheduled processing and relies on stored data, while stream processing is based on real-time processing and usually relies on continuous data. In batch processing, the data is first stored in a storage area such as a data warehouse for a period of time and is processed at a scheduled time. Micro-batch processing is a type of real-time processing in which the data is stored for a very short amount of time before getting processed. All different types are used in IoT applications. Many IoT applications require micro-batch processing.

E11.7 Supervised machine learning algorithms require training data sets to have labels. Preparing a data set with labels usually requires measurements, which are costly and time-consuming.

E11.8
a. Reinforcement learning is widely used in robotics, where robots can learn new tasks by interacting with their environment and using trial and error.
b. It is shown in Figure 11.SE8.

E11.9
a. Anomaly in the electric system of a building can happen for several different reasons such as a fault in one or more appliances or electric devices, negligence of people living in the building (leaving the window open on a cold day makes the electric heater to work continuously), or a power theft where a thief is tapping electricity directly from a distribution feeder.

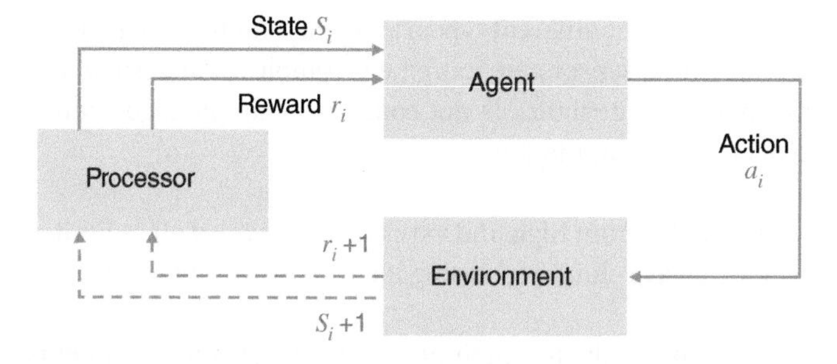

Figure 11.SE8 Reinforcement learning.

b. The output of an anomaly detection algorithm can be a binary label that indicates whether an anomaly has occurred or not. It can also show a value that indicates the level of abnormality.

E11.10 a. It is unbalanced. Most of the time the data is good and there are very small portions of the data that might be faulty. So, most of the time the data must be classified as perfect and in rare times as corrupted.

b. If the training data set is unbalanced (10 000 000 good IoT packets and 100 faulty ones), we can use several methods to make the training data set balanced. The following are some of these methods:

1. Down sample the majority class. This means to use only 100 out of 10 000 000 good packets. And use a training data set of 200 data points to train our model. This means throwing out many samples.
2. Up sampling the minority class by duplication.
3. Up sampling the minority class by making synthetic examples.
4. Design several machine learning systems, and model each using a training data set that contains the training data sets of minority class and a subset of training data sets from majority class. Then calculate the majority vote of all the outputs of these machine learning algorithms to calculate the final classification.

c. No, accuracy is not good for an unbalanced data set.

d. Recall, precision, and F-Beta are good metrics for unbalanced data sets.

E11.11 Data points a, c, and d belong to one cluster and data point b belongs to another cluster as shown in Figure 11.SE11.

E11.12 Generally speaking, in the *K*-Means clustering method, it is possible that the calculation of new cluster centers results in a cluster center that may not have any association with any data points.

E11.13 Over fitting happens when a model learns the details in the training data to the degree that it adversely impacts the model. It may perfectly work for the training set; however, when the new data arrives, the model is inefficient. Underfitting, on the other hand, makes a model that has poor performance with the training data. And therefore, it cannot be generalized to new data.

E11.14 The network has no memory in a feedforward neural network, and signals go from the input to the output. Only the current inputs are considered in this network, and there is

				Distance from		Round 1		Distance from		Round 2
	x	y	z	Centroid 1	Centroid 2	Class		Centroid 1	Centroid 2	Class
a	0	1	0	1	2	1		$\sqrt{2}/3$	$\sqrt{2}$	1
b	1	0	0	$\sqrt{3}$	$\sqrt{2}$	2		$\sqrt{4}/3$	0	2
c	0	0	1	1	$\sqrt{6}$	1		$\sqrt{6}/3$	$\sqrt{2}$	1
d	1	1	1	1	$\sqrt{2}$	1		$\sqrt{2}/3$	$\sqrt{2}$	1
Centroid 1	0	1	1	New Centroid 1		0.33	0.67	0.67		
Centroid 2	2	1	0	New Centroid 2		1	0	0		

Figure 11.SE11 Clustering of data for E11.11.

no memory of previous inputs. RNN takes prior knowledge into account, making it ideal for time-based signals.

E11.15 CNN consists of three layers. Convolutional layer, hidden layers, and pooling layers. The convolutional layer performs the convolutional operation by sliding a kernel over data, and builds the feature map. Hidden layers and connected dense layers perform operation of feature maps. A pooling layer performs down sampling, which decreases the feature map's dimensions.

E11.16 The distance of the new data point to A, B, C, and D is 6, $\sqrt{13}$,5, 6, and 7, respectively. So, we choose points A, B, and C as nearest neighbors. Based on the majority vote, this new data point belongs to a class with label 1.

E11.17 a. With $n = 2$, it is Euclidean distance, and with $n = 1$, it is a Manhattan distance.
b. $|1-5| + |2-6| + |3-7| + |4-8| = 16$

E11.18 The accuracy is 0.5 (50%).

E11.19 The minimum size of the training set is $P(M)^{D-1}$. Considering $D = 1$ means that the root is a leaf node.

E11.20 Decision trees are susceptible to overfitting and effective pruning can reduce the problem with overfitting.

E11.21 $2g(2x) - 1 = tanh\,(x)$,

$$g(2x) = \frac{e^{2x}}{1+e^{2x}} = \frac{e^x}{e^{-x}+e^x} \rightarrow 2g(2x) = \frac{2e^x}{e^{-x}+e^x}$$

$$2g(2x) - 1 = \frac{2e^x}{e^{-x}+e^x} - 1 = \frac{e^x-e^{-x}}{e^{-x}+e^x} = tanh(x)$$

E11.22

Input		Output
4	Softmax	0.643914
3		0.236883
1		0.032059
2		0.087144

E11.23 All linear functions $y = ax + b$. Composition of linear functions is linear.

E11.24 a. Loss function is $\frac{1}{2}(w_0 + \ldots + w_d - 100)^2$.
b. $\text{Grad} = \frac{1}{2} \times 2 \times (w_0 + \ldots + w_d - 100) = (2 \times (d+1) - 100) = 2d - 98$

E11.25 a. The hyperplane can be expressed as $\vec{w}.\vec{x} + b = 0$. The marginal hyperplanes are $\vec{w}.\vec{x} + b = 1$, and $\vec{w}.\vec{x} + b = -1$. Designing SVM means to determine the values of w and b in such a way that maximizes the margin. It is clear that the normal vector of $\vec{w}$ is perpendicular to the hyperplane. The unit vector in the direction of w is $\frac{\vec{w}}{\|w\|}$. Assume there is a vector $\vec{x}$ on the decision boundary. And therefore, $\vec{w}.\vec{x} + b = 0$. To get to positive region $\vec{w}.\vec{x} + b = 1$, the distance from the hyperplane is α in direction of unit vector $\frac{\vec{w}}{||w||}$. Therefore,

$$\vec{w}.\left(\vec{x} + \alpha\frac{\vec{w}}{||w||}\right) + b = 1 \rightarrow \vec{w}.\vec{x} + \alpha\frac{\vec{w}.\vec{w}}{||w||} + b = 1$$

Since $\vec{w}.\vec{x} + b = 0$, we can say that $\alpha\frac{\vec{w}.\vec{w}}{||w||} = 1$. Therefore, $\alpha = \frac{1}{||w||}$. This is the distance between the hyperplane and the positive marginal hyperplane. By a similar argument, the distance between the hyperplane and the positive marginal hyperplane is $\frac{1}{||w||}$. Therefore, the margin between two hyperplanes is $\frac{2}{||w||}$.

b. Margin $= \frac{2}{||w||} = \frac{2}{\sqrt{14}}$

c. The data point belongs to the negative cluster. $\vec{w}.\vec{x} - 10 = -5$, which is below -1.

E11.26 a. The output of the neural network is:

$$y = g(\,g(x_1 + x_2w_1 - 0.5) + g(-7x_1 + x_2w_2 - 4) - 0.5)$$

for $y = x_1 \text{OR}\, x_2$, then all the combinations of x_1, x_2 are:

$$x_1 = 0, x_2 = 0 \rightarrow y = 0$$

$$y = g(\,g(-0.5) + g(-4) - 0.5) = g(0 + 0 - 0.5) = 0$$

$$x_1 = 0, x_2 = 1 \rightarrow y = 1$$

$$y = g(g(w_1 - 0.5) + g(w_2 - 4) - 0.5)$$

$$x_1 = 1, x_2 = 0 \rightarrow y = 1$$

$$y = g(\,g(1 - 0.5) + g(-7 - 4) - 0.5) = g(1 + 0 - 0.5) = g(0.5) = 1$$

$$x_1 = 1, x_2 = 1 \rightarrow y = 1$$

$$y = g(\,g(1 + w_1 - 0.5) + g(-7 + w_2 - 4) - 0.5) = g(g(w_1 + 0.5) + g(w_2 - 11) - 0.5)$$

Therefore,

$$g(\,g(w_1 - 0.5) + g(w_2 - 4) - 0.5) = 1$$

and

$$g(g(w_1 + 0.5) + g(w_2 - 11) - 0.5) = 1$$

Both conditions are satisfied, if $w_1 > -0.5$, no matter what w_2.

b. For $y = x_1 \text{XOR}\, x_2$, then all the combinations of x_1, x_2 are:

$$x_1 = 0, x_2 = 0 \rightarrow y = 0$$

$$y = g\,(g(-0.5) + g(-4) - 0.5) = g(0 + 0 - 0.5) = 0$$

$$x_1 = 0, x_2 = 1 \rightarrow y = 1$$

$$y = g(g(w_1 - 0.5) + g(w_2 - 4) - 0.5)$$

$$x_1 = 1, x_2 = 0 \rightarrow y = 1$$

$$y = g(\, g(1 - 0.5) + g(-7 - 4) - 0.5) = g(1 + 0 - 0.5) = g(0.5) = 1$$

$$x_1 = 1, x_2 = 1 \rightarrow y = 0$$

$$y = g(\, g(1 + w_1 - 0.5) + g(-7 + w_2 - 4) - 0.5) = g(g(w_1 + 0.5) + g(w_2 - 11) - 0.5)$$

Therefore,

$$g(\, g(w_1 - 0.5) + g(w_2 - 4) - 0.5) = 1$$

and

$$g(g(w_1 + 0.5) + g(w_2 - 11) - 0.5) = 0$$

Both conditions are satisfied if

$$w_1 < -0.5 \text{ and } 4 < w_2 < 11.$$

E11.27 The convolution matrix represented by $f(x)*f(x)$ is

$$\begin{bmatrix} 0 & 0 & 0 & 0 & 0 & 0 & 0 \\ 0 & 0 & 0 & 4 & -4 & 1 & 0 \\ 0 & 0 & 0 & 0 & 0 & 0 & 0 \\ 0 & 0 & 0 & -8 & 4 & 0 & 0 \\ 0 & 0 & 0 & 0 & 0 & 0 & 0 \\ 0 & 0 & 0 & 4 & 0 & 0 & 0 \\ 0 & 0 & 0 & 0 & 0 & 0 & 0 \end{bmatrix}$$

Chapter 12

E12.1 a. The data that is collected during account creation such as age and gender, the subscription data such as payment data, smartphone application data that is used to connect the toy to the smartphone such as WiFi password, and the data that is recorded daily whenever the child is playing with the toy.

b. US Children's Online Protection Act

c. This ensures that the doll is not passively listening to conversations. The method used to provide consumers' information on the collection of data is called notice. In this case, the doll manufacturer provides notice into its product by adding a push button.

d. To transport data from the doll to the cloud, the designer can use an encrypted session and also use secure IP protocols such as IPSec and secure transport protocols such as Transport Layer Security (TLS) protocol.

E12.2 Threat modeling is a set of activities that are performed in order to enhance security by identifying potential vulnerabilities and threats, calculating the level of risk of each threat, and developing countermeasures to prevent and mitigate threats.

E12.3 In Brute Force attack, the threat actor tries to find the username and password using trial and error. To do that, the threat actor tries all the possible combinations. In most cases, these attacks are automated where a software program automatically checks possible credentials in order to login to the system. Increasing password complexity or setting a limit for the number of login failures are some of possible methods to prevent this attack.

E12.4 The threat actors use port scanning techniques to identify open ports and interfaces of a system. They can use these ports to enter the system for malicious purposes.

E12.5 An MITM or Man In The Middle is a type of attack where a threat actor intercepts communication between two people or systems. The primary intention of the MITM attack is to access confidential data.

E12.6 A botnet is a group of Internet-connected devices that is infected by malware. Threat actors can control these groups to initiate botnet attacks such as DDoS attacks, unauthorized access, or data theft.

E12.7 Social engineering is the term used when the attacker tries to fool users into revealing confidential information. For example, an attacker can pretend like a legitimate user who requests for the confidential information. The attacker may use email or text messaging to fool people into giving them their credentials or to convince people to download a malicious app that the attacker can use to access authentication information.

E12.8 Encryption of firmware images, adding hash to protect integrity of data, digitally sign firmware to protect the authenticity, and validate the digital signatures using hardware-based boot loaders.

E12.9 This is correct. Since some of the biometrics information of a user might be found through high resolution photos.

E12.10 If with the information the driver can be identified, then it is a privacy concern. If it tracks the driver's movements and the driver's habit can be linked with other information in which the driver can be identified, then there might be possible privacy issues.

E12.11 a. $n = 97 \times 83 = 8051$, $\varphi = (97-1)(83-1) = 7872$, $d = 6661 > (13 \times d)\ mod\ 7872 = 1$
b. IoT $- > 67737984$, $m_1 = 6773$, $m_2 = 7984$

$$c_1 = 6773^{13}\ mod\ 8051 = 2792$$
$$c_2 = 7984^{13}\ mod\ 8051 = 3181$$

c.

$$c_1 = 2792^{6661}\ mod\ 8051 = 6773 \qquad c_2 = 3181^{6661}\ mod\ 8051 = 7984$$

So, it decrypts it correctly.

E12.12 Denial of Sleep Attack (DoSA) is a threat to the availability of IoT systems. There is a need for low power consumption in battery-powered IoT devices. In these types of IoT systems, an IoT device usually sends a small amount of data and goes to sleep to reduce power consumption. This increases the lifetime of the battery and allows the device to operate for a long time. In the denial of sleep attack, the attacker tries to prevent the IoT device from going into sleep mode. It is actually an attack on the power supply of the device. The objective of this attack is to increase power consumption in order to reduce the lifetime of the device. A simple way to do DoSA is by using a fake IoT device. The attacker may use a fake IoT device to transmit data to "legitimate" IoT devices. The objective of using a fake device is to make IoT devices in the range to consume more energy. This reduces the lifetime of their batteries and eventually makes them become unavailable.

E12.13

a. DES uses two straight P-Boxes. The input to these P-Boxes is 64 bits and the output is also 64 bits, which makes them straight P-Boxes.
b. P-Boxes are keyless, and basically shuffle the input bits around.
c. In each round there are two P-Boxes. One is an expansion P-Box that receives 32 bits as input, and it outputs 48 bits. The other P-Box is straight since its input and output are both 32 bits. Each round uses a compression S-Box. Its input is 48 bits, and its output is 32 bits.
d. DES uses a compression P-Box. However, it has been designed so that in each round of DES, the data is not lost. For example, there is an expansion P-Box at the beginning of each round, and also, (R_i) in each round will be repeated as L_{i+1}. Therefore, data is not lost.

E12.14 Because a hash function generates a fixed size output for any arbitrary size input, it is possible that multiple inputs generate the same hash results. However, the possibility that modifying an input results in the same hash is unlikely.

E12.15 Yes. A password can be entered as the input to a hash function. In this case, instead of sending the password, the hash value can be used. For example, for web authentication, servers do not save the actual password and can just check the hash values. It is clear that from the hash value, the actual password cannot be found.

E12.16

a. The register office uses a centralized model. Marriage information that is recorded in a register office can be accessed by authorized personnel. Asking people that attended the wedding is a decentralized approach.
b. The problem with a centralized approach is that all the data is stored in one place. From a security point of view, if the threat actor can go inside the database of the register office, it may be able to change the information.
c. In a decentralized model, the information is found by asking people that attended the wedding. Each person has a record of the information. If a threat actor wants to change the data, it needs to change the data recorded by a large group of people who attended the weeding (at least more than 50%). This is very difficult and costly.
d. IoT devices may use a centralized model and get the data from a database in the cloud (centralized data center) or use blockchain technology in which the data is distributed among many nodes in the network.

E12.17 De-identified information is a type of personal information, and therefore, it is part of privacy law. Information is personal if it is possible to identify an individual using the information. Personal information can be considered de-identified if there are no possible situations to identify an individual from the information. It should be noted that the de-identified information can be used for research purposes.

E12.18 Consider a situation where an IoT device records environmental conditions including temperature and humidity that goods are subjected to while in transit. Assume that there is a smart contract on the blockchain that sets out the desired conditions based on the requirements of the regulators, senders, or customers. Once the goods are arrived at a transit point or delivered to an end customer, the data is verified against the smart contract and ensures that all of the requirements are satisfied. It can then trigger appropriate actions such as sending notifications to different parties to the contract.

E12.19 In this case, if process A randomly selects $x = 5$, then

$$R_1 = g^x mod\, p = 7^5 mod\, 37 = 9$$

If process B randomly selects y=8, then

$$R_2 = g^y mod\, p = 7^8 mod\, 37 = 16$$

Process A calculates the value of the shared key as

$$k = R_2{}^x mod\, p = 16^5 mod\, 37 = 33$$

Process B calculates the value of the shared key as

$$k = R_1{}^y\, mod\, p = 9^8\, mod\, 37 = 33$$

E12.20 The responses usually are saved in Non-Volatile Random-Access Memory (NVRAM) in an IoT device. The threat actors like to access NVRAM to find this information. PUFs are lightweight and cost-efficient solutions. PUFs can provide secure authentication without keeping any authentication data on the device. This makes PUFs especially interesting for resource-constraint IoT devices.

E12.21
a. A stronger encryption scheme does not provide any protection against this attack. The threat actor does not need to decrypt the packet or tamper with its data.
b. A digital signature does not provide any protection against this attack. The message might even be digitally signed. The threat actor does not need to change anything. It just delays or resends the same message.
c. The replay attack in smart grid applications can be disastrous. If a threat actor delays the transmission of a fault message, the protection relay will not react quickly enough, potentially causing harm to other smart grid equipment. On the other hand, threat actors could create blackouts if they replay a message containing information about a failure at a later time.
d. Adding a session ID or timestamping the message may be useful. In this case, if a message arrives late, it becomes discarded.

E12.22
a. A blacklist is a list of users, systems, or applications that are dangerous and need to be blocked from accessing data. To protect a system, blacklists need to be constantly updated. A whitelist is the inversion of a blacklist. A whitelist blocks everything except the users, systems, or applications that are on the list.

b. The central BLE node (IoT gateway) should be configured to use whitelisting and should accept the MAC addresses of peripheral nodes in its whitelist. Each peripheral BLE node should also be configured to use whitelisting. Each peripheral BLE node should accept only the MAC address of the central node in its whitelist.

E12.23 On a hot day, the attacker can turn on all the heaters inside every house in an area. This causes the heaters to consume a huge amount of energy. It also causes the air conditioning units to work nonstop and draw huge amount of power. Since each heater or air conditioning unit has its own circuit, the circuit breaker inside the house does not go off. However, the load on the area increases sharply. This can cause line failure and power outages in the area.

E12.24 Most encryption algorithms require the IoT device to generate large random numbers. These large random numbers can guarantee that the threat actors are not able to decrypt the data easily. Using small random numbers, the encryption can be decrypted in a short amount of time, and therefore, the IoT device is not very secure.

E12.25 a. MQTT provides username and password. However, to make this application very secure, we can use MQTT with TLS or MQTT Secure (MQTTS) for communication between the MQTT client A and the MQTT broker, and the communication between the MQTT broker and the MQTT client B.

b. A threat actor can play the role of MQTT client A, or MQTT broker, and sends messages to indicate the temperature is high. This can cause the air conditioning unit to work nonstop until it fails to operate.

c. Making secure communication requires the use of SSL/TLS, which demands high computational power. This makes the design of MQTT client A challenging. For example, generating long random numbers, calculating the asymmetric keys for the exchange of cipher suite, the use of symmetric key for encryption of messages, applying a compression method, and verifying the certificates with a trusted certificate authority require a substantial processing power and memory. However, the type of cipher suites can be selected in such a way that it satisfies the specification of an IoT device in terms of processing power and memory.

d. We need to introduce a root of trust for this system. Let us create a private CA and establish a sense of trust between the MQTT client A and the private CA, between the MQTT broker and the private CA, and between the MQTT client B and the private CA. We explain how trust can be established between the MQTT client A and the private CA, but the same method can be applied for the communication between the MQTT client B or the MQTT broker and the private CA.

1. MQTT client A needs to create a private key, and create a Certified Signing Request (CSR) to be signed by the private CA. The MQTT client A needs to send CSR to the private CA.
2. The private CA should have its own private key and certificate. It should receive CSR from the MQTT client A, and make an MQTT client A certificate. This certificate should be sent to the MQTT client A. This certificate is signed by the trusted CA and relates to the private key of the MQTT client A. Therefore, it can be used as the proof of identity for MQTT client A.

As mentioned earlier, the same can be down by the MQTT broker and MQTT client B. Now, when MQTT client A wants to communicate with the MQTT broker, as part of the TLS Handshake protocol, the MQTT broker sends its certificate to the MQTT client A. The certificate can be verified by the private CA to ensure the authenticity of the MQTT broker and its keys.

Chapter 13

E13.1 Since IoT requires a wide range of expertise that usually does not exist within a company.

E13.2 The purpose of the sustainment step is to maximize readiness and deliver the best possible product, systems, services, or features by solving minor problems that users or customers may face in a fully deployed scenario. Also, the organization should be able to answer the questions, educate and train the users, and help them to use proper configurations in order to maximize the performance or usage of the solution.

E13.3 The wind data that is collected in a wind park by a power generation company is valuable for the environmental agencies that want to make a better environmental model and better forecast the weather. The organization that collects the data usually owns the data; however, sometimes the organization may sell its ownership to a vendor or another organization. The data might also be shared publicly for research purposes.

E13.4 The water department might be interested in modeling the water quality in a city, and the building owners might be interested, since it may show the quality of the pipes in their buildings. Since some material in water may damage the pipes over time, the home insurance companies might also be interested in the data.

E13.5 Integration points are the interfaces among different systems.

E13.6
a. Business case document changes during the IoT solution development period, and therefore, it needs to be updated accordingly.
b. The executives of an organization, and sometimes the board of directors.
c. It contains high-level cost estimates.

E13.7 It most likely needed to receive one or more certifications for the IoT medical device from the regulatory bodies before it could start the deployment.

E13.8 System integrators, similar to vendors, are partners who help a company in its IoT solution development. A product vendor is a vendor who has a product that can be used as part of an IoT solution. A system integrator usually does not have a product and may use other vendor's products, if needed.

E13.9 The non-IoT machinery cannot usually be replaced right away and often should work next to IoT-based machines. Therefore, a solution should consider both IoT-based and non-IoT-based machineries.

E13.10 In the Agile methodology, the development of different pieces happens simultaneously. In other words, instead of planning for the entire work, the Agile methodology divides the development work into small increments that are completed in iterations. A comparison between the Waterfall and Agile methodologies is provided in Table 13.1.

E13.11 The first reason is the rapid advancements in the tools, techniques, and algorithms of an IoT solution. It would be great if there is a possibility to take advantage of these advancements in order to improve the quality of an existing IoT solution. The second reason is that both the organization and its customers may feel a need for additional new features, processes, or services when they realize the benefit of deployed IoT solutions. The third reason is related to the possibility of improving the IoT system through software updates. Therefore, we can continue updating the IoT system software in order to improve the solution with ease. The fourth reason is related to the IoT solution design itself. For example, in the Agile implementation of an IoT solution, we need to continuously improve the initial implementation.

E13.12 Ethical hackers are people employed by an organization to test the security of an IoT solution.

E13.13 The experience and the references that a vendor can provide is one of the most important evaluation criteria. It would be perfect if a vendor has done exactly similar solutions. Otherwise, it should be determined if any vendor has done similar use cases or projects in similar verticals. It is good to see if the company has or uses any IoT platform as its core business. It is important to find the partner networks of a vendor, especially those partners that can help with your solution. Many IoT projects use a combination of public and private networks, and it is good to check if the vendor has performed hybrid solutions in the past.

E13.14 In this case, the business model of the company's operation has changed. This creates a huge change, since the staff have no experience in the new operations and do not know how to react to different customer needs. Before, they sold the products, but the operation will be changed after the deployment of the IoT solution. Changing the mindset of the staff and providing education and training takes time, and it is just one of many organizational challenges that organizations are faced with. Change management plan can ensure that staff understand the reason behind the change. The staff should be consulted, and they must know that the organization has plans for their positions after the solution is deployed.

E13.15 Most companies are not ready for the Agile implementation. The Agile approach is not as straightforward as the Waterfall approach and therefore, it needs experts to develop solutions using the Agile approach.

E13.16 The protocol translation should be handled in the detailed design step.

E13.17
a. Company A and D
b. Company A and B

E13.18 Example 1: A manufacturing company wants to monitor its machines and makes an IoT solution to do so. The company uses a simple data structure for this purpose. Later, it wants to add predictive maintenance as a new feature. But the data structure used in the previous solution cannot be easily modified and the entire database needs to be changed for adding the new features.
Example 2: A company needed to add several sensors to its initial design before deployment, but there was not enough power available on the board to make this possible.

E13.19 If the organization encounters many problems in a deployment phase, it may delay the deployment of the other phases, until it can resolve the existing problems.

E13.20 In pilot testing, the organization performs testing in a live environment.

E13.21 In the Agile approach, there is a need for various teams of developers to collaborate, and therefore, communication among different teams is an important necessity for a successful implementation. The fact that there are not many interactions among different departments in this company makes the Agile implementation challenging.

Abbreviations

3GPP Third Generation Partnership Project

A

AB	Access Barring
AC	Access Control
ACB	Access Class Barring
ADMS	Advanced Distribution Management Systems
ADSL	Asymmetric Digital Subscriber Line
AES	Advanced Encryption Standard
AGNSS	Assisted Global Navigation Satellite System
AH	Authentication Header
AI	Artificial Intelligence
AMI	Advanced Metering Infrastructure
AMR	Automatic Meter Reading
ANOVA	ANalysis Of VAriance
ANSI	American National Standards Institute
AoA	Angle of Arrival
API	Application Programming Interface
AR	Augmented Reality
ARIB	Association of Radio Industries and Businesses
ARM	Advanced RISC Machines
ARP	Address Resolution Protocol
ARQ	Automatic Repeat reQuest
ASCII	American Standard Code for Information Interchange
AT	ATtention
AWS	Amazon Web Services

B

BAN	Body Area Network
BB-PLC	Broadband Power Line Communication
BCT	Building, Configuration, and Testing

Fundamentals of Internet of Things: For Students and Professionals, First Edition. F. John Dian.

BLE	Bluetooth Low Energy
BMCA	Best Master Clock Algorithm
BPSK	Binary Phase Shift Keying
BSS	Basic Service Set

C

CA	Certificate Authority
CAN	Campus Area Network
CAN	Controller Area Network
CBS	Credit Based Shaper
CCD	Charged-Coupled Device
CCSA	China Communications Standards Association
CDMA	Code Division Multiple Access
CE	Coverage Enhancement
CENELEC	Comité Européen de Normalisation ÉLECtrotechnique
CEP	Complex Event Processing
CI	Connection Interval
CID	Cell Identity
CIDR	Class Inter Domain Routing
CIoT	Cellular Internet of Things
CM	Change Management
CNN	Convolutional Neural Network
CoAP	Constrained Application Protocol
CON	CONfirmable messages
COTS	Commercial Off-The-Shelf
CP	Cyclic Prefix
CP	Control Plane
CPU	Central Processing Unit
CR	Carriage Return
CRC	Cyclic Redundancy Check
CSA	Connectivity Standards Alliance
CSIA	Cyber Security Improvement Act
CSMA	Carrier Sense Multiple Access
CSMA/CD	Carrier Sense Multiple Access/Collision Detection
CSR	Certificate Signing Request
C-DRX	Connected DRX

D

DCI	Downlink Control Indicator
DCS	Digital Cellular System
DDoS	Distributed Denial of Service
DDS	Data Distribution Service
DES	Data Encryption Standard

DHCP Dynamic Host Configuration Protocol
DHS Department of Homeland Security
DL DownLink
DLC Distribution Line Carrier
DNS Domain Name System
DoS Denial of Service
DoSA Denial of Sleep Attack
DREAD Damage, Reproducibility, Exploitability, Affected users, and Discoverability
DRX Discontinuous reception (RX)
DSL Digital Subscriber Line
DTDL Digital Twins Definition Language

E

EAB Extended Access Barring
EC European Commission
ECS Electric Charging Station
EC-GSM Extended Coverage GSM
ECID Enhanced Cell Identity
EDGE Enhanced Data rates for GSM Evolution
eDRX Extended/Enhanced Discontinuous Reception (RX)
EDT Early Data Transmission
eMBB enhanced Mobile BroadBand
EMS Energy Management System
EPRI Electric Power Research Institute
eSIM embedded SIM
E-SMLC Evolved Serving Mobile Location Center
ESP Encryption Security Payload
EtherCAT Ethernet for Control Automation Technology
ETSI European Telecom Standards Institute
E-UTRA Evolved Universal Terrestrial Radio Access
EV Electric Vehicle
EVSE Electric Vehicle Service Equipment

F

FCC Federal Communications Commission
FDA Food and Drug Administration
FDD Frequency Division Duplexing
FDMA Frequency Division Multiple Access
FEC Forward Error Correction
FFD Full Function Devices
FN False Negative
FP False Positive
FR Frequency Range

FSK Frequency Shift Keying
FTP File Transfer Protocol
FWA Fixed Wireless Access

G

GATT Generic ATTribute
GCF Global Certification Forum
GE General Electric
GEO Geostationary Earth Orbit
GLONASS GLObal Navigation Satellite System
GMLC Gateway Mobile Location Center
GNSS Global Navigation Satellite System
GPIO General Purpose Input/Output
GPRS General Packet Radio Service
GPS Global Positioning System
GSM Global System for Mobile Communications
GTS Guaranteed Time Slot

H

HaaS Hardware as a Service
HAN Home Area Network
HARQ Hybrid Automatic Repeat reQuest
HD-FDD Half-Duplex Frequency Division Duplexing
HLEN Header LENgth
HSS Home Subscriber Server
HTTP HyperText Transfer Protocol
HTTPS HTTP Secure

I

IaaS Infrastructure as a Service
IANA Internet Assigned Numbers Authority
IC Integrated Circuit
ICANN Internet Corporation for Assigned Names and Numbers
ICNIRP International Commission on Non-Ionizing Radiation Protection
IDE Integrated Development Environment
IEEE Institute of Electrical and Electronics Engineers
IETF Internet Engineering Task Force
IHD In Home Display
IIC Industrial Internet Consortium
IIoT Industrial IoT
IIRA Industrial Internet Reference Architecture

IMU	Inertial Measurement Unit
IoT	Internet of Things
IoTWF	Internet of Things Word Forum
IP	Internet Protocol
IPSec	Internet Protocol Security
iSIM	integrated SIM
ISM	Industrial, Science, and Medical
ISP	Internet Service Provider
ISO	International Standardization Organization
ITU	International Telecommunication Union
I-DRX	Idle DRX

J

JPEG	Joint Photographic Experts Group
JTAG	Joint Test Access Group
JSON	JavaScript Object Notation

K

KNN	K-Nearest Neighbor

L

LAN	Local Area Networks
LBS	Location-Based Services
LED	Light-Emitting Diode
LEO	Low-Earth Orbit
LIDAR	LIght Detection And Ranging
LNA	Low Noise Amplifier
LoRaWAN	Long Range Wide Area Network
LPP	LTE Positioning Protocol
LPPa	LTE Protocol Positioning annex
LPWAN	Low-Power Wide Area Network
LSTM	Long Short-Term Memory
LTE	Long-Term Evolution
LTE-M	LTE for Machine-Type Communications
LVQ	Last Value Queue
LWT	Last Will and Testament

M

MA	Multiple Access
MAC	Media Access Control

MAC	Message Authentication Code
MAN	Metropolitan Area Network
MB-PLC	MidBand-PLC
MCL	Maximum Coupling Loss
MCS	Modulation and Coding Scheme
MCU	Micro-Controller Unit
MCU-PT	Micro-Controller Unit-Packet Tracer
MDMS	Meter Data Management Systems
MEO	Medium Earth Orbit
MIMO	Multiple Input Multiple Output
MITM	Man In The Middle
ML	Machine Learning
MME	Mobility Management Entity
mMTC	massive Machine Type Communications
MPL	Maximum Path Loss
MQTT	Message Queue Telemetry Transport
MQTTS	MQTT Secure
MQTT-SN	MQTT for Sensor Networks
MSS	Minimum Segment Size
MTC	Machine-Type Communications

N

NAK	Negative Acknowledgment
NAL	Network Access License
NAT	Network Address Translation
NB-IoT	Narrow Band Internet of Things
NB-PLC	Narrow Band Power Line Communication
NFV	Network Function Virtualization
NIC	Network Interface Card
NIST	National Institute of Standards and Technology
NON	NON-confirmable message
NR	New Radio
NSA	Non-StandAlone
NTC	Negative Temperature Coefficient
NVRAM	Non-Volatile Random Access Memory

O

OASIS	Organization for the Advancement of Structured Information Standards
OCF	Open Connectivity Foundation
OFDM	Orthogonal Frequency Division Multiplexing
OFDMA	Orthogonal Frequency Division Multiple Access
OSI	Open Systems Interconnection
OTA	Over The Air
OTDOA	Observed Time Difference Of Arrival

P

PaaS	Platform as a Service
P-Box	Permutation-Box
PCB	Printed Circuit Board
PCS	Personal Communications Service
PIA	Privacy Impact Assessment
PLC	Power Line Communication
PoDL	Power over DataLine
PoE	Power over Ethernet
PON	Passive Optical Network
PRIME	PoweRline Intelligent Metering Evolution
PROFINET	Process FIeld NETwork
PRS	Positioning Reference Signal
PSD	Power Spectral Density
PSM	Power Saving Mode
PSS	Primary Synchronization Signal
PTCRB	PCS Type Certification Review Board
PTP	Precision Time Protocol
PTW	Paging Time Window
PU	Processing Unit
PUE	Power Usage Effectiveness
PUF	Physically Unclonable Function
PUR	Preconfigured Uplink Resource

Q

QAM	Quadrature Amplitude Modulation
QoS	Quality of Service

R

RA	Random Access
RAI	Release Assistance Indication
RAM	Random Access Memory
RAN	Radio Access Network
RAO	Random Access Opportunity
RAP	Random Access Preamble
RAR	Random Access Response
RAT	Radio Access Technology
RAW	Restricted Access Window
ReLU	Rectified Linear Unit
RF	Radio Frequency
RFD	Reduced Function Device
RFID	Radio Frequency IDentification

RISC	Reduced Instruction Set Computer
RLF	Radio Link Failure
RNN	Recurrent Neural Network
RoI	Return of Investment
RPM	Revolution Per Minute
RRC	Radio Resource Control
RSA	Rivest–Shamir–Adleman
RSRP	Reference Signal Received Power
RSRQ	Reference Signal Received Quality
RSSI	Received Signal Strength Indicator
RSTD	Reference Signal Time Difference
RTD	Real-Time Difference
RTD	Resistance Temperature Detectors
RTU	Remote Terminal Unit

S

SA	Systems Aspects
SA	StandAlone
SaaS	Software as a Service
S-Box	Substitution-Box
SBC-PT	Single Boarded Components – Packet Tracer
SCADA	Supervisory Control And Data Acquisition
SCCB	Serial Camera Control Bus
SDK	Software Development Kit
SDN	Software Defined Networking
SF	Scaling Factor
SHA	Secure Hash Algorithm
SHMS	Structural Health Monitoring System
SIG	Special Internet Group
SIM	Subscriber Identification Module
SINR	Signal to Interference-Noise Ratio
SLA	System Level Agreement
SMS	Short Messaging System
SMTP	Simple Mail Transfer Protocol
SNMP	Simple Network Management Protocol
SNR	Signal to Noise Ratio
SPE	Single Pair Ethernet
SPI	Serial Peripheral Interface
SRP	Stream Reservation Protocol
SSE	Sum of Squared Errors
SSL	Secure Socket Layer
SSS	Secondary Synchronization Signal
SUPL	Secure User Plane Location
SVM	Support Vector Machine

T

TA	Time Alignment
TA	Tracking Area
TAL	Tracking Area List
TAS	Time Aware Shaper
TAT	Time Alignment Timer
TAU	Tracking Area Update
TBS	Transport Block Size
TCP	Transmission Control Protocol
TDD	Time Division Duplexing
TDMA	Time Division Multiple Access
TELNET	TELetype NETwork
TF	TensorFlow
TFL	TensorFlow Lite
TI	Texas Instrument
TinyML	Tiny Machine Learning
TLS	Transport Layer Security
TN	True Negative
ToA	Time of Arrival
ToS	Type of Service
TP	True Positive
TSDSI	Telecommunications Standards Development Society India
TSG	Technical Specification Group
TSN	Time Sensitive Networking
TTA	Telecommunications Technology Association
TTC	Telecommunication Technology Committee
TTF	Time-To-Fix
TTL	Time to Live
TTN	The Things Network
TWT	Target Wake Time

U

UART	Universal Asynchronous Receiver/Transmitter
UDP	User Datagram Protocol
UE	User Equipment
UI	User Interface
UL	UpLink
UP	User Plane
UPnP	Universal Plug and Play
URI	Universal Resource Identifier
URLLC	Ultra-Reliable Low Latency Communications
USB	Universal Serial Bus

V

VDSL	Very high-speed Digital Subscriber Line
VIN	Vehicle Identification Number
VLAN	Virtual Local Area Network
VM	Virtual Machine
VPN	Virtual Private Network
VR	Virtual Reality

W

WAN	Wide Area Network
WG	Working Group
WUS	Wake Up Signal
WWW	World Wide Web

X

XML	eXtensible Markup Language
XMPP	Extensible Messaging and Presence Protocol

Index

q

r

Printed and bound by CPI Group (UK) Ltd, Croydon, CR0 4YY
16/04/2025
14658588-0002